T5-BAF-996

ATLA BIBLIOGRAPHY SERIES
edited by Dr. Kenneth E. Rowe

1. *A Guide to the Study of the Holiness Movement*, by Charles Edwin Jones. 1974.
2. *Thomas Merton: A Bibliography*, by Marquita E. Breit. 1974.
3. *The Sermon on the Mount: A History of Interpretation and Bibliography*, by Warren S. Kissinger. 1975.
4. *The Parables of Jesus: A History of Interpretation and Bibliography*, by Warren S. Kissinger. 1979.
5. *Homosexuality and the Judeo-Christian: An Annotated Bibliography*, by Thom Horner. 1981.
6. *A Guide to the Study of the Pentecostal Movement*, by Charles Edwin Jones. 1983.
7. *The Genesis of Modern Process Thought: A Historical Outline with Bibliography*, by George R. Lucas Jr. 1983.
8. *A Presbyterian Bibliography*, by Harold B. Prince. 1983.
9. *Paul Tillich: A Comprehensive Bibliography . . .*, by Richard C. Crossman. 1983.
10. *A Bibliography of the Samaritans*, by Alan David Crown. 1984 (see no. 32).
11. *An Annotated and Classified Bibliography of English Literature Pertaining to the Ethiopian Orthodox Church*, by Jon Bonk. 1984.
12. *International Meditation Bibliography, 1950 to 1982*, by Howard R. Jarrell. 1984.
13. *Rabindranath Tagore: A Bibliography*, by Katherine Henn. 1985.
14. *Research in Ritual Studies: A Programmatic Essay and Bibliography*, by Ronald L. Grimes. 1985.
15. *Protestant Theological Education in America*, by Heather F. Day. 1985.
16. *Unconscious: A Guide to Sources*, by Natalino Caputi. 1985.
17. *The New Testament Apocrypha and Pseudepigrapha*, by James H. Charlesworth. 1987.
18. *Black Holiness*, by Charles Edwin Jones. 1987.
19. *A Bibliography on Ancient Ephesus*, by Richard Oster. 1987.
20. *Jerusalem, the Holy City: A Bibliography*, by James D. Purvis. Vol. I, 1988; Vol. II, 1991.
21. *An Index to English Periodical Literature on the Old Testament and Ancient Near Eastern Studies*, by William G. Hupper. Vol. I, 1987; Vol. II, 1988; Vol. III, 1990; Vol. IV, 1990; Vol. V, 1992; Vol. VI, 1994; Vol. VII, 1998; Vol. VIII, 1999.
22. *John and Charles Wesley: A Bibliography*, by Betty M. Jarboe. 1987.
23. *A Scholar's Guide to Academic Journals in Religion*, by James Dawsey. 1988.

24. *An Oxford Movement and Its Leaders: A Bibliography of Secondary and Lesser Primary Sources*, by Lawrence N. Crumb. 1988; Supplement, 1993.
25. *A Bibliography of Christian Worship*, by Bard Thompson. 1989.
26. *The Disciples and American Culture: A Bibliography of Works by Disciples of Christ Members, 1866–1984*, by Leslie R. Galbraith and Heather F. Day. 1990.
27. *The Yogacara School of Buddhism: A Bibliography*, by John Powers. 1991.
28. *The Doctrine of the Holy Spirit: A Bibliography Showing Its Chronological Development* (2 vols.), by Esther Dech Schandorff. 1995.
29. *Rediscovery of Creation: A Bibliographical Study of the Church's Response to the Environmental Crisis*, by Joseph K. Sheldon. 1992.
30. *The Charismatic Movement: A Guide to the Study of Neo-Pentecostalism with Emphasis on Anglo-American Sources*, by Charles Edwin Jones. 1995.
31. *Cities and Churches: An International Bibliography* (3 vols.), by Loyde H. Hartley. 1992.
32. *A Bibliography of the Samaritans*, 2nd ed., by Alan David Crown. 1993.
33. *The Early Church: An Annotated Bibliography of Literature in English*, by Thomas A. Robinson. 1993.
34. *Holiness Manuscripts: A Guide to Sources Documenting the Wesleyan Holiness Movement in the United States and Canada*, by William Kostlevy. 1994.
35. *Of Spirituality: A Feminist Perspective*, by Clare B. Fischer. 1995.
36. *Evangelical Sectarianism in the Russian Empire and the USSR: A Bibliographic Guide*, by Albert Wardin Jr. 1995.
37. *Hermann Sasse: A Bibliography*, by Ronald R. Feuerhahn. 1995.
38. *Women in the Biblical World: A Study Guide. Vol. I: Women in the World of Hebrew Scripture*, by Mayer I. Gruber. 1995.
39. *Women and Religion in Britain and Ireland: An Annotated Bibliography from the Reformation to 1993*, by Dale A. Johnson. 1995.
40. *Emil Brunner: A Bibliography*, by Mark G. McKim. 1996.
41. *The Book of Jeremiah: An Annotated Bibliography*, by Henry O. Thompson. 1996.
42. *The Book of Amos: An Annotated Bibliography*, by Henry O. Thompson. 1997.
43. *Ancient and Modern Chaldean History: A Comprehensive Bibliography of Sources*, by Ray Kamoo. 1999.
44. *World Lutheranism: A Select Bibliography for English Readers*, by Donald L. Huber. 2000.

45. *The Christian and Missionary Alliance: An Annotated Bibliography of Textual Sources*, by H. D. (Sandy) Ayer. 2001.
46. *Science and Religion in the English-Speaking World, 1600–1727: A Bibliographic Guide to the Secondary Literature*, by Richard S. Brooks and David K. Himrod. 2001.

To Carolyn and Lola

and

In Memory of Jo Dobbs

Science and Religion in the English-Speaking World, 1600–1727

A Bibliographic Guide to the Secondary Literature

Richard S. Brooks
David K. Himrod

American Theological Library Association Bibliography Series, No. 46

The Scarecrow Press, Inc.
Lanham, Maryland, and London
2001

SCARECROW PRESS, INC.

Published in the United States of America
by Scarecrow Press, Inc.
4720 Boston Way, Lanham, Maryland 20706
www.scarecrowpress.com

4 Pleydell Gardens, Folkestone
Kent CT20 2DN, England

British Library Cataloguing-in-Publication Information Available

Library of Congress Cataloging-in-Publication Data

Brooks, Richard S., 1946–
Science and religion in the English-speaking world, 1600–1727 : a bibliographic guide to the secondary literature / Richard S. Brooks, David K. Himrod.
p. cm. — (Bibliography series / American Theological Library Association ; no. 46)
Includes bibliographical references and indexes.
ISBN 0-8108-4011-1 (alk. paper)
1. Religion and science–History–17th century–Bibliography. I. Himrod, David K., 1938– II. Title. III. ATLA bibliography series ; no. 46.
Z7844.5 .B76 2001
[BR115]
016.2615'5–dc21 Library of Congress Control Number: 2001020062

The paper used in this publication meets the minimum requirements of American National Standard for Information Sciences—Permanence of Paper for Printed Library Materials, ANSI/NISO Z39.48-1992.
Manufactured in the United States of America.

Contents

Series Editor's Foreword

The American Theological Library Association Bibliography Series is designed to stimulate and encourage the preparation of reliable bibliographies and guides to the literature of religious studies in all of its scope and variety. Compilers are free to define their field, make their own selections, and work out internal organization as the unique demands of the subject requires. We are pleased to be able to publish this annotated guide to the interplay between natural science and religion prepared by David Himrod and Richard Brooks.

Following undergraduate studies in the humanities at the University of Kansas, Richard S. Brooks completed a doctorate in religion at Northwestern University. The author of a book *The Interplay Between Science and Religion in England, 1640-1720: A Bibliographical and Historiographical Guide* (Evanston, Ill.: Garrett-Evangelical Theological Seminary Library, 1975), along with several articles and book reviews, Dr. Brooks is currently Associate Professor of Religious Studies at the University of Wisconsin-Whitewater.

David K. Himrod completed undergraduate studies in physics at the California Institute of Technology, earned a master's degree in theology at Claremont School of Theology and took the doctorate in history at the University of California at Los Angeles. The author of several articles, essays, and reviews, Dr. Himrod is Assistant Librarian for Reader Services at the United Library, Garrett-Evangelical Theological Seminary in Evanston, Illinois.

Kenneth E. Rowe
Drew University Library
Madison, New Jersey

Acknowledgments

Without the support, encouragement, and patience of many people, we would not have been able to begin, much less complete this volume.

We thank Dick's colleagues and administrators at the University of Wisconsin-Whitewater for supporting a sabbatical leave for one semester and approving a part-time leave of absence in another. Wade Dazey, David Cartwright, and Andrea Nye, especially, have been supportive. We thank Dave's colleagues at The United Library of Garrett-Evangelical Theological Seminary and Seabury-Western Theological Seminary. Al Caldwell and Dianne Robinson have been particularly helpful.

We thank the Publication Committee of the American Theological Library Association. And we greatly appreciate the patience and support of Kenneth Rowe. We began this project with a financial grant from the ATLA.

We thank our several research assistants through the years, especially Meagan O'Dowd and Penelope Johnson.

This volume is dedicated to our wives, Carolyn Brooks and Lola Himrod, and to the late Betty Jo Teeter Dobbs. Jo Dobbs, Dick's dissertation advisor, provided us with her insights into the period as well as encouragement and friendship.

Lastly, we thank Seth, Tim, Judson, and Adam who, along with Lola and Carolyn, impatiently have believed in and supported us—and yet have always reminded us where lie the true priorities of life.

Introduction

This is a work in progress. When we began it a decade ago, we believed that in a few years of concentrated effort—reading one thousand items or so—we would complete our coverage from 1600 to 1750 and then move on to later periods. However, our personal lives have since become more complicated, and the "science and religion industry" has mushroomed. Thus, we have both taken more time on the project and radically changed our originally planned format. We here present our first 2000 items. We here have a publication cutoff date of 1994—and this set does not include even all of the relevant books and articles published before that date.

The parameters that limit our topic have evolved through the years. All of our decisions have to do with limiting the overwhelming pool of books and articles. We are not claiming that the continent, historiographically speaking, should be treated separately from the English-speaking world or that there is anything special about the year 1727. But we do need to make our project bibliographically manageable. Chronologically, we have decided to change the later limit from 1750 to 1727, the year of Newton's death. We are, therefore, excluding Hume and several other important individuals whose work mainly occurred after 1727. Geographically, we are restricting our project to England, Ireland, Scotland, Wales, and the English colonies in America. Thus, we are excluding Descartes, Galileo, Leibniz, and other important individuals of the period. We also exclude, generally speaking, works that are not significant for the study of both science and religion in the period.

Our book is structured as follows: the major portion consists of an annotated bibliography of books and articles arranged alphabetically by author. This is followed by unannotated lists of bibliographies and doctoral dissertations. We are not annotating dissertations because we have not seen most of them.

We include three indexes. The first is topical, relating each work in our bibliography to one or more broad topical categories. Our understanding of each of these categories is explained in the next section after this introduction. Our topics reflect an originally planned format of chapter organization which we have since abandoned. We now wish that we had indexed the works we are reading according to fifty or more topics! But we do hope that the present topical identifications will yet remain useful. The second index is one of persons who wrote or worked in the seventeenth and early-eighteenth centuries. We provide nutshell identification, focusing on science and religion for each of 380 men and women, and we identify significant discussions about each found in the works listed in this bibliography. The final index lists the authors and editors of works cited in our bibliography.

We have considered and decided to omit a number of books and articles (now over 1,300). Reasons for omission for works with apparently relevant titles include: (1) that none or not enough of the work is focused within our chronological framework; (2) that none or not enough of the work is focused within our geographical framework; (3) that there is not enough connection to science; (4) that there is not enough connection to religion; and (5) that science and religion are treated in the work in a totally separate way without connection or interplay; (6) situated on the chronological border, we have decided to omit George Berkeley in this volume. We plan, eventually, to maintain a web site where we will share, electronically, information concerning the items that we have decided to omit (with reasons why).

Our decisions to include or omit works at times are "judgment calls." We include some items which arguably could be omitted. In some works, for example, the relationship between science and religion is not explicit. (Such a work might discuss a bishop's natural philosophy and not his religion—or Newton's theology but not his science.) We also present some historiographic essays in which internalist historians of science argue that science and religion are totally separate with no connection or interplay. We have omitted George Berkeley in this set of 2000 items.

In our choice of geographic scope, we do not intend to suggest that the English-speaking areas were isolated. Indeed, travel and correspondence were commonplace, especially between the British Isles and the European continent. Significant continental influences and debates occurred in the English-speaking areas. (Items about these are included in this bibliography.) But we are not able to cover the entire literature on Galileo, Descartes, Leibniz, and the rest of the continental situation. Thus, such topics occur in our selections and continental persons appear in our listing of names only if they were actually present in the British Isles or America.

In our annotations we try to be fair and, as a general rule, to represent most authors in their own terms. Occasionally, we make judgmental remarks having to do with our assessment of scholarly usefulness, quality of writing, out-of-date status, or historiographic adequacy. Obviously, these judgments may show our own biases, yet we hope that they help rather than hinder the usefulness of our annotations.

As regards our own historiographic position, several points should be made clear. We are convinced of the validity and of the importance of a profound historical interplay between science and religion. We think that both the "separation metaphor" and the "warfare metaphor" are inadequate. As our annotations and topical labels indicate, we see complexity and diversity in the interplay between science and religion, especially in the period under study. We believe that *religion* refers to a category of phenomena and thought wider than, yet inclusive of, Christianity. We take the *science* or natural philosophy of the period to refer to a category of phenomena and thought far wider than, yet inclusive of, some later—and narrow—understandings of natural science. Our descriptions of topical categories in the next section show where we are coming from. Even by themselves, they demonstrate the inadequacy of the separation and warfare metaphors.

We try not to be confined by disciplinary boundaries. Our own training was, respectively, in the history of science and history of religions (Himrod) and in the history of Christianity (Brooks). In this bibliography, however, we have actively sought out works produced by historians of philosophy, literary historians, historians of ideas, cultural historians, historical sociologists, and writers of general history—as well as historians of science and historians of Christianity.

Since this is a work in progress, we request feedback from the users of this work. We would like suggestions as to relevant books and articles. And we will consider rewriting or amending any annotation which our readers convince us is incomplete or not balanced. Richard Brooks' e-mail address is brooksr@mail.uww.edu; David Himrod's e-mail address is dhimrod@garrett.edu.

Descriptions of Topical Categories

Following each entry in our bibliography, a Roman numeral in italics identifies the primary or most basic topical category within which the book, article, essay, bibliography, or dissertation falls. Other topical categories to which the work is relevant are then identified by the Roman numeral codes (without italics).

The nature of these broad categories and the nature of our discipline is such that significant overlap occurs between the categories. Thus, we have made no attempt to create watertight categorical compartments; rather, these categories and codes are intended mainly to be a useful tool for purposes of identification and indexing.

Unless otherwise noted, all topics and discussions mentioned in the following summary descriptions apply to the period and geographical limits indicated by the title of our bibliography. These descriptions do not totally exhaust all the possible subtopics, but we hope that they are reasonably thorough and give the reader a good idea of each topical category.

Topic I: Historiography includes general theories about how best to study and interpret the interrelationships between science and religion. It, thus, also includes discussions concerning the methodological principles in:

the history of science; the history of religion; the history of literature; the history of ideas; the history of philosophy; cultural history; the history of ideology; and other such disciplines.

Topic I includes debates about:

"internalist" versus "externalist" approaches to the history of science;

the propriety of "intellectualist" approaches to the history of science;

the "retrospective" or "Whig" approach to history as applied to cultural history or the history of science;

the implications of Marxist historiography for the history of science and religion;

the implications of feminist historiography for the history of science; and

the proper use of sociological, anthropological, and prosopographical methods for the history of science and religion.

Topic I also includes discussions of:

the "warfare" and "separation" metaphors as applied to the historical relations between science and religion;

methodological issues connected to Topic II, Topic III, Topic IV, Topic VI, and/or Topic XI;

key terminology on which historiographic debates turn (such as "science," "scientific revolution," and "Puritanism"); and

scientific change and Kuhn's model of "paradigm."

Finally, Topic I includes specific discussions of the methods of influential scholars (such as Foster, Koyre, Merton, Metzger, and Yates).

Topic II: The Magical, Alchemical, and *Prisca* Traditions includes discussions of the following movements, especially as they have influenced science and religion:

alchemy; astrology; gnosticism; Hermeticism; Kabbala; magic; mysticism; Neoplatonism; Paracelsianism; Rosicrucianism; and witchcraft.

This topic also includes discussions of the contemporary usage and the influence of the following concepts:

the ancient theology or *prisca theologia*; invisible spirits; microcosm and macrocosm; the occult; plastic nature; and Pythagorean harmony.

Finally, Topic II includes discussions of the following subtopics:

how science and religion in our period were related to Renaissance worldviews;

nonmechanical (including organic) cosmic images and symbols;

seventeenth-century precursors to belief in paranormal phenomena; and
seventeenth-century precursors to new-age religion.

Topic III: Protestantism and the Rise of Modern Science includes all of the debates concerning:
the role of Protestant Christianity in general in the origins or rise of modern science;
the specific role of Puritanism or Calvinism in such developments; and
the so-called "Merton thesis."
In connection to these controversies, Topic III includes the related discussions of:
the significance for science of Puritan theology, Puritan practice, Puritan motivation, and/or Puritan values;
science and "ascetic Protestantism";
the relationship of Puritanism to Baconianism;
educational issues, science, and Puritanism;
the role of experiential, experimental, and empirical orientations in Puritanism and science; and
questions as to who was and who was not a Puritan.
Topic III also includes discussions of the corresponding significance for the rise-of-science of:
Anglicanism, Latitudinarianism, and/or Catholicism;
religion and the Royal Society;
religion and the universities;
the role of capitalism in these matters; and
ideas of progress and programs of reform.
Finally, Topic III includes:
historiographic debates related to these issues;
discussions of the proper role of historical sociology as regards these issues;
debates concerning the definition of science as it applies to these issues; and
debates concerning the definition of Puritanism.

Topic IV: Christianity, Social Ideals, Ideology, and Science includes general discussions and debates about:
social values, ideals, and doctrines as an area of interplay between religion and science;
how science influences and is influenced by its social context;

the relationship of various "ideological" positions to science;
politics and political views as an area of interplay between science and religion;
science and religion in connection to class structure and popular culture; and
gender issues, science, and religion.

Topic IV thus includes more specific discussions about:

the social ideals of Latitudinarians, dissenters, high-church Anglicans, Catholics, Radical Protestants, and religious freethinkers in relationship to science;
Newtonianism as a socioreligious ideology;
social values motivating and justifying the pursuit of natural philosophy;
science and religion in connection to the "ancients versus moderns" controversy;
science, religion, and the belief in progress;
science, religion, and proposals for the reform of society;
science, religion, and proposals for the advancement of learning;
science, religion, and utopian schemes;
science, religion, and jurisprudence;
the Boyle Lectures as socioreligious ideology based on the new science;
science and millennialism; and
secrecy and openness as Christian, alchemical, and scientific ideals.

Topic V: Social Institutions, Science and Christianity includes discussions and debates about:

the universities in relation to science and religion;
institutions of government in relation to science and religion;
the Royal Society and religion;
religion and precursor groups of the Royal Society;
religion and the Royal College of Physicians;
the Church of England and science;
the Boyle Lectures and the new science; and
dissenter institutions and science.

Topic V also includes discussions of institutions as carriers or disseminators of:

Baconianism;
Aristotelian conservatism;

physico-theology;
religious support for the new science; and
Newtonian ideology.

Topic VI: Religion, Technology, Architecture and the Environment includes general discussions about:

technology and religion;
medical technology and religion;
the effects on theology of technological change;
the effects on theology of mechanical metaphors;
architecture and religion; and
religion and the natural environment.

Topic VI thus includes more specific descriptions of:

religious justifications of technology and technological progress;
technology and religiously motivated public service;
religion and the technology of warfare;
the clock metaphor for God;
how telescopic and microscopic observations influenced concepts of God;
how improved astronomical and mathematical methods influenced religious chronology and dating;
the printing press, religion, and science;
the magnet as a religious metaphor;
the architecture of churches;
how architectural and gardening metaphors influenced concepts of God, Christ, and Adam; and
religious attitudes towards nature (including "man's dominion over nature").

Topic VI also includes discussions which relate or correlate:

specific religious groups to support for technology;
social class to ideology in terms of technological and religious issues;
magic, alchemy, and astrology to religious and technological topics; and
architecture to various religious cosmologies.

Finally, this topic includes debates about:

religion, technology and the contextualist approach to "pure science";
religion, utopian technological dreams, and social reform; and
Christianity and the ecological crisis.

Topic VII: Theology, Philosophy and Science includes, in general, discussions of:

proposed ways in which science and theology were philosophically integrated;
the purported separation between natural philosophy (as science) and theology;
the purported "warfare" or conflict between natural philosophy (as science) and theology;
theology and the new philosophy or the mechanical philosophy;
science, theology, and the traditional schools of natural philosophy;
the changes in theology effected by natural philosophy or science; and
the impact on scientific theories of various theological positions.

This topic thus includes discussions about:

science, theology, and Neoplatonism;
science, theology, and Aristotelianism;
science, theology, and Stoicism;
science, theology, and Epicureanism;
science, theology, and scholasticism;
the theological implications of materialism;
the theological implications of atomism and other doctrines of matter;
the theological implications of a mechanistic worldview and/or an organic worldview;
science, theology, and the Great Chain of Being;
the natural philosophic implications of various doctrines of God (including voluntarism);
God's relation to nature;
God as a scientific hypothesis or an explanatory factor in natural philosophy;
science, theology, and the relation of humans to nature;
the logical and methodological boundaries between natural philosophy and different types of theology;
the theological overtones of certain scientific concepts (such as the aether and attraction);
the carrying of religious meanings into natural philosophy by the usage of natural metaphors such as light, life, rainbow, comets, the moon, and the sun;
science and miracles;
science and divine providence;

natural philosophy and ghosts or invisible spirits;
the changes in traditional Christian theology effected by science;
the effects of science on understandings of the relationship between reason and revelation; and
mathematics as it effected theology.

Topic VII also includes discussions of the interplay between science (especially geology) and theology concerning "the theory of the earth" (the creation, the fall of man, the flood, and the apocalyptic end of the world).

Topic VII includes the interplay between natural philosophy and theology in various systems of metaphysics, cosmology, and epistemology. More specifically, it includes discussions of the conceptual interaction of theology and natural philosophy in relation to various theories of: causality; determinism; proper scientific method; transduction or non-observables; natural law or the laws of nature; space; time; the void; gravity; active principles; the spirit of nature and plastic nature; embryology; music; body, mind, and soul; and immortality.

Finally, there is major overlap between Topics VII, VIII, and IX. Most of the subtopics of Topic VIII and Topic IX are also subtopics of Topic VII.

Topic VIII: Natural Theology and Natural Philosophy includes general discussions and debates about:
the three-way relationships (including patterns of separation) between revelation, natural theology, and natural philosophy;
arguments for the existence and attributes of God based on natural philosophy;
the effects (on literature, philosophy, and religion) of natural theology based on science; and
the usage of science-based natural theology for ideological purposes.

Topic VIII thus includes discussions about or descriptions of:
natural philosophy, different kinds of theology, and theories of knowledge;
revelation, natural theology, and "the mechanical philosophy";
the new science used as a basis for the cosmological argument (or "First Cause" argument);
the new science used as a basis for the teleological argument (or "argument from design");

natural theology used as part of the defense of "the new philosophy" and of the Royal Society;
natural theology as a connecting link between natural philosophy and views of God and of Providence;
the relationship of natural theology to attitudes towards nature;
the relationship of science to the Clockmaker God, the Newtonian God, and other such phrases;
natural theology as a connecting link in historical theses about the effects of science on the development of deism and secularism;
science-based natural theological argumentation as a source of scepticism; and
Newtonian natural theology as sociopolitical ideology.

Topic IX: Heretical Christianity, Deism, and Atheism includes discussions and debates about:

the influence of science on deism;
the influence of science on atheism;
science as a source of pantheism;
science as a source of theological skepticism regarding divine revelation, miracles and providence; and
the relation of science and famous scientists to Arianism, Socinianism, and unitarianism.

Topic IX thus also includes the following subtopics:

the connections between atheism and various forms of atomism, materialism, and/or the mechanical philosophy;
the mix of natural philosophy, radical religion, radical politics, and/or radical ideology;
science, accusations of godlessness, and literature; and
science, accusations of godlessness, and social ideology.

Topic X: Science, the Bible, and Literature, in general, includes all subtopics having to do with:

science, religion, and culture;
science, religion, and literature; and
the interplay between science and the Bible.

Topic X thus includes discussions and debates concerning the religious representation in literature of the following objects, images and concepts:

natural objects such as light, life, rainbows, comets, mountains, the moon, and the sun;

macrocosm and microcosm;
space, time, and infinity;
magnetism;
the circle metaphor;
garden metaphors; and
number symbolism.

Topic X also includes the discussions of literary-scientific explications ("the physics") of: the creation; the fall of man; the flood; the Apocalypse or end of the world; heaven and hell; and angels, devils, and ghosts. The first three of the subtopics above sometimes are called "Genesis and Geology."

In addition, Topic X includes discussions of the respective roles of science and religion in the following controversies:

the ancients versus moderns debates;
the decay-of-nature controversy;
the idea and reality of progress; and
plurality-of-worlds (life-on-other-planets) debates.

Topic X includes discussions of the effects of science on religious rhetoric and religious elements in literature as this applies to:

styles of preaching and style in written sermons;
usage of typological figures;
the role of and trust in literary imagination;
usage of ancient Greek and Roman authors;
usage of ancient myths and allegories; and
history writing and theories of history.

Also concerning the effects of science on religion, literature, and culture, Topic X includes all proposed theses and related discussions about:

the bad effects of science on literature and religion; and
the "two cultures" conflict (between science and the humanities).

Concerning the Bible, this topic includes discussions of:

the effects of science on biblical interpretation;
scientific influence on and correlations to biblical chronology;
famous scientists' views of the Bible; and
literary representations of Adam, Noah, and other biblical figures in a natural philosophical way.

Finally, Topic X includes discussions about:

attempts to develop a religious and scientific universal language;
claims about the language of God and the two-books theory of God's communication through nature and through the Bible;
Royal Society religious apologetics as literature;

the effects of the telescope and microscope on religious elements in literature;
literary representations of astrology and alchemy as magic or part of religion; and
science, accusations of godlessness, and literature.

Topic XI: Religion and Medicine includes discussions of:
the relationships in general between medicine and religion;
scientific, religious, and magical understandings of healing;
the religious views and practices of physicians;
religion and the institutions of medicine;
religious motivation in medicine;
medicine as religious calling and/or service;
theories about spirits, especially "vital spirit," in medical theory;
religio-medical theories of macrocosm and microcosm;
religio-medical theories of soul and body;
religion and embryology;
religion, medicine, and the plague;
controversies over miraculous healings (especially those of Valentine Greatrakes);
religion and the controversy over the circulation of the blood;
Cotton Mather and the inoculation controversy in New England;
explanations of venereal disease as the consequences of sin;
religion and proposals for the reform of medicine;
Newtonianism and medicine;
class analysis, medicine, and religion; and
gender issues, medicine, and religion.

Topic XII: Newtonian Studies, in general, includes discussions and debates about science and religion in connection:
to Isaac Newton, the man;
to those persons who consciously were followers of Newton in one aspect of their thought or another; and
to "Newtonianism" as a historical movement theoretically constructed by later historians.

This topic overlaps with each and every other major topical category used in this bibliography. Historiographically (Topic I), scholarly debates occur both with respect to Newton and to Newtonianism. These include arguments regarding:
"internalist" versus "externalist" approaches;
the propriety of "intellectualist" approaches;

"retrospective" or "Whig" historiography;
the implications of Marxist historiography;
the "warfare" and "separation" metaphors; and
scientific change and Kuhn's model of "paradigm."

As regards overlap with Topic II, Topic XII includes discussion and debates about:

Newton, alchemy, and religion;
Newton, Neoplatonism, and religion;
Newton and mysticism;
Newton and Jewish traditions; and
Newton, religion, and the *prisca tradition*.

In overlapping Topics III, IV, V, and VI, Topic XII includes discussions and debates about:

Newton and Puritanism;
Newton as an example of "Protestantism-and-the-rise-of-modern-science";
Newtonianism as a socioreligious ideology:
Newton, religion, and the Royal Society;
Newton and the Boyle Lectures;
Newtonianism and the clock metaphor for God; and
Newtonianism and the architect metaphor for God.

The huge overlap between Topic XII and Topic VII (philosophy) includes explications of:

Newton and the theological implications of mechanism and materialism;
Newton and the theological implications of atomism and other doctrines of matter;
Newtonianism and the theological implications of a mechanistic worldview;
Newton's voluntaristic understanding of God;
God as a scientific hypothesis or an explanatory factor in Newton's and Newtonian natural philosophy;
the logical and methodological boundaries, for Newton, between natural philosophy and different types of theology;
the theological overtones of certain Newtonian concepts (such as the aether and attraction);
Newton's and Newtonian views of miracles;
Newton's and Newtonian views of divine providence;
the changes in traditional Christian theology effected by Newtonianism;
Newtonian mathematics as it affected theology.

Scholars also take up the interplay between natural philosophy and theology with respect to Newton's and Newtonian understanding of:

causality;
proper scientific method;
transduction or nonobservables;
natural law or the laws of nature;
space, time, and the void;
gravity;
active principles;
music; and
"theory of the earth."

Finally, overlapping Topic VII are discussions about:

Newton and stoicism;
Newton's differences with Leibniz; and
the Clarke-Leibniz debates.

Concerning Topic VIII (natural theology and natural philosophy), Topic XII includes discussions and debates about:

the three-way relationships in Newton's thought (including patterns of separation) between revelation, natural theology, and natural philosophy;
arguments for the existence and attributes of God based on Newton's natural philosophy;
the effects (on literature, philosophy, and religion) of Newtonian natural theology; and
the usage of Newtonian natural theology for ideological purposes.

Topical XII overlaps with Topic IX as regards the following subtopics:

the influence of Newtonian science and natural theology on deism;
the influence of Newtonian science on atheism;
Newtonian science and pantheism;
Newtonian science as a source of theological skepticism regarding divine revelation, miracles and providence; and
the relation of Newton and his followers to Arianism, Socinianism, and unitarianism.

As regards overlap with Topic X (science, the Bible, and literature), Topic XII includes the following discussions and debates:

Newtonianism and culture;
Newton and literature;
Newton and the Bible;
Newton and biblical chronology;

the bad effects of Newtonian science on literature and religion;
Newtonianism and the "two cultures";
Newton as religious cultural symbol; and
Newton's relationship to the Enlightenment.

Finally, some scholars also write about Newtonianism and medicine.

Periodical Abbreviations

Ambix	*Ambix: Journal of the Society for the History of Alchemy and Chemistry*
Am. Hist. Rev.	*American Historical Review*
Am. J. Sociol.	*American Journal of Sociology*
Am. Philos. Q.	*American Philosophical Quarterly*
Ann. Sci.	*Annals of Science: Quarterly Review of the History of Science since the Renaissance*
Arch. Gesch. Phil.	*Archiv fur Geschichte der Philosophie*
Arch. Philos.	*Archiv fur Philosophie*
Arch. Hist. Exact Sci.	*Archive for History of Exact Sciences*
Arch. Int. d'Hist. Sci.	*Archives Internationale d'Histoire des Sciences*
Brit. J. Hist. Sci.	*British Journal for the History of Science*
Brit. J. Philos. Sci.	*British Journal for the Philosophy of Science*
Brit. J. Sociol.	*British Journal of Sociology*
Bull. Hist. Med.	*Bulletin of the History of Medicine*
Cambridge Hist. J.	*Cambridge Historical Journal*
Can. J. Philos.	*Canadian Journal of Philosophy*
Church Q. Rev.	*Church Quarterly Review*
Daedalus	*Daedalus: Journal of the American Academy of Arts and Sciences*
Diogenes	*Diogenes: A Quarterly Publication of the International Council for Philosophy and Humanistic Studies*

ELH	*ELH: A Journal of English Literary History*
Eng. Lang. Notes	*English Language Notes*
Harv. Theo. Rev.	*Harvard Theological Review*
Hermathena	*Hermathena: A Dublin University Review*
Hist. Scientiarum	*Historia Scientiarum: International Journal of the History of Science Society of Japan*
Historian	*The Historian*
Hist. J.	*The Historical Journal*
Hist. Stud. Phys. Bio. Sci.	*Historical Studies in the Physical and Biological Sciences*
Hist. Stud. Phys. Sci.	*Historical Studies in the Physical Sciences*
Hist. Euro. Ideas	*History of European Ideas*
Hist. Ideas News.	*History of Ideas Newsletter*
Hist. Sci.	*History of Science*
Hist. Technol.	*History of Technology*
Hunt. Lib. Q.	*Huntington Library Quarterly*
Ideas and Production	*Ideas and Production: A Journal in the History of Ideas*
Int. Philos. Q.	*International Philosophical Quarterly*
Int. Stud. Philos. Sci.	*International Studies in the Philosophy of Science*
Isis	*Isis: International Review Devoted to the History of Science and Its Cultural Influences*
Janus	*Janus: Revue Internationale de l'Histoire des Sciences, de la Medecine et de la Technique*
J. Hist. Astron.	*Journal for the History of Astronomy*
J. Brit. Stud.	*Journal of British Studies*
J. Eccles. Hist.	*Journal of Ecclesiastical History*
J. Engl. Germ. Philol.	*Journal of English and German Philology*
J. Euro. Stud.	*Journal of European Studies*
J. Mod. Hist.	*Journal of Modern History*
J. Rel.	*Journal of Religion*
J. Rel. Ethics	*Journal of Religious Ethics*

J. Hist. Astron.	*Journal of the History of Astronomy*
J. Hist. Ideas	*Journal of the History of Ideas*
J. Hist. Med.	*Journal of the History of Medicine and Allied Sciences*
J. Hist. Philos.	*Journal of the History of Philosophy*
J. Soc. Bibliog. Natur. Hist.	*Journal of the Society for the Bibliography of Natural History*
J. Warb. Court. Inst.	*Journal of the Warburg and Courtauld Institute*
J. World Hist.	*Cahiers d'Histoire Mondial/Journal of World History (International Commission for a History of the Scientific and Cultural Development of Mankind)*
Knowledge and Society	*Knowledge and Society: Studies in the Sociology of Culture Past and Present*
Listener	*The Listener*
Locke Newsletter	*The Locke Newsletter*
Metascience	*Metascience: Annual Review of the Australian Association for the History, Philosophy, and Social Studies of Science*
Mod. Lang. Notes	*Modern Language Notes*
Mod. Lang. Q.	*Modern Language Quarterly*
Modern Schoolman	*The Modern Schoolman*
Monist	*The Monist*
New Engl. Q.	*New England Quarterly*
Notes Rec. Royal Soc. London	*Notes and Records of the Royal Society of London*
Paedagogica Europaea	*Paedagogica Europaea: The European Yearbook of Educational Research*
Pap. Bibliogr. Soc. Amer.	*The Papers of the Bibliographic Society of America*
Philol. Q.	*Philological Quarterly*
Philos. Q.	*Philosophical Quarterly*
Philos. Rev.	*The Philosophical Review*
Philos. Sci.	*Philosophy of Science*
PMLA	*PMLA: Publications of the Modern Language Association of America*

Proceed. Amer. Philos. Soc.	*Proceedings of the American Philosophical Society Held at Philadelphia for Promoting Useful Knowledge*
Proc. Mass. Hist. Soc.	*Proceedings of the Massachusetts Historical Society*
Proc. Royal Irish Acad.	*Proceedings of the Royal Irish Academy*
Proc. Royal Soc. A.	*Proceedings of the Royal Society of London, Series A*
Pub. Colonial Soc. Mass. Trans.	*Publications of the Colonial Society of Massachusetts. Transactions*
Rel. Stud. Rev.	*Religious Studies Review*
Renaiss. Mod. Stud.	*Renaissance and Modern Studies*
Renaiss. Stud.	*Renaissance Studies: Journal of the Society for Renaissance Studies*
Review (Indiana Univ.)	*Review (Indiana University, College of Arts and Sciences, Graduate School Alumni Association)*
Rev. English Stud.	*Review of English Studies*
Rev. Metaphysics	*The Review of Metaphysics*
Rev. d'Hist. Sci.	*Revue d'Histoire des Sciences*
Riv. Crit. Stor. Fil.	*Rivista Critica di Storia della Filosofia*
Scot. J. Theology	*Scottish Journal of Theology*
Seventeenth Cent.	*The Seventeenth Century*
Skeptic	*Skeptic: A Quarterly Publication of the Skeptics Society*
Smith Coll. Stud. Mod. Lan.	*Smith College Studies in Modern Languages*
Soc. Stud. Sci.	*Social Studies of Science*
Stud. Hist. Philos. Sci.	*Studies in History and Philosophy of Science*
Texas Q.	*Texas Quarterly*
Thomist	*The Thomist*
Trans. Royal Hist. Soc.	*Transactions of the Royal Historical Society*
Univ. Toronto Q.	*University of Toronto Quarterly*
Wash. Univ. Stud.	*Washington University Studies*
Wm. Mary Q.	*William and Mary Quarterly*

An Annotated Bibliography of Books and Articles

0001 Aaron, Richard I. *John Locke*. 3rd ed. Oxford: Clarendon Press, 1971. (First edition, 1937.)

This book is a systematic explication of Locke's thought by a historian of philosophy. Only minor changes have been made in the text since the first edition of 1937; the bibliographic references have been updated through the 1960s. The 1971 edition does include a revised version of Aaron's essay "The Limits of Locke's Rationalism." It includes attention both to Locke's natural philosophy and to his theology—but not specifically to their interaction or relationship in Locke's thought. Aaron sees Locke as both reacting against and influenced by Descartes; he sees Locke as both an "empiricist" and a "rationalist."

Topic: *VII* Names: Cudworth, Locke, More, Norris

0002 Aarsleff, Hans. "Some Observations on Recent Locke Scholarship." In *John Locke: Problems and Perspectives*, edited by John Yolton, 262-271. London: Cambridge University Press, 1969. See (1715).

Aarsleff condemns previous Locke scholars for their concern with pseudo-problems and their emphasis on confusion and contradiction in Locke. This essay is marginally relevant to our field; but it includes a good notice of the theological motives of Locke (along with Bacon, Boyle, and the Royal Society).

Topics: *I*, VII Names: Bacon, Boyle, Locke

0003 Aarsleff, Hans. "The State of Nature and the Nature of Man in Locke." In *John Locke: Problems and Perspectives*, edited by John W. Yolton, 99-136. London: Cambridge University Press, 1969. See (1715).

This essay mainly focuses on Locke's political philosophy and ethics—but it includes a good analysis of how the different aspects of Locke's thought are interrelated and consistent. Aarsleff also discusses Locke's understanding of natural theology and revelation (and relates this to the first two chapters of Paul's Epistle to the Romans).

Topics: *VII*, IV, VIII Name: Locke

0004 Abraham, Gary A. "Misunderstanding the Merton Thesis: A Boundary Dispute between History and Sociology." *Isis* 74(1983): 368-387. Abridged in *Puritanism and the Rise of Modern Science*, edited by Bernard I. Cohen, 233-245. New Brunswick, N.J.: Rutgers University Press, 1990. See (0246).

Abraham argues that historians: (1) have shown a disciplinary bias in their approach to the Merton thesis; (2) have harmfully changed the terms of the thesis; and (3) have failed even to understand Merton's *sociological* perspective. Abraham criticizes all of the major historical writings—both friendly to Merton and hostile—as simply missing the point. That point is that social values and cultural ideas ("institutions" in the sense used by sociologists) are what is at stake: it is a mistake to focus on theological doctrines, specific individuals, or specific writings as such. Likewise, argues Abraham, focusing on institutions (in the nonsociological sense of organizations) as a test case is also a misunderstanding of Merton's claims.

Topics: *III*, I, IV

0005 Acton, Henry. *The Religious Opinions of Milton, Locke, and Newton*. London, 1833. Published in the United States as *Religious Opinions and Example of Milton, Locke, and Newton*. Boston: Charles Bowen, 1833. Reprinted, New York: AMS Press, 1973.

This is a pamphlet which reproduces a sermon-like public lecture. Acton was a nineteenth-century Unitarian who here argues that Milton, Locke, and Newton were all devout, Christian, morally upright—and unitarian.

Topics: *VII*, X, XII Names: Locke, Milton, Newton

0006 Adams, Robert P. "The Social Responsibilities of Science in *Utopia*, *New Atlantis* and after." *J. Hist. Ideas* 10(1949): 374-398.

Adams compares Thomas More's *Utopia* and Bacon's *New Atlantis* for their respective understandings of the relationships between science, religion, and power. He finds that More has a greater contribution to make (than does Bacon) regarding the social responsibilities of science. Bacon's optimism about the goodness of the scientists, for Adams, makes him an impractical dreamer. The article concludes with a discussion of current (1949) issues resulting from the development of the atomic bomb.

Topic: *IV* Name: Bacon

0007 Agassi, Joseph. "The Ideological Import of Newton." *Vistas in Astronomy* 22(1979): 419-430. Reprinted in *Science and Society*, by Joseph Agassi, 372-387. Dordrecht: D. Reidel, 1981. See (0010).

Agassi is concerned with the question: how could Newton's absolute authority live side-by-side with the Enlightenment view of the autonomy of the rational individual? He uses Levi-Strauss' thesis that myths come in pairs of conflicting qualities and Newton thus represented both poles. Agassi also poses the corollary historiographic question about the neglect (or suppression) of Newton's alchemical and theological writings. In this case, Agassi finds an ideological choice of Newton the rational positivist over (its opposite) Newton the metaphysical genius.

Topics: *XII*, I, II, VII Name: Newton

0008 Agassi, Joseph. "The Origins of the Royal Society." *Organon* (Warsaw) 7(1970): 117-135. Reprinted in *Science and Society*, by Joseph Agassi, 352-371. Dordrecht: D. Reidel, 1981. See (0010).

This is a lengthy critique of Margery Purver's *The Royal Society: Concept and Creation* (1270). Agassi concludes that Purver's search for antecedent groups is "neither possible nor interesting." For Agassi, both the religious and the scientific backgrounds of the individual members—and hence of the organization as a whole—are more complex and confused than her radical Baconian thesis indicates.

Topics: *V*, III Names: Bacon, Boyle, Sprat, Wallis

0009 Agassi, Joseph. "Robert Boyle's Anonymous Writings." *Isis* 68(1977): 284-287.

In this note, Agassi postulates that Boyle wrote a few of his theological works anonymously because he disliked disputes (especially among Christians). Truly anonymous writings, Agassi says, begin with the scientific revolution when authorship becomes public.

Topic: *X* Name: Boyle

0010 Agassi, Joseph. *Science and Society: Studies in the Sociology of Science*. Dordrecht: D. Reidel, 1981.

This book is a collection of articles by Agassi (mainly having to do with philosophy of science). An article on Newtonian historiography (0007) and an essay review of Purver's *Royal Society* (0008) are listed separately in this bibliography.

Topic: *I*

0011 Agassi, Joseph. *Towards an Historiography of Science*. (Beiheft 2 of *History and Theory*.) The Hague: Mouton, 1963.

Agassi, a philosopher of science, harshly condemns most of the histories of science then in existence and explains why they are so bad. He sees bad philosophy of science as the cause of bad history of science. He advocates "Popper's methodology as a means of improving the present lamentable state of affairs in the field of history of science." Most of Agassi's examples are post-1720 but the work claims historiographic applicability to all periods.

Topic: *I*

0012 Aiton, Eric J. *Leibniz: A Biography*. Boston: A. Hilger, 1985.

Aiton includes brief notices having to do with Locke, Newton, Clarke, and Princess Caroline—from Leibniz's point of view. He discusses the religious differences between Leibniz and the English philosophers.

Topics: *VII*, XII Names: Caroline, Clarke, Newton

0013 Albee, Ernest. "The Philosophy of Cudworth." *Philos. Rev.* 33(1924): 245-272.

This is an explication and evaluation of Cudworth's philosophy by a writer sympathetic to modern idealism. Albee's emphasis is on

moral theory but he includes reference to Cudworth's epistemology, theology, philosophy of nature, and atomism.
Topic: *VII* Name: Cudworth

0014 Albury, W. R. "Halley's Ode on the *Principia* of Newton and the Epicurean Revival in England." *J. Hist. Ideas* 39(1978): 24-43.
Albury argues that Halley's ode was based upon passages from Lucretius which expressed skeptical theological views but that Newtonians interpreted it (and even changed Halley's lines) in terms of orthodox physico-theology.
Topics: *XII*, VII, X Name: Halley

0015 Alexander, H. G., ed. *The Leibniz-Clarke Correspondence.* Manchester, England: Manchester University Press, 1956.
The useful introduction is relevant to issues having to do with space, time, God, and natural theology.
Topics: *XII*, VII, VIII Names: Clarke, Newton

0016 Allen, Don Cameron. *Doubt's Boundless Sea: Scepticism and Faith in the Renaissance.* Baltimore: Johns Hopkins Press, 1964.
This is a survey of sixteenth- and seventeenth-century European writers on the topics of skepticism, atheism, the existence of God, the immortality of the soul, natural theology, and the value of revelation. The book reads like a series of reading notes—jumping from one author to another, with gratuitous judgments thrown in for good measure. A few seventeenth-century Englishmen are summarized, including Henry More and Charles Blount. Natural philosophy is mentioned in passing but not emphasized. Heresy and atheism are seen to follow from reasoning in general, not specifically from science.
Topics: *IX*, VII, VIII, X Names: Blount, More

0017 Allen, Don Cameron. *The Legend of Noah: Renaissance Rationalism in Art Science, and Letters.* (Illinois Studies in Language and Literature, Vol. 33.) Urbana Ill.: University of Illinois Press, 1949.
This is a study of the attempts to provide a rational explanation for the legend of Noah. Allen's primary concern is with poets and

artists who were sensitive to the artistic pleasures of the myth—but who also were aware of attempts to establish its truth by reason. One chapter concerns the new science and the question of the universality of the flood; it is preceded by lengthy discussions of the history of biblical interpretation and the relations between faith and reason.

Topics: *X*, VII Names: Bacon, Boyle, Browne, T., Burnet, T., Donne, Hale, Milton

0018 Allen, Don Cameron. *The Star-Crossed Renaissance: The Quarrel about Astrology and Its Influence in England*. Durham, N.C.: Duke University Press, 1941.

This is a survey of astrology in literature from fifteenth-century Italy to seventeenth-century England. It was original in its time, opening up the topic and showing that most Renaissance thinkers and writers (more or less) believed in astrology. In the chapters on Elizabethan and Jacobean England, Allen proposes that most literary figures held a moderate position regarding astrology. He also suggests that theological discussions concerning free will and the providence of God formed the basis for questions about astrology. Allen seems to treat the opinions throughout the period as essentially static.

Topics: *II*, X, XI Names: Bacon, Carleton, Chamber, Heydon, Melton

0019 Allen, Phyllis. "Medical Education in Seventeenth-Century England." *J. Hist. Med.* 1(1946): 115-143.

Allen summarizes the roles of various institutions in educating physicians in the period. She implies relationships between medicine and religion in her discussions of licensing by the Church of England, Laud's Caroline Code at Oxford, and the Dissenting academies after the Restoration.

Topics: *XI*, V

0020 Allen, Phyllis. "Scientific Studies in the English Universities of the Seventeenth Century." *J. Hist. Ideas* 10(1949): 219-253.

Allen discusses the ups and downs of science in the universities in terms of its acceptance and success at different periods during the century. She sees a noncontinuous but finally successful development of science in gaining a place in the universities. Generally,

Allen ascribes opposition to the new philosophy to Aristotelianism and other types of conservatism not specifically religious. She sees a positive role in the support of scientific studies for Puritanism.
Topics: *V*, III, IV

0021 Almond, Philip C. "Henry More and the Apocalypse." *J. Hist. Ideas* 54(1993): 189-200.
Almond describes More as using reason to interpret Biblical texts in his apocalyptic writings. In so doing, he reads the Bible in opposition to radical millennialists: for More, Christ's millennial reign is a heavenly one. Thus, More emphasizes the advancement of knowledge in a stable society—with the apocalypse in the distant future.
Topics: *X*, IV Name: More

0022 Anderson, Fulton H. *Francis Bacon, His Career and Thought.* Los Angeles: University of Southern California Press, 1962.
Roughly 240 pages are devoted to apologetic biographical chapters and 110 pages to Bacon's thought. The presentation of Bacon's thought centers on his natural philosophic work with much summary of his writings. Anderson emphasizes the separation of natural philosophy from revealed religion for Bacon. This book is aimed at a nonscholarly audience.
Topic: *VII* Name: Bacon

0023 Anderson, Fulton H. *The Influence of Contemporary Science on Locke's Method and Results.* (University of Toronto Studies in Philosophy, Vol. 2.) Toronto: The University Library, 1923.
This is a thirty-one page dissertation summary. It emphasizes the influence of Boyle and Sydenham on Locke. It has been superseded by later works on Locke, including Yolton (1715).
Topic: *VII* Names: Boyle, Locke, Sydenham

0024 Anderson, Fulton H. *The Philosophy of Francis Bacon.* Chicago: University of Chicago Press, 1948.
Anderson presents Bacon's philosophy and his views of other philosophers by outlining and summarizing Bacon's writings. He organizes the ideas topically and does not try to assess Bacon's positions. He classifies Bacon as a rationalist and realist—not a minor empiricist. For our purposes, the chapters entitled "Clas-

sification of the Sciences Respecting God and Nature" and "Suing for Science: Attack upon the Universities" are most useful. Anderson emphasizes Bacon's separation of humanly discoverable truth from the "dogmas of revealed theology." This was an important book in its time.

Topics: *VII*, IV Name: Bacon

0025 Anderson, Paul Russell. *Science in Defense of Liberal Religion: A Study of Henry More's Attempt to Link Seventeenth Century Religion with Science*. New York: G. P. Putnam's Sons, 1933.

Anderson discusses More's important views on matter, space, Cartesian physics, and God as an appropriation of science to defend religion. He discusses the influence of religion on More's natural philosophy and the influence of seventeenth-century science on religion in general. In a way similar to E. A. Burtt (0160), Anderson shows irritation at the "bad" influence of science on religion. He briefly addresses the topic of Royal Society apologetics.

Topics: *VII*, V, VIII Name: More

0026 Andrade, Edward Neville da Costa. *Isaac Newton*. London: M. Parrish, 1950.

This is a slightly different version of (0027).

Topic: *XII* Name: Newton

0027 Andrade, Edward Neville da Costa. *Sir Isaac Newton: His Life and Works*. London: Collins, 1954.

This is a good popular and short biography. It is not particularly strong on Newton's religion but Andrade at least mentions its importance.

Topic: *XII* Name: Newton

0028 Anselment, Raymond A. "Seventeenth-Century Pox: The Medical and Literary Realities of Venereal Disease." *Seventeenth Cent*. 4(1989): 189-211.

Anselment argues that the seventeenth-century attitude to venereal disease was neither a lighthearted acceptance nor a neutral medical analysis. Rather, there was a social and moral judgment involved which commonly included the view that venereal disease was a divine punishment for sexual sin.

Topics: *XI*, X

0029 Ardolino, Frank. "The Saving Hand of God: The Significance of the Emblematic Frontispiece of the *Religio Medici.*" *Engl. Lang. Notes* 15(1977): 1923.
Ardolino indicates that Browne retained (throughout his editions of the *Religio Medici*) the image of a man falling from the rock and of God's hand reaching down to save him from the sea. The man, rock, sea and hand of God each have symbolic meaning. The hand represents the redeeming, healing providence of God.
Topics: *XI*, X Name: Browne, T.

0030 Armistead, J. M. "The Occultism of Dryden's 'American' Plays in Context." *Seventeenth Cent.* 1(1986): 127-152.
Armistead describes the impact of science on the representation of the occult in Restoration drama generally—and particularly with respect to Dryden's early plays. He argues that *The Indian Queen* (1664) and *The Indian Emperor* (1665) contain the first genuine supernatural episodes in Restoration drama. One can link Dryden's poetic and dramatic practice to "the curious blend of science and pseudo-science in Restoration thought." The audience, influenced by empirical science, lacked sympathy for the cruder forms of magic and demonology but retained confidence in divine Providence and the spirit world. Spiritual intelligences were accepted "as part of 'scientifically' definable reality." Armistead also argues that the demonology of Dryden's plays can be related to the writings of Henry More and Joseph Glanvill.
Topics: *X*, II Names: Dryden, Glanvill, More

0031 Armitage, Angus. "Rene Descartes and the Early Royal Society." *Notes Rec. Royal Soc. London* 8(1950): 1-9.
In this survey, Armitage summarizes some of the writings of Descartes and briefly notes the reactions to Descartes of selected members of the Royal Society. This essay has been superseded by many later authors, including Gabbey (0521-0523) and Pacchi (1190).
Topics: *VII*, XII Name: Newton

0032 Armstrong, Robert L. "The Cambridge Platonists and Locke on Innate Ideas." *J. Hist. Ideas* 30(1969): 187-202.
Armstrong argues that the Cambridge Platonists used the doctrine of innate ideas to support their moral and religious values (they feared the new science as materialistic and subversive of moral and

religious values). Locke, on the other hand, was not worried about any threat to morality and religion from the new science. He attacked the doctrine of innate ideas and, indeed, later philosophers saw it as having been refuted by Locke.
Topic: *VII* Names: Cudworth, Locke

0033 Armytage, W. H. G. "The Early Utopists and Science in England." *Ann. Sci.* 12(1956): 247-254.
This is a rambling essay on utopian writers, natural philosophers, alchemists, and religious and social reformers. It has been superseded by Webster (1611-1614; 1616; 1620; 1625; 1627) and Hill (0690; 0699).
Topic: *IV*

0034 Ashcraft, Richard. "Faith and Knowledge in Locke's Philosophy." In *John Locke: Problems and Perspectives*, edited by John W. Yolton, 194-223. London: Cambridge University Press, 1969. See (1715).
Ashcraft argues that Locke's philosophy (and especially his epistemology) must be read within Locke's seventeenth-century context. "Bringing together man's ignorance and God's wisdom" is Locke's aim and thus produces not a modern philosophy but a worldview characteristic of his era. Locke's treatment of innate ideas, knowledge, faith, reason, and probability drew criticism then and appears to be logically inconsistent now. But if we understand Locke's consistent theological ends, we can understand why he wrote what he did. This is an excellent essay.
Topics: *VII*, VIII Name: Locke

0035 Ashcraft, Richard. "John Locke's Library: Portrait of an Intellectual." *Transactions of the Cambridge Bibliographical Society* 5(1969): 47-60. Reprinted in *A Locke Miscellany*, edited by Jean S. Yolton, 226-245. Bristol: Thoemmes, 1990. See (1712).
This analysis is based upon and goes beyond Harrison and Laslett's *The Library of John Locke*. Ashcraft compares Locke's holdings to a survey of eighty other contemporary personal libraries. He discusses Locke's holdings in religion and in science—as well as in philosophy, politics, economics, and the arts. Ashcraft assumes a rather rigid compartmentalization but the essay is still useful.
Topic: *VII* Name: Locke

0036 Ashworth, William B., Jr. "Catholicism and Early Modern Science." In *God and Nature*, edited by David C. Lindberg and Ronald Numbers, 136-166. Berkeley and Los Angeles: University of California Press, 1986. See (0954).

Ashworth restricts his discussion to the continent but it still is relevant to the Protestantism and the rise of science debates.

Topic: *III*

0037 Aspelin, Gunnar. *John Locke: Tankaren och Upplysningsmannen*. Lund: Gleerup, 1948.

This is written in Swedish ("John Locke: Thinker and Enlightenment Man.") It primarily is a study of the *Essay Concerning Human Understanding*. The chapters on the intellectual background and on Locke as thinker place him in the context both of the natural philosophy and of the theological issues of the time.

Topic: *VII* Name: Locke

0038 Aspelin, Gunnar. "Locke and Sydenham." *Theoria* 15(1949): 29-37.

Aspelin shows that the ideas expressed in Locke's *Essay* are strikingly similar to the conception of scientific method expressed by the London physician, Thomas Sydenham. "And behind this attitude, we find, in Sydenham as well as in Locke, a religious view of life that is typical for English scientists during the seventeenth century."

Topics: *VII*, XI Names: Locke, Sydenham

0039 Aspelin, Gunnar. "The Polemics in the First Book of Locke's Essay." *Theoria* 6(1940): 109-122.

Aspelin argues that the object of Locke's polemics (against the theory of innate ideas) is the conservative academic tradition in philosophy, religion, and natural philosophy. Locke is trying to free the situation for the new philosophy and true religious fulfillment. Locke, like Glanvill, believes that innatism and the philosophy of the schools lead to meaningless phrases and intellectual laziness. But God gave us our reasoning faculties in order that we gain the kingdom of knowledge through diligence and industry. Aspelin sees Locke's theory of knowledge as related to "his Puritan valuation of labor as the calling of man."

Topics: *VII*, III, IV, V Names: Glanvill, Locke

0040 Aspelin, Gunnar. "Ralph Cudworth's Interpretation of Greek Philosophy." Translated by Martin S. Allwood. *Goteborgs Hogskolas Arsskrift* 49(1943): 1-47.

This is a short monograph which takes up the whole issue of Volume 49, No. 1, of the journal. Aspelin demonstrates Cudworth's conviction that religious and scientific truth was manifest in the beginning of human history, was then distorted, but now is being restored again to its original purity. The author follows Cudworth's views of both the history of cosmology and the history of religion through the Greeks to their Jewish and Egyptian sources. Thus, Aspelin places Cudworth in a Renaissance Neoplatonic tradition which aimed at a harmony between the new science and Christian philosophy.

Topics: *VII*, II Name: Cudworth

0041 Atherton, Margaret, ed. *Women Philosophers of the Early Modern Period*. Indianapolis: Hackett, 1984.

This anthology is designed to be a supplementary text for courses in the history of philosophy. Atherton's introductions to selections from seventeenth-century women philosophers are good, brief summaries of their lives and thought.

Topic: *VII* Names: Cavendish, Cudworth, Masham

0042 Atkinson, A. D. "William Derham, F.R.S. (1657-1735)." *Ann. Sci.* 8(1952): 368-392.

This is a pleasant biographical essay on the author of *Physico-Theology* and *Astro-Theology*. It is based on primary sources.

Topics: *VIII*, VII Name: Derham

0043 Attfield, Robin. "Christian Attitudes to Nature." *J. Hist. Ideas* 44(1983): 369-386.

Attfield summarizes and critiques the theses of Lynn White (in a famous article focusing on the medieval era), John Passmore (1207), and William Coleman (0250) regarding the Christian roots of the ecological crisis. Attfield finds much more complexity and diversity among Christian attitudes to nature. He directs specific attention to the examples of seventeenth- and early-eighteenth-century natural theology used by Passmore and Coleman.

Topics: *VI*, VII, VIII Names: Derham, Grew, Hale, Ray

0044 Attfield, Robin. "Clarke, Collins, and Compounds." *J. Hist. Philos.* 15(1977): 45-54.

Attfield provides a summary and philosophical analysis of the debate between Samuel Clarke and Anthony Collins in 1706-1708. The debate was largely over the materiality or immateriality of the soul and the correspondence ran to over 400 pages.

Topics: *VII*, IX Names: Clarke, Collins, A.

0045 Attfield, Robin. *The Ethics of Environmental Concern*. Oxford: Blackwell, 1983.

Portions of Chapters 3 and 4 are drawn from Attfield's article, (0043)—including his critiques of White, Passmore, and Coleman. He aims to show that the Christian tradition has resources that can be used to develop an ethic of stewardship. In Chapter 5, he suggests that Locke's revision of the doctrine of human sin forms a basis both for the subsequent belief in progress and for ethical activity.

Topics: *VI*, VII, VIII Names: Bacon, Hale, Locke, Ray

0046 Austin, William H. "Isaac Newton on Science and Religion." *J. Hist. Ideas* 31(1970): 521-42.

Austin discusses what he calls "Newton's maxim"—which is the principle that science and theology should be kept separate.

Topics: *XII*, VII, VIII Name: Newton

0047 Axtell, James L. Introduction to *The Educational Writings of John Locke*, edited by James L. Axtell, 1-104. Cambridge: Cambridge University Press, 1968.

In the section on "Locke and Scientific Education," Axtell describes Locke's views on the order of subject areas which an educated person should follow in study. Locke recommends "the study of metaphysics, and of the nature and qualities of spirits, before that of natural philosophy." And metaphysics includes natural theology based on the phenomena of nature. Axtell also discusses Locke's participation in the Royal Society, Locke's education, and his roles in Lord Shaftesbury's household.

Topics: *IV*, VII, VIII, XI Names: Coste, Locke

0048 Axtell, James L. "Locke, Newton, and the Elements of Natural Philosophy." *Paedagogica Europaea* 1(1965): 235-244.

Axtell argues that Locke was truly interested in science and worked with Newton on his philosophy of scientific education.
Topics: *VII*, XII Names: Locke, Newton

0049 Axtell, James L. "Locke, Newton and the Two Cultures." In *John Locke: Problems and Perspectives*, edited by John W. Yolton, 165-182. London: Cambridge University Press, 1969. See (1715).

Taking C. P. Snow's distinction between the two intellectual cultures (scientific and literary), Axtell presents Locke as an intellectual polymath who was the first "Newtonian philosopher without the help of Geometry." He sees Locke as an early and important popularizer of Newton's *Principia* to nonspecialists not skilled in mathematics. Religion is treated as barely relevant to either culture.
Topics: *VII*, X, XII Names: Locke, Newton

0050 Ayers, Michael R. "Mechanism, Superaddition, and the Proof of God's Existence in Locke's Essay." *Philos. Rev.* 90(1981): 210-251.

This is an excellent article. Ayers explains Locke's understanding of "mechanism" and "superaddition" both in the context of the "thinking matter" debates and in the context of his proof of the existence of God. Although he does analyze weaknesses in Locke's arguments, the focus consistently is on Locke's meaning (rather than on whatever we would mean if we uttered selected sentences from Locke's writings). Ayers argues that, for Locke, "superaddition" does not entail the supernatural or miraculous. This article sheds light on Locke's natural theology, on his natural philosophic "agnosticism," and even on his understanding of the ontological status of gravity.
Topics: *VII*, VIII Name: Locke

0051 Aylmer, G. E. "Unbelief in Seventeenth-Century England." In *Puritans and Revolutionaries*, edited by D. Pennington and K. Thomas, 22-46. Oxford: Clarendon Press, 1978. See (1219).

Aylmer finds few actual atheists (but several men who might later have been called agnostics). Yet he finds many anti-atheist Christian apologists. He distinguishes between popular scoffers, genuine philosophic doubters, and silent cynics. Treating the

subject by chronological eras, Aylmer finds religious diversity and atomism as two perceived causes of atheism among the Christian writers.
Topics: *IX*, IV, VII

0052 Backscheider, Paula R., ed. *Probability, Time, and Space in Eighteenth-Century Literature*. New York: AMS Press, 1979.
This is a collection of thirteen papers produced by a Modern Language Association group which focuses on "Problems in Eighteenth-Century Evidence." The essays by Beck (0074), LeClerc (0940), Milic (1069), and Osler (1180) are relevant to our field and are annotated individually in this bibliography.
Topic: *X*

0053 Bainton, Roland H. "Comment on R. Hooykaas' Science and Reformation." *J. World Hist.* 3(1957): 140-141.
Bainton generally agrees with Hooykaas but wants to modify his position. He makes the interesting point (for the 1950s) that the idea of the millennium (leading to the concept of the Holy Commonwealth) was an important factor in Puritan social thought.
Topics: *III*, IV

0054 Baker, Herschel. *The Wars of Truth: Studies in the Decay of Christian Humanism in the Earlier Seventeenth Century*. Cambridge, Mass.: Harvard University Press, 1952.
Baker aims "to indicate the intellectual and emotional pressures which shaped men's conception of 'truth' and of their capacity to attain it, and to suggest some of the consequences for literature." He assesses the place of religion, natural philosophy, and Renaissance humanism on many issues of a socio-ideological kind: the ancients vs. moderns debate, history and providence, and the idea of progress. The book is written in a judgmental, sometimes sarcastic, style.
Topics: *X*, IV Names: Bacon, Hobbes, Milton

0055 Baker, John Tull. "The Emergence of Space and Time in English Philosophy." In *Studies in the History of Ideas*, Vol. 3, 273-296. New York: Columbia University Philosophy Department, 1935. Reprinted, New York: AMS Press, 1970.

This is a schematic discussion (without notes or bibliography) in the history of philosophy. After sketching the Neoplatonic, Aristotelian, Galilean, and Cartesian positions, Baker discusses the views of More and Barrow. He then argues that Newton rejected the Cartesian positions and developed that of Galileo, More, and Barrow. This article overlaps with but is by no means the same as his book/dissertation.

Topics: *VII*, XII Names: Barrow, Locke, More, Newton

0056 Baker, John Tull. "Henry More and Kant." *Philos. Rev.* 49(1937): 298-306.

Baker here argues that Kant, concerning space, was influenced by More through Newton. He compares More and Kant and notes metaphysical and logical deficiencies. And he connects More's doctrine of space to his theologically motivated attack on Cartesian materialism.

Topics: *VII*, XII Names: More, Newton

0057 Baker, John Tull. *A Historical and Critical Examination of English Space and Time Theories from H. More to Bishop Berkeley*. Bronxville, N.Y.: Sarah Lawrence College, 1930.

This is Baker's doctoral dissertation and is ninety pages long. It includes chapters on More, Barrow, Newton, Locke, and Clarke (as well as Berkeley). Baker describes the views of each thinker and includes the theological dimension. A student of John Herman Randall, Baker is more descriptive, less retrospective and less judgmental than his mentor.

Topics: *VII*, XII Names: Barrow, Clarke, Locke, More, Newton, Smith

0058 Baker, John Tull. "Space, Time, and God: A Chapter in Eighteenth-Century English Philosophy." *Philos. Rev.* 41(1932): 577-593.

For the pre-1720 period, Baker discusses Samuel Clarke and the minor philosophical theologians, Samuel Colliber and William King. He sees Clarke as separate from Newton and treats Clarke as an example "of the assumption by theology of the results of science and philosophy."

Topic: *VII* Names: Clarke, Colliber, King

0059 Ballard, Keith Emerson. "Leibniz's Theory of Time and Space." *J. Hist. Ideas* 21(1960): 49-65.
Ballard analyzes the Leibniz-Clarke debate. He gives special attention to the Newtonian side and to the rival understandings of God.
Topics: *VII*, XII Names: Clarke, Newton

0060 Barbour, Ian G. *Issues in Science and Religion.* Englewood Cliffs, N.J.: Prentice Hall, 1966.
This is a presentist survey of topics related to the title. It oversimplifies intellectual history using a retrospective historiography similar to (and possibly based on) Baumer (0068), Burtt (0160), Randall (1285-1288), and the early Westfall (1650). But Barbour provides a valuable typological scheme of five key issues by which to analyze the historical relations between science and religion and to compare historical eras. This scheme is: (1) methods in science; (2) the character of nature; (3) methods in theology; (4) God and his relation to nature; and (5) man and his relation to nature.
Topics: *VII*, VIII, XII Name: Newton

0061 Barnouw, Jeffrey. "The Separation of Reason and Faith in Bacon and Hobbes, and Leibniz's *Theodicy*." *J. Hist. Ideas* 42(1981): 607-628. Reprinted in *Philosophy, Religion and Science in the Seventeenth and Eighteenth Centuries*, edited by John W. Yolton, 206-227. Rochester, N.Y.: University of Rochester Press, 1990. See (1720).
This is mainly directed to understanding Leibniz's *Theodicy* but also includes significant discussion of the Englishmen. Barnouw sees Bacon and Hobbes as Ockhamists who strictly separated natural and revealed knowledge (science and religion). Such is opposite to the Christian virtuosi of the Royal Society. On this point, says Barnouw, the Royal Society was *unBaconian*.
Topics: *VII*, V, VIII Names: Bacon, Hobbes

0062 Baron, Hans. "The *Querelle* of the Ancients and the Moderns as a Problem for Renaissance Scholarship." *J. Hist. Ideas* 20(1959): 3-22.
Baron sees the quarrel as *not* originating in religion or science or technology. Rather, he finds the origins of the quarrel in Renaissance humanism in France and Italy. Baron writes against Bury (0163) and Richard Jones (0835).
Topics: *IV*, III, X Names: Bacon, Hakewill

0063 Barr, James. "Why the World was Created in 4004 B.C.: Archbishop Ussher and Biblical Chronology." *Bulletin of the John Rylands University Library of Manchester* 67(1985): 575-608.

This is the best detailed study of Ussher's comprehensive biblical chronology. Barr emphasizes the complexity of the historical problems and the key places where Ussher's handling of Old Testament evidence is questionable.

Topic: *X* Name: Ussher

0064 Batten, J. Minton. *John Dury: Advocate of Christian Reunion.* Chicago: University of Chicago Press, 1944.

Batten's biography focuses on Dury's irenic activity and his ideas about Christian reunion. Batten touches on Dury's interest in organizing scientific research and discusses his proposals for educational reforms.

Topics: *IV*, V Names: Comenius, Dury, Hartlib, Milton, Roe

0065 Baumer, Anne. "Christian Aristotelianism and Atomism in Embryology (William Harvey, Kenelm Digby, Nathaniel Highmore)." In *Science and Religion/Wissenschaft und Religion*, edited by Anne Baumer and Manfred Buttner, 16-24. Bochum, Germany: Universitatsverlag N. Brockmeyer, 1989. See (0066).

Baumer briefly argues that seventeenth-century "embryologists" had theological motives for their observations and new theories. In particular, they were motivated to observe the development of embryos and proposed in various ways that God was the primary cause of development.

Topic: *VII* Names: Digby, Harvey, Highmore

0066 Baumer, Anne, and Manfred Buttner, eds. *Science and Religion/Wissenschaft und Religion: From the Proceedings of the XVIIIth International Congress of History of Science at Hamburg-Munich, 1-9 August 1989.* (Abhandlungen zur Geschichte der Geowissenschaften und Religion, 3.) Bochum, Germany: Universitatsverlag N. Brockmeyer, 1989.

The essays by Baumer (0065), Force (0490), and Mason (0995) are annotated separately.

Topic: *VII*

0067 Baumer, Franklin L. *Modern European Thought: Continuity and Change in Ideas, 1650-1950*. New York: Macmillan, 1977.

This is a survey history of ideas in modern Europe. Five perennial questions—concerning God, nature, man, society, and history—provide the structure while the change from an emphasis on being to a sense of becoming provides Baumer's main theme. Science and religion are discussed only in passing but in a provocative way.

Topic: *VII*

0068 Baumer, Franklin L. *Religion and the Rise of Scepticism*. New York: Harcourt Brace, 1960.

Baumer discusses the rise of religious skepticism in the seventeenth century. He sees the new science (not as the only cause but) as "probably the most important single source of skepticism in the early modern period." Using a retrospective historiography, Baumer argues that the traditional conceptions of God, providence, miracles, and religious knowledge were all undermined by science; and that the likes of Bacon, Boyle, and Newton were unknowingly taking the positions of Voltaire, Hume, and Condorcet.

Topics: *IX*, VII, VIII, XII Names: Bacon, Boyle, Newton

0069 Beall, Otho T., Jr. "Cotton Mather's Early 'Curiosa Americana' and the Boston Philosophical Society of 1683." *Wm. Mary Q.* 18(1961): 360-372.

Beall draws evidence of the existence and sophisticated interest of the Boston Philosophical Society (1683-1688?) from an analysis of Mather's "Curiosa Americana"—letters to the Royal Society of London from 1712 to 1724. The scientific content of these letters, Beall also notes, indicate a transition from an association with a theology of nature to a greater objectification.

Topics: *VII*, V Name: Mather, C.

0070 Beall, Otho T., Jr., and Richard H. Shryock. *Cotton Mather: First Significant Figure in American Medicine*. Baltimore: Johns Hopkins Press, 1954.

This deals primarily with Mather's "The Angel of Bethesda" and the inoculation controversy. Beall and Shryock also discuss the medical background—both placing Mather in his early eighteenth-century context and noting his contributions in the general history

of medicine. The authors present Mather as "the only American example of a complete medico-theological synthesis."
Topic: *XI* Names: Boylston, Mather, C.

0071 Bechler, Zev, ed. *Contemporary Newtonian Research*. Dordrecht: D. Reidel, 1982.
This anthology contains four relevant articles which are annotated separately in this bibliography. They are by Bechler (0073), McGuire (1031), Rogers (1344), and Westfall (1646).
Topics: *XII*, VII Name: Newton

0072 Bechler, Zev. "The Essence and Soul of Seventeenth-Century Scientific Revolution." *Science in Context* (1)1987: 87-101.
This is a response to Schaffer's "Godly Men and Mechanical Philosophers" (1402). Bechler argues from the intellectualist position that scientific theories "belong to different ontological strata of the world" from psychological, social, or biological states. But he does find God and spirits to be essential to seventeenth-century versions of the Platonic philosophy of nature.
Topics: *VII*, I, XII Name: Newton

0073 Bechler, Zev. "Introduction: Some Issues of Newtonian Historiography." In *Contemporary Newtonian Research*, edited by Zev Bechler, 1-20. Dordrecht: D. Reidel, 1982. See (0071).
Bechler discusses Newton's philosophy of science and scientific method—and defends the role of metaphysical assumptions in both. Specifically, he asserts the importance of "God's place in his (Newton's) system of science" and pursues the related historiographic implications of this point. He suggests that even the invention of the infinitesimal calculus is related to the theological context.
Topics: *XII*, I, VII Name: Newton

0074 Beck, Lewis White. "World Enough, and Time." In *Probability, Time, and Space in Eighteenth-Century Literature*, edited by Paula R. Backscheider, 113-139. New York: AMS Press, 1979. See (0052).
This is a philosophical (and ahistorical) analysis of selected eighteenth-century concepts of time. About half of the essay is given over to Newton, Locke, and Leibniz. Beck's conclusion is that eternity was "shaken out of the philosophy of nature and the

philosophy of man"—its place taken by human time—but "time was quite enough."
Topics: *VII*, XII Names: Locke, Newton

0075 Becker, George. "Pietism and Science: A Critique of Robert K. Merton's Hypothesis." *Am. J. Sociol.* 89(1984): 1065-1090.
Becker challenges Merton's subthesis concerning German pietism and suggests that "the more inclusive ascetic Protestantism-science thesis" is suspect. Becker's examples are totally continental and many are post-1720.
Topic: *III*

0076 Bedford, R. D. *The Defense of Truth: Herbert of Cherbury and the Seventeenth Century*. Manchester, England: Manchester University Press, 1979.
For deism, see especially the last chapter.
Topic: *IX* Name: Herbert, E.

0077 Beier, Lucinda McCray. *Sufferers and Healers: The Experience of Illness in Seventeenth-Century England*. London: Routledge and Kegan Paul, 1987.
Beier examines the topic expressed by her subtitle: what did people do and believe when they became ill? In the chapter on approaches to illness, she divides the people (anachronistically, she admits) into religious and secular. By "religious" she means a passive trust in God and God's healers; by "secular" she means an active attack on the symptoms. Included in the former category are magical healers as well as literature on "moral medicine." The case studies which Beier provides in the second half of the book indicate the complexity of the actual "history of suffering" in the seventeenth century.
Topic: *XI* Names: Hooke, Josselin, R., Josselyn, J., Pepys

0078 Ben-David, Joseph. "Puritanism and Modern Science: A Study in the Continuity and Coherence of Sociological Research." In *Comparative Social Dynamics*, edited by Erik Cohn, 207-223. Boulder, Colo., 1985. Reprinted in *Puritanism and the Rise of Modern Science*, edited by Bernard I. Cohen, 246-261. New Brunswick, N.J.: Rutgers University Press, 1990. See (0246).

A sociologist argues that the Merton thesis was on the right track but needed to be modified and refined in its details. Ben-David believes that Puritanism was a major factor in the institutionalization of science and in a "value shift" (which conferred religious significance to scientific endeavor) in English society. As regards social-psychological factors, however, Ben-David argues that the Merton thesis needs to be modified by considering the key religious factor to have been "Protestantism" (and not "Puritanism").
Topic: *III*

0079 Ben-David, Joseph. *The Scientist's Role in Society: A Comparative Study*. Englewood Cliffs, N.J.: Prentice Hall, 1971.
This book includes two chapters in which Ben-David analyzes seventeenth and early-eighteenth century England. Ben-David's own philosophy of science is determinative for many of his historical conclusions.
Topics: *V,* III, IV

0080 Bercovitch, Sacvan. "Empedocles in the English Renaissance." *Studies in Philology* 65(1968): 67-80.
Bercovitch argues that a Neoplatonic interpretation of the philosophy and poetry of Empedocles had an appreciable effect on sixteenth- and seventeenth-century English writers.
Topics: *X*, VII Name: Cudworth

0081 Berkeley, Edmund, and Dorothy Smith Berkeley. "Biographical Sketch." In *The Reverend John Clayton, a Parson with a Scientific Mind: His Scientific Writing and Other Related Papers*, edited by Edmund Berkeley and Dorothy Smith Berkeley, xv-lxiii. Charlottesville, Va.: University Press of Virginia, 1965.
The authors describe Clayton's varied scientific interests and church activities in England, Ireland, and Virginia. While rector of a series of parishes Clayton: practiced medicine; developed and tested experimental devices with Boyle; assisted in agricultural development; and made botanical and geological observations. He did all this as well as observing Indian customs in Virginia.
Topics: *VII*, VI, XI Name: Clayton

0082 Berman, David. "Anthony Collins and the Question of Atheism in the Early Part of the Eighteenth Century." *Proc. Royal Irish Acad.* Section C, 75(1975): 85-102.

Berman argues that Collins was not a deist but truly was an atheist. He sees Collins as believing in "the atheist argument from the eternity of matter."

Topic: *IX* Name: Collins, A.

0083 Berman, David. "Anthony Collins: Aspects of His Thought and Writings." *Hermathena* 119(1975): 49-70.

Berman's "primary aim" in this article "is to fill up gaps in, and to supplement, O'Higgins' pioneer work." Included is an assessment of Collins' place in the history of philosophy "between Locke and Berkeley."

Topics: *VII*, IX Name: Collins, A.

0084 Berman, David. "Enlightenment and Counter-Enlightenment in Irish Philosophy." *Arch. Gesch. Phil.* 64(1982): 148-165.

Berman discusses Toland and the Irish bishops Browne, King, and Berkeley, as constituting an Irish philosophical tradition from the 1690s to the 1750s.

Topics: *VII*, IX Names: Browne, P., King, Toland

0085 Berman, David. *A History of Atheism in Britain: From Hobbes to Russell*. London: Croom Helm, 1988.

This basically is in the mode of history of philosophy—treating several of the key concepts, arguments, and logical implications of atheism through the past three centuries. It includes no direct discussion of science or natural philosophy. But there is some relating of atheism (and theism) to philosophic theories of matter during our period. Also, this book addresses some important historiographic issues—especially those having to do with covert atheism in the seventeenth and eighteenth centuries.

Topics: *IX*, I Names: Collins, A., Hobbes

0086 Berman, Morris. *The Reenchantment of the World*. Ithaca, N.Y.: Cornell University Press, 1981.

This is a controversial and stimulating "new age" book. Berman believes that something is fundamentally wrong with our twentieth-century worldview. The loss of meaning has had various destruc-

tive social and psychological consequences. Berman argues that "the disenchantment of the world" (Weber's phrase) has been caused by an alienating scientific consciousness. He proposes a metaphysics of participation and reenchantment. Concerning the seventeenth century, Berman relates science to the rise of capitalism and presents alchemy as "the last great coherent expression of participating consciousness." He pictures Newton as a tragic man of his own time and of ours. Berman draws mostly on secondary sources in his historical and academically-based theses.
Topics: *VII*, I, II, III, V, XII Names: Bacon, Newton

0087 Bernal, John D. "Comenius' Visit to England (1641), and the Rise of Scientific Societies in the Seventeenth Century." In *The Teacher of Nations*, edited by Joseph Needham, 27-34. Cambridge: Cambridge University Press, 1942.
Bernal points out that the religious mission of peace and justice was the basis for Comenius' proposals for educational and scientific reform. He suggests that the same message needs to be heard again in the twentieth century.
Topic: *IV* Name: Comenius

0088 Bernal, John D. *Science in History*. 2 vols. London: Watts, 1954.
This is a survey history of science written by a Marxist.
Topic: *IV*

0089 Bernstein, Howard R. "Leibniz and the *Sensorium Dei*." *J. Hist. Philos*. 15(1977): 171-182.
Bernstein discusses Leibniz's side of the debate with Clarke. He focuses on Leibniz's response to Clarke's use of the "Sensorium Dei" analogy—analyzing the crucial differences concerning the relationships between God, extension, space, and perception.
Topics: *VII*, XII Names: Clarke, Newton

0090 Bethell, S. L. *The Cultural Revolution of the Seventeenth Century*. London: D. Dobson, 1951.
This work by a literary historian focuses on the influence of science on religion. Bethell puts special emphasis on Anglican conceptions of "reason" and "faith." He argues that "the Elizabethan world picture" was pushed aside by "the Locke-Newton universe" (in which "God was on his way out" and "man as a

spiritual being was bound to follow"). Bethell writes in the mid-century historiographic tradition of noting the "bad" effects of science on literature and religion.

Topics: *VII*, VIII, X, XII Names: Donne, Locke, Newton, Tillotson, Vaughan, H.

0091 Biarnais, Marie-Francoise. *De la Gravitation ou Les Fondements de la Mecanique Classique*. Paris: Les Belles Lettres, 1985.

This is a translation into French of Newton's *De Gravitatione* with about 100 pages of notes and commentary. Biarnais discusses the role of God in Newton's early thought, his opposition to Descartes, and his relationship to Henry More. She notes Newton's views on gravitation, absolute space, absolute time, mechanics, and metaphysics. On all of these topics, both natural philosophy and theology are treated as relevant to Newton's approach. But Newton is also seen as maintaining a certain separation between science and religion.

Topics: *XII*, VII Names: More, Newton

0092 Biarnais, Marie-Francoise. *Les Principia de Newton: Genese et Structure des Chapitres Fondamentaux avec Traduction Nouvelle*. Paris: Centre de Documentation Sciences Humaines, 1982.

Biarnais provides a translation into French of key portions of the *Principia* with about 150 pages of notes and commentary. In the commentary on the General Scholium, Biarnais describes Newton's understanding of the relation between natural philosophy and natural theology. She sees Newton as basing a natural theology upon but as otherwise keeping theology separate from experimental philosophy.

Topics: *XII*, VIII Name: Newton

0093 Biddle, John C. "Locke's Critique of Innate Principles and Toland's Deism." *J. Hist. Ideas* 37(1976): 411-422. Reprinted in *Philosophy, Religion and Science in the Seventeenth and Eighteenth Centuries*, edited by John W. Yolton, 140-151. Rochester, N.Y.: University of Rochester Press, 1990. See (1720).

Biddle discusses Locke's anti-deist intentions and Toland's deistic use of Locke's epistemology. Natural philosophy is involved only indirectly.

Topic: *IX* Names: Locke, Toland

0094 Bienvenu, Richard T., and Mordechai Feingold, eds. *In the Presence of the Past: Essays in Honor of Frank Manuel* (International Archives of the History of Ideas, 118). Dordrecht: Kluwer Academic, 1991.

The essays by Feingold (0455) and Ryan (1385) are annotated separately.

Topics: *IV*, V

0095 Birkett, Kirsten, and David Oldroyd. "Robert Hooke, Physico-Mythology, Knowledge of the World of the Ancients and Knowledge of the Ancient World." In *The Uses of Antiquity: The Scientific Revolution and the Classical Tradition*, edited by Stephen Gaukroger, 145-170. Dordrecht: Kluwer Academic, 1991. See (0530).

The authors show how Hooke made use of scripture and classical texts to defend his theory of the geological history of the earth. His approach to the writings of Plato and Ovid, for example, was a kind of euhemerism which the authors call physico-mythology. Birkett and Oldroyd also discuss Hooke's explanation of creation and of Noah's flood.

Topic: *X* Name: Hooke

0096 Black, Robert C., III. *The Younger John Winthrop*. New York: Columbia University Press, 1966.

This is a biography of the personable and versatile son of the governor of the Massachusetts Bay colony and himself governor of Connecticut. Black finds Winthrop to be indifferent to theological controversies but enthusiastic at being the first North American member of the Royal Society. Following Wilkinson, Black argues that this Winthrop probably was the noted alchemist "Eirenaeus Philalethes."

Topics: *V*, II Name: Winthrop

0097 Blau, Joseph L. *The Christian Interpretation of the Cabala in the Renaissance*. New York: Columbia University Press, 1944.

This is an older but still useful study of the background of the *prisca theologia* and seventeenth-century magic by a literary historian.
Topics: *II*, X Name: Browne, T.

0098 Blay, Michel. "Leon Bloch et Helen Metzger: La Quete de la Pensee Newtonienne." *Corpus: Revue de Philosophie* 8/9 (1988): 67-84. Reprinted in *Etudes sur/Studies on Helene Metzger*, edited by Gideon Freudenthal, 67-84. Leiden: Brill, 1990. See (0513).
Blay compares Bloch's positivistic approach to Newton's approach to natural philosophy. He adopts Metzger's emphasis on the relation of universal attraction to religious inspiration and natural theology in Newton's thought.
Topics: *XII*, I, VII, VIII Name: Newton

0099 Blaydes, Sophia B. "Nature Is a Woman: The Duchess of Newcastle and Seventeenth-Century Philosophy." In *Man, God, and Nature in the Enlightenment*, edited by Donald C. Mell, Jr. and others, 51-64. East Lansing, Mich.: Colleagues Press, 1988.
The author summarizes Margaret Cavendish's writings. Blaydes indicates that Cavendish held to a fundamental materialism and a skepticism toward experimental science. Yet she was motivated by a hope for immortality and an identification with nature.
Topics: *VII*, VIII Name: Cavendish

0100 Bloch, Ernst. *The Principle of Hope*. 3 vols. Translated by Neville Plaice and others. Cambridge, Mass.: MIT Press, 1986.
The first section of Chapter 37 contains philosophical reflections on historic technological utopias. Alchemy and Bacon are among the topics discussed. Bloch, a Marxist, says little directly about religion. Yet religious hopes and dreams are implied in his insights about the transformation of nature.
Topics: *IV*, II, VI Name: Bacon

0101 Bloch, Leon. *La Philosophie de Newton*. Paris: F. Allen, 1908.
Bloch interprets Newton in a positivistic manner typical of turn-of-the-century historiography.
Topics: *XII*, VII Name: Newton

0102 Boas, George. *The History of Ideas: An Introduction*. New York: Scribner, 1969.

Boas wrote this book toward the end of his teaching career. It is possibly the best discussion of the basis for the "history of ideas" approach. In the first section, Boas examines in clear language the elements of the history of ideas (e.g., basic metaphors, how ideas change). In the second section, he sketches out three examples: the people; monotheism; and microcosm. Although ideas in the seventeenth century are not discussed, both his theory and his latter two sketches are relevant to the historiography of the relations between science and religion.

Topic: *I*

0103 Boas, Marie. "The Establishment of the Mechanical Philosophy." *Osiris* 10(1952): 412-541.

See Marie Boas Hall, (0612).

Topic: *VII*

0104 Bodemer, Charles W. "Materialism and Neoplatonic Influences in Embryology." In *Medicine in Seventeenth Century England*, edited by Allen G. Debus, 183-213. Berkeley and Los Angeles: University of California Press, 1974. See (0349).

Bodemer traces theories of embryology from the middle to the end of the seventeenth century by discussing various writers from Digby to Boyle. In his examination of Neoplatonic influences, he touches on the interactions of religion and science—especially in the writings of Thomas Browne, the Cambridge Platonists, and Boyle.

Topics: *XI*, VII Names: Boyle, Browne, T., Cudworth, Charleton, Digby, Harvey, Highmore, More

0105 Bolam, Jeanne. "The Botanical Works of Nehemiah Grew, F. R. S. (1641-1712)." *Notes Rec. Royal Soc. London* 27(1973): 219-231.

This essay appears to be aimed at a general audience. In addition to the main discussion indicated by the title, Bolam briefly discusses Grew's dissenting religious background, his religious motives, and his teleological arguments.

Topic: *VII* Name: Grew

0106 Bond, Donald F. "Distrust of Imagination in English Neoclassicism." *Philol. Q.* 14(1935): 54-69. Reprinted in *Essential Articles: For the Study of English Augustan Backgrounds*, edited by Bernard N. Schilling, 281-301. Hamden, Conn.: Archon Books, 1961. See (1411).

Bond argues distrust of imagination in the late-seventeenth and early-eighteenth centuries was a matter of degree and not absolute. The reaction against religious enthusiasm and the excesses of preachers did combine with a positive valuation of the new science to cause such distrust: but this should be noted in a balanced way. Moreover, Bond goes on to argue, such developments affected prose style more than poetry.

Topic: *X*

0107 Borges, Jorge Luis. "The Analytical Language of John Wilkins." In *Other Inquisitions, 1937-1952*, edited by Jorge Luis Borges, 51-64. Austin, Texas: University of Texas Press, 1964.

In this imaginative essay, Borges reflects on the impossibility of penetrating "God's secret dictionary." Yet, Borges writes, this impossibility did not dissuade Wilkins from outlining a provisional and conjectural analytic language within which each letter is meaningful and each word scientifically defines itself.

Topic: *X* Name: Wilkins

0108 Bowles, Geoffrey. "John Harris and the Powers of Matter." *Ambix* 22(1975): 21-38.

Bowles analyzes in detail the discussions of the powers of matter found in the two volumes of Harris' *Lexicon Technicum* (1704, 1710)—an encyclopedia aimed at popularizing the breadth of human knowledge. In Harris' second volume, Bowles finds his theology to have played a significant role in his description of Newtonian active powers in matter (a view which differed from that of Clarke and others). For Bowles, this illustrates that the reception of new notions (including those of Newtonianism) is explicable not only in terms of the influence of the originator's conception—but also in terms of the nonscientific preconceptions of its transmitters.

Topic: *VII* Name: Harris

0109 Bowles, Geoffrey. "Physical, Human and Divine Attraction in the Life and Thought of George Cheyne." *Annals of Science* 31(1974): 473-488.

The author delineates Cheyne's theories which combine Newtonian attractions and a Neoplatonic theology. Bowles also correlates Cheyne's developing views of attraction to his life circumstances—implying some causal efficiency to Cheyne's personal experiences.

Topics: *VII*, XI, XII Name: Cheyne

0110 Boylan, Michael. "Henry More's Space and the Spirit of Nature." *J. Hist. Philos.* 18(1980): 395-405.

This is a philosophical analysis of the relations between God, space, and the spirit of nature in More's writings. Boylan concludes that, for More, space is "an ontological condition *sine qua non* of existence" of any thing. And the spirit of nature serves as a ubiquitous "tool of God for the governance of physical events."

Topic: *VII* Name: More

0111 Bradish, Norman C. "John Sergeant: A Forgotten Critic of Descartes and Locke." *Monist* 39(1929): 571-628. Reprinted as a pamphlet, *John Sergeant: A Forgotten Critic of Descartes and Locke*. Chicago: Open Court, 1929.

This is an introduction to Sergeant and some of his work. Sergeant was a Roman Catholic apologist who entered into theological controversies with Anglicans and attacked the philosophy of Locke. He tried to develop a mathematics-like philosophy which describes the universe ordered according to preexisting ideas in the mind of God. Bradish also reproduces "Non Ultra" (one of Sergeant's pamphlets).

Topic: *VII* Names: Locke, Sergeant

0112 Brandt, Frithiof. *Thomas Hobbes' Mechanical Conception of Nature*. Translated by Vaughan Maxwell and Annie I. Fausboll. Copenhagen: Levin and Munksgaard, 1928. Originally published in Danish as *Den Mekaniske Naturpfattelse hos Thomas Hobbes*. Copenhagen: Levin and Munksgaard, 1921.

Brandt treats Hobbes as a natural philosopher with little reference to theology or religion. But he does see Hobbes as a materialistic atheist.

Topic: *IX* Name: Hobbes

0113 Brasch, Frederick E., ed. *Sir Isaac Newton: A Bicentenary Evaluation of His Work*. Baltimore: Williams and Wilkins, 1928.

This is an outdated collection of essays—two of which (Brett on religion and Newell on alchemy) are nonetheless listed in this bibliography: see (0119) and (1109).

Topic: *XII* Name: Newton

0114 Brauer, Jerald C. "Puritan Mysticism and the Development of Liberalism." *Church History* 19(1950): 151-170.

Internal developments within Puritanism and the new science both are seen as important factors in the rise of liberal theology in England. Articles such as this raise complicating questions as to the exact role of science in the rise of deism.

Topics: *IX*, III Names: Locke, Tillotson

0115 Bredvold, Louis I. "Dryden, Hobbes, and the Royal Society." *Modern Philology* 25(1928): 417-438.

Bredvold notes that Dryden went to both the old science and the new science for sources of poetic imagery and that certain characters in his plays reflect Hobbesian materialism. But, on deeper investigation, one sees that Dryden was interested in and sympathetic to the new science of the Royal Society. The apologetics of Sprat, Glanvill, and Boyle (argues Bredvold) represent Dryden's views on science and religion.

Topics: *X*, V Names: Dryden, Hobbes

0116 Bredvold, Louis I. *The Intellectual Milieu of John Dryden: Studies in Some Aspects of Seventeenth-Century Thought*. Ann Arbor: University of Michigan Press, 1934.

Bredvold puts forth a complex argument about Dryden's skepticism, his religious consistency (despite converting to Roman Catholicism), and his interest in the new science. Dryden was inclined to skepticism by temperament and he eventually combined Pyrrhonism with fideism: this accounts both for his conversion to Catholicism and for his Tory political conservatism. His reading of Montaigne and Thomas Browne, his contacts with the Royal Society, the tradition of Roman Catholic apologetics in England, and the biblical criticism of Richard Simon all sustained, clarified, and developed Dryden's skeptical outlook. Bredvold's theses about

Dryden both have been influential and have been contested. His chapter-long treatments of "The Traditions of Scepticism," "The Crisis of the New Science," and seventeenth-century Roman Catholic apologetics are excellent. (This reads like Richard Popkin twenty-five years before its time!) His discussion of the complexities of religious rationalism also is good.

Topics: *X*, VII, IX Names: Boyle, Browne, T., Dryden, Glanvill, Hobbes

0117 Bredvold, Louis I. "The Religious Thought of Donne in Relation to Medieval and Later Traditions." In *Studies in Shakespeare, Milton, and Donne*, 190-232. New York: Macmillan, 1925.

Bredvold describes Donne as a Renaissance man of great curiosity whose poetry and philosophy were influenced by the new science of Copernicus, Kepler, and Galileo. But the mature Donne, argues Bredvold, developed a philosophical skepticism and saw a certain opposition between reason and faith. Donne's mature religion was mystical—one in which God and the spiritual life were beyond the power of reason to comprehend. Thus, while his poetry uses images from both medieval and Renaissance natural philosophy, his poetry (ultimately) is not intended to expound any particular natural philosophy.

Topics: *X*, VII Name: Donne

0118 Breitwieser, Mitchell Robert. "Cotton Mather's Pharmacy." *Early American Literature* 16(1981): 42-49.

For Mather, smallpox inoculation was understood as a theological metaphor—analogous to salvation. Thus medicine, particularly homeopathy, was not an incongruous addition to theology. Rather, it was one instance of humble submission to the archetypical creation of God.

Topic: *XI* Name: Mather, C.

0119 Brett, George S. "Newton's Place in the History of Religious Thought." In *Sir Isaac Newton: A Bicentenary Evaluation of His Work*, edited by Frederick E. Brasch, 259-273. Baltimore: Williams and Wilkins, 1928. See (0113).

Although written by a scientist, this is like a sermon; today, it is of little use.

Topic: *XII* Name: Newton

0120 Brett, R. L. "Thomas Hobbes." In *The English Mind: Studies in the English Moralists Presented to Basil Willey*, edited by Hugh Sykes Davies and George Watson, 30-54. Cambridge: Cambridge University Press, 1964.

For Hobbes, Brett says, empiricist psychology explains poetry. Epic poetry, accordingly, loses its supernatural references and its claims to knowledge. In this, Hobbes was opposed by Milton but followed by Dryden. Brett also touches upon Hobbes' separation of faith from scientific knowledge.

Topics: *X*, VII Names: Dryden, Hobbes, Milton

0121 Brewster, Sir David. *The Life of Sir Isaac Newton*. London: J. Murray, 1831.

This work was superseded by Brewster's later biography of Newton (0122).

Topic: *XII* Name: Newton

0122 Brewster, Sir David. *Memoirs of the Life, Writings, and Discoveries of Sir Isaac Newton*. 2 vols. Edinburgh: T. Constable, 1855. Reprinted, New York: Johnson Reprint, 1965.

In his work on Newton, Brewster was embarrassed by the religious (and alchemical) issues. He admitted Newton's unorthodoxy but did not pursue it; and, of course, he did not address the connections between Newton's science and religion. For many years, this was the best available biography and some scholars still mine Brewster for useful tidbits. But, for most purposes, this work is now outdated.

Topics: *XII*, VII Name: Newton

0123 Bricker, Philip, and R. I. G. Hughes, eds. *Philosophical Perspectives on Newtonian Science*. Cambridge, Mass.: MIT Press, 1990.

Essays on Newton's understanding of God, space, and time by Carriero (0190) and McGuire (1029) are annotated separately.

Topics: *XII*, VII Name: Newton

0124 Briggs, E. R. "English Socinianism around Newton and Whiston." *Studies on Voltaire and the Eighteenth Century* 216(1983): 48-50.

Briggs briefly outlines Socinian theology; he treats Whiston with

a favorable bias. (But there is almost nothing about Newton.)
Topics: *IX*, XII Name: Whiston

0125 Briggs, Katherine M. Introduction to *The Last of the Astrologers: Mr. William Lilly's History of His Life and Times from the Year 1608 to 1681*, edited by Katherine M. Briggs, vii-xvii. London: Folklore Society, 1974.

Briggs' introduction to Lilly's autobiography provides a brief sketch of Lilly's adventures and an explanation of his astrological terms.

Topic: *II* Name: Lilly

0126 Broad, C. D. "John Locke." *Hibbert Journal* 31(1933): 249-267. Reprinted in *A Locke Miscellany*, edited by Jean S. Yolton, 1-24. Bristol: Thoemmes, 1990. See (1712).

This is a brief summary of Locke's life and work. It is addressed to a general audience. Broad is insightful in discussing Locke's religious views and their relation to his theory of knowledge.

Topic: *VII* Name: Locke

0127 Broad, C. D. "Leibniz's Last Controversy with Newtonians." *Theoria* 12(1946): 143-168.

This article is an interpretive summary and philosophical analysis of the Leibniz-Clarke controversy over space, time, and God. Broad sympathizes with and notes the contemporary implications of Leibniz's position.

Topic: *XII* Names: Clarke, Newton

0128 Broad, C. D. *The Philosophy of Francis Bacon*. Cambridge: Cambridge University Press, 1926.

This was a public lecture outlining Bacon's classification of knowledge. For Bacon, "religion and morality have little to hope and nothing to fear from the advance of natural philosophy." Broad concludes that Bacon has been overrated.

Topic: *VII* Name: Bacon

0129 Brock, C. Helen. "The Influence of Europe on Colonial Massachusetts Medicine." In *Medicine in Colonial Massachusetts, 1620-1820*, edited by Philip Cash and others, 101-143. Boston: The Society, 1980. See (0193).

Brock describes the European influence through books, travel, and Galenic and Paracelsian ideas. She also notes the lack of institutions (particularly hospitals and professional organizations) in the colonies. An appendix contains a useful list of persons in Massachusetts who received some medical training in Europe from 1620 to 1800.
Topic: *XI*

0130 Brooke, John Hedley. "The God of Isaac Newton." In *Let Newton Be!*, edited by John Fauvel and others, 169-183. Oxford: Oxford University Press, 1988. See (0449).
This essay is the best treatment of the topic yet to be published. It is an excellent summary of Newton's theological views (aimed at a general audience but good for specialists as well). In a well-organized way, Brooke treats a broad range of issues—including parallels between Newton's scientific and religious quests and the influence of Newton's theology on his interpretation of nature.
Topics: *XII*, VII, X Name: Newton

0131 Brooke, John Hedley. "Natural Law in the Natural Sciences: the Origins of Modern Atheism?" *Science and Christian Belief* 4(1994): 83-103.
Taking examples from the seventeenth and nineteenth centuries, Brooke stresses the ambivalence and flexibility of the metaphor "laws of nature." He thereby argues against Michael Buckley's thesis that emphasis on natural theology and law invited its own refutation. (He likewise shows the flexibility of the use of the law-of-nature metaphor in politics.) And, Brooke says, the real issues lie in the "domain behind the laws," i.e., in the doctrines of God.
Topics: *IX*, IV, VII, VIII, XII Names: Newton, Whiston

0132 Brooke, John Hedley. *Science and Religion: Some Historical Perspectives*. Cambridge: Cambridge University Press, 1991.
This survey of the sixteenth-through-nineteenth centuries is good. It is fairly advanced for an introductory survey—with an emphasis on both the complexity of the issues and the historiographic complications. About fifty pages of the book deal directly with seventeenth and early-eighteenth century England; and about 200 pages are relevant in some way to our field. Brooke emphasizes the importance of natural theology in the seventeenth-to-early nineteenth centuries (and especially in England). For some readers,

Brooke's emphasis on complexity may render his presentation of "Protestantism and the Rise of Science" rather disjointed and confusing. The book includes a good bibliographic essay.
Topics: *VII*, I, III, VIII, XII Names: Boyle, Newton, Wilkins

0133 Brooke, John Hedley. "Science and the Fortunes of Natural Theology: Some Historical Perspectives." *Zygon* 24(1989): 3-22.
Brooke shows the ironic ease with which natural theological arguments for divine providence could graduate into secular arguments against Christianity—and vice-versa. In outlining various strategies used from the sixteenth through the nineteenth centuries, Brooke proposes that science has been seen as a "cultural resource" for "higher level assumptions" (such as religious ones) which determine the positions of specific historical figures.
Topics: *VIII*, I, VII, IX, X, XII Names: Burnet, T., Newton

0134 Brooke, John Hedley. "Why did the English Mix their Science and their Religion?" In *Science and Imagination in XVIIIth-Century British Culture*, edited by Sergio Rossi, 57-78. Milan: Edizioni Unicopli, 1987. See (1362).
Brooke offers several conjectures as to why the tradition of natural theology strongly persisted in England from the late seventeenth into the nineteenth century. First, he shows its new prominence in the late seventeenth century and proposes some possible reasons for this. (And he takes polite exception to the thesis that English natural theology led to deism.) Second, he considers the social and political context. Finally, Brooke analyzes the various levels of interaction between science and religion that allowed a flexible and resilient natural theological tradition to be maintained. He concludes that natural theology allowed natural philosophers "to demonstrate that science was not subversive," especially in times of social or political tension.
Topics: *VIII*, I, IV

0135 Brooke, John L. *The Refiner's Fire: The Making of Mormon Cosmology, 1644-1844*. Cambridge: Cambridge University Press, 1994.
The thesis here is that the antecedents of Mormon cosmology are to be found in the hermetic dispensationalism of elements in the

Radical Reformation. The cosmologies of seventeenth-century English sects are usefully summarized in the first few chapters. Brooke emphasizes the affinities between the theology of the restoration of the primal Adam on the one hand and, on the other, "the occult cultures of popular conjuring and esoteric hermeticism."
Topic: *II*

0136 Brooks, David. "The Idea of the Decay of the World in the Old Testament, the Apocrypha, and the Pseudepigrapha." In *The Light of Nature*, edited by J. D. North and J. J. Roche, 383-404. Dordrecht: M. Nijhoff, 1985. See (1145).
Brooks mainly analyzes texts as indicated by the title. He argues that the development of the decay-of-the-world theme in Hellenistic Judaism and early Christianity made it potentially available to all later Christians. Reference is made to the seventeenth-century debate in England and its relation to the new science; but these topics are not pursued. This is an interesting background article for our period.
Topic: *X*

0137 Brooks, Richard S. *The Interplay Between Science and Religion in England, 1640-1720: A Bibliographical and Historiographic Guide*. Evanston, Ill.: Garrett-Evangelical Theological Seminary Library, 1975.
This 145-page work is the predecessor of the current annotated bibliography. The original includes nine historiographic essays (ranging in length from four to eleven pages each). The bibliographic entries and annotations of the original are now superseded by the present work.
Topics: *I*, II, III, IV, V, VII, VIII, IX, XII Name: Newton

0138 Brown, K. C. "Hobbes's Grounds for Belief in a Deity." *Philosophy* 37(1962): 336-334.
Brown argues: (1) that Hobbes was not an atheist; and (2) that, while he rejected the First Mover argument, Hobbes sincerely accepted the Argument from Design.
Topics: *IX*, VIII Name: Hobbes

0139 Brown, Stuart. "Leibniz and More's Cabbalistic Circle." In *Henry More (1614-1687) Tercentenary Studies*, edited by Sarah Hutton, 77-95. Dordrecht: Kluwer, 1990. See (0782).

Brown takes up the question of the influence of More, Conway, and F. van Helmont on Leibniz. Rather than claiming a direct influence, Brown proposes a theory of "convergence." Leibniz, he says, used a similar kind of solution to the same traditional philosophical problems because he and they had shared perspectives. Brown's two examples are: the doctrine that every particle of matter contains a world of infinitely many creatures; and the theory that the universe, including matter, emanates from God.
Topics: *VII*, I Names: Conway, Helmont, More

0140 Brown, Theodore M. "The College of Physicians and the Acceptance of Iatromechanism in England, 1665-1695." *Bull. Hist. Med.* 44(1970): 12-30.
This article is not directly relevant to religion: but it does show the interplay between professional conservatism and the new science. The point is made that iatromechanism provided a progressive-conservative compromise for the professional physicians.
Topics: *XI*, V Name: Willis

0141 Buchdahl, Gerd. "Explanation and Gravity." In *Changing Perspectives in the History of Science: Essays in Honour of Joseph Needham*, edited by Mikulas Teich and Robert Young, 167-203. London: Heinemann, 1973. See (1535).
In this philosophical essay, Buchdahl discusses divine activity as one possible explanation (the "ontological anchor") that could make intelligible the concept of action-at-a-distance for Newton and his contemporaries. He proposes that our understanding of Newton's "force" should occur on three levels—inductive, conceptual, and architectonic.
Topics: *VII*, XII Name: Newton

0142 Buchdahl, Gerd. "Gravity and Intelligibility: Newton to Kant." In *The Methodological Heritage of Newton*, edited by Robert E. Butts and John W. Davis, 74-102. Toronto: University of Toronto Press, 1969. See (0172).
A few pages are given over to God as an explanatory factor in the attempts to understand universal gravitation. Buchdahl describes this as a "counter-move" on the part of Newtonians in answering Leibniz and other critics.
Topics: *VII*, XII Names: Locke, Newton

0143 Buchdahl, Gerd. "History of Science and Criteria of Choice." In *Historical and Philosophical Perspectives of Science*, edited by Roger H. Struewer, 204-230 and 239-245. Minneapolis: University of Minnesota Press, 1970. See (1514).

This is a difficult essay in the philosophy of science. Buchdahl argues for "three components that appear to determine the acceptability of scientific hypotheses." These are "conceptual explication" (or conventional terminology), "constitutive articulation" (presentation and interpretation), and "architectonic determination." The last component is the "deeper level" of explanation—which in the seventeenth century included God. See also the "Comment" by Laurens Lauden (0936) and Buchdahl's response (pp. 239-245.)

Topics: *I*, VII, XII Names: Locke, Newton

0144 Buchdahl, Gerd. *Metaphysics and the Philosophy of Science: The Classical Origins, Descartes to Kant*. Oxford: Blackwell; Cambridge, Mass.: MIT Press, 1969.

This book is oriented towards present-day philosophy of science (and away from theology). But it is of some use for historians of our field—especially with respect to Locke.

Topic: *VII* Name: Locke

0145 Buchholtz, Klaus-Dietwardt. *Newton als Theologe*. Witten, Germany: Luther, 1965.

Even in 1965, this book added nothing that was not already written in English. It relies on now outdated secondary sources.

Topics: *XII*, VII Name: Newton

0146 Buckley, George T. *Atheism in the English Renaissance*. Chicago: University of Chicago Press, 1932.

This book mainly deals with the sixteenth century and without reference to science or natural philosophy. But, oriented towards English literature, it is of some use as background for our field of study.

Topics: *IX*, X Names: Davies (1565), Davies (1569), Greville, Ralegh

0147 Buckley, Michael J. *At the Origins of Modern Atheism*. New Haven, Conn.: Yale University Press, 1987.

In his basically ahistorical approach, Buckley argues that modern atheism logically developed from a strand of early modern thinking which made Christian theology subservient to philosophical reason. About seventy pages are devoted to analyzing the thought of Isaac Newton and of Samuel Clarke. Buckley sees these English Christians as making philosophical (including natural philosophical) arguments which are at the origins of modern atheism. Materialism, absolute space and time, and natural theology are all central to Buckley's discussion.

Topics: *IX,* VII, VIII, XII Names: Clarke, Newton

0148 Buckley, Michael J. "God in the Project of Newtonian Mechanics." In *Newton and the New Direction in Science*, edited by G. V. Coyne and others, 85-105. Vatican City: Specola Vaticana, 1988. See (0281).

This is a good summary and philosophical analysis of the natural theology found in the *Principia* and in the *Opticks*. Buckley argues that Newton's understanding of God is not logically separate from but, in fact, logically follows from his mechanics and from his understanding of light. This essay is incorporated into Buckley's book *At the Origins of Modern Atheism* (0147).

Topics: *VIII*, VII, XII Name: Newton

0149 Buckley, Michael J. *Motion and Motion's God: Thematic Variations in Aristotle, Cicero, Newton, and Hegel*. Princeton, N.J.: Princeton University Press, 1971.

Part III (pp. 159-204) is an excellent study of Newton's natural theology and its philosophical context.

Topics: *XII*, VII, VIII Name: Newton

0150 Buckley, Michael J. "The Newtonian Settlement and the Origins of Atheism." In *Physics, Philosophy and Theology*, edited by Robert John Russell and others, 81-102. Vatican City: Vatican Observatory, 1988. See (1381).

A concise version of parts of (0147). The "Newtonian Settlement" refers to the relation between science and religion in which mechanics is the foundation for the credence of theology. Buckley argues skillfully that Enlightenment atheism emerged dialectically out of the arguments of Newtonian natural theology (with the development of a dynamic conception of mass). For Buckley,

religion betrayed itself by allowing that it needed other resources for dealing with the question of the existence of God.
Topics: *IX*, VII, XII Name: Newton

0151 Budick, Sanford. *Dryden and the Abyss of Light: A Study of "Religio Laici" and "The Hind and the Panther."* New Haven, Conn.: Yale University Press, 1970.
This book focuses on the context or milieu of Dryden's religious poetry. Although Budick intends to write the definitive work on this topic, his book supplements rather than supplants Bredvold and Harth. Budick emphasizes the Cambridge Platonists as an important part of the intellectual milieu (though he throws in gratuitous negative judgments of them). He mentions the new science in passing but does not emphasize it. Budick seems very sympathetic to Dryden's Catholicism.
Topics: *X*, VII, IX Names: Dryden, More

0152 Bukharin, N. I., and others. *Science at the Cross Roads: Papers Presented to the International Congress of the History of Science and Technology held in London from June 29th to July 3rd, 1931 for the Delegates of the U.S.S.R.* London: Kniga, 1931. Reprinted with a Foreword by Joseph Needham and an Introduction by Paul Gary Werskey. London: F. Cass, 1971.
This is a famous collection of Marxist interpretations which take economic determinism as presupposed. It includes the especially well-known essay by Boris Hessen (0676).
Topics: *I*, IV

0153 Bullough, Geoffrey. "Bacon and the Defense of Learning." In *Seventeenth Century Studies presented to Sir Herbert Grierson*, 1-20. Oxford: Clarendon, 1938. See (1424). Reprinted in *Essential Articles for the Study of Francis Bacon*, edited by Brian Vickers, 93-113. Hamden, Conn.: Archon Books, 1968. See (1588).
This essay is a meandering discussion of Bacon's defence of learning in Book I of *The Advancement of Learning*. The author sees the new science as part of what Bacon was defending; and he sees the religious objections to learning as one of the forces against which Bacon was fighting.
Topics: *X*, IV Names: Bacon, Greville, Ralegh

0154 Bullough, Geoffrey. Introduction to *Philosophical Poems of Henry More, comprising Psychozoia and Minor Poems*, edited by Geoffrey Bullough, xi-lxxxi. Manchester, England: Manchester University Press, 1931.

This is a good introduction and summary of More's *Psychozoia* and other poems. The place of natural philosophy is recognized but treated as incidental. Bullough explicates the background influences on More's Platonism and on his poetic style. Complementing Bullough's introduction are his notes and commentary on the poems (pp. 175-250).

Topics: *X*, II, VII Names: More, Vaughan, T., Whichcote

0155 Burke, Henry R. "Sir Isaac Newton's Formal Conception of Scientific Method." *New Scholasticism* 10(1936): 93-115.

This is a rather interesting analysis for the 1930s. Burke discusses Newton's empiricism, his view of hypotheses, and his four *regulae philosophandi*. For Newton, he notes, the metaphysical foundation for the intelligibility and order of nature (as well as the rules) was God. Moreover, "science was to be valued because it brought men to a knowledge of God."

Topics: *XII*, VII Name: Newton

0156 Burke, John G. Introduction to *The Uses of Science in the Age of Newton*, edited by John G. Burke, xi-xxii. Berkeley and Los Angeles: University of California Press, 1983. See (0157).

Burke discusses several "contextualist" types of history of science and questions the validity of contextualist assumptions and generalizations.

Topic: *I*

0157 Burke, John G., ed. *The Uses of Science in the Age of Newton*. Berkeley and Los Angeles: University of California Press, 1983.

The essays by Burke (0156), Miner (1075), and Westfall (1648) address the social and intellectual context of seventeenth and early-eighteenth century science.

Topic: *I*

0158 Burnham, Frederic B. "The More-Vaughan Controversy: The Revolt Against Philosophic Enthusiasm." *J. Hist. Ideas* 35 (1974): 33-49.

Philosophic enthusiasm was a label used against Hermetic and magical philosophy; Burnham describes the Latitudinarian revolt against this outlook.

Topics: *VII*, II, III, IV Names: More, Vaughan, T.

0159 Burns, R. M. *The Great Debate on Miracles from Joseph Glanvill to David Hume*. Lewisburg, Pa.: Bucknell University Press, 1981.

Burns proposes that it was the seventeenth-century natural philosophers (moderate empiricists and liberal Anglicans) who began stressing the importance of miracles as evidence of revelations. He also argues that it was the deists who were first responsible for the negative theological response. Thus, he sees science and religion existing "symbiotically" in the leading natural philosophers of the period even on the question of miracles (thereby contradicting Westfall).

Topics: *VII*, IX Names: Boyle, Clarke, Glanvill, Locke, Toland, Wilkins, Wollaston, Woolston

0160 Burtt, Edwin Arthur. *The Metaphysical Foundations of Modern Physical Science*. London: Paul, Trench, Trubner, 1925. Revised Edition, Garden City, N.Y.: Doubleday, 1931.

Burtt provides a philosophical and historical analysis of seventeenth-century science. He discusses many of the theological and philosophical issues raised explicitly and implicitly by Hobbes, Boyle, Newton, and important continental philosophers. The book includes extensive coverage of Newton and Newtonianism. It includes thorough discussion of the teleological and cosmological arguments as used in seventeenth-century England. Burtt describes the fundamental assumptions inherent in organicism, mechanism, and empiricism. He sees the new science as significant in the rise of deism and sees the new theories of matter as leading to atheism. Although Burtt is overly retrospective and harps on the "bad" consequences of the new science for religion, his book is a classic. It is well-written and, in its time, it truly broke new ground. The book has been reprinted many times by several publishers.

Topics: *VII*, VIII, IX, XII Names: Barrow, Bentley, Boyle, Hobbes, More, Newton

0161 Burtt, Edwin Arthur. *The Metaphysics of Sir Isaac Newton; an Essay on the Metaphysical Foundations of Modern Science.* London: N.p., 1925.
This work is identical to Burtt's *The Metaphysical Foundations of Modern Physical Science* (0160)—except for the changed title and changed thesis note.
Topic: *XII* Name: Newton

0162 Burtt, Edwin Arthur. "Method and Metaphysics in Newton." *Philosophy of Science* 10(1943): 57-66.
Burtt defends his *Metaphysical Foundations of Modern Physical Science* (1924 and 1932) against E. E. Strong's *Procedures and Metaphysics* (1936). Burtt again emphasizes the relevance of metaphysical and theological ideas to Newton's natural philosophy.
Topics: *XII*, I VII Name: Newton

0163 Bury, J. B. *The Idea of Progress*. London: Macmillan, 1920.
This is a survey in which the author assumes an inherent opposition between science and Christianity. The idea of progress, for Bury, goes hand in hand with the developments of science. It definitely is "Whig" history: but this book is important for many later writers who discuss progress, utopias, and the quarrel between the ancients and moderns.
Topic: *IV*

0164 Bush, Douglas. *Science and English Poetry: A Historical Sketch, 1590-1950*. New York: Oxford University Press, 1950.
This survey provides an introduction to science, religion, and poetry during our period.
Topics: *X*, IV Name: More

0165 Bush, Douglas. "Science and Literature." In *Seventeenth Century Sciences and the Arts*, edited by Hedley Howell Rhys, 29-62. Princeton, N.J.: Princeton University Press, 1961. See (1328).
This is an introductory essay about science, literature, and the surrounding culture. As regards literature, Bush focuses on the effects of science on belief and sensibility—arguing that scientific progress "entailed far greater losses than gains" for literature.
Topics: *X*, IV Names: Donne, Milton

0166 Bush, Mary Delaney. "Rational Proof of a Deity from the Order of Nature." *ELH* 9(1942): 288-319.

Marred by the author's retrospective value judgments, this concerns the design argument in early eighteenth-century poetry.

Topics: *VIII*, X Name: Addison

0167 Butler, Eliza M. *Ritual Magic*. Cambridge: Cambridge University Press, 1949.

Most of this book deals with magicians in times and places outside our scope. But in Chapter 5, "The Art in England," Butler discusses Reginald Scot's *Discoverie of Witchcraft* (1584) and an "anti-Scot" addition to the third edition of the same work (1665). The amended edition favorably pictured various magic rituals. Butler's book serves both as a reminder of the persistence of ritual magic and as a good background for the philosophical debates among clergy and natural philosophers.

Topics: *II*, VII

0168 Butler, Ian Christopher. *Number Symbolism*. London: Routledge and K. Paul, 1970.

This is a useful survey—from Greek symbolism, to medieval Biblical exegesis, to some modern developments. Chapter 4 has to do with cabalism and magic. Chapter 6 discusses the use of number symbolism both for metaphors and as determinants of the structure of selected poetry in the seventeenth century.

Topics: *X*, II

0169 Butler, Jon. "Magic, Astrology and the Early American Religious Heritage, 1600-1760." *Am. Hist. Rev.* 84(1979): 317-346. Revised as "Magic and the Occult," in *Awash in a Sea of Faith: Christianizing the American People*, by Jon Butler, Chapter 2. Cambridge, Mass.: Harvard University Press, 1990.

Butler proposes that European occult practices and beliefs survived and even flourished in the American colonies as "noninstitutional popular religion" until about 1720. The author uses as evidence: (1) occult books on library lists; (2) the astrological and other occult information found in contemporary almanacs; (3) church and court records; (4) the practicing alchemists respected in English alchemical circles; (5) a German Christian-Rosicrucian community;

and (6) contemporary protests against occult practices. Butler also suggests causes for the decline of occult traditions after 1720.
Topic: *II* Name: Teackle

0170 Butterfield, Herbert. *The Origins of Modern Science, 1300-1800*. London: G. Bell and Sons, 1949. Revised Edition. New York: Macmillan, 1958.
This is a well-written survey by a generalist historian famous for his early opposition to "Whig" history. As a cultural historian, Butterfield is well aware of the importance of Christianity in seventeenth-century thought. He emphasizes intellectual elements and rationalism in his approach to both science and religion. He tends to see neither a conflict nor causal relations between science and religion. Instead, he sees a rather easy reconciliation between science, the mechanical view of nature, and Christianity. Once a classic, this work is now a bit outdated.
Topic: *VII*

0171 Butterfield, Herbert. *The Whig Interpretation of History*. London: G. Bell and Sons, 1931.
In what is now a classic little book, Butterfield criticizes those historians who write in terms of the "friends and enemies of progress." The book concerns English political historiography but also applies to our field. Although Butterfield does not mention them, most historians of science and historians of religion before 1930 wrote history in this way. Even now, it continues to be an issue, although now "Whig interpretations" are done with more subtlety.
Topic: *I*

0172 Butts, Robert E., and John W. Davis, eds. *The Methodological Heritage of Newton*. Toronto: University of Toronto Press, 1969.
These articles written by scholars who basically are interested in the philosophy of science. The essays by Buchdahl (0142), John Davis (0329), and Priestley (1263) are entered separately in this bibliography.
Topics: *XII*, VII Name: Newton

0173 Cachemaille, Ernest Peter. *Sir Isaac Newton on the Prophetic Symbols. Symbolism of the Visions of Revelation*. (Prophecy

Investigation Society Aids to Prophetic Study, No. 8.) London: Thynne, 1916.

This is a pamphlet which includes five pages summarizing Newton's *Observations upon the Prophecies of Daniel and the Apocalypse of St. John*. Basically, it is useless for the historical study of our period.

Topics: *XII*, X   Name: Newton

0174 Cafiero, Luca. "Robert Fludd e la Polemica con Gassendi." *Rivista di Storia della Filosophia* 19(1964): 367-410.

Cafiero examines Fludd's writings in his controversy with Gassendi. He presents his work as a supplement to Yates' discussion of the Fludd-Kepler debate.

Topic: *II*   Name: Fludd

0175 Cairns, John. *Unbelief in the Eighteenth Century: as Contrasted with Its Earlier and Later History*. Edinburgh: Adam and Charles Black, 1881.

This theological and judgmental treatment of seventeenth- and eighteenth-century free thought includes a section on English deism. It is of little use today.

Topic: *IX*

0176 Cajori, Florian. "Sir Isaac Newton's Early Study of the Apocalypse." *Popular Astronomy* 34(1926): 75-78.

This short note now is of little use.

Topics: *XII*, X   Name: Newton

0177 Callebaut, D. K. "Comparison between the Philosophies of Newton and Einstein." In *Newton and the New Direction in Science*, edited by G. V. Coyne and others, 193-201. Vatican City: Specola Vaticana, 1988. See (0281).

Callebaut makes broad generalizations comparing Newton and Einstein—including the topics of God, determinism, space, and cosmology. There is nothing original here (at least as far as Newton is concerned).

Topics: *XII*, VII   Name: Newton

0178 Cantor, Geoffrey N. "Eighteenth Century Materialism." *Hist. Sci.* 23(1985): 201-206.

This is an essay review of John Yolton's *Thinking Matter: Materialism in Eighteenth-Century Britain* (1722). Cantor questions Yolton's method and raises related historiographic questions for the history of philosophy.

Topics: *VII*, I

0179 Cantor, Geoffrey N. "Revelation and the Cyclical Cosmos of John Hutchinson." In *Images of the Earth*, edited by L. J. Jordanova and Roy S. Porter, 3-22. Chalfont St. Giles, England: British Society for the History of Science, 1979. See (0847).

Cantor describes Hutchinson's mechanistic, biblically-based cosmology, epistemology, and theory of the Hebrew language. He also discusses Hutchinson's serious opposition to the Newtonians and the influence of Hutchinson later in the eighteenth century.

Topics: *X*, VII, XII Name: Hutchinson

0180 Capek, Milic. "The Conflict Between the Absolutist and the Relational Theory of Time Before Newton." *J. Hist. Ideas* 48(1987): 595-608.

Capek's essay is not focused on our period or on England; and there is no direct discussion of theology in relation to theories of time. But God does keep popping up in his analysis; and the implicit connection between the Christian understanding of God and the development of what came to be the "Newtonian" understanding of time is rather obvious.

Topics: *VII*, XII Names: Barrow, Newton

0181 Capkova, Dagmar. "The Comenian Group in England and Comenius' Idea of Universal Reform." *Acta Comeniana* 1(1969): 25-34.

The author describes the intellectual exchange between Comenius and his friends in England during 1641-1642. Comenius' pansophic emphasis on the whole of mankind and his concrete proposals for scientific and educational reform are noted by Capkova.

Topic: *IV* Names: Comenius, Hartlib, Hubner

0182 Capp, Bernard. *English Almanacs, 1500-1800: Astrology and the Popular Press*. Ithaca, N.Y.: Cornell University Press, 1979. Published in England as *Astrology and the Popular Press: English Almanacs, 1550-1800*. London, 1979.

Capp's study shows that astrological practitioners and the compilers of popular almanacs participated in political and religious controversies. This work includes chapters on the relation of almanacs to politics, religion, and science over the span of time indicated by the title. But Capp focuses especially on the period during and after the mid-seventeenth century. Regarding religion, Capp indicates that the compilers defended themselves against clerical opposition by attempting to harmonize astrology with Christianity. Likewise, they saw themselves as proponents of the new science. Capp concludes with very useful biographical notes and a bibliography of almanacs.

Topics: *II*, X Names: Booker, Buckmaster, Coley, Culpepper, Gadbury, Heydon, Lilly, Partridge, Saunders, Tanner, Wharton (1617), Wing

0183 Capp, Bernard. "The Status and Role of Astrology in Seventeenth-Century England: The Evidence of the Almanac." In *Scienze, Credenze Occulte, Livelli di Cultura*, 279-290. Florence: L. S. Olsehki, 1982.

Capp here summarizes his monograph (0182). He deals first with the almanacs as instruments in the popularization of social, religious, and scientific ideas. Then he considers several questions regarding astrology, its popularity, and its demise (at least among intellectuals).

Topics: *II*, X

0184 Carlini, Armando. *La Filosofia di G. Locke*. 2 vols. Florence: Vallecchi, 1920. 2nd ed., Florence: Vallechi, 1929.

In its time, this was a thorough Italian-language classic by an historian of philosophy. It touches on science and religion only tangentially.

Topic: *VII* Names: Bacon, Cudworth, Locke, More, Stillingfleet

0185 Carpenter, Edward F. *Thomas Tenison, Archbishop of Canterbury, His Life and Times*. London: S.P.C.K., 1948.

This is the standard biography of this archbishop and patron of the early Boyle lectures. Carpenter generally deemphasizes the new science and Tenison's views on natural philosophy. The book does include a description of Tenison's handling of the Whiston affair.

Topics: *VII*, V Names: Tenison, Whiston

0186 Carpenter, S. C. *Eighteenth Century Church and People*. London: Murray, 1959.

Carpenter discusses deism and the social and intellectual context of Anglicanism in the early eighteenth century. The new philosophy and the leading scientists of the period are seen as factors in the rise of deism. But deism is considered a plausible development (not a logical or necessary one) of the natural religion of the scientists.

Topics: *IX*, IV, VIII Names: Burnet, G., Tillotson

0187 Carre, Meyrick H. "New Philosophy and the Divines." *Church Q. Rev.* 156(1955): 33-44.

Focusing on Anglican clerics and theologians of the first quarter of the seventeenth century, Carre emphasizes the *variety* of responses to the new astronomy and to related issues in natural philosophy.

Topic: *VII*

0188 Carre, Meyrick H. *Phases of Thought in England*. Oxford: Clarendon Press, 1949.

In this broad survey, Carre discusses philosophy, religion, and (for the seventeenth and early-eighteenth centuries) the new science.

Topic: *VII* Names: Bacon, Boyle, Cudworth, Hobbes, Locke, More

0189 Carre, Meyrick H. "Ralph Cudworth." *Philos. Q.* 3(1953): 342-351.

This essay includes a summary description of *The True Intellectual System of the Universe*. Carre emphasizes Cudworth's criticism of atheism and of ancient and Hobbesian materialism. He argues that "Our author's overriding purpose is the vindication of natural theology."

Topics: *VIII*, IX Names: Cudworth, Hobbes

0190 Carriero, John. "Newton on Space and Time: Comments on J. E. McGuire." In *Philosophical Perspectives on Newtonian Science,* edited by Philip Bricker and R. I. G. Hughes, 109-133. Cambridge, Mass.: MIT Press, 1990.

Carriero gives a philosophical analysis of Newton's views concerning: (1) space and time as real beings; and (2) space and time as common affections of divine being. He aims to correct McGuire's interpretation of Newton on these matters. (Carriero seems not to be challenging McGuire's fundamental interpretation as to the importance of God in Newton's metaphysics and natural philosophy.)

Topics: *XII*, VII Names: Clarke, Newton

0191 Carroll, James W. "Merton's Thesis on English Science." *American Journal of Economics and Sociology* 13(1954): 427-432. Excerpted in *Puritanism and the Rise of Modern Science*, edited by Bernard I. Cohen, 203-205. New Brunswick, N.J.: Rutgers University Press, 1990. See (0246).

Carroll argues with and insults Merton. Carroll, a sociologist, argues on both sociological and historical grounds. (In turn, Merton criticizes Carroll in two responses).

Topic: *III*

0192 Carroll, Robert Todd. *The Common-Sense Philosophy of Religion of Bishop Edward Stillingfleet, 1635-1699*. The Hague: Nijhoff, 1975.

Carroll's book adds to—and is an extension of—the works of Popkin and Van Leeuwen. In this genre of intellectual history, Carroll focuses on Stillingfleet's defense both of the reasonableness of Christianity and of the reasonableness of natural religion. Stillingfleet approved of the new philosophy of the Royal Society; and he defended his views of true religion and true philosophy against his understandings of Locke, Toland, Descartes, Hobbes, Spinoza, and Roman Catholicism. All this, of course, involved him in a variety of epistemological and metaphysical issues.

Topics: *VII*, VIII Names: Chillingworth, Locke, Stillingfleet

0193 Cash, Philip, and others, eds. *Medicine in Colonial Massachusetts, 1620-1820*. (Publications of the Colonial Society of Massachusetts, Vol. 57). Boston: The Society, 1980.

This volume includes essays by Brock (0129) and Gifford (0539) which are listed separately in this bibliography.

Topic: *XI*

0194 Casini, Paolo. "Newton, a Sceptical Alchemist?" In *Reason, Experiment, and Mysticism in the Scientific Revolution*, edited by M. L. Righini Bonelli and William R. Shea, 233-238 and 316. New York: Science History, 1975. See (1329).

This is a commentary on R. S. Westfall's article (1649) in the same volume. Casini proposes that Newton was a "sceptical alchemist" and that "Newton's alchemical research was a rational attempt to conquer the irrational world of the occult."
Topics: *XII*, II Name: Newton

0195 Casini, Paolo. "Newton, le Lois de la Nature et la 'Grand Ocean de la Verite'." In *Proceedings of the XVth International Congress of the History of Science*, 278-285. Edinburgh: Edinburgh University Press, 1978.
Casini here focuses on the unity or synthesis of the various dimensions of Newton's thought (science, theology, metaphysics, alchemy). Newton found himself at the crossroads of the Hermetic and mechanical traditions—both opposing and putting forward metaphysical hypotheses. From the time of his youthful *De Gravitatione* to the end of his life, Newton showed an agnostic modesty towards "the great ocean of truth" but also maintained his faith. He also concluded that natural theology must be a posteriori (it can not be deductive).
Topics: *XII*, II, VIII Name: Newton

0196 Casini, Paolo. "Newton: The Classical Scholia." *Hist. Sci.* 22(1984): 1-57.
Casini attempts to correct what he considers to be the exaggerated viewpoints of McGuire, Rattansi, and Dobbs regarding Newton's use of *prisca philosophia* or *prisca theologia*. He does this by recovering the whole manuscript of "the classical scholia" and by "the recension of the sources cited by Newton." The last 34 pages of this article constitute Casini's critical edition of the scholia manuscripts and the first 23 pages constitute his introduction. Casini understands the classical scholia as reflecting a particular variant of the tradition of the *prisca*: that is, it was a creative way to vindicate the validity of cosmological models different from the geostatic system and a way to add an "aura of truth" to Newton's own mathematical analysis.
Topics: *XII*, II, VII Names: Gregory, Newton

0197 Casini, Paolo. "Toland e l'attivita della materia." *Riv. Crit. Stor. Fil.* 22(1967): 24-53.

Casini here discusses Toland's philosophical and theological opposition to Newtonianism.

Topics: *IX*, XII Name: Toland

0198 Casini, Paolo. *L'Universo-Macchina: Origini della Filosofia Newtoniana*. Bari, Italy: Laterza, 1969.

This breaks no new ground but it still is a good book.

Topics: *XII*, VIII Name: Newton

0199 Cassirer, Ernst. "Newton and Leibniz." *Philos. Rev.* 52(1943): 366-391.

Cassirer stresses the impossibility of eliminating the metaphysical and theological aspects from the natural philosophy of this period.

Topics: *VII*, XII Name: Newton

0200 Cassirer, Ernst. *The Philosophy of the Enlightenment*. Translated by Fritz C. A. Koelin and James P. Pettegrove. Princeton, N.J.: Princeton University Press, 1951.

Although this work is devoted mostly to continental Europe in the eighteenth century, it includes many references to Newton and to Newton's influence. The Enlightenment, Cassirer argues, was permeated by the conviction that the secrets of nature were opened to the light of independent human knowledge. This meant that "it was above all necessary to sever the bond between theology and physics once and for all."

Topics: *VII*, XII Name: Newton

0201 Cassirer, Ernst. *Die Platonische Renaissance in England und die Schule von Cambridge*. Leipzig, Germany: B. G. Teubner, 1932. Translated by James P. Pettegrove as *The Platonic Renaissance in England*. Austin, Texas: University of Texas Press, 1953.

Besides the focus indicated by the title, this work addresses: the theological implications of mechanism and materialism; concepts of space; and epistemological and methodological issues. Cassirer writes in an "I'm-telling-you-so" style—but he has some good insights and the book is worth reading.

Topics: *VII*, VIII Names: Cudworth, More, Whichcote

0202 Castillejo, David. *The Expanding Force in Newton's Cosmos: as Shown in His Unpublished Papers.* Madrid: Ediciones de Arte y Bibliofilia, 1981.

Castillejo attempts to correlate Newton's alchemical, biblical, and chronological writings on the basis of "an expanding force in Newton's cosmos" which is complementary to the contracting force of gravitation. Moreover, the author claims, the pattern which he finds can be found in Newton's published scientific writings. Castillejo's writing style is rather incoherent—perhaps because his theses are not persuasive. The book is interlaced with lengthy (and valuable) quotations from Newton's unpublished manuscripts.

Topics: *XII*, II, VII, X Name: Newton

0203 Centore, F. F. "Mechanism, Teleology, and Seventeenth-Century English Science." *Int. Philos. Q.* 12(1972): 553-571.

Centore discusses teleology and its place in the thought of seventeenth-century mechanical philosophers. His purpose is to show that mechanism and teleology have not been incompatible—either for seventeenth-century natural philosophers or for twentieth-century scientists.

Topics: *VIII*, VII, XII Names: Bacon, Boyle, Hooke, Newton

0204 Chalmers, Gordon Keith. "Sir Thomas Browne, True Scientist." *Osiris* 2(1936): 28-79.

Chalmers discusses Browne's work and argues for the propriety of considering him a true scientist.

Topics: *VII*, XI Name: Browne, T.

0205 Chappell, Vere, ed. *Seventeenth-Century Natural Scientists.* (Essays on Early Modern Philosophers, 7.) New York: Garland, 1992.

This is a good anthology of reprinted articles. See the separate annotations on works by Dobbs (0388), Henry (0665), Home (0713), McGuire (1023), O'Toole (1188), Rattansi (1291), Rogers (1344), and Shanahan (1425).

Topics: *VII*, XII Name: Newton

0206 Chitnis, Anand. *The Scottish Enlightenment: A Social History.* Totowa, N.J.: Rowman and Littlefield, 1976.

Chitnis mainly focuses on the period after 1720. But in background discussions, he does describe some of our period. His coverage is especially good for social and institutional topics. He notes the developments in medicine and science as part of the Scottish Enlightenment. Chitnis postulates the relevance of church, religion, and theology in the seventeenth century but he emphasizes the importance of secularization in the eighteenth century.

Topics: *IV*, V

0207 Christianson, Gale E. *In the Presence of the Creator: Isaac Newton and His Times*. New York: Free Press, 1984.

This biography of Newton is written in a good story-telling style. It reads like a historical novel (one could compare it to the books of Barbara Tuchman). It is a useful and readable synthesis of recent scholarship: Christianson seems especially influenced by Westfall and Manuel. As the title indicates, Newton's understanding of God is emphasized as the background to all of his life and work. The book is strong on the context of science, alchemy, and natural philosophy—but not strong on the content (which largely goes unexplained). Sometimes Christianson waxes eloquent on "what must have been" when the evidence is lacking. The work is recommended as a scholarly-based biography for a general audience.

Topics: *XII*, II, V, IX Names: Barrow, Fatio, Newton, Whiston

0208 Christianson, Gale E. "Newton, the Man—Again." In *Newton's Scientific and Philosophical Legacy*, edited by P. B. Scheurer and G. Debrock, 3-21. Dordrecht: Kluwer Academic, 1988. See (1409).

This is Christianson's book (0207) in a nutshell.

Topics: *XII*, VII Name: Newton

0209 Christie, John R. R. "The Origins and Development of the Scottish Scientific Community, 1680-1760." *Hist. Sci.* 12(1974): 122-141.

A few pages of this article relate the Scottish Kirk and anti-episcopalian ideology to the University of Edinburgh and to early developments in Scottish science.

Topics: *V*, IV

0210 Christie, John R. R. "The Rise and Fall of Scottish Science." In *The Emergence of Science in Western Europe*, edited by Maurice P. Crosland, 111-126. New York: Science History, 1976. See (0300).

Christie sees the rise of Scottish science as a part of the Scottish Enlightenment. He argues that it had its roots in the social and cultural background of an "improving tradition" which was a response to the disorientation caused by the Union of Parliaments in 1707. This tradition included both individual improvement (moral and intellectual) and collective improvement (economic and technological). Christie sees this thesis as a better explanation of the social dynamics of Scottish science than Trevor-Roper's proposal (1565) that liberal Calvinists and Jacobin-Anglicans imported the new science.

Topics: *IV*, VI

0211 Churchill, Mary S. "*The Seven Chapters*, with Explanatory Notes." *Chymia* 12(1967): 29-57.

This is a transcription of Newton's manuscript copy of *The Seven Chapters* (attributed to Hermes Trismegistus) with an introduction by Churchill. In discussing *The Seven Chapters*, she notes the religious significance for Newton as well as a description of its place among Newton's manuscripts.

Topics: *XII*, II Name: Newton

0212 Clagett, Marshall, ed. *Critical Problems in the History of Science: Proceedings of the Institute for the History of Science at the University of Wisconsin*. Madison, Wisc.: University of Wisconsin Press, 1959.

This collection of articles is not concerned specifically with religion and science; but it is important with respect to the 1950s historiography of the history of science. See also Clark (0217) and Crombie (0297).

Topic: *I*

0213 Clark, George N. *A History of the Royal College of Physicians of London*. Vol. 1. Oxford: Clarendon Press, 1964.

This work is of reference value concerning a conservative social factor.

Topics: *XI*, IV, V

0214 Clark, George N. *Science and Social Welfare in the Age of Newton*. Oxford: Clarendon Press, 1937. 2nd ed. Oxford: Clarendon Press, 1949.

Clark makes a sustained argument against Hessen (0676) and the Marxist interpretation of Newton's science.

Topics: *IV*, I, III, XII Name: Newton

0215 Clark, J. Kent. *Goodwin Wharton*. Oxford: Oxford University Press, 1984.

This is a fascinating biography of Goodwin Wharton and his partner, Mary Parish (based on his manuscript autobiography). It shows that a leading public figure in the late seventeenth century could combine Christianity, occult activities, technical invention, and public service. Wharton's occult activities included alchemy and (through Parish) communication with fairies, spirits, and angels.

Topics: *II*, VI Names: Parish, Wharton (1653)

0216 Clark, Jon, Celia Modgil, and Sohan Modgil, eds. *Robert K. Merton: Consensus and Controversy*. New York: Falmer Press, 1990.

The three relevant articles by Cohen (0244), Hall (0600), and Rattansi (1295) are annotated separately.

Topic: *III*

0217 Clark, Joseph T. "The Philosophy of Science and the History of Science." In *Critical Problems in the History of Science*, edited by Marshall Clagett, 103-140. Madison, Wisc.: University of Wisconsin Press, 1959. See (0212).

Clark recommends that historians of science use current conceptions of science as a frame of reference for doing their history. He wants to *judge* works of the past in terms of these current conceptions. The goal of history writing seems to be "historical appraisal."

Topic: *I*

0218 Clark, Robert E. D. "Newton's God and Ours." *Hibbert Journal* 37(1939): 425-434.

This older essay is of little use to historians.

Topic: *XII* Name: Newton

0219 Clark, Stuart. "The Scientific Status of Daemonology." In *Occult and Scientific Mentalities in the Renaissance*, edited by Brian Vickers, 351-374. Cambridge: Cambridge University Press, 1984. See (1591).

Though mainly dealing with sixteenth-century Europe, this article also examines the views of a few seventeenth-century Englishmen. Clark raises the issue of the causal status of demonic effects: that is, what criteria were used to even include demonology as part of natural philosophy? He argues that eccentric or unusual phenomena allowed natural philosophers to present the devil as "a supremely gifted natural magician."

Topic: *II* Names: Bacon, Cotta, Glanvill

0220 Clarke, Larry R. "The Quaker Background of William Bartram's View of Nature." *J. Hist. Ideas* 46(1985): 435-448.

In an introductory style, Clarke connects early Quakerism with a negative or "apophatic" theology and a positive view of the study of nature. About five pages are given over to the pre-1720 period.

Topic: *VII* Name: Fox

0221 Clauss, Sidonie. "John Wilkins' Essay toward a Real Character: Its Place in the Seventeenth-Century Episteme." *J. Hist. Ideas* 43(1982): 531-553.

This is an examination of Wilkins' attempt to solve the political, religious, linguistic, and scientific problems and divisions of his time through his proposed universal language. He sought to match things with unequivocal (quasi-scientific and perhaps prelapsarian) terms. Though the most important work of its kind and time, Wilkins' essay is often "accounted forgettable" because political and religious divisions were insurmountable and because natural philosophers developed their own partial solutions within separate disciplines.

Topic: *X* Name: Wilkins

0222 Clericuzio, Antonio. "The Internal Laboratory: The Chemical Reinterpretation of Medical Spirits in England (1650-1680)." In *Alchemy and Chemistry in the Sixteenth and Seventeenth Centuries*, edited by Piyo M. Rattansi and Antonio Clericuzio, 51-83. Dordrecht: Kluwer Academic, 1994. See (1299).

Clericuzio discusses the role of "spirits" in the debate over the adoption of chemistry as the basis of physiology. In replacing Galenic medicine, several physicians and natural philosophers "had recourse to spirits in explaining the main functions of the human as well as animal bodies." Clericuzio also demonstrates a division among these writers over whether or not the spirit of life received a "divine illumination."

Topics: *II*, VII, XI, XII Names: Boyle, Glisson, Helmont, Newton, Thomson, Willis

0223 Clifford, James L. "Swift's *Mechanical Operation of the Spirit*." In *Pope and His Contemporaries*, edited by James L. Clifford and Louis A. Landa, 135-146. Oxford: Clarendon Press, 1949. See (0224).

Clifford discusses the reception of the last part of Swift's *A Tale of a Tub* (1704)—in which Swift satirizes religious enthusiasm as a physiological effect.

Topic: *X* Names: Pope, Swift

0224 Clifford, James L., and Louis A. Landa, eds. *Pope and His Contemporaries: Essays Presented to George Sherburn*. Oxford: Clarendon Press, 1949.

The essays by Clifford (0223) and R. F. Jones (0837) are annotated in this bibliography.

Topic: *X* Name: Pope

0225 Clucas, Stephen. "In Search of 'The True Logik': Methodological Eclecticism among the 'Baconian Reformers'." In *Samuel Hartlib and Universal Reformation*, edited by Mark Greengrass, 51-74. Cambridge: Cambridge University Press, 1994. See (0573).

Clucas shows that Hartlib and his circle were influenced by a variety of European thinkers, including some Calvinists, and that they sought a "fluid methodological outlook." Above all, "true logik" was a way to Christ. And a true logician was "a good theologian, politician, and philosopher." Thus true logic underlay all reform; yet perhaps it was too much to expect fallen man to achieve perfect logic.

Topics: *VII*, IV Names: Dury, Hartlib

0226 Clucas, Stephen. "Poetic Atomism in Seventeenth-Century England: Henry More, Thomas Traherne, and 'Scientific Imagination'." *Renaiss. Stud.* 5(1991): 327-340.

Clucas shows how atomism served as a moral metaphor in seventeenth-century religious poetry. At first, it was seen as a symbol of cosmic, political, and social disintegration. But Traherne and More unified their atomic worlds through the "cohesive coherence" of God's will and presence.

Topic: *X*, IV Names: Browne, T., Donne, More, Traherne

0227 Clucas, Stephen. "Samuel Hartlib's *Ephemerides*, 1635-59, and the Pursuit of Scientific and Philosophical Manuscripts: The Religious Ethos of an Intelligencer." *Seventeenth Century* 6(1991): 33-55.

Hartlib's *Ephemerides* is a collection of memoranda and notes (Hartlib's detailed working diary) relating to his scientific, technological, philosophical, and religious interests. Clucas reviews the contents of the *Ephemerides* and emphasizes the Protestant Christian impulses which informed Hartlib's work as an international collector, processor, and disseminator of information. This reads like a chapter in Charles Webster's *The Great Instauration* (1616).

Topics: *V*, III Name: Hartlib

0228 Coffin, Charles Monroe. *John Donne and the New Philosophy.* (Columbia University Studies in English and Comparative Literature, No. 126.) New York: Columbia University Press, 1937.

By "new philosophy" Coffin means Copernican astronomy. In a long-winded way, Coffin presents Donne as quite knowledgeable in the new astronomy even though Donne used poetic metaphors from both the new and the old world systems. Coffin notes the Christian elements involved in Donne's poetry but presents them as rather separate from his philosophical views. The last chapter ("The Two Lights") discusses Donne on the respective spheres of reason and faith in the tradition of Duns Scotus and William of Occam. Coffin's use of other secondary literature is surprisingly outdated even for 1937.

Topics: *X*, VII Name: Donne

0229 Cogley, Richard W. "Survey Article: Seventeenth-Century English Millenarianism." *Religion* 17(1987): 379-396.

This is an introductory analysis of the terms used by millenarians throughout the century. A main thesis is that millenarianism was an historiographical mode used by seventeenth century Englishmen to make history intelligible. There is nothing here on science, but the article is useful for background.

Topic: *X*

0230 Cohen, H. Floris. *The Scientific Revolution: A Historiographical Inquiry*. Chicago: University of Chicago Press, 1994.

This Dutch historian of science attempts to present a critical analysis of the vast literature on the Scientific Revolution. He does so by dividing his analysis into two parts: defining the nature of the Scientific Revolution and searching for its causes. The former, again, is broken into "the great tradition" of historiography (from the nineteenth century through Koyre, the Halls, and Westfall) and "the new science in a wider setting." Portions of this last chapter on the new science and the "old magic" along with the chapter on "religion and the rise of early modern science" in Part II are most relevant to this bibliography. Beyond these particular sections, Cohen summarizes and comments on almost every major historian of the Scientific Revolution. He finds value in the persistence of the hermetic tradition (and its historians) in that it forms a continuing undercurrent that "aims at putting into words man's [sic] stance in the cosmos." In the section on religion, Cohen deals primarily with Hooykaas (to whom he dedicates the book) and the Merton thesis and its commentators. He decides that historians of science have misunderstood Merton's original problem: "The explanation, through changes in social values, of shifts in vocational interest in seventeenth-Century England." Yet, although he has a brief section on the social setting of science, he indicates explicitly that he is not sympathetic to social history of science or to "social constructivism."

Topics: *I*, II, III

0231 Cohen, I. Bernard. *Franklin and Newton: An Inquiry into Speculative Newtonian Experimental Science and Franklin's Work in Electricity as an Example Thereof*. Philadelphia: American Philosophical Society, 1956. Cambridge, Mass.: Harvard University Press, 1966.

This early work by Cohen includes much discussion of the philosophical and some discussion of the theological context of Newton's science. There is an extended discussion of Newton's method, his use of "hypotheses," and why God is not an hypothesis for Newton.
Topics: *XII*, VII Names: Boyle, Newton

0232 Cohen, I. Bernard. "Galileo, Newton, and the Divine Order of the Solar System." In *Galileo, Man of Science*, edited by Ernan McMullin. New York: Basic Books, 1967. See (1040).
Cohen works out a series of equations to show that Newton's cosmological theory was superior to the "Galileo-Plato" hypothesis of the formation of the Copernican solar system. Cohen briefly explains why Newton believed the Copernican system (combined with universal gravitation) both requires and shows God's creativity.
Topics: *XII*, VII Name: Newton

0233 Cohen, I. Bernard. General Introduction to *Isaac Newton's Papers & Letters On Natural Philosophy*, edited by I. Bernard Cohen, 1-24. 2nd ed. Cambridge, Mass.: Harvard University Press, 1978. See (0245).
Cohen includes some brief references to God and religion. He makes the intriguing point that Richard Bentley's Boyle Lectures were the first general account of Newtonian principles for the non-mathematical reader.
Topics: *XII*, VII Names: Bentley, Newton

0234 Cohen, I. Bernard. "Introduction: The Impact of the Merton Thesis." In *Puritanism and the Rise of Modern Science: The Merton Thesis*, edited by I. Bernard Cohen, 1-111. New Brunswick, N.J.: Rutgers University Press, 1990. See (0246).
This section includes a long, rambling, autobiographical description of the topic which overlaps with Cohen's other essays on Merton. It includes an excellent annotated bibliography on books and articles related to the Merton thesis. Cohen also discusses broader historiographic issues (internalism vs. externalism) in the history of science. He praises Merton (the man) but is cautious towards "the Merton Thesis."
Topics: *III*, I, VI

0235 Cohen, I. Bernard. *Introduction to Newton's "Principia."* Cambridge: Cambridge University Press, 1971.

This is an exegetical commentary on the variorum edition of Newton's *Philosophiae Naturalis Principia Mathematica* edited by Koyre, Cohen, and Anne Whitrow (and published separately in two volumes). It also includes a discussion of previous Newtonian scholarship.

Topics: *XII*, VII Name: Newton

0236 Cohen, I. Bernard. *Isaac Newton's Papers & Letters On Natural Philosophy; and Related Documents*. Cambridge, Mass.: Harvard University Press, 1958. 2nd ed. Cambridge, Mass.: Harvard University Press,1978.

This first edition of (0245) sometimes is listed in footnotes and bibliographies without Schofield's name.

Topics: *XII*, VII Name: Newton

0237 Cohen, I. Bernard. "Isaac Newton's *Principia*, the Scriptures, and the Divine Providence." In *Philosophy, Science, and Method: Essays in Honor of Ernest Nagel*, edited by Sidney Morgenbesser and others, 523-548. New York: St. Martin's Press, 1969.

Cohen provides a detailed account of the God passages in various editions and drafts of the *Principia*. He argues that Newton's concern with God and divine providence was a continuing feature in all editions (even though, for Cohen, such concern amounts to straying from science and meandering through theological metaphysics). This article also includes some useful manuscript sources.

Topics: *XII*, VIII Name: Newton

0238 Cohen, I. Bernard. "The Many Faces of the History of Science." In *The Future of History*, edited by Charles F. Delzell, 65-110. Nashville, Tenn.: Vanderbilt University Press, 1977.

In this long essay, Cohen ranges across historiographic approaches to the history of science—both of the past and in the 1970s. He is quite clear about his own favored positions. For example, while admiring those historians who relate science to its social and religious context in particular instances, he denies the direct influence of hermetic ideas on seventeenth-century physics. Above all, he says, the history of science is about "the actual contents of the scientific record."

Topic: *I*

0239 Cohen, I. Bernard. "Newton in the Light of Recent Scholarship." *Isis* 51(1960): 489-514.

This wide-ranging bibliographic and historiographic discussion covers the study of Newton's religious, chronological, and alchemical interests as well as his mathematical and scientific ones. Cohen calls for the systematic publication of *all* of Newton's manuscripts and for further research into *all* facets of Newton's personality and interests. Cohen's early call, of course, has since been somewhat fulfilled.

Topics: *XII*, I, VII, X Name: Newton

0240 Cohen, I. Bernard. "A Note on Harvey's 'Egg' as Pandora's 'Box'." In *Changing Perspectives in the History of Science*, edited by Mikulas Teich and Robert Young, 233-249. London: Heinemann, 1973. See (1535).

Cohen explores the background and meaning of the frontispiece to Harvey's *De Generatione Animalium*. It pictures Jove opening an egg from which small animals emerge. Study of its predecessors indicate a shift from Pandora to Jove. This also implies a continuity in the significance of the myth: "a letting loose in the world of 'goods' and 'evils,' in short, *omnia*."

Topics: *X*, XI Name: Harvey

0241 Cohen, I. Bernard. Preface to *Opticks: or A Treatise of the Reflections, Refractions, Inflections & Colours of Light*, by Isaac Newton, ix-lviii. New York: Dover, 1952.

Cohen's wide-ranging introduction to the *Opticks* is directed to a general audience. He discusses various reactions to the work over the intervening centuries; and he compares reactions towards the *Opticks* to those towards the *Principia*. Cohen mentions the theological passages of the Queries but does not pursue their implications.

Topics: *XII*, VII Name: Newton

0242 Cohen, I. Bernard. "The Publication of *Science, Technology and Society*: Circumstances and Consequences." *Isis* 79(1988): 571-582.

This essay has been superseded by Cohen's longer essays "Introduction: The Impact of the Merton Thesis" (0234) and "Some Documentary Reflections on the Dissemination and Reception of the 'Merton Thesis'" (0244).
Topic: *III*

0243 Cohen, I. Bernard. "*Science, Technology and Society in Seventeenth Century England* by Robert K. Merton." *Scientific American* 288(1973): 117-120.
This review, in which Cohen argues that Merton's theses concerning technology should be given equal billing to his theses concerning Puritanism, has been superseded by Cohen's 1990 "Introduction," (0234).
Topics: *III*, VI

0244 Cohen, I. Bernard. "Some Documentary Reflections on the Dissemination and Reception of the Merton Thesis." In *Robert K. Merton: Consensus and Controversy*, edited by Jon Clark and others, 307-348. Philadelphia: Falmer Press, 1990. See (0216).
This is a shorter version of Cohen's "Introduction: The Impact of the Merton Thesis" (0234). It includes some material not published elsewhere—including longer discussions of the works relevant to the Merton thesis written by J. B. Conant, R. F. Jones, M. H. Nicolson, Joseph Needham, and Walter Pagel.
Topics: *III*, I, VI

0245 Cohen, I. Bernard, ed., assisted by Robert E. Schofield. *Isaac Newton's Papers & Letters On Natural Philosophy; and Related Documents*. Cambridge, Mass.: Harvard University Press, 1958. 2nd ed. Cambridge, Mass.: Harvard University Press, 1978.
This work has several relevant introductory essays, including those by: Gillespie (0543); Miller (1071); and, in the second edition, Cohen (0233).
Topics: *VII*, XII Name: Newton

0246 Cohen, I. Bernard, ed., assisted by K. E. Duffin and Stuart Strickland. *Puritanism and the Rise of Modern Science: The Merton Thesis*. New Brunswick, N.J.: Rutgers University Press, 1990.

This anthology has 24 sections, four of which (totaling about 160 pages) appear to be new. Many of the reprinted portions are abridged selections. Cohen provides a rambling autobiographical introduction with a good annotated bibliography (0234). An appendix by Duffin and Strickland (0404) gives an analytical synopsis of the principal theses of Merton's *Science, Technology, and Society in Seventeenth Century England* (1057). See also: Abraham (0004); Ben-David (0078); Carroll (0191); Cook (0258); de Candolle (0363); Gillespie (0544); Hall (0602); Merton (1053), (1055), and (1061); Stimson (1504); and Webster (1624).
Topics: *III*, I, IV

0247 Cohen, I. Bernard, and Rene Taton. "Hommage a Alexandre Koyre." In *Melanges Alexandre Koyre*, Vol. 1, xix-xxv. Paris: Hermann, 1964. See (1046).
In their biographical introduction, the authors note especially Koyre's conviction of the unity of thought. To him, it seems impossible to separate philosophical from religious thought. Koyre thus studies the structure of scientific ideas in the context of "ideas transscientifiques, philosophiques, metaphysiques, religienses."
Topics: *I*, VII

0248 Cohen, Leonora D. "Descartes and Henry More on the Beast-Machine. A Translation of their Correspondence pertaining to Animal Automation." *Ann. Sci.* 1(1936): 48-61.
In about eight pages of commentary, Cohen notes the biological, metaphysical, and theological differences between Descartes and More.
Topic: *VII* Name: More

0249 Cole, Stephen. "In Defense of the Sociology of Science." *The G. S. S. Journal* (1965): 30-38.
This is a critique of Hall's "Merton Revisited" (0602).
Topics: *III*, I

0250 Coleman, William. "Providence, Capitalism, and Environmental Degradation: English Apologetics in an Era of Economic Revolution." *J. Hist. Ideas* 37(1976): 27-44.
Regarding the contribution of Christian ethics to the environmental crisis, Coleman takes Lynn White's thesis concerning doctrine and

technology in the middle ages and shifts it to natural theology and economic activity in the late seventeenth and early eighteenth centuries. Coleman argues that Christian apologetic arguments about man the maker and entrepreneur sanctified unrestrained economic individualism. His primary example is William Derham.

Topics: *VI*, IV, VIII Name: Derham

0251 Colie, Rosalie L. *Light and Enlightenment: A Study of the Cambridge Platonists and the Dutch Arminians*. Cambridge: Cambridge University Press, 1957.

Colie discusses controversies concerning mechanism, determinism, atheism, and deism. Her emphasis is on selected philosophical theologians. She argues that theological platonism in England and Holland presented an important antimechanistic tradition in the seventeenth and eighteenth centuries. She also discusses Royal Society apologetics.

Topics: *VII*, V, IX Names: Cudworth, Hobbes, More

0252 Colie, Rosalie L. "Some Paradoxes in the Language of Things." In *Reason and the Imagination*, edited by Joseph Anthony Mazzeo, 93-128. New York: Columbia University Press, 1962. See (1004).

This essay is good for seeing how the connotations of several concepts from religious and poetic thinking carried over to that of the natural philosophers.

Topics: *X*, IV, VII, X Names: Bacon, Boyle, Swift

0253 Colie, Rosalie L. "Spinoza and the Early English Deists." *J. Hist. Ideas* 20(1959): 23-46.

Colie sees deism as a complex philosophical movement with important social and political overtones. She treats science as *not* causally significant in the rise of deism. Rather, men such as Blount, Collins, Tindal, and Toland used the formulations of natural science for their own philosophical ends.

Topic: *IX* Names: Blount, Collins, A., Tindal, Toland

0254 Collier, Katherine Brownell. *Cosmogonies of Our Fathers: Some Theories of the Seventeenth and Eighteenth Centuries*. New York: Columbia University Press, 1934.

This book is about cosmological theories in the period. Collier emphasizes the changing interpretations of Genesis in reaction to new scientific theories. She first treats individual writers and then topics.

Topics: *X*, II, VII Names: Burnet, T., Derham, Dickinson, Fludd, Grew, Ray, Stillingfleet, Warren, Whiston, Woodward

0255 Colligan, J. Hay. *The Arian Movement in England*. Manchester, England: Manchester University Press, 1913.

Although an older work, this book provides background for the late seventeenth- and eighteenth-century theological controversies inside and outside of the established church. Colligan stresses the biblical bases of the controversies. He mentions several scientists but only in passing.

Topic: *IX*

0256 Collingwood, R. G. *The Idea of Nature*. Oxford: Clarendon Press, 1945.

This is not a history of the idea of nature—but rather a logical analysis of the relations between cosmologies, doctrines of God, and natural philosophy. Included is an early discussion of the connection between the voluntaristic doctrine of God and the natural philosophy of our period.

Topic: *VII*

0257 Conant, James B. "The Advancement of Learning during the Puritan Commonwealth." *Proc. Mass. Hist. Soc.* 66(1936-1941): 3-31.

Conant discusses, among other issues, the Puritanism-and-the-rise-of-science thesis. He acknowledges the general value of the arguments of Merton, Stimson, and R. F. Jones. But he wants to correct the hypothesis by laying stress on what he calls "moderate Puritans."

Topics: *III*, IV

0258 Cook, Harold J. "Charles Webster on Puritanism and Science." In *Puritanism and the Rise of Science: The Merton Thesis*, edited by I. Bernard Cohen, 265-268. New Brunswick, N.J.: Rutgers University Press, 1990. See (0246).

Cook describes the career and works of Webster as regards the "Merton Thesis" and the topic more generally of Puritanism and science. He presents Webster as a master of case studies and of nuanced consideration of complicated historical contexts.
Topic: *III*

0259 Cook, Harold J. "Living in Revolutionary Times: Medical Change under William and Mary." In *Patronage and Institutions: Science, Technology and Medicine at the European Court, 1500-1750*, edited by Bruce T. Moran, 111-135. Rochester, N.Y.: Boydell Press, 1991.
Cook shows that the way Mary and especially William chose their court physicians and surgeons led to a decline in the power of the medical establishment, of Oxford- and Cambridge-educated physicians, and (thereby) of the Anglican Church.
Topic: *XI*, V Names: Bidloo, Blackmore, Hutton, Mary II, Radcliffe, William III

0260 Cook, Harold J. "Medical Innovation or Medical Malpractice? or, A Dutch Physician in London: Joannes Groenevelt, 1694-1700." *Tractrix* 2(1990): 63-91.
This essay has been superseded by Cook's book. See the annotation for (0261).
Topic: *XI* Name: Groenevelt

0261 Cook, Harold J. *Trials of an Ordinary Doctor: Joannes Groenevelt in Seventeenth Century London*. Baltimore: Johns Hopkins University Press, 1994.
This is a well-written and thoroughly researched biography. It demonstrates the difficulties that a Dutch Calvinist physician, trained in the methods of the new science, had with the Anglican-dominated College of Physicians. The centerpiece of Cook's work is the prosecution of Groenevelt by the censors of the College, Groenevelt's imprisonment for malpractice, and his determined appeal to the courts and to public opinion.
Topics: *XI*, V Name: Groenevelt

0262 Cooney, Brian. "John Sergeant's Criticism of Locke's Theory of Ideas." *Modern Schoolman* 50(1973): 143-158.

Cooney discusses Sergeant's "immediate realism" as found in his *Solid Philosophy*. Sergeant's criticism of Locke's imposition of ideas between things and the knower led to his postulation of a spiritual nature (being) in humans and the intervention of God in the epistemological process. Locke heavily annotated Sergeant's book but did not respond publicly.

Topic: *VII* Names: Locke, Sergeant

0263 Cope, Jackson I. "Appendix B. Aftermath: Stubbe's Attacks on the Royal Society." In *History of the Royal Society by Thomas Sprat*, edited by Jackson I. Cope and Harold Whitmore Jones, 68-74. St. Louis, Mo.: Washington University, 1958. See (0268).

Cope briefly describes the debate between Glanvill and Stubbe (after the publication of *Stubbe's History*). He indicates the internal contradictions of each. He concludes by emphasizing Stubbe's scripture-based attack on the natural theological defense of the apologists for the Royal Society.

Topics: *V*, VIII, X Names: Glanvill, Stubbe

0264 Cope, Jackson I. "Evelyn, Boyle, and Dr. Wilkinson's Mathematico-Chymico-Mechanical School." *Isis* 50(1959): 30-32.

This is a short note relating the Puritan Henry Wilkinson to the program of reforming college education along "Baconian" lines.

Topics: *IV*, III Names: Boyle, Evelyn, Wilkins

0265 Cope, Jackson I. Introduction to *History of the Royal Society by Thomas Sprat*, edited by Jackson I. Cope and Harold Whitmore Jones, xii-xxxii. St. Louis, Mo.: Washington University, 1958. See (0268).

Cope proposes that Sprat emphasized: (1) the transplantation of the glorious destiny of England from theology to the new utilitarian science; (2) the rational religion of the latitudinarian Church of England (against the enthusiastic Puritans); and (3) the "mathematical plainness" of language as an instrument of calming theological passions.

Topics: *V*, IV Name: Sprat

0266 Cope, Jackson I. *Joseph Glanvill: Anglican Apologist*. St. Louis, Mo.: Washington University, 1956.

In this biography, Cope deals extensively with the interplay between science and religion. And Glanvill (1636-1680) is an excellent biographical subject for introducing beginning students to the field. Some topics include: (1) apologetics not only for Christianity but also for the new science and the Royal Society; (2) relationships between the natural and the supernatural, especially with respect to witchcraft; and (3) skepticism and the limits of human reasoning.

Topics: *VII*, II, V, VIII, IX Names: Bacon, Baxter, Boyle, Glanvill, Hobbes, More, Stubbe, White

0267 Cope, Jackson I. "Science, Christ, and Cromwell in Dryden's Heroic Stanzas." *Mod. Lang. Notes* 71(1956): 483-485.

Cope here argues that Dryden applies to Oliver Cromwell "language which had become familiar in the argument from design as it was used by the pious virtuosi in defense of the new science." Cope thus sees science, religion, and political views brought together in Dryden's poem.

Topics: *X*, IV Names: Cromwell, Dryden

0268 Cope, Jackson I., and Harold Whitmore Jones, eds. *History of the Royal Society by Thomas Sprat*. St. Louis, Mo.: Washington University, 1958.

See Cope's "Introduction" (0265) and his "Appendix B. Aftermath: Stubbe's Attacks on the Royal Society" (0263).

Topic: *V* Name: Sprat

0269 Copenhaver, Brian P. "Natural Magic, Hermetism, and Occultism in Early Modern Science." In *Reappraisals of the Scientific Revolution*, edited by David C. Lindberg and Robert S. Westman, 261-301. Cambridge: Cambridge University Press, 1990. See (0957).

Copenhaver emphasizes the complexity of the occult traditions from the late middle ages through the Renaissance and into the seventeenth century. He argues for precise understandings of the terms "natural magic," "occultist," and "hermetic." He begins and ends his essay with statements of admiration for Frances Yates' imaginative exploration of areas of history otherwise ignored.

Topic: *II* Name: Fludd

0270 Copleston, Frederick C. *A History of Philosophy*. Vols. 4 and 5. Westminster, Maryland: Newman Press, 1958-1959.

Although outdated, Copleston remains a solid introductory history.

Topic: *VII*

0271 Cornish, D. "Time, Space, and Freewill: The Leibniz-Clarke Correspondence." In *The Study of Time III: Proceedings of the Third Conference of the International Society for the Study of Time, Alpbach—Austria*, edited by J. T. Fraser and others, 634-655. Berlin: Springer, 1978.

This is a summary and philosophical analysis of selected aspects of the Leibniz-Clarke correspondence. Cornish logically correlates Clarke's position defending absolute time and space with: divine freewill; the mathematical quantification of temporal and spatial points; and (yet) the impossibility of consistent scientific explanation of events in the world. He logically correlates Leibniz's relational understanding of time and space with: the principle of sufficient reason; no freewill (not even divine); the possibility of scientific explanation of events in the world; and the inability to quantify temporal and spatial points. Cornish then asks, retrospectively, "Who won the argument?" and denies that Leibniz did so.

Topics: *XII*, VII Names: Clarke, Newton

0272 Corrigan, John. *The Prism of Piety: Catholick Congregational Clergy at the Beginning of the Enlightenment*. New York: Oxford University Press, 1991.

This is a detailed study of the group of Massachusetts ministers who were influenced by the Cambridge Platonists, English natural philosophers, and Latitudinarians—while also retaining the Puritan covenant of grace. Their theological view combined the unified, reasonable order of nature with the positive value of the affections; and it contributed to a redefinition of the purpose of New England. Corrigan sees them thus as instrumental in the beginning of the Enlightenment in America.

Topics: *VII*, X Names: Brattle, W., Coleman, Pemberton, Wadsworth

0273 Costello, W. J. *The Scholastic Curriculum at Early Seventeenth Century Cambridge*. Cambridge, Mass.: Harvard University Press, 1958.

Costello analyzes the framework and the content of the scholastic curriculum in great detail. He uses manuscripts of student theses and notebooks as well as printed materials. Of particular interest are his chapters on the undergraduate sciences (Aristotelian metaphysics, mathematics, and physics) and graduate studies (theology and medicine). Costello clarifies the meaning of terms (e.g., "nature") as they were used at the time. In the section on theology, he demonstrates both its scholastic structure and its Puritan orientation—especially at Emmanuel College.
Topics: *V*, VII, XI Names: Duport, Holdsworth

0274 Coudert, Allison P. *Alchemy: The Philosopher's Stone*. Boulder, Colo.: Shambhala, 1980.
This book is a well-written, cross-cultural, popular introduction. It combines historical scholarship with Jungian thought. See Chapter 8, "The End of the Quest," for most of the discussion of seventeenth-century Englishmen.
Topics: *II*, VII, XII Names: Ashmole, Boyle, Newton

0275 Coudert, Allison P. "A Cambridge Platonist's Kabbalist Nightmare." *J. Hist. Ideas* 36(1975): 633-652.
Coudert provides a good description of F. M. von Helmont's and Lady Anne Conway's positive view of Jewish Kabbala and Henry More's rejection of it. Her discussion is only indirectly related to natural philosophy.
Topics: *II*, VII Names: Conway, Helmont, More

0276 Coudert, Allison P. "Forgotten Ways of Knowing: The Kabbalah, Language, and Science in the Seventeenth Century." In *The Shapes of Knowledge from the Renaissance to the Enlightenment*, edited by Donald R. Kelley and Richard H. Popkin, 83-99. Dordrecht: Kluwer Academic, 1991. See (0872).
In this study of F. M. van Helmont and C. K. von Rosenroth, Coudert argues that there was a "fluid line" between scientific and occult mentalities in the seventeenth century. Van Helmont's study of the Kabbala, especially the *Zohar*, led him to search for a universal language and to have faith in universal reform and the improvement of the environment. Coudert also discusses van Helmont's influence on Leibniz.
Topics: *II*, IV, X Name: Helmont

0277 Coudert, Allison P. "Henry More and Witchcraft." In *Henry More (1614-1687) Tercentenary Studies*, edited by Sarah Hutton, 77-95. Dordrecht: Kluwer, 1990. See (0782).

Coudert discusses More's views on witchcraft in the context of the new experimental science and the Glanvill-Webster debate. She argues that the protagonists' positions on witchcraft cannot be neatly correlated to their scientific views or to their religious positions. (And, likewise, they can not be categorized in terms of their sociopolitical ideologies.) Underlying the witchcraft debate, Coudert says, were fundamental questions about the authority of Christian revelation, the physical constitution of the universe, the basis of morality, and what constitutes valid knowledge. Because different writers could agree on their answers to one of these issues but disagree on other issues, the situation was far more complex than previous writers have indicated. Coudert shows that More's target was atheistic materialism and that he was an expedient skeptic who used an experimental method awkwardly.

Topics: *II*, IV, VII, IX Names: Boyle, Glanvill, More, Webster

0278 Coudert, Allison P. "Henry More, the Kabbalah, and the Quakers." In *Philosophy, Science, and Religion in England 1640-1700*, edited by Richard Kroll and others, 31-67. Cambridge: Cambridge University Press, 1992. See (0917).

Coudert explores the reasons why More was first attracted to the teachings in the Kabbalah—as expressed by Quakers—and then rejected them. Although he found them similar to his Neoplatonism, useful in his refutation of Descartes, and possibly relevant to his concept of space, More came to disagree with his friend F. M. van Helmont in these matters. More rejected van Helmont's perfectionism inherent in his Lurianic Kabbalah and alchemy. Coudert concludes that More's rejection was related to "his political and social conservatism and Calvinist training."

Topics: *II*, IV Names: Conway, Helmont, More

0279 Coulter, Harris L. *Progress and Regress: J. B. Van Helmont to Claude Bernard*. Vol. 2 in *Divided Legacy: A History of the Schism in Medical Thought*. Washington, D.C.: Wehawken Book, 1977.

Coulter surveys the history of medicine by dividing the profession into "Empirics" and "Rationalists." British and continental figures

in the seventeenth and early-eighteenth centuries are discussed in the first third of this volume. His style includes much stringing together of quotations. References to God are thus included and religious motives are thus acknowledged.
Topic: *XI* Names: Bacon, Cheyne, Sydenham, Willis

0280 Cowling, T. G. *Isaac Newton and Astrology*. (The Eighteenth Selig Brodetsky Memorial Lecture.) Leeds, England: Leeds University Press, 1977.
This is a lecture, aimed at a general audience, which has as its aim the refutation of modern astrologers and their references to Newton. It is of little scholarly use.
Topics: *XII*, II, VII Name: Newton

0281 Coyne, G. V., M. Heller, and J. Zycinski, eds. *Newton and the New Direction in Science*. Vatican City: Specola Vaticana, 1988.
The essays by Buckley (0148), Callebout (0177), Hoskin (0736), Sokolowski (1462), Wallace (1602), and Zycinski (1735) are annotated separately in the present bibliography.
Topics: *VII*, XII Name: Newton

0282 Cragg, Gerald R. *The Church and the Age of Reason, 1648-1789*. Hammondsworth, England: Penguin Books, 1960.
Cragg here deals only in broad generalities about the effect of the new science on theology and the Christian mentality.
Topic: *VII*

0283 Cragg, Gerald R. *Freedom and Authority: A Study of English Thought in the Early Seventeenth Century*. Philadelphia: Westminster Press, 1975.
Cragg relates the intellectual concerns of the first forty years of the seventeenth century to the theme of freedom and authority (and the tension between new approaches and old authorities). The first chapter contains a good summary of the complexities of the issues—pessimism and progress, certainty and skepticism, differing views of the occult. The second chapter portrays Bacon as an advocate of the new scientific outlook and contains a useful summary of his views on faith, reason, and natural theology.

Cragg concludes that, in the early seventeenth century, important questions were clarified (if not always solved).
Topics: *VII*, II, VIII Names: Bacon, Browne, T.

0284 Cragg, Gerald R. *From Puritanism to the Age of Reason: A Study of Changes in Religious Thought Within the Church of England, 1660 to 1700*. Cambridge: Cambridge University Press, 1950.
Cragg here makes general remarks concerning the influence of science on religion.
Topics: *VII*, VIII Names: Cudworth, More, Smith, Whichcote

0285 Cragg, Gerald R. Introduction to *The Cambridge Platonists*, edited by Gerald R. Cragg, 3-31. New York: Oxford University Press, 1968.
This is a good introduction to the philosophy of the Cambridge Platonists and its relation to theology and natural philosophy. Cragg specifically addresses: (1) the theological implications of mechanism and materialism; and (2) the relationship between reason and revelation.
Topics: *VII*, VIII Names: Cudworth, More, Smith, Whichcote

0286 Cragg, Gerald R. *Reason and Authority in the Eighteenth Century*. Cambridge: Cambridge University Press, 1964.
In this book, Cragg discusses the effects of science on the theological doctrines of reason and revelation in eighteenth-century England. He includes a discussion of the theological implications of mechanism and materialism. His work is characterized by an older historiography which saw the new science—and especially that of Newton and Newtonianism—as replacing the God of the Bible with a distant, deistic, metaphysical projection of the creation.
Topics: *VII*, IX, XII Name: Newton

0287 Craig, William Lane. *The Historical Argument for the Resurrection of Jesus during the Deist Controversy*. Text and Studies in Religion, Vol. 23. Lewiston, N.Y.: Edwin Mellen Press, 1985.
As the title implies, science is not addressed in this treatment of deism and Christianity. It, however, provides much evidence for

those who would argue for little or no connection between science and deism.
Topics: *IX*, X

0288 Crane, Ronald S. "Suggestions toward a Genealogy of the ‹Man of Feeling›." *ELH* 1(1934): 205-230. Reprinted in *The Idea of the Humanities*, by Ronald S. Crane, Vol. 1, 188-213. Chicago: University of Chicago Press, 1967.
This is a commonly cited but rather unreliable discussion of Latitudinarianism as applied to eighteenth-century English literature. It does not directly relate to science.
Topic: *X*

0289 Cranston, Maurice. *John Locke: A Biography*. London: Longmans, Green, 1957.
At the time it was written, this was an outstanding biography of Locke. It covers all aspects of Locke's life, activities, and writings—including religion and natural philosophy—in a sympathetic way. Without putting it forward as an actual thesis, Cranston tends to present Locke's religion and theology as separate from his medicine and science. He sees Locke as a Socinian who would not admit it (even to himself) and thus simultaneously a Latitudinarian Anglican. The book is now a bit outdated but is still useful.
Topics: *VII*, IX Names: Collins, A., Locke, Molyneux, W.

0290 Cranston, Maurice. *Locke*. (Bibliographic Supplements to the *British Book News*, No. 35.) London, 1961.
This is a thirty-page popularization of the main points of Cranston's major biography of Locke. See above, (0289).
Topic: *VII* Name: Locke

0291 Craven, J. B. *Count Michael Maier: Doctor of Philosophy and of Medicine, Alchemist, Rosicrucian, Mystic, 1568-1622*. Kirkwall, England: W. Peace, 1910. Reprinted, London: Dawsons, 1968.
Primarily, this is a descriptive bibliography with summaries of Maier's writings. It includes those on alchemy, Hermeticism, and the Egyptian and Greek gods. The brief introduction discusses Maier's visit to England and the dedication of his first book to prominent Englishmen.
Topics: *II*, XI Name: Maier

0292 Craven, J. B. *Doctor Robert Fludd: The English Rosicrucian.* Kirkwall, England: W. Peace, 1902.

This work contains useful biographical material and detailed summaries of Fludd's writings and controversies. Craven describes him as devoted to the Church of England and as an ardent Rosicrucian. He describes Fludd as "the last of a long and wondrous procession" of persons who saw God's presence in everything. (Surprisingly, Hermes is not included in the procession.) Craven writes in a way naively favorable to Fludd (as one might expect from a 1902 publication on a Rosicrucian). An appendix includes an analytic bibliography of some of Fludd's writings.

Topic: *II* Names: Fludd, Foster, W.

0293 Craven, Kenneth. *Jonathan Swift and the Millennium of Madness: The Information Age in Swift's "A Tale of a Tub."* (Brill's Studies in Intellectual History, 30.) Leiden: Brill, 1992.

Craven argues that Swift challenged, and continues to challenge, the optimistic and seductive "scientific millenarian myth." That myth, says Craven, accommodated the Judeo-Christian God, Christian Neoplatonism, and Epicurean atomism. But Swift juxtaposed against it the "sober truth" of the myth of Kronos-Saturn, the god of satire. Craven's analyses of *A Tale of a Tub* and *A Tritical Essay* demonstrate how Swift used and satirized major figures who exemplified various scientific and religious traditions and institutions. There are chapters on Toland (Swift's major satiric victim), Narcissus Marsh and Peter Browne, Milton, Harrington, William Temple, Paracelsus, and Newton. Craven's last chapter is a Swiftian critique of "scientific millennialism" as discussed and assumed by historians of science and literary critics. Craven's perspective is that Swift has something to teach us about the superficiality of modern information systems: "If moderns insist on scientific progress to achieve the good life, Kronos will soon deny them enduring art and life itself."

Topics: *X*, II, IV, V, IX, XI, XII Names: Browne, P., Harrington, Locke, Marsh, Milton, More, Newton, Swift, Temple, Toland

0294 Cristofolini, Paolo. *Cartesiani e Sociniani: Studio su Henry More.* Urbino, Italy: Argalia, 1974.

Cristofolini emphasizes the importance of More in the introduction of Cartesianism into England. He analyzes More's philosophy in relation to various philosophical traditions, the new mechanical philosophy, natural religion, and Christian revelation. And he discusses More's qualified opposition to Descartes, Spinoza, and Socinianism. The book includes long reproductions of More's correspondence.

Topic: *VII* Name: More

0295 Crocker, Robert. "Henry More: A Biographical Essay." In *Henry More (1614-1687) Tercentenary Studies*, edited by Sarah Hutton, 1-17. Dordrecht: Kluwer, 1990. See (0782).

This is a condensed and scholarly intellectual biography which also serves as an introductory chapter to a collection of other essays. Crocker summarizes More's writings, taking them in chronological order. He emphasizes More's devotional mystical theology as it evolved from his Puritan beginnings to the Neoplatonism of his mature years.

Topics: *VII*, II Name: More

0296 Crocker, Robert. "Mysticism and Enthusiasm in Henry More." In *Henry More (1614-1687) Tercentenary Studies*, edited by Sarah Hutton, 137-155. Dordrecht: Kluwer, 1990. See (0782).

Crocker argues that More chastised "enthusiasts," including alchemists and Quakers, because their illuminism was similar to his own. But More differentiated himself from them through the role of "irradiated" reason, which discriminated between soul and body. Thus, seeing the issue as both a spiritual and a physiological one, he understood himself as a physician ideally placed to cure their disease. This is one of the better discussions of More's mystical theology.

Topics: *XI*, II Name: More

0297 Crombie, Alistair C. "Commentary on the Papers of Rupert Hall and Giorgio de Santillana." In *Critical Problems in the History of Science*, edited by Marshall Clagett, 66-78. Madison, Wisc.: University of Wisconsin Press, 1959. See (0212).

Crombie argues for a continuity of "principles and attitudes" between: (1) the theologians, philosophers, craftsmen, and cathedral architects of the middle ages; and (2) the natural philosophers of the scientific revolution. This essay complements

a longer paper by Crombie (on medieval discussions of scientific method) in the same volume.
Topic: *I*

0298 Crombie, Alistair C. Introduction to *Scientific Change*, edited by Alistair C. Crombie, 1-11. London: Heinemann, 1963. See (0299).

Crombie outlines the most important goals, problems, and approaches to pursue in the history of science as he describes the organization of the symposium held at Oxford in 1961. He raises questions both about the whole lives of past scientists and about the social contexts of their motives and opportunities.
Topic: *I*

0299 Crombie, Alistair C., ed. *Scientific Change: Historical Studies in the Intellectual, Social and Technical Conditions for Scientific Discovery and Technical Invention, from Antiquity to the Present*. London: Heinemann, 1963.

For our field, the importance of this collection of essays lies mainly in the historiographic debates (Part Nine, pp. 797-878). See annotations for Crombie's introduction (0298), and essays by Guerlac (0581), and Koyre (0898).
Topic: *I*

0300 Crosland, Maurice P., ed. *The Emergence of Science in Western Europe*. New York: Science History, 1976.

The articles by Christie (0210) and Rattansi (1296) are listed separately in this bibliography.
Topic: *I*

0301 Crosland, Maurice P., ed. *Historical Studies in the Language of Chemistry*. London: Heinemann, 1962.

This book is written on an introductory level but it includes good observations on the language of alchemy.
Topic: *II*

0302 Crous, Ernst. *Die Grundlagen der Religionsphilophischen Lehren Lockes*. Halle, Germany: Karras, 1909.

This work was superseded by (0303).
Topic: *VII* Name: Locke

0303 Crous, Ernst. *Die Religionsphilosophischen Lehren Lockes, und ihre Stellung zu dem Deismus seiner Zeit*. Halle, Germany: M. Niemeyer 1910. Reprinted, Hildesheim, Germany: G. Olms, 1980.

Crous focuses on epistemology and Locke's distinction between reason and faith. He relates Locke to the "deism" of Herbert, Chillingworth, and Hobbes.

Topics: *VII*, IX Names: Chillingworth, Herbert, E., Hobbes, Locke

0304 Cunnar, Eugene R. "Donne's 'Valediction: Forbidding Mourning' and the Golden Compasses of Alchemical Creation." In *Literature and the Occult: Essays in Comparative Literature*, edited by Luanne Frank, 72-110. Arlington, Texas: University of Texas at Arlington, 1977.

Cunnar proposes that Donne found in alchemy "a symbolic paradigm for his creative imagination." That is, he used spiritual alchemy as a basis for self-understanding and as a means to interconnect his poetry with his theology. Cunnar explicates his view in a detailed analysis of the poem indicated in the title.

Topics: *X*, II Name: Donne

0305 Cunningham, Andrew. "Thomas Sydenham: Epidemics, Experiment and the 'Good Old Cause'." In *The Medical Revolution of the Seventeenth Century*, edited by Roger French and Andrew Wear, 164-190. Cambridge: Cambridge University Press, 1989. See (0510).

In this clear and well-organized essay, Cunningham explores the "what?," "why?," and "how?" of Sydenham's attempts to reform medicine. The author considers Sydenham in his own context rather than as the hero remembered in later centuries. Above all, he shows that Sydenham was motivated by the religious and political convictions of his early years as a "commonwealth man": that was the "Good Old Cause" which he translated into his medical practice during the Restoration. Possibly because of his background, he practiced primarily among the poor. (Yet he lived next to Boyle's sister and near Locke and the Earl of Shaftesbury.) Among the poor he could use Boyle's experimental methods and with Locke make empirical histories of bedside observations.

Topics: *XI*, III, IV Names: Boyle, Coxe, Locke, Sydenham

0306 Curry, Patrick. *Prophecy and Power: Astrology in Early Modern England*. Princeton, N.J.: Princeton University Press, 1989.

Curry examines the history of astrology from its "heyday" in the Interregnum through its crisis after the Restoration and into its survival in the eighteenth century. Historiographically, Curry both emphasizes the understanding of this occult science in its own terms and treats the topic as social history with class distinctions. Churchmen and natural philosophers play roles in this history—both as supporters and as critics of astrology. For example, William Lilly appears as an important astrologer while also a church warden during the interregnum; and some of the Boyle lecturers briefly revise "high astrology" even though the Royal Society had opposed judicial astrology.

Topics: *II*, IV Names: Ashmole, Gadbury, Goad, Lilly, Partridge

0307 Curry, Patrick. "Revisions of Science and Magic." *Hist. Sci.* 33(1985): 299-325.

This is an essay review of *From Paracelsus to Newton* edited by Charles Webster (1615) and of *Occult and Scientific Mentalities in the Renaissance* edited by Brian Vickers (1591). Curry points out the historiographic background for each of the volumes. Most of the review is a commentary on the Vickers collection. Vickers' article and introduction are severely criticized as an unhistorical attempt to protect the purity of modern science. Curry approves of Westfall's article on Newton and alchemy and concludes with some historiographic suggestions concerning viable approaches to the question implicit in the title of the essay review.

Topics: *II*, I, VII, XII Name: Newton

0308 Curry, Walter Clyde. *Milton's Ontology, Cosmogony, and Physics*. Lexington: University of Kentucky Press, 1957.

Curry's book is good for showing that (generally and for Milton) there was no simple relationship between the older thought and the "new philosophy." Milton's theology was an important factor in his views on these natural philosophic matters.

Topics: *X*, II, VII Name: Milton

0309 Curry, Walter Clyde. "Some Travels of Milton's Satan and the Road to Hell." *Philol. Q.* 29(1950): 225-235.

Curry here discusses both God and Satan in *Paradise Lost*. He analyzes Milton's views of infinity, space and time, cosmic chaos, and creation. This article is incorporated into and superseded by Curry's 1957 book.
Topic: *X* Name: Milton

0310 Curtis, L. P. *Anglican Moods of the Eighteenth Century*. New Haven, Conn.: Archon Books, 1966.
This book includes a discussion of the milieu of Latitudinarianism. Curtis sees some influence of science on religion but sets it within the total movement of rational theology.
Topic: *VII* Names: Locke, Tillotson

0311 Dahm, John J. "Science and Apologetics in the Early Boyle Lectures." *Church History* 39(1970): 172-186.
Dahm discusses the theological and philosophical implications of mechanistic materialism as the issue to which the physico-theologians addressed themselves. This essay is a helpful but not complete discussion of the apologetic context of natural theology in the years between 1690 and 1720.
Topics: *VIII*, IV, V, IX, XII Names: Bentley, Boyle, Clark, Derham, Newton

0312 Dampier, William C. *A History of Science and its Relations with Philosophy and Religion*. Cambridge: Cambridge University Press, 1929. 4th ed. Cambridge: Cambridge University Press, 1948.
This survey now is outdated because of the way in which Dampier imposes on the subject matter an anachronistic distinction between science and religion.
Topic: *VII*

0313 Daniel, Stephen H. *John Toland: His Methods, Manners, and Mind*. Kingston, Canada: McGill-Queens University Press, 1984.
This book presents systematically the diverse (even disparate) interests and activities of Toland. Daniel argues for an overall unity in Toland's approach which is rooted in a common methodological pattern. Toland's pantheistic deism and materialistic natural philosophy are relayed to each other within this larger

whole which includes social, political, polemical, and other dimensions. Toland's metaphysics, natural philosophy, and theology are all seen as greatly influenced by Bruno and as running counter to certain aspects of Newtonianism.

Topic: *IX* Name: Toland

0314 Darst, David H., with Steven L. Jeffers. "Wizards and Magicians: In the King James Old Testament." *Seventeenth Cent.* 6(1991): 1-10.

This is a study of how the English words "wizard" and "magician" were given new, diabolical, meaning with the publication of the King James Bible. For social and political reasons, as much as for lexical ones, these words were used to translate Hebrew words which had no clear or exact English equivalents. Darst and Jeffers see this as an attempt to attack folk and Catholic subcultures by using words which previously had connotations of wisdom and white magic and giving them the connotation of heresy and of working with the devil. Although the authors do not argue the point, they also note that this change helps lay "the intellectual groundwork for the rational, concrete, non-animistic, mechanized world of the forthcoming scientific revolution in Britain."

Topics: *X*, II, IV

0315 Davie, Donald. *The Language of Science and the Laguage of Literature*. London: Sheed and Ward, 1963.

This is a sustained scholarly polemic against the thesis that science and mechanistic philosophy produced in the early eighteenth century an atmosphere inimical to poetry (and to its companion, religion). Davie is effective in his literary analyses but seems not to understand Locke and Newton. (He treats Descartes, Hobbes, Locke, Newton, and early eighteenth-century science all as a single movement.) This short book has the length and feel of a long pamphlet.

Topic: *X* Names: Mandeville, Pope, Swift

0316 Davies, Godfrey. *The Early Stuarts, 1603-1660*. Oxford: Clarendon Press, 1937. 2nd ed. Oxford: Clarendon Press, 1959.

This well-written survey is part of the old Oxford History of England. It includes sections on science and religion—but the section on science is rather badly outdated.

Topics: *VII*, IV, V

0317 Davies, Gordon L. "The Concept of Denudation in Seventeenth-Century England." *J. Hist. Ideas* 27(1966): 278-284.

Denudation is an interesting geological principle which had important theological implications in the seventeenth century. This article has been incorporated into Davies' book (0319).

Topic: *VII*

0318 Davies, Gordon L. "Early British Geomorphology 1578-1705." *Geographical Journal* 132(1966): 252-262.

This essay has been superseded by Davies' book (0319).

Topic: *VII*

0319 Davies, Gordon L. *The Earth in Decay: A History of British Geomorphology, 1578-1878*. London: McDonald, 1969.

About 100 pages of this survey history are given over to our period. Davies is a historian of science whose interpretive approach is retrospective and judgmental. He postulates and emphasizes a harmful "bibliolatry" which "tended to cramp, stultify, and warp the infant subject" of geomorphology. But he has mastered the primary sources and he does treat science and religion as an integrated topic. He is thorough with the "Genesis and geology" theme—covering the sacred theory of the earth and hypotheses about the creation, the fall, the flood, and the end of the world. Davies is good at noting the theological implications of the various geomorphological theories (despite his gratuitous judgments). He also touches upon natural theology in its relations with geological subjects. The book provides some background for the ancients and moderns debate—although it treats this theme only tangentially.

Topics: *VII*, IV, VIII, X Names: Burnet, T., Hooke, Ray, Whiston, Woodward

0320 Davies, Gordon L. "From Flood and Fire to Rivers and Ice—Three Hundred Years of Irish Geomorphology." *Irish Geography* 5(1964): 1-16.

This essay has been superseded by Davies' book (0319).
Topic: *VII*

0321 Davies, Gordon L. "Robert Hooke and his Conception of Earth History." *Proceedings of the Geological Association, London* 75(1964): 493-498.
This article also has been superseded by Davies' book (0319).
Topic: *VII* Name: Hooke

0322 Davies, Horton. *Worship and Theology in England from Andrewes to Baxter and Fox, 1603-1690*. (Worship and Theology in England, Vol. 2.) Princeton, N.J.: Princeton University Press, 1975.
Chapters 1 and 2 describe the theologies of church architecture and the achievements of Anglican, Roman Catholic, Puritan, and Quaker architects. Davies argues that differences in theology and social position are reflected in differences in church architecture.
Topic: *VI* Names: Andrewes, Baxter, Charles I, Laud, Wren

0323 Davies, Marie-Helene. *Reflections of Renaissance England: Life, Thought and Religion Mirrored in Illustrated Pamphlets, 1535-1640*. Allison Park, Pa.: Pickwick, 1986.
This work contains summaries of pamphlets—including some on science, magic, astrology, and theology. It also contains illustrations and a good listing of pamphlets.
Topics: *X*, II

0324 Davis, Edward B., Jr. "The Anonymous Works of Robert Boyle and the 'Reasons Why a Protestant Should not Turn Papist'." *J. Hist. Ideas* 55(1994): 611-629.
Thanks to Davis' detective work, the problem of the last anonymous book in Fulton's bibliography of Boyle's work is solved. Davis finds convincing evidence that Boyle did not write *Reasons why a Protestant Should not Turn Papist*. Rather, it was written by David Abercromby, a Scottish physician and former Jesuit. (Abercromby was a member of Boyle's circle and translated several of Boyle's works into Latin.) Davis sets his solution of this case in the context of a review of Boyle's anonymous and pseudonymous writings (both theological and scientific).
Topic: *X* Names: Abercromby, Boyle

0325 Davis, Edward B., Jr. "Blessed are the Peacemakers: Rewriting the History of Christianity and Science." *Perspectives on Science and Christian Faith* 40(1988): 47-52.

This is an essay review of three surveys of the historical relations between science and religion: *The Galileo Connection* by Charles E. Hummel (Downers Grove, Ill.: Intervarsity Press, 1986); *Crosscurrents* by Colin A. Russell (1376); and *God and Nature*, edited by David Lindberg and Ronald Numbers (0954). While Davis sees them as various attempts to overcome the "warfare" thesis, he points out that their purposes differ. Whereas Hummel and Russell have written Christian apologetic works, *God and Nature* is a more sophisticated collection of academic studies. Concerning the seventeenth century, Davis is critical of Russell and likes Deason's essay (0333) in the Lindberg and Numbers volume.

Topics: *I*, VII

0326 Davis, Edward B., Jr. "Newton's Rejection of the 'Newtonian World View': The Role of Divine Will in Newton's Natural Philosophy." *Fides et Historia* 22(1990): 6-20. Reprinted in *Sci. Christ. Belief* 3(1991): 103-117.

Davis argues that the "textbook" version of Newton as "the paragon of Enlightenment deism" is absolutely false. He focuses on Newton's doctrine of the dominion of the "free and powerful God" who creates, preserves, and (more importantly) acts providentially in and on nature.

Topics: *XII*, VII, X Names: Clarke, Newton

0327 Davis, Edward B., Jr. " Parcere Nominibus : Boyle, Hooke and the Rhetorical Interpretation of Descartes." In *Robert Boyle Reconsidered*, edited by Michael Hunter, 157-175. Cambridge: Cambridge University Press, 1994. See (0759).

Davis analyzes Boyle's rhetoric in response to Descartes on the issue of final causes. He proposes that Hooke assisted Boyle as he tried to understand and respond to Cartesianism. He finds that Boyle's theological voluntarism and desire to encourage piety motivated his emphasis on teleological rather than ontological arguments. Yet Boyle's irenical purpose led him to avoid *ad hominem* attacks against Christian Cartesians (in other words, "*parcare nominibus*, to go little by names").

Topic: *VII* Names: Boyle, Hooke

0328 Davis, J. C. *Utopia and the Ideal Society: A Study of English Utopian Writing, 1516-1700*. Cambridge: Cambridge University Press, 1981.

After distinguishing between utopias and other forms of writing about ideal societies, Davis devotes chapters to particular utopian writings of the sixteenth and seventeenth centuries. For Davis, the distinguishing mark of these utopias is the acceptance of human deficiencies (not here called sinfulness) which is disciplined by the totally ordered society. Davis usefully outlines Bacon's *New Atlantis* as a drama and examines each scene. He finds a tolerant but necessary Christian faith as well as the all-knowing yet isolated natural philosophers. He concludes that Bacon's unresolved tension is still with us: his optimism about the new science in part undoes the consequences of the fall; yet there is utopian pessimism in his assumption of the corruptibility of his scientists. Davis also discusses the *Macaria* by Gabriel Plattes and others in the Hartlib circle. He does so in the context of "full employment" utopias. This economic, state-centered approach was associated with small Christian societies and with those who advocated technological and agricultural innovations.

Topics: *IV*, IV, X Names: Bacon, Hartlib, Plattes

0329 Davis, John W. "Berkeley, Newton, and Space." In *The Methodological Heritage of Newton*, edited by Robert E. Butts and John W. Davis, 57-73. Toronto: University of Toronto Press, 1969. See (0172).

Davis analyzes the differences between Newton and the young Berkeley as regards space and its theological implications. He challenges the theses of several other scholars and argues that Berkeley had no influence on Newton.

Topics: *VII*, XII Name: Newton

0330 Dear, Peter. "Miracles, Experiments, and the Ordinary Course of Nature." *Isis* 81(1990): 663-683.

In this intricately argued essay, Dear correlates sociocultural patterns with scientific practice in France and England. The English experimental philosophy had a focus on description of discrete historical events and detailed experiments while the French emphasis on legitimizing scientific laws put a focus on mathematical models. The different practices in the physical sciences in the two countries, argues Dear, can be correlated to the respective (French Catholic and Anglican) understandings of miracles and the ordinary course of nature.

Topics: *VII*, IV Names: Boyle, Sprat

0331 Deason, Gary B. "John Wilkins and Galileo Galilei: Copernicanism and Biblical Interpretation in the Protestant and Catholic Traditions." In *Probing the Reformed Tradition: Historical Studies in Honor of Edward A. Dowey, Jr.*, edited by Elsie Anne McKee and Brian G. Armstrong, 313-338. Louisville, Ky.: Westminster/John Knox Press, 1989.

Comparing Galileo and Wilkins on biblical interpretation, Deason revises the popularly accepted view that Protestant literalism was an obstacle to Copernicanism. The application of the "accommodation theory" to the whole sense of the text allowed Wilkins to separate completely the new astronomy from matters of faith and morals—that is, the study of nature from the interpretation of scripture. Galileo, while using the principle of accommodation for specific texts, had to grapple with the unity of Catholic truth, traditional interpretations, and the multiple meanings of scriptural passages.

Topic: *X* Name: Wilkins

0332 Deason, Gary B. "The Protestant Reformation and the Rise of Modern Science." *Scot. J. Theology* 38(1985): 221-240.

Deason argues for an indirect (noncausal) relationship between Protestantism and science. He considers both sixteenth-century continental reformers and seventeenth-century English natural philosophers. His main argument is that "Protestantism contributed to the rise of the mechanical philosophy largely by helping to undermine Aristotelian philosophy." It did this in three ways: by transforming the locus of authority, by emphasizing empirical methods; and by asserting the sovereignty of God. Bacon, Boyle, and Newton are used briefly as examples.

Topics: *III*, VII, XII Names: Bacon, Boyle, Newton

0333 Deason, Gary B. "Reformation Theology and the Mechanistic Conception of Nature." In *God and Nature*, edited by David C. Lindberg and Ronald Numbers, 167-189. Berkeley and Los Angeles: University of California Press, 1986. See (0954).

Deason here develops the connection between the Protestant emphasis on "the radical sovereignty of God" and the mechanistic conception of nature (with its fundamental view that matter is passive).
Topics: *VII*, III

0334 Debus, Allen G. "Alchemy and the Historians of Science." *Hist. Sci.* 6(1967): 128-138.
This is an essay review of C. H. Josten's edition of *Elias Ashmole* (0849). In criticizing Josten, Debus argues in favor of a more contextual historiography of the Hermetic tradition. He notes that the alchemists and natural magicians "sought a universal science" and not just a secret mystical brotherhood.
Topics: *II*, I Name: Ashmole

0335 Debus, Allen G. "The Chemical Debates of the Seventeenth Century: The Reaction to Robert Fludd and Jean Baptiste van Helmont." In *Reason, Experiment, and Mysticism in the Scientific Revolution*, edited by M. L. Righini Bonelli and William R. Shea, 19-47 and 291-298. New York: Science History, 1975. See (1329).
Debus here gives an account of the scope and issues of the chemical debates of the early seventeenth century. He summarizes the views of the Paracelsians on such topics as chemistry and alchemy, macrocosm and microcosm, religion and natural philosophy, the divine light and the vital spirit, and natural magic. He argues for the relevance of Hermetic influences to our understanding of the Scientific Revolution.
Topics: *II*, VII, XI Name: Fludd

0336 Debus, Allen G. "The Chemical Dream of the Renaissance." In *Churchill College Overseas Lecture Number 3*, 7-40. Cambridge: Heffer, 1968.
In this early lecture, Debus examines those men who advocated a reform of learning outside of the universities and based on chemistry. For them, chemistry had a divine significance—the creation itself was interpreted as a chemical process. Debus also relates the writings of the Rosicrucian movement to these advocates and to their understanding of chemistry.
Topics: *II*, IV, V Names: Fludd, Ward, Webster

0337 Debus, Allen G. "The Chemical Philosophers: Chemical Medicine from Paracelsus to Van Helmont." *Hist. Sci.* 12(1974): 243-259.

This is a brief survey of what later became Debus' two-volume work. Here, he summarizes the alchemically- and biblically-based chemical philosophies of Fludd and J. B. van Helmont.

Topics: *XI*, II, X Name: Fludd

0338 Debus, Allen G. *The Chemical Philosophy: Paracelsian Science and Medicine in the Sixteenth and Seventeenth Centuries*. 2 vols. New York: Science History, 1977.

Debus' magnum opus is a thorough examination of the topic described in the subtitle. In the chapters on the seventeenth-century chemical philosophy, Fludd and J. B. van Helmont are the central figures. But others, in England and elsewhere, are discussed. Throughout, Debus shows the centrality of the chemical interpretation of Genesis and the importance of the macrocosm-microcosm analogy.

Topics: *XI*, II, VII, X Names: Boyle, Charleton, Fludd, Ward, Webster

0339 Debus, Allen G. *Chemistry, Alchemy, and the New Philosophy, 1550-1700: Studies in the History of Science and Medicine*. (Collected Studies Series, 249.) London: Variorum Reprints, 1987.

This is a collection of previously published papers—two of which, (0355) and (0358), are annotated separately in this bibliography.

Topic: *II*

0340 Debus, Allen G. "Chemistry and the Quest for a Material Spirit of Life in the Seventeenth Century." In *Spiritus*, edited by M. Fattori and M. Bianchi, 245-263. Rome: Edizioni dell' Ateneo, 1984. See (0446).

Debus shows the deep interest in the search for a material spirit of life by chemical philosophers in the seventeenth century. Traditionally, this spirit was understood to be divine in origin and necessary for all life. Debus uses a wheat experiment by Fludd and analyses of blood by van Helmont and Boyle as his primary case studies. He indicates that such investigations had both medical and cosmological implications.

Topics: *VII*, XI Names: Boyle, Fludd

0341 Debus, Allen G. *The English Paracelsians*. London: Oldbourne, 1965.

This is a study of the thought and practice of Englishmen in the period before 1640. For theological (as well as other) reasons, these men transmitted and advocated Paracelsian chemical and medical philosophy. Of particular interest, for our purposes, are certain clergymen and Robert Fludd. Debus finds that their medical remedies were more popular than their cosmology.

Topics: *XI*, II Names: Fludd, Tymme

0342 Debus, Allen G. "Iatrochemistry and the Chemical Revolution." In *Alchemy Revisited*, edited by Z. R. W. M. von Martels, 51-66. Leiden: Brill, 1990.

Against the general view of the chemical revolution as a late-eighteenth century "delayed" revolution, Debus proposes a series of phases dating from the sixteenth century. The first phase was primarily a Paracelsian medical development in which a mystical cosmology and religio-political influences were important. Debus' revisionist approach requires a reconsideration of the historiographic separation of the histories of science, medicine, and religion.

Topics: *XI*, I, II Names: Boyle, Fludd

0343 Debus, Allen G. Introduction. In *Theatrium Chemicum Britannicum*, by Elias Ashmole, ix-xlix. New York: Johnson Reprint, 1967.

Debus discusses Ashmole's life and this text in the context of alchemy in England. The "understanding of nature" as a whole (not just transmuting metals into gold) was the true object of the alchemist for Ashmole. Therefore, alchemy led to "a perfect knowledge of the works of God." Debus also describes the thirty works by English alchemists which Ashmole compiled in the volume and adds a bibliographic note.

Topic: *II* Name: Ashmole

0344 Debus, Allen G. Introduction to *Robert Fludd and His Philosophical Key; being a Transcription of the Manuscript at Trinity College, Cambridge*, edited by Allen G. Debus, 1-58. New York: Science History, 1979.

This essay contains a good summary of Fludd's life and works—as well as an analysis of the *Philosophical Key*. Fludd is important,

Debus argues, because he engaged in both scientific and religious debates.
Topics: *II*, X Name: Fludd

0345 Debus, Allen G. "John Webster and the Educational Dilemma of the Seventeenth Century." In *Science et Philosophie XVIIe et XVIIIe Siecles*, 15-23. Paris: Albert Blanchard, 1971. See (1418).
Debus summarizes Webster's *Academiarum Examen*. He points out its rejection of Aristotle and its advocacy of a combination of Baconian experimentation, Paracelsian chemistry, and scriptural principles. He thereby shows that the then current (1968) division between ancients and moderns was outmoded and proposes a more complex formulation of the philosophical divisions in the mid-seventeenth century.
Topics: *II*, I, IV, VII Name: Webster

0346 Debus, Allen G. "Key to Two Worlds: Robert Fludd's Weather-glass." *Annali dell'Instituo e Museo di Storia della Scienza di Firenze* 7(1982): 109-144.
This is a detailed, illustrated discussion of Fludd's use of a thermoscope to confirm his own "mystical world view." Debus first points out the role of the divine spirit of life in Fludd's hermetic philosophy. Then he shows how, for Fludd, the weatherglass was important not only for measuring temperature and contributing to human health but also for discovering fundamental dichotomies of the macrocosm in a microcosmic setting.
Topics: *VI*, II, XI Name: Fludd

0347 Debus, Allen G. *Man and Nature in the Renaissance*. Cambridge: Cambridge University Press, England, 1978.
This is an introductory textbook on the history of science and medicine from the mid-fifteenth century to the mid-seventeenth century. Although he also deals with astronomy and the new methods, Debus focuses his primary attention on chemistry, alchemy and medicine. He clearly acknowledges that natural magic and the religious quest for divine truths were a part of the science of the period. His conclusion restates the thesis of his inaugural lecture: there was a lengthy dialogue and debate in the seventeenth

century over how best to found a natural philosophy that would account for the entire divinely created cosmos.
Topics: *XI*, II, IV, VII Names: Fludd, Gilbert, Harvey

0348 Debus, Allen G. "Mathematics and Nature in the Chemical Texts of the Renaissance." *Ambix* 15(1968): 1-28.
Debus demonstrates that the two mathematical paths recommended by Cusanus for learning about the Creator and creation were taken respectively by Robert Fludd and by J. B. van Helmont. Fludd believed that mathematics was a mystical science whose symbols were the keys to the cosmic mysteries (of his alchemical universe). For van Helmont, mathematics assisted the philosopher in his determination of experimental weights. Debus also outlines the Neoplatonic background and their use of biblical prooftexts.
Topics: *II*, VII Name: Fludd

0349 Debus, Allen G., ed. *Medicine in Seventeenth-Century England: A Symposium Held at UCLA in Honor of C. D. O'Malley*. Berkeley and Los Angeles: University of California Press, 1974.
The essays by Bodemer (0104) and Debus (0353) are annotated separately in this bibliography.
Topic: *XI*

0350 Debus, Allen G. "Motion in the Chemical Texts of the Renaissance." *Isis* 64(1973): 5-17.
Debus here proposes that the chemical philosophers of the early seventeenth century opposed both the mathematicization of nature and the Aristotelian view of motion. They argued against the latter because it required God to be immovable while they believed in the complete freedom and power of the Creator. Indeed, they understood motion to be the result of the divine spirit within bodies. Debus also suggests that these views influenced later natural philosophers.
Topic: *VII* Names: Fludd, Webster

0351 Debus, Allen G. "Myth, Allegory, and Scientific Truth: An Alchemical Tradition in the Period of the Scientific Revolution." *Nouvelles de la Republique des Lettres* 1(1987): 13-35.
Debus illustrates the importance of myth and allegory in sixteenth- and seventeenth-century texts of alchemists, chemical philosophers,

and Bacon. He shows how their search for hidden meanings in ancient myths related to their understandings of nature, the biblical creation story, and scientific method. Thus, he argues for an expanded, interdisciplinary approach to the history of science.
Topics: *X*, I, II Names: Bacon, Fludd

0352 Debus, Allen G. "Paracelsian Doctrine in English Medicine." In *Chemistry in the Service of Medicine*, edited by F. N. L. Poynter, 5-26. London: Pitman, 1963.
In this early paper, Debus finds that prior to 1650 English "Paracelsian" physicians generally accepted the practical applications of chemical remedies but rejected Paracelsus' "mysticism." (He notes Fludd as an exception.) During the Interregnum, Paracelsian physicians became more radical and partisan. All, however, interpreted creation as a chemical separation.
Topics: *II*, I, IV, VII, XI Names: Fludd, Tymme

0353 Debus, Allen G. "Paracelsian Medicine: Noah Biggs and the Problem of Medical Reform." In *Medicine in Seventeenth Century England*, edited by Allen G. Debus, 33-48. Berkeley and Los Angeles: University of California Press, 1974. See (0349).
Debus summarizes the contents of the book *Mataeotechnia Medicinae Praxews: The Vanity of the Craft of Physick* (1651). It was a plea to Parliament to reform the education of physicians by a Puritan iatrochemist who believed that his union of medicine and natural philosophy was a divine calling.
Topics: *XI*, IV, V Name: Biggs

0354 Debus, Allen G. "Renaissance Chemistry and the Work of Robert Fludd." *Ambix* 14(1967): 42-59. An earlier version was published (as a Clark Library paper) in *Alchemy and Chemistry in the Seventeenth Century*, by Allen G. Debus and Robert P. Multheuf, 1-29. Los Angeles: University of California, 1966.
This is an excellent discussion of the interrelationships between alchemy, philosophy, Christianity, and the new mechanical philosophy. The Paracelsians and Fludd developed a philosophy based upon a mystical chemical account of creation. Debus argues that Fludd and others had the same goals as later (more modern) chem-

ists. And, historiographically, they must be taken seriously if we are to understand the development of seventeenth-century science.

Topics: *II*, VII Names: Fludd, LeFevre

0355 Debus, Allen G. "Robert Fludd and the Circulation of the Blood." *J. Hist. Med.* 16(1961): 374-393. Reprinted in *Chemistry, Alchemy and the New Philosophy*, Chapter 11. London: Variorum Reprints, 1987. See (0339).

In this study of the controversy between Gassendi and Fludd over the circulation of the blood, Debus shows that both differed in their respective views from Harvey. While Fludd believed in a kind of circulation on metaphysical and "mystical" grounds, Gassendi rejected circulation of the blood until very late in his life.

Topics: *XI*, II Names: Fludd, Harvey

0356 Debus, Allen G. *Science and Education in the Seventeenth Century: The Webster-Ward Debate*. London: Macdonald, 1970.

Debus here shows that the Webster-Ward controversy does not make sense when put in the ancients-versus-moderns pattern. He discusses the magical tradition (represented by Webster) in relation to "Baconianism" and experimental philosophy.

Topics: *IV*, II, III Names: Ward, Webster

0357 Debus, Allen G., ed. *Science, Medicine and Society in the Renaissance: Essays to Honor Walter Pagel*. 2 vols. New York: Science History, 1972.

See the bibliography by Winder (1812) and the essays by Poynter (1258), Rattansi (1293), Ravetz (1305), Rossi (1359), Walker (1597), Westfall (1644), and Wilkinson (1679).

Topic: *XI*

0358 Debus, Allen G. "Science vs. Pseudo-Science: The Persistent Debate." Inaugural Lecture, the University of Chicago, 1978. The Morris Fishbein Center for the Study of the History of Science and Medicine. Publication No. 1. Chicago: University of Chiacgo, 1979. Reprinted in *Chemistry, Alchemy and the New Philosophy*, Chapter 1. London: Variorum Reprints, 1987. See (0339).

Debus traces what he calls the debate between the "mechanistic" and the "mystical" viewpoints from the recovery of ancient texts

in the fifteenth and sixteenth centuries through the nineteenth century. His point is that, although members of the "mystical" party may not be found among twentieth-century scientists, one cannot understand the history of science without observing this persistent debate among "scientists" in the past.
Topics: *II*, I

0359 Debus, Allen G. "The Significance of the History of Early Chemistry." *Cahiers d'Histoire Mondiale* 9(1965): 39-58.
This article is a bibliographical essay concerning research on the history of chemistry, primarily in the early modern period. It contains a good discussion of writings up to the mid-1960s. Debus concludes that the more recent research links alchemy to philosophical and religious thought.
Topics: *I*, II

0360 Debus, Allen G. "Some Comments on the Contemporary Helmontian Renaissance." *Ambix* 19(1972): 145-150.
This is an essay review of the facsimile reprint of J. B. van Helmont's *Aufgang der Artzney-Kunst* (with an introduction by Walter Pagel). Debus discusses both recent studies (especially by Pagel) and van Helmont's thought (especially his combination of "mysticism" and science). The essay contains a good brief discussion of van Helmont's emphasis of "man" and his rejection of the macrocosm-microcosm universe. Although basically continental, this has relevance to the English scene as well.
Topics: *VII*, I

0361 Debus, Allen G. "The Sun in the Universe of Robert Fludd." In *Le Soleil a le Renaissance: Sciences et Mythes*, 259-277. Brussels: Presses Universitaires de Bruxelles, 1965.
Debus shows that the sun, for Fludd, was the "tabernacle of the Lord" and, hence, was the source of the divine light necessary for all life. Fludd rejected the Copernican model and retained a geocentric universe, among other reasons, because it housed God's spirit and was the source of all motion.
Topics: *II*, X Name: Fludd

0362 Debus, Allen G. "Thomas Sherley's *Philosophical Essay* (1672): Helmontian Mechanism as the Basis of a New Philosophy." *Ambix* 27(1980): 124-135.

Debus analyzes Sherley's book and shows that it combines the writings of Boyle and van Helmont on the doctrine of the elements and the generation of human stones. This union included a final proof of Helmontian principles from the first chapter of Genesis. Sherley's work indicates that seventeenth-century persons did not make the same sort of historiographical distinctions (e.g., modern versus mystic) that twentieth-century historians tend to make.
Topics: *II*, X Names: Boyle, Sherley

0363 de Candolle, Alphonse. *Histoire des Sciences et des Savants depuis Deux Siecles*. Geneva: H. Georg, 1873. 2nd ed. Geneva-Bale: H. Georg, 1885. Excerpt in English translation appears in *Puritanism and the Rise of Modern Science: The Merton Thesis*, edited by I. Bernard Cohen, 373-389. New Brunswick N.J.: Rutgers University Press, 1990. See (0246).
Alphonse de Candolle was the first scholar to make a statistical argument connecting Protestantism and science. He lumped together the whole period from 1666 to his own time. His work has been superseded by Merton (1055 and 1057).
Topic: *III*

0364 De Grazia, Margreta. "The Secularization of Language in the Seventeenth Century." *J. Hist. Ideas* 41(1980): 319-329.
DeGrazia sees the "deverbalization of God's message and the dissociation of language from God" as two parallel seventeenth-century developments in the relation between human language and God. She suggests that: the "book of nature" lost its traditional verbal form; the second book, the Bible, came to be interpreted through science; and a third book, the "tablet of the soul" intuited through each person, became more authoritative. Finally, she says, words and language (having become human conventions) lost their place as a mark of humankind's resemblance to God.
Topic: *X*

0365 De Morgan, Augustus. *Essays on the Life and Work of Newton*. Edited by Philip E. B. Jourdain. Chicago: Open Court, 1914.
DeMorgan made miscellaneous contributions to Newton's biography. He treated Newton's science as separate from his religion—but he did recognize the importance and legitimacy of his religion. His was the best nineteenth-century treatment of Newton's

heterodoxy. His work still is worth reading for advanced Newton scholars.
Topics: *XII*, IX Name: Newton

0366 De Morgan, Augustus. "Newton." *The Cabinet Portrait Gallery of British Worthies* 11(1846): 78-117. Reprinted in *Essays on the Life and Work of Newton*, edited by Philip E. B. Jourdain, 3-63. Chicago: Open Court, 1914. See (0365).
See (0365).
Topic: *XII* Name: Newton

0367 De Morgan, Augustus. "Review of Brewster's *Memoirs of the Life, Writings, and Discoveries of Sir Isaac Newton.*" *North British Review* 23(1855): 307-338. Reprinted in *Essays on the Life and Work of Newton*, edited by Philip E. B. Jourdain, 119-182. Chicago: Open Court, 1914. See (0365).
See (0365).
Topic: *XII* Name: Newton

0368 de Pauley, W. C. *The Candle of the Lord: Studies in the Cambridge Platonists*. London: Society for Promoting Christian Knowledge, 1937.
This book includes nine freestanding chapters with a focus on theology. It consists of basic summaries of the writings and sermons of eight individuals (including Whichcote, Cudworth, More, and Stillingfleet). Science is not directly relevant but epistemology, philosophy of nature, and natural religion are covered for these various theologians.
Topics: *VII*, VIII Names: Cudworth, Culverwel, More, Smith, Stillingfleet, Whichcote

0369 de Santillana, Giorgio. *Reflections on Men and Ideas*. Cambridge, Mass.: M.I.T. Press, 1968.
Several of the essays in this collection deal with the spirit of "scientific rationalism" from the Greeks, through the seventeenth century, to the present. For de Santillana, scientific rationalism is "a faith and an implicit doctrine," a sense of symbolic unity, and the myth of science.
Topics: *VII*, XII Name: Newton

0370 Dessauer, Friedrich. "Galileo and Newton: The Turning Point in Western Thought." In *Spirit and Nature: Papers from the Eranos Yearbooks*, edited by Joseph Campbell, 288-321. New York: Pantheon Books, 1954. Translated by Ralph Manheim. The German original was published in *Eranos-Jahrbucher* 14(1946).

Dessauer argues that Newton held "the two spheres" of spirit and nature together. The division which has since occurred has happened in spite of Newton's intentions. This was an interesting insight for the 1940s and 1950s.

Topics: *XII*, VII Name: Newton

0371 Dessauer, Friedrich. *Weltfahit der Erkenntis: Leben und Werk Isaac Newtons*. Zurich: Rascher, 1945.

This is a biography with selected commentaries on Newton's life and work. It focuses especially on Newton's optics and mathematics. The role of God and Newton's theological writings are briefly noted and accepted—but not emphasized. The book has been superseded by Westfall, Manuel, and Christianson.

Topics: *XII*, VII Name: Newton

0372 Dick, Oliver Lawson. "The Life and Times of John Aubrey." In *Aubrey's Brief Lives*, edited by Oliver Lawson Dick, xiii-cvi. Ann Arbor, Mich.: University of Michigan Press, 1949.

In his biographical introduction to Aubrey's *Brief Lives*, Dick explicates at length Aubrey's wide-ranging interests in medicine, natural history, and philosophy. Dick weaves Aubrey's own words into the narrative, so that it appears to be as autobiographical as it is biographical. Dick's ahistorical and judgmental biases are clear—and make the quotations more valuable than the analysis.

Topics: *X*, VII, XI Name: Aubrey

0373 Dick, Steven J. "The Origins of the Extraterrestrial Life Debate and Its Relation to the Scientific Revolution." *J. Hist. Ideas* 41(1980): 3-27.

Dick shows that metaphysical principles, interpretations of scripture, and natural theology all were crucial factors in the seventeenth-century debate among European natural philosophers. These factors "reinforced, extended, or undermined" the Copernican theory as well as empirical observations of the moon and planets.

Thus, he argues that theological issues were part of the seventeenth-century scientific revolution in astronomy.
Topics: *VII*, X Name: Wilkins

0374 *The Dictionary of National Biography*. Edited by Leslie Stephen and Sidney Lee. London: Smith, Elder, 1885-1900. (Various reprints have appeared.)
This classic still is useful for basic biographical facts and primary references.
Topics: *VII*, II, X, XI, XII

0375 *The Dictionary of the History of Ideas*. Edited by Philip P. Wiener. 5 vols. New York: Scribner, 1968-1973.
This helpful resource includes article-length discussions on several relevant topics. The equivalent of about 200 normal pages are relevant to our field and era. Some of the topics and the scholars who have written the essays are: alchemy (Debus); atomism (Kargon); Baconianism (Rossi); certainty in the seventeenth century (Van Leeuwen); deism (Emerson); design argument (Ferre); nature (Boas); Newtonian topics (Guerlac and Nicolson); and skepticism (Popkin). The essays are thorough, up-to-date (as of 1970), and generally characterized by the unit-idea historiography suggested by the title of the dictionary. Volume 5 contains a thorough index of names and topics.
Topics: *VII*, II, VIII, IX, XII Names: Bacon, Newton

0376 Dijksterhuis, E. J. *The Mechanization of the World Picture*. Translated by C. Dikshoorn. Oxford: Clarendon Press, 1961.
This is a classic survey of the development of "mechanicism" by a Dutch historian of science. It extends from the ancient Greek philosophers to Newton. As well as their physics, Dijksterhuis includes brief summaries of the theological views of his chosen figures. He concludes with Newton and the seeming necessity for the separation of religion and science.
Topics: *VII*, VIII, XII Names: Bacon, Boyle, Newton

0377 Dillenberger, John. *Protestant Thought and Natural Science*. Garden City, N.Y.: Doubleday, 1960.
In this survey, Dillenberger addresses the following topics: conflicts between the new philosophy and older philosophical

traditions; the cosmological and teleological arguments as used in our period; changes in traditional Christianity brought about by early modern science; reason and revelation during our period. Dillenberger sees the new science as contributing to the rise of deism and, from a theological perspective, passes a negative value judgment on this development. As an introductory survey, this book has been superseded by Kaiser (0855).

Topics: *VII*, III, VIII, IX, XII Names: Boyle, Newton

0378 Dobbs, Betty Jo Teeter. *Alchemical Death and Resurrection: The Significance of Alchemy in the Age of Newton*. Washington, D. C.: Smithsonian Institution Libraries, 1990.

This is a revised and illustrated lecture originally delivered at the Smithsonian. Dobbs draws on her expertise in Newton's writings and on the sixteenth-century Ripley scrolls. She proposes "to reconstruct by historical methods some of the meaning of alchemical texts and illustrations." Alchemists believed that in alchemy lay the secrets of life, death, and resurrection. The symbolic details of the illustrations indicate that Christian alchemy was a way of finding Christ, the active agent, and therefore was a way of achieving both personal and the world's salvation.

Topics: *II*, VII, XII Name: Newton

0379 Dobbs, Betty Jo Teeter. "Alchemische Kosmogonie und Arianische Theologie bei Isaac Newton." In *Die Alchemie in der Europaischem Kulturund Wissenschaftsgeschichte*, edited by Christoph Meinel, 137-150. Wiesbaden, Germany: Harrassowitz, 1986.

Dobbs focuses her attention on Newton's commentary upon the *Emerald Tablet* attributed to Hermes Trismegistus. She shows how Newton: (1) used analogical reasoning to draw parallels between cosmogony and alchemy; (2) used a typological exegesis on the text to interpret Hermes as God's divine agent; and (3) believed that the Christ of his Arian theology is the omnipresent creative logos as well as the alchemical spirit. Newton's Christ is forever transmitting God's will into action but is united with God only in a "unity of dominion." (For Newton, Christ is not united with God in a metaphysical union.) But also, for Newton, Christ's continuous activity refutes ideas of a fully mechanical system of the world.

Topics: *XII*, II, VII, IX Name: Newton

0380 Dobbs, Betty Jo Teeter. *The Foundations of Newton's Alchemy: "The Hunting of the Greene Lyon."* Cambridge: Cambridge University Press, 1975.

In this early work, based on her dissertation research, Dobbs focuses on Newton's alchemical studies between 1668 and 1675. She intends the word "Foundations" in the title simultaneously to refer to: (1) the origins of Newton's alchemy; (2) the experimental and scholarly bases of Newton's alchemical studies; and (3) alchemy as a pillar which supported Newton's mature science. Thus, Dobbs provides a survey of the natural philosophical background; she describes and interprets Newton's alchemical experiments and manuscripts; and she indicates possible later influences of the early experiments and speculations. Dobbs presents Newton as striving to integrate alchemical and Hermetic ideas with the mechanical philosophies of his day. Concerning religion, Dobbs treats religion and alchemy: (1) as intertwined in the tradition on which Newton draws; but (2) to be rather separate for Newton. She includes an interesting Jungian analysis of "the older alchemy and its soteriological function." But she states that Newton's "alchemical manuscripts do not deal with religious matters." (In her later writings, of course, Dobbs came to see a unity of thought in Newton's mind—and religion and alchemy were intertwined for him.) In terms of the seventeenth-century background, Dobbs discusses the Hartlib circle (alchemy, mechanism, and reform), as well as Barrow and More (chemistry and alchemy at Cambridge). She sees the influence of Barrow and More in Newton's adherence to the *prisca* tradition, sympathy for the Neoplatonic philosophic position, and possibly even his very interest in alchemy. Dobbs emphasizes the importance of *prisca sapientia* in Newton's natural philosophy and believes that "any real understanding of Newton's alchemy is precluded if his adherence to the *prisca sapientia* is ignored." As regards the influence of alchemy on Newton's mature thought, Dobbs discusses Newton's theory of matter, his speculations about the aether, his creation of a new concept of force, and his mature chemical thought. But she does not (in this book) connect these topics to religion.

Topics: *XII*, II, VII Names: Barrow, Boyle, Digby, More, Newton

0381 Dobbs, Betty Jo Teeter. "From the Secrecy of Alchemy to the Openness of Chemistry." In *Solomon's House Revisited: The Organization and Institutionalization of Science*, edited by

Tore Frangsmyr, 75-94. Canton, Mass.: Science History, 1990.

This is an historical essay on how "an originally occult and secret millenarianism subtly but surely transformed into a public and social utilitarianism." Alchemy was assimilated into Christianity both as an analogical parallel to the original creation and as a typological prefiguration of human salvation history. Regarding the latter, Dobbs traces the shift in the image of the actor in this history from an individual seeker to a secret Hermetic society to an open community of seekers. That society's utopian task and hope was the "relief of man's estate." Yet she also notes the irony that, when "entangled with nationalism and war," such a hope brings humanity to the brink of apocalypse.

Topics: *II*, IV, V, X

0382 Dobbs, Betty Jo Teeter. "Gravity and Alchemy." In 1579, *The Scientific Enterprise*, edited by Edna Ullman-Margalit, 205-222. Dordrecht: Kluwer Academic, 1992. See (1579).

Dobbs here presents a chronological study of the changes in Newton's thought in the 1660s and 1670s concerning: the relationship between the active spirit of alchemy and the cause of gravity; and, indirectly, his doctrine of God.

Topics: *XII*, II, VII Name: Newton

0383 Dobbs, Betty Jo Teeter. *The Janus Face of Genius: The Role of Alchemy in Newton's Thought*. Cambridge: Cambridge University Press, 1992.

Dobbs' brilliant volume is an attempt to understand Newton and his writings as a whole. She does this by presenting "a religious interpretation of all his work." Newton, she argues, believed in "the unity of Truth, guaranteed by the unity and majesty of God." Thus she openly revises her previous view of Newton's alchemy as being without religious significance. She now pictures Newton as continuously searching for the bases both for alchemy and for gravity. The answers varied, yet were always related to his Arian theology. All of Newton's diverse studies, she shows, were motivated by his desire to learn about God's activity in every area—in micromatter, in the cosmic order, in history. Thus, after two conceptual chapters on "Vegetability and Providence," and "Cosmogony and History," she presents four chronological chapters on "modes of divine activity in the world." Newton's views of the manner in which God acts changed before, during, and after the period of the *Principia*. In these six chapters she incorporates material drawn from her previously published articles (and annotated elsewhere). The epilogue summarizes her present interpretation and briefly notes that it may reinforce the thesis that a Christian doctrine of creation motivated men to study nature. Fifty pages of transcriptions of manuscripts and a bibliography conclude the volume. We believe that this is the most significant discussion of the interrelations between Newton's theology, his alchemy, and his natural philosophy (i.e., the unity of his thought) yet available.

Topics: *XII*, II, IV, VII, X Names: Boyle, Burnet, T., Clarke, Fatio, Halley, More, Newton

0384 Dobbs, Betty Jo Teeter. "Newton and Stoicism." *Southern Journal of Philosophy* 23 Supplement(1985): 109-123.

Dobbs tentatively proposes that Newton saw in Stoicism fragments of the "ancient wisdom" and that he used the Stoic doctrines regarding the end of the world, total blending, and cohesion. The Stoic idea of an all-pervading creative divine pneuma was particularly important in Newton's discussions of God as the cause of gravity.

Topics: *XII*, II, VII Name: Newton

0385 Dobbs, Betty Jo Teeter. "Newton as Alchemist and Theologian." In *Standing on the Shoulders of Giants*, edited by Norman J. W. Thrower, 128-140. Berkeley and Los Angeles: University of California Press, 1990. See (1545).

Dobbs argues that scholars should see Newton as "one whole and historical human being." His diverse studies constituted a unified plan for obtaining truth, the unity of which was guaranteed by God. Dobbs understands Newton to be a voluntarist in theology—stressing the will of God. Hence empirical examination was the method by which to study how God actually decided to institute the world. In Newton's understanding of providence, comets and the "vegetable spirit" of alchemy were God's agents in creation.

Topics: *XII*, II, VII Name: Newton

0386 Dobbs, Betty Jo Teeter. "Newton as Final Cause and First Mover." *Isis* 85(1994): 633-643.

In one of her last papers, Dobbs declared that she intends "to undermine one of our most hallowed explanatory frameworks, that of the Scientific Revolution." After exploring the history of the use of the metaphor, she perceives the basic problem to be in our interpretation of our scientific heroes. "We choose for praise the thinkers that seem to us to have contributed to modernity, but we unconsciously assume that their thought patterns were fundamentally like ours." Hence we become uncomfortable when we discover that they took seriously astrology, alchemy, magic, divine providence, and salvation history. Not unexpectedly, Dobbs uses Newton as her prime example—not only because of her own research but because he appears as the "final cause" and "first mover" in many histories. She suggests in conclusion that perhaps Newton should be seen not as the first mover of modern science but as "one of history's great losers" in "a titanic battle between the forces of religion and the forces of irreligion."

Topics: *I*, VII, XII Name: Newton

0387 Dobbs, Betty Jo Teeter. "Newton's Alchemy and his 'Active Principle' of Gravitation." In *Newton's Scientific and Philosophical Legacy*, edited by P. B. Scheurer and G. Debrock, 55-80. Dordrecht: Kluwer Academic, 1988. See (1409).

This is a detailed analysis of Newton's search in the writings of ancient philosophers and church fathers for fragments of the *prisca sapientia* which would provide insight into the cause of gravity. In this search, Philo became central—his Platonized Stoicism providing a key to Newton's formulation of an omnipotent spiritualized divine agent. Thus, Dobbs corrects herself and Westfall in deciding that (between 1684 and 1710) ancient cosmic thought was more important than alchemical ideas in Newton's conception of a cause for gravity.

Topics: *XII*, II, VII Name: Newton

0388 Dobbs, Betty Jo Teeter. "Newton's Alchemy and His Theory of Matter." *Isis* 73(1982): 511-528. Reprinted in *Seventeenth-Century Natural Scientists*, edited by Vere Chappell, 227-244. New York: Garland, 1992. See (0205).

Dobbs traces the historical process by which Newton produced his published theory of matter. In that theory, alchemy and corpuscularianism interacted. Activity, however, required the divinity. Dobbs concludes with an argument that, for Newton, the Arian Christ was God's viceroy—and that forces, active principles, spirits, virtues, etc., were ultimately subsumed under the providential activity of Christ.
Topics: *XII*, II, VII Name: Newton

0389 Dobbs, Betty Jo Teeter. "Newton's *Commentary* on the *Emerald Tablet* of *Hermes Trismegistus*: Its Scientific and Theological Significance." In *Hermeticism and the Renaissance*, edited by Ingrid Merkel and Allen G. Debus, 182-191. Washington, D.C.: Folger Shakespeare Library; London: Associated Universities Presses, 1988. See (1051).
This is a brief article which focuses on Newton's search for an activating spiritual agent that acts on matter. In his commentary on the *Emerald Tablet*, Dobbs finds hints that Newton may have believed that Christ is the alchemical agent. She stresses the unity of Newton's thought regarding matter theory, alchemy, and theology.
Topics: *XII*, II, VII Name: Newton

0390 Dobbs, Betty Jo Teeter. "Newton's Copy of 'Secret's Reveal'd' and the Regimens of the Work." *Ambix* 26(1979): 145-169.
This is a detailed analysis of major portions of Newton's manuscript commentary on Eirenaeus Philalethes' *Secrets Revealed*. Dobbs points out that Newton's interest in alchemy had a theological basis: alchemical work mirrored the work of God in creation. The manuscript also indicates Newton's continuing search for the activating agent in the process.
Topics: *XII*, II, VII Name: Newton

0391 Dobbs, Betty Jo Teeter. "Newton's Rejection of a Mechanical Aether: Empirical Difficulties and Guiding Assumptions." In *Scrutinizing Science: Empirical Studies of Scientific Change*, edited by A. Donavan and others, 69-83. Dordrecht: Kluwer Academic, 1988.
In this case study, Dobbs argues that Newton rejected a mechanical aether as the cause of gravitation in late 1684 or early 1684/85.

The question was basically one in physics, but Dobbs suggests that Newton could temporarily leave the issue unresolved because of his voluntarist theology.

Topics: *XII*, VII Name: Newton

0392 Dobbs, Betty Jo Teeter. "Stoic and Epicurean Doctrines in Newton's System of the World." In *Atoms, Pneuma, and Tranquility*, edited by Margaret J. Osler, 221-238. Cambridge: Cambridge University Press, 1991. See (1179).

Here, Dobbs analyzes Newton's adoption of ancient philosophical ideas after his early Cartesian phase. Despite his atomism, Dobbs argues, Newton rejected Epicureanism both for theological and for scientific reasons. She describes a "Platonizing Stoicism" which made sense to Newton in terms of his theological and historical views. And she discusses how this same ancient outlook solved some of his natural philosophical problems (such as how to understand gravity, the void, the cohesion of particles, and"active principles").

Topics: *XII*, VII Name: Newton

0393 Dobbs, Betty Jo Teeter. "Studies in the Natural Philosophy of Sir Kenelm Digby: Part I." *Ambix* 18(1971): 1-25.

Dobbs pictures Digby as a private gentleman and virtuoso: he had an energetic, eclectic interest in all aspects of life. Although this article focuses almost exclusively on natural philosophy and the "sympathetic powder" controversy, she acknowledges Digby's Catholic Aristotelianism.

Topics: *XI*, VII Name: Digby

0394 Dobbs, Betty Jo Teeter. "Studies in the Natural Philosophy of Sir Kenelm Digby: Part II. Digby and Alchemy." *Ambix* 20(1973): 143-163.

Dobbs presents a chronological survey of Digby's alchemical activities and theories in the context of his mentors and friends. (At this early point in her career, she used a Jungian approach to alchemy). She concludes that Digby's work–by rationalizing and chemicalizing the language of alchemy–had aided the separation of the older psychological/religious aspects from the material aspects of alchemy.

Topics: *II*, VII Name: Digby

0395 Dobbs, Betty Jo Teeter. "Studies in the Natural Philosophy of Sir Kenelm Digby: Part III. Digby's Experimental Alchemy—The Book of *Secrets*." *Ambix* 21(1974): 1-28.

In this installment, Dobbs analyzes particular recipes and processes from Digby's posthumously published collection entitled *Secrets* (1682). They include examples of processes influenced by Christian trinitarian doctrine and by Neoplatonism (as well as alchemy in Aristotelian terms and alchemy by strict recipe).

Topics: *II*, VII Names: Digby, LeFevre

0396 Dobbs, Betty Jo Teeter. "‹The Unity of Truth›: An Integrated View of Newton's Work." In *Action and Reaction*, edited by S. G. Brush and others, 105-122. Newark: University of Delaware Press, 1993.

This is an edited excerpt from *The Janus Face of Genius* (0383). In it, Dobbs uses gravity as a case study both to show Newton's flexibility and to argue for the unity of purpose in his theology, alchemical studies, and mechanics.

Topics: *XII*, II, VII Name: Newton

0397 Dorsey, Stephen P. *Early English Churches in America, 1607-1807*. New York: Oxford University Press, 1952.

This work includes good black and white photographs of seventeenth- and eighteenth-century churches. It also includes lists of churches still standing. There is no architectural analysis.

Topic: *VI*

0398 Downes, Kerry. *Christopher Wren*. London: Allen Lane, 1971.

This is an illustrated study of Wren's personality and architectural style. Downes emphasizes Wren's originality and his preference for visible results. Downes suggests that, for Wren, the relationship between science and art (and, by implication, religion as well) was complex but integrated.

Topic: *VI* Names: Hawksmoor, Wren

0399 Downes, Kerry. *Sir Christopher Wren: the Design of St. Paul's Cathedral*. London: Trefoil, 1988.

This book includes a comprehensive description and a beautiful catalogue of the drawings of St. Paul's made by Wren and his assistants. The introduction describes the sequence of designs he

proposed for rebuilding the cathedral and includes a brief biography.
Topic: *VI* Name: Wren

0400 Drake, Stillman. "Galileo in English Literature of the Seventeenth Century." In *Galileo: Man of Science*, edited by Ernan McMullin, 415-431. New York: Basic Books, 1967. See (1040).
This is a bibliographical essay of works published by seventeenth-century Englishmen and Scots who either translated or wrote about Galileo. Drake takes "literature" to mean any nonscientific publication written for the educated public. He discusses (among others) Digby, Donne, Milton, Salusbury, and Wilkins. He notes religious issues in passing.
Topics: *X*, VII Names: Digby, Donne, Milton, Salusbury, Wilkins

0401 Draper, John W. *History of the Conflict between Religion and Science*. New York: D. Appleton, 1874.
The warfare metaphor tells it all for Draper. This survey now is basically worthless as a secondary source for our period of study. (But it is a good and interesting primary source for nineteenth-century historiography.)
Topics: *VII*, I

0402 Drennon, Herbert. "Newtonianism: Its Method, Theology, and Metaphysics." *Englische Studien* 68(1934): 397-409.
Drennon argues for the influence on eighteenth-century society of the combined scientific and religious worldview of Newtonianism. He discusses natural philosophic natural theology in the service of religious apologetics.
Topics: *XII*, IV, VIII Name: Newton

0403 Duchesneau, Francois. *L'Empirisme de Locke*. The Hague: M. Nijhoff, 1973.
Duchesneau focuses on Locke's theory of knowledge and philosophy of science by charting seventeenth-century empirical epistemology and by setting these within the context of Locke's medical interests. Locke's philosophy of nature is seen as affected by his view that God is the creator of a rationally ordered world.
Topic: *VII* Name: Locke

0404 Duffin, K. E., and Stuart W. Strickland. "Appendix: The Principal Theses of *Science, Technology and Society in Seventeenth Century England*: An Analytical Synopsis." In *Puritanism and the Rise of Modern Science: The Merton Thesis*, edited by I. Bernard Cohen, 373-389. New Brunswick, N.J.: Rutgers University Press, 1990. See (0246).

This is a good, concise section-by-section summary of Merton's famous monograph (1057). It includes no evaluative comments by Duffin and Strickland.

Topic: *III*

0405 Duffy, Eamon. "Valentine Greatrakes, the Irish Stroker: Miracle, Science and Orthodoxy in Restoration England." In *Religion and Humanism*, edited by Keith Roberts, 251-273. Oxford: Basil Blackwell, 1981.

Duffy sets Greatrakes' healings in the historical context of apologetic theology during the Restoration. He sees Greatrakes' activity as posing a dilemma for Latitudinarian virtuosi. They recognized the reality of his healings—yet they also emphasized natural order and minimized miracles. Duffy provides a good survey of Greatrakes' life, the events of 1666, and responses to his healings. This article should be supplemented by the article on Greatrakes by Nicholas Steneck (1494).

Topics: *XI*, VII Names: Boyle, Conway, Greatrakes, Lloyd, D., More, Stubbe

0406 Duffy, Eamon. " Whiston's Affair : The Trials of a Primitive Christian, 1709-1714." *J. Eccles. Hist.* 27(1976): 129-150.

Duffy describes the persecution and the attempted persecution of Whiston by high church Anglicans. There is no direct connection to science—but the article is good as an antidote to simple generalizations about the success of "Newtonian ideology" in our period.

Topics: *IV*, V, IX, XII Name: Whiston

0407 Duncan, Edgar H. "Satan-Lucifer: Lightning and Thunderbolt." *Philol. Q.* 30(1951): 441-443.

Duncan extends Svendsen's thesis that Milton relied on popular schooltexts by paralleling passages in *Paradise Lost* with Comenius' textbook explanation of thunderbolts.

Topic: *X* Names: Comenius, Milton

0408 Dunn, John. *Locke*. Oxford: Oxford University Press, 1984.
This work is aimed at a general audience; it consists mainly of philosophical analysis and is largely ahistorical. The chapter entitled "Knowledge, belief and faith" is a summary and analysis of Locke's views with much focus on religion and some focus on natural science.
Topic: *VII* Name: Locke

0409 Dunn, William P. *Sir Thomas Browne: A Study in Religious Philosophy*. Menasha, Wisc.: G. Banta, 1926. Revised ed. Minneapolis: University of Minnesota Press, 1951.
Dunn emphasizes Browne's religious philosophy and not his scientific ideas.
Topics: *VII*, IV, X, XI Name: Browne, T.

0410 Dupre, Louis. "The Modern Idea of Culture: Its Opposition to Its Classical and Christian Roots." In *Modernity and Religion*, edited by Ralph McInerny, 1-18. Notre Dame, Ind.: University of Notre Dame Press, 1994.
This is a philosophical essay about the cosmological issues underlying the historiography of our period. Dupre sees the seventeenth and eighteenth centuries as the critical time in the transformation from the classical and Christian worldviews to the modern. The latter is characterized by an opposition between nature and culture, the "instrumentalization of nature," the "vanishing of man" as self, and the "disappearance of transcendence."
Topics: *I*, VII

0411 Eamon, William. "From the Secrets of Nature to Public Knowledge." In *Reappraisals of the Scientific Revolution*, edited by David C. Lindberg and Robert S. Westman, 333-365. Cambridge: Cambridge University Press, 1990. See (0957).
Eamon links the ideology of openness in science to the vision of the unity of religion in the ideal of pansophia of the Hartlib circle. He presents this as one of many factors over a long period of time concerning the development of science as public knowledge.
Topic: *IV* Name: Hartlib

0412 Eamon, William. "Technology as Magic in the Late Middle Ages and the Renaissance." *Janus* 70(1983): 171-212.

Toward the end of this interesting survey of the association between technology and magic are a few pages on Robert Fludd. "For Fludd the magus and the engineer . . . were the same person." Generally, Eamon argues that the "technological dream" of the late Middle Ages and the Renaissance was largely a product of a magical worldview.

Topics: *VI*, II Name: Fludd

0413 Easlea, Brian. *Science and Sexual Oppression*. London: Weidenfeld and Nicholson, 1981.

In the chapter on "Male Sexism and the Seventeenth Century Revolution," Easlea argues that the men who spearheaded the scientific and medical revolutions used metaphors for nature (feminine) and God (masculine) to reinforce the power of active males over passive nature (and women). His *ad hominem* argument proposes that a "problematic" sexuality (e.g. of Newton) correlates with oppression of both women and nature.

Topics: *IV*, XI, XII Names: Bacon, Newton, Oldenburg

0414 Easlea, Brian. *Witch Hunting, Magic and the New Philosophy: An Introduction to the Debates of the Scientific Revolution, 1450-1750*. Brighton, England: Harvester Press; Atlantic Highlands, N.J.: Humanities Press, 1980.

This introductory text attempts to correlate ideas expounded by natural philosophers with social and sexual stratification. In Easlea's general picture, witchcraft and natural magic lost out to the Christianized mechanical philosophy because it became the religious natural philosophy of the male establishment.

Topics: *IV*, II, XII Names: Bacon, Boyle, Newton, Winstanley

0415 Edel, Abraham. "Levels of Meaning and the History of Ideas." *J. Hist. Ideas* 7(1946): 355-360.

Edel proposes that ideas have "levels of meaning" and he indicates possible ways these levels can be related to their sociohistorical context. Although he discusses the issues abstractly, his thesis is useful in understanding the various approaches reflected in this bibliography. Edel was a member of the "history of ideas" school.

Topic: *I*

0416 Edelin, Georges. "Joseph Glanvill, Henry More, and the Phantom Drummer of Tedworth." *Harvard Library Bulletin* 10 (1956): 186-192.

This is a brief description of Glanvill's correspondence with More concerning their common interest in ghosts. Edelin describes the haunted house at Tedworth and Glanvill's "Baconian investigations" of it.

Topics: *VII*, II Names: Glanvill, More

0417 Eiseley, Loren. *Francis Bacon and the Modern Dilemma*. Lincoln, Neb.: University of Nebraska Press, 1962.

Eiseley writes a wonderful essay on the ambiguity of the study of nature and on the necessarily complex relationships between nature and culture, between inductive science and mystical experience. As seen by Eiseley, Bacon was one of the first of modern men. He was a paradoxical visionary who both foresaw the advancement of science and yet could not anticipate all the directions in which humans have taken science.

Topics: *IV*, VII Name: Bacon

0418 Eisenstein, Elizabeth L. *The Printing Press as an Agent of Change: Communications and Cultural Transformation in Early Modern Europe*. 2 vols. Cambridge: Cambridge University Press, 1979.

These volumes are a significant examination of the printing press as *an* (Eisenstein's emphasis) agent of change. The author first outlines the historiography of the history of printing and then defines some features of print culture. She highlights dissemination, standardization, reorganization, data collection, amplification and reinforcement, and (especially) preservation as such features. Eisenstein discusses at length how the printing revolution reoriented classical and Christian traditions during the Renaissance and the Reformation. In the second volume, she shows how natural philosophy ("the book of nature") was transformed. Although much of her discussion focuses on continental and sixteenth-century materials, Eisenstein does mention Newton and the Royal Society almost as much as Luther or Galileo. Her concluding chapter, "Scripture and Nature Transformed," is especially relevant to the interplay of technology, religion, and science. She also contributes

to the Protestantism-and-the-rise-of-science discussion (including the issue of the censorship of printed materials).
Topics: *VI*, III, X, XII

0419 Eisenstein, Elizabeth L. *The Printing Revolution in Early Modern Europe*. Cambridge: Cambridge University Press, 1983.
In this abridgement of *The Printing Press as an Agent of Change*, the primary theses are perhaps more clear. Citations have been eliminated but the important conclusion is essentially retained. This version of Eisenstein's work generally is easier to obtain.
Topics: *VI*, III, X, XII

0420 Eisenstein, Elizabeth L. "Some Conjectures about the Impact of Printing on Western Society and Thought: A Preliminary Report." *J. Mod. Hist.* 40(1968): 1-56.
This essay has been superseded by Eisenstein's books.
Topic: *VI*

0421 Eliade, Mircea. *The Forge and the Crucible*. Translated by Stephen Corrin. New York: Harper, 1971.
Although this phenomenological study does not treat our time period directly, it is an important background study of the motifs of alchemy. Eliade speculates that the intentions of alchemists—to transform nature and to supersede time—have been secularized but are continued in modern science and technology.
Topic: *II*

0422 Eliade, Mircea. *A History of Religious Ideas*. Vol. 3. Translated by Alf Hiltebeitel and Diane Apostolos-Cappadona. Chicago: University of Chicago Press, 1985.
Chapter 38 is entitled "Religion, Magic, and Hermetic Traditions before and after the Reformation." It contains a brief discussion of the integration of Christianity, the Hermetic tradition, and the natural sciences from the perspective of this influential historian of religions.
Topic: *II*

0423 Elmer, Peter. "Medicine, Religion and the Puritan Revolution." In *The Medical Revolution of the Seventeenth Century*, edited by Roger French and Andrew Wear, 10-45. Cambridge: Cambridge University Press, 1989. See (0510).

Elmer argues that the medical reformers after 1640 were not Puritans, but rather they were more radical sectarians. Mainstream Puritans, he points out, denounced Paracelsian or iatrochemical reforms both before and after 1640. Furthermore, the "essential thread of religious belief which runs through the writings of the reformers" may be characterized as a spirit of hermetic irenicism (that is, of religious reconciliation and reunification).

Topics: *XI*, II, III Name: Hart

0424 Emerson, Roger L. "Heresy, the Social Order, and English Deism." *Church History* 37(1968): 389-403.

In arguing that one "must pay attention not only to the ideas constituting Deism but to the motives and circumstances of those accepting the ideas," Emerson emphasizes social, political, and heretical factors in the rise of English deism. The acceptance of a rational religion of nature in the deist movement was a social, political, and religious affair—and science was merely coincidental.

Topics: *IX*, IV

0425 Emerson, Roger L. "Latitudinarianism and the English Deists." In *Deism, Masonry, and the Enlightenment*, edited by J. A. Leo Lemay, 19-48. Newark, Del.: University of Delaware Press, 1987.

Emerson challenges those scholars (Westfall, M. C. Jacob, O'Higgins, Sullivan) who connect deism either to ideology or to liberal Anglicanism. He argues that deism should be defined in a European-wide context and mainly in terms of religious beliefs. This is good for historiographic purposes even though there is only small reference to science.

Topics: *IX*, II, IV Names: Blount, Tillotson

0426 Emerson, Roger L. "Science and Moral Philosophy in the Scottish Enlightenment." In *Studies in the Philosophy of the Scottish Enlightenment*, edited by M. A. Stewart, 11-36. Oxford: Clarendon Press, 1990.

Emerson traces a pattern which begins with ethics based on Christian revelation and ends with moral philosophy as an independent (secular) discipline. He postulates that science or natural philoso-

phy was central in this process: thus, natural philosophy in this period was a secularizing force in Scotland. Emerson is mainly concerned with the post-1720 period but he begins his discussion in the mid-seventeenth century.
Topic: *VII*

0427 Emerson, Roger L. "Science and the Origins and Concerns of the Scottish Enlightenment." *Hist. Sci.* 26(1988): 333-366.
Emerson argues that the Scottish Enlightenment of the mid-to-late eighteenth century had its origins in Scottish, English, and continental developments of the period 1690-1720. A major factor was the new science. Emerson presents the earlier period as an era when Scots believed that science bolstered religion. (He presents the period *after* 1720 as a time when science contributed to a growing religious skepticism and secularity of outlook in Scotland and elsewhere.)
Topics: *VII*, V, IX   Names: Pitcairne, Sibbald

0428 Emerton, Norma E. "The Argument from Design in Early Modern Natural Theology." *Sci. Christ. Belief* 1(1989): 129-147.
Emerton presents a popular survey of natural theology in relation to natural philosophy from Cicero to Paley. The seventeenth century receives most attention; and she briefly discusses each of its major figures.
Topics: *VIII*, XII   Names: Bacon, Boyle, Newton

0429 Emerton, Norma E. "Creation in the thought of J. B. van Helmont and Robert Fludd." In *Alchemy and Chemistry in the Sixteenth and Seventeenth Centuries*, edited by Piyo M. Rattansi and Antonio Clericuzio, 85-101. Dordrecht: Kluwer Academic, 1994. See (1299).
Emerton contrasts the interpretations of the Biblical creation story by van Helmont and Fludd. She shows that their theologies affected how each used different chemical theories to explain the account in Genesis. On the one hand, van Helmont viewed the creation story from an Augustinian and literalistic perspective. On the other, Fludd's account was a composite in which he inserted a Hermetic creation story within the framework of Genesis 1.
Topics: *X*, II   Name: Fludd

0430 Empson, William. "A Deist Tract by Dryden." *Essays in Criticism* 25(1975): 74-100.

Empson argues that a deist tract published by Charles Blount in 1693 was written by Dryden. The connection to science is only indirect.

Topics: *X*, IX Names: Blount, Dryden

0431 Empson, William. "Dryden's Apparent Skepticism." *Essays in Criticism* 20(1970): 172-181 and 21(1971): 111-115.

Empson argues that Dryden was a deist (and judges this to be good). He sees Dryden's conversion as a pragmatic matter (and unfortunate). The whole argument is based on selected passages and is not convincing.

Topics: *X*, IX Name: Dryden

0432 *Encyclopedia of Religion*. 16 vols. Mircea Eliade, Editor-in-Chief. New York: Macmillan, 1987.

These are articles of various lengths directed to a general readership. Both science and the seventeenth century seem rather unimportant to the editors of this encyclopedia. There are a total of about 25 relevant pages (in the 16 volumes). These include short notices on Bacon, Newton, deism, physics and religion, science and religion, and Renaissance alchemy.

Topics: *VII*, II, IX, XII Names: Bacon, Newton

0433 *The Encyclopedia of Unbelief*. Edited by Gordon Stein. 2 vols. Buffalo, N.Y.: Prometheus Books, 1985.

This encyclopedia includes articles by well-known scholars on Atheism, Deism, Collins, Herbert, Hobbes, Tindal, and Toland.

Topic: *IX* Names: Collins, A., Herbert, E., Hobbes, Tindal, Toland

0434 Engdahl, Sylvia Louise. *The Planet-Girded Suns: Man's View of Other Solar Systems*. New York: Atheneum, 1974.

This is a popular (high-school-level) book covering the period from the late-sixteenth century to the twentieth century. Engdahl's use of quotes from primary sources is good and shows how theological issues have played a major role in discussions about this aspect of cosmological theory.

Topic: *VII* Names: Bentley, Burnet, T., Derham

0435 Epstein, Julia L., and Mark L. Greenberg. "Decomposing Newton's Rainbow." *J. Hist. Ideas* 45(1984): 115-140.

This is a brief survey of the literary history of rainbow symbolism and the effects of Newton's *Opticks* on rainbow imagery. After Newton explained the rainbow in scientific terms, "a large portion of its meaning" was displaced "forever from the sacred realm into the profane." Approximately two-thirds of this article is relevant to the period through the 1720s.

Topics: *X*, VII, XII Name: Newton

0436 'Espinasse, Margaret. "The Decline and Fall of Restoration Science." *Past and Present* 14(1958): 71-89. Reprinted in *The Intellectual Revolution of the Seventeenth Century*, edited by Charles Webster, 347-368. London: Routledge and Kegan Paul, 1974. See (1618).

'Espinasse argues that science generally and applied science in particular were depreciated during and after the Restoration period in England. She uses literary and social evidence to support her argument. She treats religion as part of this overall ideological trend.

Topics: *IV*, III, V, VI, X, XII Names: Addison, Boyle, Dryden, Newton

0437 'Espinasse, Margaret. *Robert Hooke*. London: W. Heinemann, 1956. Reprinted, Berkeley and Los Angeles: University of California Press, 1962.

This is an apologetic survey of Hooke's person and work. Generally, it is in the style of intellectual history (and influenced by Jacob Bronowski and Basil Willey). 'Espinasse defends Hooke against Newton, Oldenburg, and what she considers an unfair tradition of nineteenth- and early twentieth-century scholars. She acknowledges the importance of religion for Hooke and his contemporaries—but then largely ignores it. She assumes that science and religion are quite separate (but Hooke and his peers just did not realize this). She includes Hooke's interests and work in technology and architecture as well as in "pure" science.

Topics: *VII*, VI, XII Names: Aubrey, Boyle, Hooke, Newton, Oldenburg, Wren

0438 Eurich, Nell. *Science in Utopia: A Mighty Design*. Cambridge, Mass.: Harvard University Press, 1967.

Large sections of this book are devoted to England in our period. On the one hand, the book suffers from the author's totally utilitarian conception of science; on the other hand, this emphasis allows her to treat technological visions and practices. Yet she also underrates the complexity of seventeenth-century religious thought.

Topics: *IV*, VI Names: Bacon, Cowley, Glanvill, Hartlib, Petty, Winstanley

0439 Evans, Robert Rees. *Pantheisticon: The Career of John Toland*. New York: P. Lang, 1991.

This was written in the 1960s and it shows no substantial alterations reflecting the historiographic developments between then and the 1991 publication date. Evans argues: that *Pantheisticon* is the key to all of Toland's thought and writings; and that Toland's pantheism is both an ideological and a philosophical system. He sees Toland's political views, his philosophy of nature, and his radical theology as complementary aspects of an intellectual revolt against medieval, authoritarian, hierarchical, and clerical traditionalism. *Pantheisticon*, Evans says, helps reveal to us the points of relation between the scientific and anticlerical movements. Moreover, Toland's writings are presented as a simultaneous attack upon the two equally stupendous errors of "continental rationalism" and "English empiricism"—especially as these were used in defense of Christianity. Spinoza and Newton are the "two philosophical villains." Central both to Toland's natural philosophy and to his theology, Evans believes, was his dynamic metamorphic biologism or his circulating biological universe. This book includes many references to the new science, the rise of science, and Newtonian physics—but with no explication of any of these. Evans describes his approach as being on three levels: biographical; revolutionary; and with respect to the Enlightenment. This last level is a strongly retrospective (almost teleological) historiography in which he sees Toland's ideas as "meeting the onrushing eighteenth century" and as "motivated by the approaching French Enlightenment." Although characterized by eccentric terminology and by an annoying (no-one-understands-Toland-but-me) style, Evans' book still is provocative and useful.

Topics: *IX*, IV, VII, XII Names: Locke, Newton, Toland

0440 Ewan, Joseph, and Nester Ewan. *John Bannister and His Natural History of Virginia, 1678-1692*. Urbana, Ill.: University of Illinois Press, 1970.

About one-third of this volume is devoted to a description of: Bannister's life; his activities as an Oxford-trained naturalist and clergyman; and the use of his work after his death. In the bulk of the volume, the Ewans have produced a critical edition of Bannister's manuscripts. Bannister's life in Virginia (1678-1692) clearly indicates that, for him, natural history was his vocation while theology was only a profession.

Topic: *VII* Names: Bannister, Byrd, Clayton, Ray

0441 Fabro, Cornelio. *God in Exile. Modern Atheism: A Study of the Internal Dynamic of Modern Atheism, from Its Roots in the Cartesian Cogito to the Present Day*. Translated and edited by Arthur Gibson. Westminster, Maryland: Newman Press, 1968. (Originally published as *Introduzione all' Ateismo Moderno*. Rome: Studium, 1964.)

This is a philosophical and theological critique of modern atheism which devotes about one hundred pages to More, Toland, Collins, Locke, and English deism. It is ahistorical in its approach: Fabro's judgmental analysis is based upon Thomistic Catholic theology. As regards seventeenth-century English deism and atheism, Fabro puts a heavy emphasis on theories of matter and philosophical materialism.

Topic: *IX*, VII Names: Collins, A., Locke, More, Toland

0442 Fairchild, Hoxie Neale. *Religious Trends in English Poetry. Vol. 1: 1700-1740*. New York: Columbia University Press, 1939.

Fairchild includes commentary on the influence of natural philosophy (as well as religion) on poetry in the period.

Topics: *X*, IV Names: Blackmore, Norris, Pope.

0443 Farr, James. "The Way of Hypothesis: Locke on Method." *J. Hist. Ideas* 48(1987): 51-72. Reprinted in *Philosophy, Religion and Science in the Seventeenth and Eighteenth Centuries*, edited by John W. Yolton, 284-305. Rochester, N.Y.: University of Rochester Press, 1990. See (1720).

Farr takes the position that Locke approved of "the hypothetical method" both theoretically and substantively. In agreeing with Mandelbaum and Laudan (against Yost and Yolton), Farr brings much evidence and a strong argument to bear on this debate. He also argues that, for Locke, there are no clear lines between natural philosophy, metaphysics, natural religion, political theory and other of *our* categories of human understanding. Several of Farr's examples and emphases are explicitly theological. This is a good article.

Topics: *VII*, VIII Name: Locke

0444 Farrell, Maureen. *William Whiston*. New York: Arno Press, 1981.

This is a reprint of a 1973 doctoral dissertation. Farrell gives basic biographical details and summaries of Whiston's writings. Her descriptions of historical context are rather simple and she stays away from the complexities of Whiston's personal and controversial network. (She hardly mentions Newton or Newtonianism with respect to Whiston's theological views.) She states that Whiston's science and theology cannot really be separated; but then she generally treats his science and religion separately. Her book is useful but needs to be completed by Force's book, (0497).

Topics: *VII*, XII Names: Burnet, T., Newton, Whiston, Woodward

0445 Farrington, Benjamin. *Francis Bacon, Philosopher of Industrial Science*. New York: H. Schuman, 1949.

This volume is one of a series of semipopular life-and-thought biographies of men and movements in the history of science. (Farrington also sought to use this book to correct the British tendency of the time to see Bacon as a minor logician of empiricism.) The author argues that Bacon's primary idea was "that science ought to be applicable to industry, that men ought to organize themselves as a sacred duty to improve and transform the conditions of life." This idea had a religious sanction: knowledge should restore to humankind its God-given dominion over nature.

Topics: *VII*, VI Name: Bacon

0446 Fattori, M., and M. Bianchi, eds. *Spiritus*. Rome: Edizioni dell' Ateneo, 1984.

The essays by Debus (0340), Rees (1318), and Walker (1598) are annotated separately.
Topics: *XI*, VII

0447 Faur, Jose. "Newton, Maimonides, and Esoteric Knowledge." *Cross Currents* 40(1990-91): 526-538.
Faur renews and gives evidence for Keynes' thesis that Newton had views similar to (and probably was influenced by) Maimonides. The connections have to do with a hermeneutical tradition concerning esoteric knowledge about God through natural philosophic and revealed sources. Faur cites good primary sources—though he makes no reference to recent secondary literature and none to Popkin! Compare Popkin's articles (1239) and (1246).
Topics: *XII*, VII Name: Newton

0448 Fauvel, John, Raymond Flood, Michael Shortland, and Robin Wilson. Introduction to *Let Newton Be!*, edited by John Fauvel and others, 1-21. Oxford: Oxford University Press, 1988. See (0449).
This is a good summary description of Newton's life and a good popular introduction to the importance of Newton—as well as an introduction to the essays which follow. Newton is presented *not* as some sort of pure scientist or as strangely unscientific in his other concerns—but as a multifaceted individual living in the seventeenth century. Newton's theological concerns are seen as part of the whole man and a man of the time.
Topics: *XII*, VII Name: Newton

0449 Fauvel, John, Raymond Flood, Michael Shortland, and Robin Wilson. *Let Newton Be! A New Perspective on His Life and Works*. Oxford: Oxford University Press, 1988.
This is an extremely good anthology which focuses on Newton as a whole man (rather than a compartmentalized one). It summarizes recent historiography in Newtonian studies for the general audience. The introduction (0448) and twelve chapters include cross-references to each other and constitute a unified and systematic presentation. Newton's religion and his interests in alchemy, ancient wisdom, and cosmic harmony are all given significant attention. The introduction and six of the chapters are relevant to science and religion and are annotated separately in this bibliogra-

phy. See also: Brooke (0130), Gjertsen (0547), Golinski (0553), Gouk (0557), Henry (0664), and Rattansi (1292).
Topics: *XII*, II, VII Name: Newton

0450 Favre, Antoine. "Rosicruciana." *Revue de l'Histoire des Religions* 190(1976): 73-88 and 157-180.
This is a positive review of Yates' work (including *The Rosicrucian Enlightenment*). Yates is commended for thoroughly examining the Rosicrucian tradition and for setting it within its political context. Favre also discusses works by Craven (on Maier), Hutin (on Fludd), and Montgomery (on Andrea).
Topic: *II*

0451 Feilchenfeld, W. "Leibniz und Henry More: Ein Beitrag zur Entwicklungsgeschichte der Monadologie." *Kant-Studien* 28(1923): 323-334.
Feilchenfeld argues that Leibniz, in his *Monadology*, was influenced by More.
Topic: *VII* Name: More

0452 Feingold, Mordechai, ed. *Before Newton: The Life and Times of Isaac Barrow*. Cambridge: Cambridge University Press, 1990.
This volume, as a whole, illustrates Barrow's involvement both in natural philosophy and in divinity. It contains separate essays on his theory of optics, his mathematics, his classical studies, and his sermons. Two essays, by Feingold (0454) and Gascoigne (0527), attempt to provide overviews of Barrow's various interests in his university context. (These latter two essays are annotated separately.)
Topic: *VII* Name: Barrow

0453 Feingold, Mordechai. "A Friend of Hobbes and an Early Translator of Galileo: Robert Payne of Oxford." In *The Light of Nature*, edited by J. D. North and J. J. Roche, 265-280. Dordrecht: M. Nijhoff, 1985. See (1145).
This is a brief biography of Payne (1596-1651). Besides what is noted in the title, Payne was an Anglican (Royalist) clergyman, an amateur scientist, and a friend of the Cavendishes and various bishops.
Topic: *VII* Names: Hobbes, Payne

0454 Feingold, Mordechai. "Isaac Barrow: Divine, Scholar, Mathematician." In *Before Newton: The Life and Times of Isaac Barrow*, edited by Mordechai Feingold, 1-104. Cambridge: Cambridge University Press, 1990. See (0452).

Through this detailed biographical essay, Feingold presents Cambridge University life as Barrow might have experienced it. The curriculum, interest in Cartesianism, the effects of Puritanism on learning, a revision of the image of the "conservatism" of university education, and university politics all are discussed from Barrow's perspective. Perhaps the subtitle should read "Scholar, Mathematician, Divine," since this is the sequence of Barrow's life and university career that Feingold portrays.

Topics: *VII*, III, V Names: Barrow, Duport

0455 Feingold, Mordechai. "John Seldon and the Nature of Seventeenth-Century Science." In *In the Presence of the Past: Essays in Honor of Frank Manuel*, edited by Richard T. Bienvenu and Mordechai Feingold, 55-78. Dordrecht: Kluwer Academic, 1991. See (0094).

In this discussion of Seldon's activities and interests in the science of his day, Feingold advances the broader thesis that "general scholars" played important roles in the "collaborative atmosphere" of seventeenth-century natural philosophy. More specifically, he shows that Seldon's learning extended from medicine, mathematics and astronomy to chronology, ancient history, Arabic, and a cautious reading of astrology. This was in addition to his activities in government and law. Overall, following his interpretation of Roger Bacon and Maimonides, Seldon advanced the metaphor of the "two books of God."

Topics: *V*, II, XI Name: Seldon

0456 Feingold, Mordechai. *The Mathematician's Apprenticeship: Science, Universities and Society in England, 1560-1640*. Cambridge: Cambridge University Press, 1984.

In this revisionist history of science, Feingold hopes to correct the standard view of a division between the (conservative, Aristotelian) universities and the (new) scientific communities after 1640. He examines the lives, activities, and diverse worldviews of students and teachers. He finds that most of them were "general scholars" and clerics who were accepted as respectable members of the scientific community. Thus, Feingold argues for continuity between scientific activity in the second half of the century and this "incub atory period of English science."

Topics: *V*, I Names: Allen, Bacon, Bainbridge, Briggs, Camden, Digby, Gellibrand, Gilbert, Hariot, Harvey, Lydiat, Mede, Oughtred, Savile, Ussher

0457 Feingold, Mordechai. "Partnership in Glory: Newton and Locke Through the Enlightenment and Beyond." In *Newton's Scientific and Philosophical Legacy*, edited by P. B. Scheurer and G. Debrock, 291-308. Dordrecht: Kluwer Academic, 1988. See (1409).

Here, Feingold discusses the senses in which Newton and Locke had a "shared vision"—both actually (in their own time) and symbolically (among selected writers in the eighteenth and early-nineteenth centuries). Both natural philosophic and theological elements were part of this shared vision. About half of the article is relevant to the period before 1720.

Topics: *XII*, VII, X Names: Locke, Newton

0458 Feingold, Mordechai. "The Universities and the Scientific Revolution." In *New Trends in the History of Science*, edited by R. P. W. Visser and others, 29-48. Amsterdam: Rodopi, 1989.

Feingold argues against the "anachronistic" thesis that the Anglican universities were reactionary while the new academies and societies were enlightened. He does this by describing the actual research that went on in them. He outlines the chronology of the writings of the new natural philosophers—and suggests that the Puritan Revolution "merely coincided" with their publication and dissemination.

Topics: *V*, III

0459 Ferguson, James P. *An Eighteenth-Century Heretic: Dr. Samuel Clarke*. Kineton, England: Roundwood Press, 1976.

This is not a biography. It consists mainly of summaries of Clarke's writings and of those of the various controversialists in the religious debates triggered by Clarke's works. Ferguson provides some context for the publications and debates. He relies totally on the primary sources and makes no reference at all to any secondary literature. He presents Clarke as a polymath and men-

tions Clarke's interest in natural philosophy. But Clarke's theology and Newtonian natural philosophy are not shown to have much influence on each other. Clarke's theology is presented here as being based mainly on a biblical literalism and on an *a priori* metaphysics. Occasionally, Ferguson also gives his own theological critiques of Clarke's positions. The book does provide detailed evidence of how Clarke (as well as Newton and Whiston) derived their unorthodox views concerning the Trinity on biblical grounds (not natural philosophical ones). There is significant overlap between Ferguson's two books.
Topics: *IX*, XII Names: Clarke, Whiston

0460 Ferguson, James P. "The Image of the Schoolmen in Eighteenth-Century English Philosophy, with Reference to the Philosophy of Samuel Clark." In *Actes du 4e Congres International de Philosophie Medievale*, 1199-1206. Paris: Librairie Philosophique J. Vin, 1969.
Ferguson uses Clarke's 1704 Boyle lectures to show the negative image of the medieval schoolmen in the early eighteenth century.
Topics: *VII*, VIII, XII Name: Clarke

0461 Ferguson, James P. *The Philosophy of Dr. Samuel Clarke, and Its Critics*. New York: Vantage Press, 1974.
In this Vantage Press publication, Ferguson summarizes, analyzes, and evaluates the specifically philosophical writings of Clarke and of his critics. The approach is rather ahistorical: he treats the issues as if all parties from Hobbes to Kant were dialoguing together. There is virtually no reference to other secondary analyses or commentaries. Absolute space and time, materialism, causality, and determinism are all discussed. But the emphasis is on metaphysics and philosophy of religion; natural philosophy and the new science come in only tangentially. Doctrines of God are discussed mainly as various unsuccessful attempts to prove the existence of God. Clarke's Arianism and other views based on biblical revelation are barely discussed. Likewise, Clarke's annotations on Rohault's *Physics* is barely mentioned.
Topics: *VII*, VIII, XII Names: Clarke, Collins, A.

0462 Ferguson, Moira. "Margaret Lucas Cavendish." In *Women Writers of the Seventeenth Century*, edited by Katharina M. Wilson and Frank J. Warnke, 305-340. Athens, Ga.: University of Georgia Press, 1989.

The first half of this article consists of a good brief biography and a summary of Cavendish's writings. Ferguson focuses on Cavendish's attitude toward women—but also mentions her natural philosophy. The last half of the article consists of primary texts and a good bibliography.

Topics: *VII*, X Name: Cavendish

0463 Ferreira, M. Jamie. *Scepticism and Reasonable Doubt: The British Naturalist Tradition in Wilkins, Hume, Reid and Newman*. Oxford: Clarendon Press, 1986.

In Chapter 2, Ferreira differentiates between Wilkins and Locke on reasonable doubt and certainty. Locke refused to use the term certainty, whereas Wilkins noted three kinds of certainty—physical, mathematical, and moral. Ferreira thus challenges Van Leeuwen's thesis that Locke's position was the culmination of that of his predecessors.

Topic: *VII* Names: Locke, Wilkins

0464 Feuer, Lewis S. *Jews in the Origins of Modern Science and Bacon's Utopia: The Life and Work of Joachim Gannse, Mining Technologist and First Recorded Jew in English-Speaking North America*. Cincinnati: American Jewish Archives, 1987.

This brochure is part of a series published by the American Jewish Archives. It outlines the scientific endeavors of and the theological controversy surrounding Joachim Gannse in the ill-fated Roanoke colony in Virginia. Feuer proposes that Gannse was the model for Joabin, the Jewish intermediary, in Bacon's *New Atlantis*.

Topics: *VI*, IV

0465 Feuer, Lewis S. *The Scientific Intellectual*. New York: Basic Books, 1963.

Feuer argues (circularly) that no religious outlook has ever exerted a positive influence on the progress of science. He disagrees especially and specifically with Merton and Stimson.

Topic: *III*

0466 Field, J. V., and Frank A. J. L. Janes, eds. *Renaissance and Revolution: Humanists, Scholars, Craftsmen and Natural Philosophers in Early Modern Europe*. Cambridge: Cambridge University Press, 1993.

One essay by Figala and Petzold (0470) and one by Hunter (0750) are annotated separately.

0467 Fiering, Norman. "The First American Enlightenment: Tillotson, Leverett, and Philosophical Anglicanism." *New Eng. Q.* 54(1981): 307-344.

Fiering proposes that Tillotson and More (rather than Newton and Locke) were the most influential writers in "the first American Enlightenment" at the beginning of the eighteenth century. He finds this Enlightenment among clergy associated with the Brattle Street Church (Boston) and among those trained by John Leverett at Harvard. Leverett, President of Harvard and a fellow of the Royal Society, followed More and Tillotson in emphasizing the use of reason in religion and nature as a revelation of God.

Topic: *VII* Names: Leverett, More, Tillotson

0468 Figala, Karin. "Newton as Alchemist." Translated by Marianne Wilder. *Hist. Sci.* 15(1977): 102-137.

In a mixed review of Dobbs' *The Foundations of Newton's Alchemy*, Figala makes detailed criticisms and goes on major digressions. There is not too much connection to religion but (sprinkled through pp. 118-127 and 134-137) Figala does suggest connections between Newton's alchemical theories and the book of Genesis.

Topics: *II*, XII Name: Newton

0469 Figala, Karin. "Newtons Rationales System der Alchemie." *Chemie in Unserer Zeit* 12(1978): 101-110.

Figala summarizes her interpretation of Newton's alchemy and theory of matter. She takes Newton's view into his late years and finds it transformed in the chemistry described in the *Opticks*. Figala concludes that, for Newton, his position explains the orderly creation by the omnipotent and omnipresent God.

Topics: *XII*, II, VII Name: Newton

0470 Figala, Karin, and Ulrich Petzold. "Alchemy in the Newtonian Circle: Personal Acquaintances and the Problem of the Late Phase of Isaac Newton's Alchemy." In *Renaissance and Revolution*, edited by J. V. Field and Frank A. J. L. James, 173-191. Cambridge: Cambridge University Press, 1993. See (0466).

The authors provide evidence that Newton continued to be interested in alchemy during his London years. They thus raise the possibility that alchemy played a role in Newton's concept of matter.

Topics: *II*, VII, XII Names: Newton, Yworth

0471 Finch, Jeremiah S. *A Catalogue of the Libraries of Sir Thomas Browne and Dr. Edward Browne, His Son*. Leiden: E. J. Brill, 1986.

This work includes a twenty-page introduction, sixty-two pages of notes, and a thirty-three page index. There is no direct discussion of science and religion but the book applies to the topic in indirect ways.

Topics: *XI*, VII, X Names: Browne, E., Browne, T.

0472 Finch, Jeremiah S. *Sir Thomas Browne: A Doctor's Life of Science and Faith*. New York: Schuman, 1950.

This is a rather popular biography but it is based solidly on the primary sources and scholarly literature. Finch presents Browne as a literary figure and physician as well as emphasizing his "life of science and faith." The historiography is one of appreciating and defending Browne in terms of mid-twentieth-century American values.

Topics: *X*, V, VII, XI Names: Browne, E., Browne, T., Digby, Ross

0473 Finocchiaro, Maurice A. "Science and Society in Newton and Marx." *Inquiry* 31(1988): 103-121.

About half of this essay is given over to a review of Freudenthal's *Atom and Individual in the Age of Newton*. Concerning the relationships of Newtonian natural philosophy to Hobbesian social philosophy and of both of these to sociopolitical conditions, Finocchiaro presents his own additional theses and radical revisions of Freudenthal's theses. Finocchiaro mentions the importance of

Newton's religion but, like Freudenthal, he does not explicate or emphasize its significance.

Topics: *IV*, I, VII, XII Names: Hobbes, Newton

0474 Fisch, Harold. "Bacon and Paracelsus." *Cambridge Journal* 5(1952): 752-758.

In this brief essay, Fisch sees the two authors and their traditions as diametrically opposed. He proposes that while Baconianism leads to a "simple-minded . . . faith in power," Paracelsianism has as its end the glory of and fear of God. Fisch places Boyle between them—a Baconian without a complete separation between science and theology.

Topics: *VII*, II Names: Bacon, Boyle

0475 Fisch, Harold. *Jerusalem and Albion: The Hebraic Factor in Seventeenth Century Literature*. London: Routledge and K. Paul, 1964.

The setting for Fisch's work is the literary issues stimulated by T. S. Eliot's thesis that "in the seventeenth century a dissociation of sensibility " occurred in English poetry. In response, Fisch argues for the importance of the "Hebrew factor" in the writings of literary, scientific, and theological figures. While Milton is central to his argument, he also discusses Bacon, Boyle, Browne, and the Cambridge Platonists. Fisch concludes by discussing Hobbes and Locke—for whom Blake's image fits: "Jerusalem has departed from Albion." This book is well-written and full of sometimes-surprising insights.

Topics: *X*, III, VII Names: Bacon, Boyle, Browne, T., Hobbes, Locke, More

0476 Fisch, Harold. "The Scientist as Priest: A Note on Robert Boyle's Natural Theology." *Isis* 44(1953): 252-265.

Fisch compares Boyle's views on natural philosophy to those of the Cambridge Platonists and to alchemical traditions. He finds that Boyle was convinced of "the indivisibility of the spheres of Science and Religion" and believed that man "must be the priest" of a unified nature.

Topics: *VIII*, II Names: Boyle, Browne, T., Cudworth

0477 Fisher, H. A. L. "The Real Oxford Movement." In *Pages from the Past*, by H. A. L. Fisher, 130-147. Oxford: Clarendon Press, 1939.

This is a now-outdated, but gracefully written, narrative lecture on the Oxford group in the 1650s. It was originally delivered in 1929 by one of England's leading historians.

Topic: *V* Names: Evelyn, Hooke, Petty, Wilkins, Wren

0478 Fisher, Mitchell Salem. *Robert Boyle: Devout Naturalist. A Study in Science and Religion in the Seventeenth Century.* Philadelphia: Oshiver Studio Press, 1945.

Originally a Columbia University dissertation, this book contains readable summaries of Boyle's writings in appropriately organized chapters. As indicated by the title, Fisher stresses the unity of Boyle's two great concerns. Though superseded by later scholars, Fisher's work is interesting and relevant in its breadth of view. It also contains a brief but useful bibliography of primary and secondary literature to 1944.

Topic: *VII* Name: Boyle

0479 Fleck, Ludwik. *Genesis and Development of a Scientific Fact.* Edited by Thaddeus J. Trenn and Robert K. Merton. Translated by Fred Bradley and Thaddeus J. Trenn. Chicago: University of Chicago Press, 1979.

Originally published in 1935 in Polish, this little case study of the history of the Wasserman reaction and its relation to syphilis has taken its place as an important historiographical monograph. As the title indicates, its primary thesis is that scientific "facts" have histories. These histories take place with "thought styles" and "thought collectives." Public science is a collective enterprise of various esoteric experts and their more exoteric publics—with mutual interchanges between them. While religion appears only in the form of an ethical "proto-idea" in the case of syphilis, Fleck does discuss in this context the anthropology of religion and the sociology of science. The notions both of "proto-idea" and of the scientific collective can be useful in the historiography of the relations between science and religion. The book is hard to follow but Trenn concludes the translation with a useful "Descriptive Analysis."

Topic: *I*

0480 Fleischmann, Wolfgang Bernard. *Lucretius and English Literature, 1680-1720*. Paris: A. G. Nizet, 1964.

This is a useful book which emphasizes the complexity and diversity of responses to Lucretius during our period. The author distinguishes between: Lucretius as an atomist; Lucretius as atheist and forerunner of the freethinkers; Lucretius as a descriptive naturalist; and Lucretius as a poet. He notes that writers and poets of our period reacted both positively and negatively to these different aspects of Lucretius. And, because acceptance or praise of Lucretius was almost always partial, his influence was limited.

Topics: *X*, IX Names: Addison, Bentley, Blackmore, Cheyne, Cudworth, Dryden, Garth, Hobbes, More, Pope, Prior, Swift

0481 Fleming, Donald. "The Judgment Upon Copernicus in Puritan New England." In *Melanges Alexandre Koyre*, Vol. 2, 160-175. Paris: Hermann, 1964. See (1046).

Fleming responds to the question: Why did educated New Englanders proclaim their allegiance to Copernicanism even though no empirical evidence differentiated the competing theories? The basic reasons he gives are: (1) their realistic Ramist logic assumed an harmony between the mind of man and that of God; and (2) the new astronomy appeared to them as an example of the recovery of Adam's innocence. Fleming also notes that Calvin's accommodation hermeneutic of scripture allowed them to suspect simple observation and to disregard passages contrary to Copernican theory.

Topics: *VII*, III, X Name: Wing

0482 Flew, Anthony G. N. "Hobbes." In *A Critical History of Western Philosophy*, edited by Daniel J. O'Connor, 153-169. New York: Free Press, 1964. See (1159).

This is a summary, explication, analysis, and evaluation of the philosophy of Thomas Hobbes. It includes significant discussions of Hobbes' natural philosophy (matter, motion, and metaphysics) and of the theological consequences of his philosophy.

Topics: *VII*, IX Names: Bramhall, Hobbes

0483 Forbes, Eric G., ed. *Human Implications of Scientific Advance: Proceedings of the XVth International Congress of the History of Science, Edinburgh 10-15 August 1977*. Edinburgh: Edinburgh University Press, 1978.

The articles by Ravetz (1306) and Rossi (1358) are annotated separately.

0484 Forbes, Eric G. "The Library of the Rev. John Flamsteed, F.R.S., First Astronomer Royal." *Notes Rec. Royal Soc. London* 28(1973): 119-143.

There is an interesting mix of scientific and religious titles in Flamsteed's book collection.

Topic: *VII* Name: Flamsteed

0485 Force, James E. "Biblical Interpretation, Newton and English Deism." In *Scepticism and Irreligion in the Seventeenth and Eighteenth Centuries*, edited by Richard H. Popkin and Arjo Vanderjagt, 282-305. Leiden: E. J. Brill, 1993. See (1250).

Force effectively demonstrates that Newton was not a deist. He correlates close readings of Newtonian texts to deist ones in order to prove that Newton's view of the Bible was Christian (although not literalist). Likewise, Force argues that Newton's Arian Christology, while not orthodox, was nonetheless Christian. Moreover, Newton's natural philosophic methodology and epistemology show a voluntaristic theology and a cautious view of the powers of human reason which was anything but deistic. Force directs his essay specifically against the views of Richard Westfall. This is a sophisticated argument with much scholarly apparatus.

Topics: *XII*, IX, X Names: Clarke, Newton, Tindal, Whiston

0486 Force, James E. "The Breakdown of the Newtonian Synthesis of Science and Religion: Hume, Newton, and the Royal Society." In *Essays on the Context, Nature, and Influence of Isaac Newton's Theology*, edited by James E. Force and Richard H. Popkin, 143-163. Dordrecht: Kluwer Academic, 1990. See (0499).

By "Newtonian Synthesis," Force means a combination of theological and natural philosophical elements in the apologetics of the early Royal Society through the early-to-mid-eighteenth century. (The "breakdown" occurred in the middle third of the century and is especially noted in the work of Hume.) This is a slightly altered version of (0489).

Topics: *V*, VIII, XII Name: Newton

0487 Force, James E. "The God of Abraham and Isaac (Newton)." In *The Books of Nature and Scripture*, edited by James E. Force and Richard H. Popkin, 179-200. Dordrecht: Kluwer Academic, 1994. See (0498).

As regards the physics and metaphysics of death and resurrection, Force argues that Newton was a "Christian Mortalist." In doing this, Force also discusses Newton's heterodoxy, his principles of Biblical interpretation, his voluntarist doctrine of God, and the synthetic unity of Newton's thought.

Topics: *XII*, VII, IX, X Names: Newton, Whiston

0488 Force, James E. "Hume and Johnson on Prophecy and Miracles: Historical Context." *J. Hist. Ideas* 43(1982): 463-475. Reprinted in *Philosophy, Religion and Science in the Seventeenth and Eighteenth Centuries* edited by John W. Yolton, 127-139. Rochester, N.Y.: University of Rochester Press, 1990. See (1720).

In dealing with the background to Hume and Johnson, about half of the article focuses on our earlier period. Miracles and prophecy were closely connected not only to Christianity and the Bible—but also to questions concerning "the laws of nature."

Topics: *VII*, IX, XII Names: Collins, A., Newton, Whiston

0489 Force, James E. "Hume and the Relation of Science to Religion Among Certain Members of the Royal Society." *J. Hist. Ideas* 45(1984): 517-536. Reprinted in *Philosophy, Religion and Science in the Seventeenth and Eighteenth Centuries* edited by John W. Yolton, 228-247. Rochester, N.Y.: University of Rochester Press, 1990. See (1720).

This is identical in content and only slightly different in form from (0486).

Topics: *V*, VIII, XII Name: Newton

0490 Force, James E. "Newton and Deism." In *Science and Religion/Wissenschaft und Religion*, edited by Anne Baumer and Manfred Buttner, 2-13. Bochum, Germany: Universitatsverlag N. Brockmeyer, 1989. See (0066).

Force argues directly against Westfall's perception of Newton having "a frankly deistic position." Force, rather, understands Newton as articulating a euhemeristic "third way"; Newton was neither a biblical literalist nor a (negative) deist. For Newton, the Bible contained a prophetic core of divine truth and, with his cautious empiricism, some divine truths exceed the grasp of reason.

Topics: *XII*, VII, IX, X Names: Newton, Whiston

0491 Force, James E. "The Newtonians and Deism." In *Essays on the Context, Nature, and Influence of Isaac Newton's Theology*, edited by James E. Force and Richard H. Popkin, 43-73. Dordrecht: Kluwer Academic, 1990. See (0499).

Force argues that Newton, Clarke, and Whiston theologically pursued "a third or middle way" between the deistic rejection of biblical revelation and a biblical extremism which rejected natural theology. This Newtonian position included a logical synthesis of biblical studies, natural philosophy, and certain historical understandings. This position was heretical with respect to traditional Christianity but strongly opposed to deism.

Topics: *XII*, IX, X Names: Clarke, Newton, Whiston

0492 Force, James E. "Newton's God of Dominion: The Unity of Newton's Theological, Scientific, and Political Thought." In *Essays on the Context, Nature, and Influence of Isaac Newton's Theology*, edited by James E. Force and Richard H. Popkin, 75-102. Dordrecht: Kluwer Academic, 1990. See (0499).

This essay, summarized by its title, is well-argued. Force explicates Newton's (and Whiston's) voluntaristic doctrine of God as a metaphysical and epistemological foundation of Newtonian natural philosophy. Force argues that Newton's understanding of God and nature influenced his political ideology but did not commit him to any specific party position or outlook.

Topics: *XII*, VII, IV Names: Newton, Whiston

0493 Force, James E. "Newton's 'Sleeping Argument' and the Newtonian Synthesis of Science and Religion." In *Standing on the Shoulders of Giants*, edited by Norman J. W. Thrower, 109-127. Berkeley and Los Angeles: University of California Press, 1990. See (1545).

Force argues persuasively that the Newtonian synthesis of science and religion consisted not just in the design argument based on

natural philosophy but also included a biblical argument based on history and prophecy. The God of dominion (characterized both by general providence and special providences) in Newtonian thought is found in both nature and history. The "argument for a Deity" concerning which Newton decided "to let it sleep" is, in Force's view, precisely the argument from prophecy and history. This synthesis is explicitly to be found in the writings of William Whiston.
Topics: *XII*, VII, X Names: Newton, Whiston

0494 Force, James E. "The Origins of Modern Atheism." *J. Hist. Ideas* 50(1989): 153-162.
This is a review article of Michael Buckley's *At the Origins of Modern Atheism* (0147). Force emphasizes the historical context (biblical criticism, deistic controversies, sociopolitical considerations) as opposed to the logical implications of philosophic texts in addressing the development of modern atheism. He strongly criticizes Buckley's approach.
Topics: *IX*, XII Names: Clarke, Newton

0495 Force, James E. "Secularization, the Language of God and the Royal Society at the Turn of the Seventeenth Century." *Hist. Euro. Ideas* 2(1981): 221-235.
Force here argues both against Westfall's internal tragic view and against DeGrazia's linguistic view of secularization. He focuses on the rise of specific deists within the Royal Society around 1700 (including Martin Folkes, who was Vice President and later President).
Topics: *IX*, V, VII, XII Names: Folkes, Newton, Whiston

0496 Force, James E. "Sir Isaac Newton, 'Gentleman of Wide Swallow'?: Newton and the Latitudinarians." In *Essays on the Context, Nature, and Influence of Isaac Newton's Theology*, edited by James E. Force and Richard H. Popkin, 119-141. Dordrecht: Kluwer Academic, 1990. See (0499).
This essay is a comparative analysis of Newton's theological views and those of the Anglican Latitudinarians. Force notes some similarities; but he emphasizes Newton's differences—in the direction of his heretical but biblical and voluntaristic theology. Force briefly explicates the related point that Newton was no deist.
Topics: *XII*, V, VII, X Names: More, Newton, Whiston

0497 Force, James E. *William Whiston: Honest Newtonian*. Cambridge: Cambridge University Press, 1985.

Force is a historian of philosophy who strongly emphasizes a contextual historiography in the history of philosophy. In this book, he aims at providing "a fuller comprehension of the rapprochement between science and religion in the Newtonian context." Force argues that, in the early eighteenth century, "Newtonianism" included not just mathematics and natural philosophy—but also certain principles of biblical interpretation and millennialism. And he shows that Whiston was "Newtonian in a strong sense" in all of these terms—as well as in terms of Arian theology and in his epistemology. (In so doing, Force presents detailed interpretations of Newton's thought as well as Whiston's.) Whiston is called an "Honest Newtonian" because he stated publicly some Newtonian ideas and principles which Newton himself kept hidden. Amongst the philosophical and theological topics which Force covers are: natural theology; biblical revelation; Genesis and geology; miracles and different types of divine providence; Arianism and heretical theology; deism; millennialism; and the relationship of Newtonianism to Whig political ideology. Although a few of Force's conclusions are debatable, this is an excellent book.

Topics: *XII*, IV, VII, VIII, IX, X Names: Burnet, T., Chubb, Keill, Newton, Swift, Whiston

0498 Force, James E., and Richard H. Popkin, eds. *The Books of Nature and Scripture: Recent Essays on Natural Philosophy, Theology, and Biblical Criticism in the Netherlands of Spinoza's Time and the British Isles of Newton's Time*. Dordrecht: Kluwer Academic, 1994.

Annotated separately are the essays by Force (0487), Goldish (0551), Henry (0666), Hutton (0784), Iliffe (0788), Kochavi (0890), and Mandelbrote (0980).

Topics: *XII*, VII, X Name: Newton

0499 Force, James E., and Richard H. Popkin. *Essays on the Context, Nature, and Influence of Isaac Newton's Theology*. Dordrecht: Kluwer Academic, 1990.

Eight of the ten essays from this collection are listed individually elsewhere in our bibliography. Most of these essays are found only in this collection. Major emphases include: Newton's biblicism;

Newton's voluntaristic doctrine of God; the religious and philosophical context of Newton and those who followed him; the point that Newton was no deist. This is an important book for Newtonian studies. See (0486), (0491), (0492), (0496), (1227), (1241), (1243), and (1246).

Topics: *XII*, VII, IX, X Names: Clarke, Newton, Whiston

0500 Foster, Herbert D. "The Political Theories of Calvinists before the Puritan Exodus to America." *Am. Hist. Rev.* 21(1915-1916): 481-503.

Although basically tangential, the last few pages are indirectly relevant to the Puritanism-and-science debate. Foster may be the earliest scholar to refer to the "scientific spirit" of Calvinists.

Topic: *III*

0501 Foster, Michael. "The Christian Doctrine of Creation and the Rise of Modern Natural Science." *Mind* 43(1934): 446-468. Reprinted in *Creation: The Impact of an Idea*, edited by Daniel O'Connor and Francis Oakley, 29-53. New York: Scribner, 1969. See (1161). Reprinted in *Creation, Nature and Political Order in the Philosophy of Michael Foster*, edited by Cameron Wybrow, 99-117. Lewiston, N.Y.: E. Mellen Press, 1992. See (1703).

See the following entry for a joint annotation.

Topics: *VII*, III

0502 Foster, Michael. "Christian Theology and Modern Science of Nature." Two parts. *Mind* 44(1935): 439-466 and *Mind* 45(1936): 1-27. Reprinted in *Creation, Nature and Political Order in the Philosophy of Michael Foster*, edited by Cameron Wybrow, 119-147. Lewiston, N.Y.: E. Mellen Press, 1992. See (1703).

In his articles, Foster provides a philosophical analysis of the relation between various doctrines of God and their respective philosophies of nature. He specifically connects the voluntaristic doctrine of God with early modern natural philosophy (empiricism). Foster does not deal specifically with English thought during our period. However, his articles are the basis of some later works which do apply his analysis to seventeenth-century England.

Topics: *VII*, III

0503 Fox, Sanford J. *Science and Justice: The Massachusetts Witchcraft Trials*. Baltimore: Johns Hopkins Press, 1968.

Fox's main purpose is to describe the part that science played in the legal prosecution of witches. He finds an uncritical over-reliance on the views of scientists whose acceptance of witchcraft was due to their attempt to harmonize religion and science. Yet he finds the legal system itself evenhanded in its administration. He relates his findings to contemporary concerns about the scientific ignorance of the legal profession.

Topic: *II*

0504 Fox Bourne, H. R. *The Life of John Locke*. 2 vols. New York: Harper, 1876.

This work was important at one time, but it now is outdated. It especially has been superseded by Yolton (1715).

Topic: *VII* Name: Locke

0505 Frank, Robert G., Jr. "Science, Medicine, and the Universities of Early Modern England: Background and Sources." *Hist. Sci.* 11(1973): 194-216 and 239-269.

Frank details the educational context of developing scientists. He discusses the institutional structures, endowed chairs, and "private" instruction. He is aware of the religious element, noting the clergy among the virtuosi. This is a lengthy and excellent article.

Topics: *V*, III, IV, XI Name: Glisson

0506 Fraser, Alexander Campbell. *Locke*. Edinburgh: W. Blackwood and Sons, 1890. Reprinted, Port Washington, N.Y.: Kennikat Press, 1970.

This late-nineteenth-century work was written for the general reader. It has been superseded, of course, but its summaries of Locke's philosophy of science and of his theology are not bad.

Topic: *VII* Name: Locke

0507 Fraser, Alexander Campbell. "Prolegomena: Biographical, Critical, and Historical." In *An Essay Concerning Human Understanding by John Locke*, edited by Alexander Campbell Fraser, Vol. 1, xi-cxl. Oxford: Clarendon Press, 1894. Reprinted, New York: Dover, 1959.

This is a condensed version of (0506), *Locke*, with special focus on the *Essay*.
Topic: *VII* Name: Locke

0508 Freeman, Edmund L. "Bacon's Influence on John Hall." *PMLA* 42(1927): 385-399.
Freeman suggests that Hall's essays reflect Bacon's influence in several areas—including that regarding religion and science.
Topics: *X*, VII Names: Bacon, Hall(1627)

0509 French, Roger. *William Harvey's Natural Philosophy*. Cambridge: Cambridge University Press, 1994.
This is the definitive scholarly study on Harvey's natural philosophy, its sources, and responses to it. Harvey's lectures on anatomy and the structure and rhetoric of *De Motu Cordis* are examined in detail. In the last two-thirds of the book, French considers the natural philosophies of Harvey and of those (doubters and believers, in England and on the continent) who responded to his proposal that blood circulates. The complexity of the interactions between theology and natural philosophy are readily apparent. Whereas Harvey did not use a religious justification for the results of his research program, both critics and supporters did. (On the one hand, Harvey's reliance on Aristotle's nature books precluded a theological justification. On the other, Fludd and Glisson accepted Harvey's thesis by beginning with their respective doctrines of God's relation to the world.) French also is concerned with how consensus about circulation of the blood was reached among educated persons. Although Harvey "took an expressly Christian view of the creator as the ultimate cause" of nature, he relied on experiments and arguments about "matters of fact" in order to persuade persons of different theological positions. When the common thread of various interpretations was seen, circulation "came to be seen as true because it was accepted."
Topics: *XI*, V Names: Fludd, Glisson, Harvey, Primrose, Read

0510 French, Roger, and Andrew Wear, eds. *The Medical Revolution of the Seventeenth Century*. Cambridge: Cambridge University Press, 1989.
This is a useful collection of papers—many of which deal with changes in medicine in England. Emphasis is placed on the

relationship between medicine and religion. The articles by Cunningham (0305), Elmer (0423), Guerrini (0584), Harley (0627), and Henry (0663) are annotated separately in this bibliography.

Topic: *XI*

0511 Freudenthal, Gideon. *Atom and Individual in the Age of Newton: On the Genesis of the Mechanistic Worldview*. (Boston Studies in the Philosophy of Science, 88.) Translated by Peter McLaughlin. Dordrecht: D. Reidel, 1986.

Freudenthal appears to be a neo-Marxist historian and philosopher. He aims to give a sociohistorical explanation of Newtonian mechanics. But he also wants to establish it as a general historiographic principle that: (1) science is influenced by philosophy; and (2) science and philosophy are influenced by social relations. Here is what Freudenthal claims: Newton's proof of the existence of absolute space can be traced to the assumption that inertia as well as other properties are attributable to a single particle independently of the existence of other particles (that is, even in an otherwise empty space); next, the origins of the assumptions about inertia and single particles can be found in bourgeois social philosophy and such assumptions correspond to the concept of a society composed of independent private proprietors. Moreover, the social-philosophical assumption about independent individuals in society has its origins in the actual social relations which developed in early modern England. Thus, Newton's theory of space (and his mechanics more generally) can be shown to have a mediated dependency on social relations. By "mediated dependency," Freudenthal means not a direct, conscious or explicit connection, but an indirect and yet real one. In his analysis, Freudenthal analyzes the following aspects of Newton's natural philosophy: absolute space; the essential properties of particles; inertia; gravitation; passive and active principles; the mechanical philosophy; the relation of the will to the body in animals and humans; epistemological issues; and the analytic-synthetic method. Social philosophers whom he discusses include: Thomas Aquinas; the Levellers; Hobbes; Locke; Leibniz; Adam Smith; Rousseau; and, by implication, Marx. Newton's natural philosophy and philosophical presuppositions are compared to those of Leibniz—who stands for taking both the material world and society as organic wholes. Freudenthal gives an extended analysis of the "clock" metaphor (equating the view of Newton and Clarke with the "artisan's clock" and Leibniz's with the "scientist's watch"); this is applied both to Newton's understanding of the analytical-synthetic method and to the respective understandings of God's relation to the world. He discusses both Newton's and Leibniz' concepts of God—treating these as being effected by social conditions and by philosophical presuppositions; but he never treats either man as a religious person or either man's ideas about God as affecting his natural philosophy. Throughout his analysis, Freudenthal downplays or even ignores the Hermetic and *prisca* traditions. Freudenthal's evidence and arguments come to at least a brilliant and provocative analysis of parallels; the connections which he claims to have made remain open to question. The book is written in what even Freudenthal calls a "Germanic" style: theses, argumentation, taxonomic labels, and selected examples follow in a dense (sometimes hard to follow) prose. The translation is awkward with many grammatical errors.

Topics: *IV*, I, IV, VI, VII, XII Names: Clarke, Hobbes, Locke, Newton

0512 Freudenthal, Gideon. *Atom und Individuum im Zeitalter Newtons: Zur Geneseder Mechanistischen Naturund Sozialphilosophie*. Frankfurt am Main: Suhrkamp, 1982.

This is an earlier edition of (0511). The English edition has a number of minor changes with more explicit discussions in Chapter 13 and in the "Afterword."

Topics: *IV*, I, VI, VII, XII

0513 Freudenthal, Gideon, ed. *Etudes sur/Studies on Helene Metzger*. (Collection de Travaux de L'Academie Internationale d'Histoire des Sciences No. 32.) Leiden: Brill, 1990.

This is a collection of studies on the work of Helene Metzger, a pioneer in considering seventeenth- and eighteenth-century science in its own context. Several letters of Metzger and a complete bibliography of her writings also are included. The papers by Blay (0098) and Schmitt (1414) are annotated separately in this bibliography. The articles originally were published in *Corpus* 8/9(1988).

Topic: *I*

0514 Froom, Le Roy Edwin. *The Prophetic Faith of Our Fathers*, Vol. 2. Washington, D.C.: Review and Herald, 1948.

Froom summarizes many writers, from the Renaissance through the eighteenth century, on the prophetic books of the Bible. He includes discussions of More, Burnet, Whiston, and a ten-page summary of Newton's work on prophecy. Froom was a Seventh Day Adventist who believed in the prophetic interpretation of history about which he was writing.

Topics: *X*, XII Names: Burnet, More, Newton, Whiston

0515 Fulton, John F. "Some Aspects of Medicine Reflected in Seventeenth-Century Literature with Special Reference to the Plague of 1665." In *The Seventeenth Century*, edited by Richard F. Jones and others, 198-208. Stanford, Calif.: Stanford University Press, 1951. See (0845).

In a rambling style, Fulton discusses the intermingling of science and medicine with religion and magic in connection with the plague of 1665 in London.

Topics: *X*, IV, XI Names: Evelyn, Pepys

0516 Funkenstein, Amos. "The Body of God in Seventeenth Century Theology and Science." In *Millenarianism and Messianism in English Literature and Thought, 1650-1800*, edited by Richard H. Popkin, 149-175. Leiden: Brill, 1988. See (1238).

Funkenstein discusses the transformation in European thought from a hierarchical universe and a symbolic (equivocal) sense of the presence of God to a homogeneous universe and an unequivocal language about God. As part of this development, God's relation to the world was given a concrete physical meaning, "the body of God." That meaning varied from Descartes' transcendent first cause to Henry More's "sensorium."

Topics: *VII*, X Name: More

0517 Funkenstein, Amos. "Revolutionaries on Themselves." In *Revolutions in Science*, edited by William Shea, 157-163. Canton, Mass.: Science History, 1988.

Funkenstein finds two seventeenth-century sources for the eighteenth-century sense of "revolution." On the one hand, the faith of activist utopians allowed them to hope to build a new world on the ruins of the old one. On the other hand, the commitment to the

new science of seventeenth-century natural philosophers was phrased in terms of religious conversion (not revolution).
Topics: *IV*, I

0518 Funkenstein, Amos. *Theology and the Scientific Imagination from the Middle Ages to the Seventeenth Century*. Princeton, N.J.: Princeton University Press 1986.
In this excellent and important study, Funkenstein traces three theological themes and the shifts they underwent from the Hellenistic and scholastic periods into seventeenth-century Europe and England. Reflecting his themes are the chapters entitled: "God's Omnipresence, God's Body, and Four Ideals of Science"; "Divine Omnipotence and Laws of Nature"; and "Divine Providence and the Course of History." These incorporate previously published articles and essays. In the chapter on God's omnipresence, Funkenstein discusses scientific "ideals": the homogeneity of nature; univocal language; mathematization and mechanization. For him, nominalist and Protestant theology discouraged equivocation and Renaissance thinkers sensed homogeneity. These converged into one world picture in the seventeenth century. Likewise, with Newton especially, there appeared "a fusion of theology and physics into almost one science." In the chapter on divine omnipotence, Funkenstein traces the history of God's absolute and ordained power, and of ideal experiments. Laws of nature (ideal-abstract conditions) became guaranteed by God. In sections on the exegetical principle of accommodation, Funkenstein shows his background as a historian of medieval Jewish thought (as well as Christian). While he does not refer directly to English natural philosophers here, it clearly provides the background for their interpretations of scripture. In a new chapter, "Divine and Human Knowledge: Knowing by Doing," he explores the new ideal of knowledge in the seventeenth century. Funkenstein also proposes the thesis that the theology which emerged among seventeenth-century natural philosophers was a "secular theology" in two senses. It was conceived by laymen for laymen, and it was oriented toward the world.
Topics: *VII*, III, X, XII Names: Hobbes, More, Newton

0519 Gabbey, Alan. "Anne Conway et Henry More: Letters sur Descartes (1650-1651)." *Arch. Philos*. 40(1977): 379-404.

This article includes four primary sources with philosophical and historical commentary.
Topic: *VII* Names: Conway, More

0520 Gabbey, Alan. "Cudworth, More and the Mechanical Analogy." In *Philosophy, Science, and Religion in England 1640-1700*, edited by Richard Kroll and others, 109-127. Cambridge: Cambridge University Press, 1992. See (0917).
Gabbey argues that Cudworth and More, among other Cambridge Platonists, were perceived by contemporaries both as Latitudinarians and as introducing the new mechanical philosophy into English intellectual life. He shows, however, that each qualified "pure mechanism" by postulating a "spirit of nature" or "plastic nature." This spirit, Cudworth and More preached, is analogous to the "quickening spirit" in the souls of humans that allows them to follow the essential "Law of the Gospel."
Topics: *VII*, X Names: Cudworth, More, Patrick

0521 Gabbey, Alan. "The English Fortunes of Descartes." *Brit. J. Hist. Sci.* 11(1978): 159-163.
This is an essay review of Pacchi's *Cartesio in Inghilterra* (1190). It includes Gabbey's view of Henry More's ambivalent response to Descartes.
Topic: *VII* Name: More

0522 Gabbey, Alan. "Henry More and the Limits of Mechanism." In *Henry More (1614-1687) Tercentenary Studies*, edited by Sarah Hutton, 19-35. Dordrecht: Kluwer, 1990. See (0782).
Gabbey discusses More's objections to Cartesian mechanical philosophy. He describes More's use of the "Spirit of Nature" as an explanatory hypothesis; and he suggests that (comparable to "the God of the gaps") More's concept functions as a "Spirit of the Causal Gaps." Gabbey argues that More's area of competence (theology) conflicted with Descartes' areas of competence (mathematics and natural philosophy) and had the effect of limiting More.
Topic: *VII* Names: Hyrne, More

0523 Gabbey, Alan. "Philosophia Cartesiana Triumphata: Henry More (1646-1671)." In *Problems of Cartesianism*, edited by

Thomas M. Lennon and others, 170-250. Kingston, Ontario: McGill-Queens University Press, 1982. See (0943).

This is a detailed examination of Henry More's complex responses to Descartes. Gabbey argues *against* the received tradition (Nicolson, Lamprecht, Lichtenstein) that More went through stages from an enthusiastic supporter, to a critic, to a hostile opponent. Gabbey gives detailed evidence from almost all of More's writings from 1646 through 1671 that More consistently was ambivalent towards Descartes. He emphasizes that More's change before and after 1660 actually was a change from philosophical to purely religious and theological priorities. In so doing, Gabbey discusses More's views on the mechanical philosophy, atomism, cosmology, the *prisca* tradition, natural religion, and the defense of Christianity.

Topics: *VII*, II, VIII, IX Name: More

0524 Garner, Barbara C. "Francis Bacon, Natalis Comes and the Mythological Tradition." *J. Warb. Court. Inst.* 33(1970): 264-291.

Garner details the influence of Comes' interpretation of ancient myths and fables on Bacon. She stresses the role of the Fall in Bacon's conception of the history of philosophy. But ancient wise men, in Bacon's view, almost restored humankind's knowledge of and control over nature—and embodied their secret knowledge in fables. Bacon's goal was to recover that ancient knowledge and control. And the truths hidden in fables and myths were identical to those to be found by using his new scientific method.

Topics: *X*, VII Name: Bacon

0525 Garrod, H. W. "Phalaris and Phalarism." In *Seventeenth Century Studies Presented to Sir Herbert Grierson*, 360-371. Oxford: Clarendon Press, 1938. See (1424).

Garrod discusses the episode in the 1690's of the ancients versus moderns controversy.

Topic: *X* Names: Bentley, King

0526 Gascoigne, John. "From Bentley to the Victorians: The Rise and Fall of British Newtonian Natural Theology." *Science in Context* 2(1988): 219-255.

Gascoigne relates the career of Newtonian natural theology to the social and political situations of popular natural theologians and their critics. This survey is a response to two questions: (1) why did Christianity in England achieve a rapprochement with science in the early period? and (2) why did they separate in the nineteenth century? Gascoigne argues that in Newton's time natural theologians had developed a balance between images of an orderly God of general providence and a voluntaristic God of special providences—and that the former was becoming a part of the dominant social ideology.

Topics: *VIII*, IV, VII, XII Names: Bentley, Newton

0527 Gascoigne, John. "Isaac Barrow's Academic Milieu: Interregnum and Restoration Cambridge." In *Before Newton: The Life and Times of Isaac Barrow*, edited by Mordechai Feingold, 149-180. Cambridge: Cambridge University Press, 1990. See (0452).

In great detail, Gascoigne presents Barrow's life and associations in Cambridge from 1642 to his death in 1677. He notes that from his student days to his time as Master of Trinity, Barrow lived through three purges of the colleges. (Yet Cambridge never was theologically monolithic.) Gascoigne outlines sequentially Barrow's relations with his colleagues as well as his broad academic interests—including mathematics, natural philosophy, and theology. Barrow's Aristotelianism, his careful anti-Cartesianism, his Arminianism, and his Royalist background are highlighted.

Topics: *V*, VII Names: Barrow, More, Wilkins

0528 Gascoigne, John. "‹The Wisdom of the Egyptians› and the Secularization of History in the Age of Newton." In *The Uses of Antiquity*, edited by Stephen Gaukroger, 171-212. Dordrecht: Kluwer Academic, 1991. See (0530).

Gascoigne argues that an emphasis on ordinary providence, rather than direct intervention, is reflected in and underlies both the natural philosophy and the historical studies of the period. Specifically, this tendency is reflected in the controversy over whether the Hebrews or the Egyptians were the originators of the original wisdom (*prisca theologia*). Newton's views about Noah and his ambivalence toward the Egyptians are described at length. Gascoigne concludes that such an emphasis effectively reduced the role of the biblical narrative in explaining the development of the ancient

world; this, he believes, led to the autonomy (or secularization) of both natural philosophy and history in the next century.
Topics: *VII*, II, X, XII Names: Burnet, T., Newton, Woodward

0529 Gaukroger, Stephen. "Introduction: The Idea of Antiquity." In *The Uses of Antiquity*, edited by Stephen Gaukroger, ix-xvi. Dordrecht: Kluwer Academic, 1991. See (0530).
Gaukroger emphasizes the "Christian-allegorical reading of antiquity" as a feature of pre-Enlightenment thought (including that of scientists and philosophers).
Topics: *VII*, II

0530 Gaukroger, Stephen, ed. *The Uses of Antiquity: The Scientific Revolution and the Classical Tradition*. Dordrecht: Kluwer Academic, 1991.
This is a useful and interesting collection of essays on the image of antiquity in the seventeenth century and of the role of antiquity in debates concerning natural philosophy, religion, and history. Gaukroger's introduction (0529) and the articles by Birkett and Oldroyd (0095), Gascoigne (0528), A. Jacob (0792), Kassler (0860), Sutton (1517), Thiel (1540) and Trompf (1570) are annotated separately.
Topics: *X*, VII

0531 Gay, John H. "Matter and Freedom in the Thought of Samuel Clarke." *J. Hist. Ideas* 24(1963): 85-105.
This article concerns Clarke's doctrine of God and various concepts of matter, spirit, and causality. Gay's retrospective approach overemphasizes Clarke's "unfortunate blindness to the future" but his article still is helpful.
Topics: *VII*, XII Name: Clarke

0532 Gelbart, Nina Rattner. "The Intellectual Development of Walter Charleton." *Ambix* 18(1971): 149-168.
Gelbart traces the "meandering route" by which Charleton arrived at atomism. Yet even his atomism, according to the author, retained elements of the hermetical and alchemical tradition.
Topics: *II*, VII Name: Charleton

0533 Genuth, Sara Schechner. "Devil's Hells and Astronomer's Heavens: Religion, Method, and Popular Culture in Speculations about Life on Comets." In *The Invention of Physical Science: Intersections of Mathematics, Theology and Natural Philosophy Since the Seventeenth Century. Essays in Honor of Erwin N. Hiebert*, edited by Mary Jo Nye and others, 3-26. Dordrecht: Kluwer Academic, 1992.

In the early eighteenth century, natural theology and scientific views combined to produce the idea that comets are "wandering hells." Later, a different natural theology produced a more benevolent view. Gennuth describes these ideas and also relates them to popular culture and a secularization trend. In this context, she discusses Whiston, Derham, and Cotton Mather.

Topics: *VIII*, VII, X Names: Derham, Mather, C., Whiston

0534 Genuth, Sara Schechner. "Newton and the Ongoing Teleological Role of Comets." In *Standing on the Shoulders of Giants*, edited by Norman J. W. Thrower, 299-311. Berkeley and Los Angeles: University of California Press, 1990. See (1545).

Genuth argues that Newton and Halley assimilated traditional comet lore into their natural philosophy. For them, God uses comets as a natural means to conserve, renovate, and reform the cosmos.

Topics: *XII*, VII Names: Halley, Newton

0535 George, Charles H. "A Social Interpretation of English Puritanism." *J. Mod. Hist.* 25(1953): 327-342.

George tries to delineate the major components of and types of Puritanism. This essay has been used by some defenders of the "Merton thesis."

Topic: *III*

0536 George, Charles H., and Katherine George. *The Protestant Mind of the English Reformation, 1570-1640*. Princeton, N.J.: Princeton University Press, 1961.

The Georges discuss and try to disprove the Weber thesis concerning Protestantism and the rise of capitalism. This work includes long passages relating Protestantism to the culture but totally ignores science. It includes a good bibliography.

Topic: *III*

0537 George, Edward A. *Seventeenth-Century Men of Latitude: Forerunners of the New Theology*. New York: C. Scribner's Sons, 1908.

This is a collection of brief biographical studies of the life and thought of various Anglican theologians—including More, Whichcote, and Thomas Browne.

Topics: *VII*, X Names: Browne, T., More, Whichcote

0538 Gibson, James. *Locke's Theory of Knowledge and Its Historical Relations*. Cambridge: Cambridge University Press, 1917. Reprinted, Cambridge: Cambridge University Press, 1960.

This work is a systematic commentary on Locke's *Essay concerning Human Understanding*. In addition to explicating Locke's views, Gibson discusses "the historical relations between Locke's doctrine" and each of the following: scholasticism; Descartes; Hobbes; the Cambridge Platonists; Boyle; Newton; Leibniz; and (beyond our period) Kant. Gibson is fully aware of Locke's theological purposes and views; and he is aware of the relevance of this, for Locke, to natural philosophy. Although epistemology is central to the book, Gibson also discusses metaphysics, the nature of matter, the nature of space and time, and the idea of God. This is an old classic, superseded in many ways, but also still useful.

Topics: *VII*, XII Names: Boyle, Hobbes, Locke, More, Newton

0539 Gifford, George E. "Botanic Remedies in Colonial Massachusetts, 1620-1820." In *Medicine in Colonial Massachusetts, 1620-1820*, edited by Philip Cash and others, 263-288. Boston: Colonial Society of Massachusetts, 1980. See (0193).

For the pre-1720 period, Gifford discusses the Paracelsian "doctrine of signatures." These were God-given signs marking plants and minerals for particular uses "for the good of man." Gifford also notes that this theory was similar to that of the Native Americans in the region. Thus, colonists had no difficulty drawing upon native healers' knowledge of the medicinal value of local animal and vegetable products.

Topic: *XI*

0540 Gilbert, Allan H. "Milton's Text Book of Astronomy." *PMLA* 38(1923): 297-307.

Gilbert describes the fifteenth-century Ptolemaic textbook which Milton used. He explains why, in Milton's view of the "great architect," it does not matter which astronomical hypotheses are correct.

Topic: *X* Name: Milton

0541 Gilbert, Allan H. "The Outside Shell of Milton's World." *Studies in Philology* 20(1923): 444-447.

Gilbert analyzes Milton's use of the Ptolemaic system and of several Church Fathers in the picture of the universe of *Paradise Lost*. For Milton, heaven and hell are invisible to humans and beyond the outside shell of the world.

Topic: *X* Name: Milton

0542 Gillespie, Charles Coulston. *The Edge of Objectivity*. Princeton, N.J.: Princeton University Press, 1960.

This is a well-written survey of "the historical development of the objectivity of modern science." Gillespie is an "internalist" historian of science who, by definition, sees almost no connection between science and religion. (He sees a minor relationship in that Christianity had to adjust itself to science.) As for Newton, he writes "it is as if there were two Newtons speaking in turn" and it is right "to let Newton the scientist, rather than . . . Newton the theologian, have the last word."

Topics: *VII*, XII Name: Newton

0543 Gillespie, Charles Coulston. "Fontenelle and Newton." In *Isaac Newton's Papers & Letters On Natural Philosophy*, edited by I. Bernard Cohen, 427-443. Cambridge, Mass.: Harvard University Press, 1958. 2nd ed., Cambridge, Mass.: Harvard University Press, 1978. See (0245).

This is a description of eighteenth- and nineteenth-century biographies of Newton with historiographic commentary. Gillespie describes the ambivalence towards Newton on the part of eighteenth-century continental scientists and philosophers. He includes an insightful critique of those historians in the mid-twentieth century who connect Newton's natural philosophy to the soulless, deterministic world picture found in France later in the eighteenth century.

Topics: *XII*, I, VII Name: Newton

0544 Gillespie, Charles Coulston. "Mertonian Theses." *Science* 184(1974): 656-660. Reprinted in *Puritanism and the Rise of Modern Science: The Merton Thesis*, edited by I. Bernard Cohen, 132-141. New Brunswick, N.J.: Rutgers University Press, 1990. See (0246).

This is an essay review of several of Merton's writings in the sociology of science and of their relevance to history of science. Merton is at his best and deepest—Gillespie says—not in statistical analysis, but "in his insight into motivation and behavior, individual and collective."

Topics: *I*, III

0545 Giorello, Giulio. "Introduction." In *Science and Imagination in XVIIIth-Century British Culture*, edited by Sergio Rossi, 13-34. Milan: Edizioni Unicopli, 1987. See (1362).

The topics included in this collection of papers are well summarized by the title. Giorello introduces the volume by raising the issue of the relation of human scientific knowledge to God's creation and God's knowledge. Tested against the latter, all human authority may be questioned—hence "new" scientific knowledge conceived. And, thus, undergirding the imaginations of scientists of the past was a "theological optimism."

Topic: *VII* Name: Milton

0546 Gjertsen, Derek. *The Newton Handbook*. London: Routledge and Kegan Paul, 1986.

This work is an encyclopedic guide to the life, science, and selected aspects of Newton and Newtonians. Individual entries are found for every published work and every major manuscript (including biblical, theological, and chronological titles). It includes some well-done entries on selected Newtonian scholars as well. Newton's mathematical and scientific accomplishments are treated thoroughly and in a rather internalist fashion. Newton's theological interests and related topics are treated in some detail but mostly as separate entries within the alphabetical arrangement. Gjertson, somewhat in the tradition of Rupert Hall, generally assumes a separation between science and religion. ("Gravity," for example, is a four-page-plus discussion with no mention of God or the theological issues.) Coverage of Newton's alchemical interests is brief and unsympathetic. "Newtonianism" generally is treated as a scientific heritage and the ideological issues are noticeably downplayed. But Gjertsen's book is an excellent reference source for factual, biographical, and manuscript information.

Topics: *XII*, VII Name: Newton

0547 Gjertsen, Derek. "Newton's Success." In *Let Newton Be!*, edited by John Fauvel and others, 23-41. Oxford: Oxford University Press, 1988. See (0449).

Gjertsen discusses Newton's success in terms of various meanings of the word. He notes, among other things, the influence of Newtonian concepts on theology; he also notes the religious usage of the phrase "on the shoulders of giants." This is a good popular introduction to the topic.

Topics: *XII*, VII Name: Newton

0548 Glacken, Clarence James. *Traces on the Rhodian Shore: Nature and Culture in Western Thought from Ancient Times to the End of the Eighteenth Century*. Berkeley and Los Angeles: University of California Press, 1967.

This is a survey of three "problems": the idea of an earth created by design; the influence of the environment on mankind; and the influence of mankind on the environment. In his chapter on "Physico-Theology: Deeper Understanding of the Earth as a Habitable Planet," Glacken studies the natural theology of the seventeenth and early-eighteenth centuries. He explores the following topics: the origin of the earth; critiques of the theory that nature was decaying; the ancients vs. moderns controversies; the "plastic nature" of the Cambridge Platonists; human design and dispersion. Above all, he appreciates the works of Ray and Derham for their arguments for the wisdom of God in the "living interrelationships in nature." In a chapter on the control of nature, Glacken explores the optimism of the seventeenth-century writers.

Topics: *VI*, IV, VIII Names: Bacon, Burnet, T., Cudworth, Derham, Evelyn, Hakewill, Ray, Whiston, Woodward

0549 Glover, Willis B. "God and Thomas Hobbes." *Church History* 29(1960): 275-297. Reprinted in *Hobbes Studies*, edited by K. C. Brown, 141-168. Cambridge, Mass.: Harvard University Press, 1965.

Glover argues that Hobbes, though a materialist and a determinist, was not an atheist. He sees Hobbes as a nominalist and a Scotist whose theology may have some inconsistencies but is nonetheless sincere.
Topic: *IX* Name: Hobbes

0550 Godwin, Joscelyn. *Robert Fludd: Hermetic Philosopher and Surveyor of Two Worlds*. Boulder, Colo.: Shambhala, 1979.
Though aimed at a popular audience, this book is a good source for Fludd's images and plates; it also includes good commentaries on the images. Godwin stresses the universality of Fludd's vision, his independence from "sectarian Christian theology," and his acceptance of a universalistic esoteric syncretism.
Topic: *II* Name: Fludd

0551 Goldish, Matt. "Newton on Kabbalah." In *The Books of Nature and Scripture*, edited by James E. Force and Richard H. Popkin, 89-103. Dordrecht: Kluwer Academic, 1994. See (0498).
The author begins by finding Hutin's "identification of Newton as a Christian kabbalist" to be "most questionable." Instead, Goldish finds Newton to be interested in the Kabbalah in order to place it in the history of the corruption of the true original faith. Goldish suggests that Newton placed it in the branch of mistaken, heathen theology which led to gnosticism. It lies, for Newton, in a metaphysical tradition that presents creation as an emission.
Topics: *XII*, II, X Name: Newton

0552 Goldman, Stephen L. "On the Interpretation of Symbols and the Christian Origins of Modern Science." *J. Rel.* 62(1982): 1-20.
Goldman shifts and expands the Protestant-origins argument by analyzing religious images, metaphors, and symbols. He compares Christian interpretations to Jewish (and, briefly, to other religions). He argues that medieval and Renaissance Christian thought generally featured an "ontological interpretation" of symbols: that is, they were interpreted "as being ikons of a reality existing external to the mind and beyond the senses." This Christian view is contrasted with Jewish attitudes and proposed as a partial answer to the question of why Christian natural philosophy uniquely made the

transition to modern science. Most of the examples discussed are outside the chronological scope of this bibliography—but the proposal is relevant to our period.
Topics: *III*, VII

0553 Golinski, Jan. "The Secret Life of an Alchemist." In *Let Newton Be!*, edited by John Fauvel and others, 147-167. Oxford: Oxford University Press, 1988. See (0449).
Aimed at a general audience, this is a description of Newton's chemistry and alchemy in seventeenth-century context. Golinski emphasizes: Newton's aims and methods; the important element of secrecy; and, as a separate issue, Newtonian chemistry as public doctrine. The essay includes a good subsection on "Alchemy and theology."
Topics: *XII*, II, VII Name: Newton

0554 Gordon, Maurice Bear. *Aesculapius Comes to the Colony: The Story of the Early Days of Medicine in the Thirteen Original Colonies*. Ventnor, N.J.: Ventnor, 1949.
Organized by colony, this study is useful for its brief biographies of physicians. It is clear that many of them also were members of the clergy.
Topic: *XI*

0555 Gosselin, Edward A. "The 'Lord God's' Sun in Pico and Newton." In *Renaissance Society and Culture: Essays in Honor of Eugene F. Rice, Jr.*, edited by John Monfasani and Ronald G. Musto, 51-58. New York: Italica, 1991.
Gosselin first discusses the religious philosophy of Pico della Mirandola in terms of Renaissance philosophical mysticism and sees it as part of "the Solar Age." He then speculates—but not very persuasively—that for Newton the sun "was the Piconian-type representative of the Lord God."
Topics: *XII*, II, VII Name: Newton

0556 Gough, John Wiedhofft. *The Superlative Prodigall: A Life of Thomas Bushell*. Bristol: J. W. Arrowsmith, 1932.
This is a biography of an interesting character, a Royalist mining engineer and promoter. Bushell was a youthful favorite of Bacon and later used this connection in his publicity. He once proposed

setting up a mining venture, modeled on Solomon's House, in which "men excellently qualified in theology, morality and humanity" would teach and convert a workforce of felons.
Topics: *VI*, IV Names: Bacon, Bushell, Charles I

0557 Gouk, Penelope M. "The Harmonic Roots of Newtonian Science." In *Let Newton Be!*, edited by John Fauvel and others, 101-125. Oxford: Oxford University Press, 1988. See (0449).
Gouk defines harmonic science as the study of ratios corresponding to musical intervals and briefly describes its place in the history of philosophy. She argues that aspects of harmonic science were directly related to central topics of Newton's scientific work—mathematics, the nature of light, and gravitational theory. Showing Newton's attitude toward the *ancient theology*, Gouk concludes that "Newton might well be regarded as a seventeenth-century Pythagorean, his life being devoted to the study of universal harmony."
Topics: *XII*, II, VII Name: Newton

0558 Gouk, Penelope. "Newton and Music: From the Microcosm to the Macrocosm." *Int. Stud. Philos. Sci.* 1 (1986): 36-59.
Gouk describes various ways in which music played a part in Newton's research programme. She relates his ideas concerning music to the *prisca sapientia* tradition and shows how his music theory, natural philosophy, theology, and cosmology all were interconnected.
Topics: *XII*, VII Name: Newton

0559 Gould, Stephen Jay. "The Godfather of Disaster." *Natural History* 96(1987): 20-29.
This is a popular and semi-humorous treatment of William Whiston's combining Newtonian physics with biblical exegesis. Gould gives a basically sympathetic rendering of how Whiston used comets to explain both salvation history and natural history from the creation to the apocalypse.
Topics: *VII*, XII Name: Whiston

0560 Gould, Stephen Jay. *Time's Arrow, Time's Cycle: Myth and Metaphor in the Discovery of Geological Time*. Cambridge, Mass.: Harvard University Press, 1987.

This is an intelligent lecture-style discussion of the dichotomy between the metaphors of time used in the title of the book. In particular, Gould explicates the texts of Thomas Burnet—whose biblical geology is seen as the first example of the tension between the two complementary views of time. Illustrations help the reader to understand Burnet's worldview.

Topics: *VII*, X Name: Burnet, T.

0561 Grant, Douglas. *Margaret the First: A Biography of Margaret Cavendish, 1623-1673*. London: Hart-Davis, 1957.

Margaret Cavendish, Duchess of Newcastle, is a minor figure for historians of science. She is of interest, however, for those studying atomism and its development in England. Grant's treatment of her also is relevant to issues concerning the possible connection of atomism to atheism.

Topics: *VII*, IX Name: Cavendish

0562 Grant, Edward. "Medieval and Seventeenth-Century Conceptions of an Infinite Void Space beyond the Cosmos." *Isis* 60(1969): 39-60.

Grant discusses ancient, medieval, and early modern concepts of God and infinite void space. The development of ideas concerning infinite space (and God) in the seventeenth century should be understood at least partially as a legacy of the late Middle Ages.

Topics: *VII*, XII Name: Newton

0563 Graubard, Mark. *Witchcraft and the Nature of Man*. Lanham, Md.: University Press of America, 1985.

Graubard argues for an awareness of our own assumptions; and he argues that science and religion were integrally related to the witchcraft controversy of the seventeenth century. But, unfortunately, the book otherwise is not very useful.

Topic: *II*

0564 Greaves, Richard L. "John Bunyan and Covenant Thought in the Seventeenth Century." *Church History* 36(1967): 151-169.

This essay is relevant to the problem of defining Puritanism.

Topic: *III*

0565 Greaves, Richard L. *The Puritan Revolution and Educational Thought: Background for Reform*. New Brunswick, N.J.: Rutgers University Press, 1969.

This is a useful discussion of various issues relating religious reformers to educational ideas and institutions in the mid-seventeenth century. Greaves emphasizes the role of science in these movements. He distinguishes sectarian from moderate Puritanism—discussing both.

Topics: *IV*, III, V Names: Bacon, Comenius, Dell, Hall(1627), Hartlib, Sprigg, Ward, Webster, Wilkins, Winstanley

0566 Greaves, Richard L. "Puritanism and Science: The Anatomy of a Controversy." *J. Hist. Ideas* 30(1969): 345-368.

This is an argumentative historiographic essay and review of the literature regarding the Puritanism-and-the-rise-of-science thesis. Greaves disagrees with Merton (1057), Stimson (1504), Jones (0835), Hill (0686) and others who take similar positions. Oddly, Greaves makes no mention of Hooykaas (0723), even though much use is made of the Rabb article (1278) which appeared in the same journal as that of Hooykaas. Greaves admits that, after the 1630s, the cultural environment in Protestant countries was more conducive to scientific progress than in Catholic areas. But he attributes this to "the restrictive effects of the Counter Reformation" and, in England, to "revolutionary ferment." This is a widely cited article.

Topics: *III*, IV, V

0567 Green, A. Wigfall. *Sir Francis Bacon*. New York: Twayne, 1966.

This book contains a brief biography and summaries of Bacon's writings. The summaries are arranged topically, with some discussion of context and sources. Portions of the book are edited excerpts from Green's *Sir Francis Bacon: His Life and Works*, a longer and more biographical work. Both books are outdated, even as introductions.

Topic: *VII* Name: Bacon

0568 Green, Henry. *Sir Isaac Newton's Views on Points of Trinitarian Doctrines; His Articles of Faith, and the General Coincidence of His Opinions with Those of John Locke: A Selection of Authorities, with Observations*. London: E. T. Whitfield, 1856.

Newton and Locke are both seen as unitarians as well as having similar views towards the Bible and scientific inquiry.
Topics: *XII*, IX, X Names: Locke, Newton

0569 Greene, Donald. "Latitudinarianism and Sensibility: The Genealogy of the 'Man of Feeling' Reconsidered." *Modern Philology* 75(1977): 159-183.
Greene emphasizes the traditional and orthodox elements in Latitudinarian thought. The essay is written as a sustained refutation of Crane's 1934 interpretation of the same movement (see 0288). There is no direct relationship to science here but the essay is historiographically valuable for our field. Greene answers not only misinterpretations of Latitudinarianism by literary scholars but also those by historians of science and by historians of the deist movement.
Topics: *X*, I, IX Names: Barrow, Glanvill, Tillotson

0570 Greene, Robert A. "Henry More and Robert Boyle on the Spirit of Nature." *J. Hist. Ideas* 23(1962): 451-474.
Greene sees the "Spirit of Nature" idea in late seventeenth-century England as part of the reaction against materialistic and overly mechanistic philosophies.
Topic: *VII* Names: Boyle, More

0571 Greene, Robert A. "Whichcote, the Candle of the Lord, and Synderesis." *J. Hist. Ideas* 52(1991):617-644.
Greene discusses the metaphor "candle of the Lord" (Proverbs 20:27) as used by Bacon, Whichcote, various English Calvinist theologians, Locke, and others. The phrase, when taken to refer to a natural human faculty, could be used to support confidence in human reasoning—but only Bacon applied this directly to natural philosophy. Even for those who applied it to moral reasoning, however, this discussion is useful. Greene also addresses the variety of views in seventeenth-century England concerning the relation between revelation and natural (including natural philosophic) reasoning.
Topic: *VII* Names: Bacon, Locke, Whichcote

0572 Greene, Robert A. "Whichcote, Wilkins, 'Ingenuity' and the Reasonableness of Christianity." *J. Hist. Ideas* 42(1981): 227-252.

Greene traces the course of several words in late-seventeenth-century England: "ingenuity"; "ingenious"; "ingenuous"; their Latin antecedents; and their English derivatives. He relates the development of these words to an ever-increasing confidence in human reason and experience (as opposed to a Calvinist emphasis on total depravity). He sees this development as supporting human science and technology—and the later understanding of the word "genius."

Topics: *X*, VII Names: Whichcote, Wilkins

0573 Greengrass, Mark, ed. *Samuel Hartlib and Universal Reformation: Studies in Intellectual Communication*. Cambridge: Cambridge University Press, 1994.

See the separate annotations for articles by Clucas (0225), Newman (1112), Oster (1186), Pumfrey (1269), and Webster (1612).

Topic: *IV* Name: Hartlib

0574 Greenlee, Douglas. "Locke and the Controversy Over Innate Ideas." *J. Hist. Ideas* 33(1972): 251-264.

Greenlee sets the controversy in the social, philosophical, and theological context of seventeenth-century England.

Topic: *VII* Name: Locke

0575 Gregory, Joshua C. "Cudworth and Descartes." *Philosophy* 7(1933): 454-467.

Gregory summarizes *The True Intellectual System* and discusses Cudworth's views on the relation between God and "Plastic Nature." His recognition of the Stoic background is interesting for the time, but the work basically is now outdated.

Topic: *VII* Name: Cudworth

0576 Gregory, T. S. "John Locke: 'The Order of Nature'." *Listener* 52(1954): 805-807. Reprinted in *A Locke Miscellany*, edited by Jean S. Yolton, 34-45. Bristol: Thoemmes, 1990. See (1712).

Gregory relates both Locke's philosophy of science and his religion to a larger native English practical worldview. The lecture is addressed to a general audience but it is provocative for scholars too.
Topic: *VII* Name: Locke

0577 Griffin, Martin I. J., Jr. *Latitudinarianism in the Seventeenth-Century Church of England*. Edited by Lila Freedman and annotated by Richard H. Popkin. Leiden: Brill, 1992.
This book is good for background although it has little direct reference to science or natural philosophy. Griffin provides a good, thorough approach to defining Latitudinarianism—both as to the key themes and the key members of the movement. It is based on research done in 1958-1962, but Popkin's annotations somewhat bring it up-to-date.
Topics: *VII*, VIII Names: Burnet, G., Glanvill, Patrick, Stillingfleet, Tillotson, Wilkins

0578 Grinnell, George. "Newton's *Principia* as Whig Propaganda." In *City and Society in the Eighteenth Century*, edited by Paul Fritz and David Williams, 181-192. Toronto: Hakkert, 1973.
For Grinnell, "science, theology and social thought are inseparable." He argues that Newton's motivation in all three areas was his hatred of that "whore of Babylon." Hence, the *Principia* is disguised Protestant propaganda aimed against the Catholic Thomistic worldview and constructed to counter the ascension of James II and Catholic attacks on Cambridge. For example, absolute space as God's sensorium is Protestant, Aristotelian "place" and the hierarchical division of heaven and earth are Thomistic.
Topics: *IV*, XII Name: Newton

0579 Gruner, Rolf. "Science, Nature and Christianity." *Journal of Theological Studies* N.S. 26(1975): 55-81. Reprinted in *Creation, Nature and Political Order in the Philosophy of Michael Foster*, edited by Cameron Wybrow, 213-243. Lewiston, N.Y.: E. Mellen Press, 1992. See (1703).
This is a philosophical critique of the thesis that modern science is a consequence of Christianity. While referring to Bacon and others in the history of Western thought, Gruner's argument is primarily a logical one. It finally comes down to his assumptions: (1) that modern science has no religious significance today; and (2) that

there is no "necessary development" in history and no "inner logic" in Christianity which produced science. Thus, one should neither praise nor blame the Christian faith for science.

Topics: *VII*, III Name: Bacon

0580 Guerlac, Henry. *Newton et Epicure.* (Conference donnee au Palais de la Decouverte le 2 Mars 1963.) Paris: Universite de Paris, 1963. Reprinted in *Essays and Papers in the History of Modern Science*, by Henry Guerlac, 82-106. Baltimore: Johns Hopkins University, 1977.

Guerlac relates Newton's atomism to the Epicurean revival of the seventeenth century. This is a pioneering work in relating seventeenth-century science to its social and historical context.

Topics: *XII*, VII Names: Charleton, Cudworth, Newton

0581 Guerlac, Henry. "Some Historical Assumptions of the History of Science." In *Scientific Change*, edited by Alistair C. Crombie, 797-812 and 875-876. London: Heinemann, 1963. See (0299).

Guerlac warns against two tendencies amongst historians of science: (1) excessive specialization; and (2) isolating the history of scientific thought from its social and historical context.

Topic: *I*

0582 Guerlac, Henry. "Theological Voluntarism and Biological Analogies in Newton's Physical Thought." *J. Hist. Ideas* 44(1983): 219-229. Reprinted in *Philosophy, Religion and Science in the Seventeenth and Eighteenth Centuries*, edited by John W. Yolton, 406-416. Rochester, N.Y.: University of Rochester Press, 1990. See (1720).

Guerlac here summarizes recent explications of Newton's theological voluntarism. He connects Newton's theological position to his emphasis on "active principles" and the deadness of matter in Newton's natural philosophy. Guerlac argues that Newton understood God's active will through an analogy with human will and animal volition. The importance of nonmechanical agency (which includes God) in Newton's natural philosophy is found in his published work but it is especially noticeable in his unpublished manuscripts.

Topics: *XII*, VII Name: Newton

0583 Guerlac, Henry, and Margaret C. Jacob. "Bentley, Newton, and Providence: (The Boyle Lectures Once More)." *J. Hist. Ideas* 30(1969): 307-318.

The authors discuss the social and political context of the first set of Boyle lectures. They argue that Newton personally took a directing role in the selection of Bentley.

Topics: *VIII*, IV, V, XII Names: Bentley, Newton

0584 Guerrini, Anita. "Isaac Newton, George Cheyne and the 'Principia Medicinae'." In *The Medical Revolution of the Seventeenth Century*, edited by Roger French and Andrew Wear, 10-45. Cambridge: Cambridge University Press, 1989. See (0510).

Guerrini traces Newton's influence on the medical theory and practice of physicians—and particularly that of George Cheyne. She notes, for example, how revisions of Newton's matter theory in various editions of the *Opticks* affected Cheyne's adoption of a theory of spirits. This, along with Newtonian mechanical terms, becomes a dominant theme in Cheyne's writings on natural theology, physiology, and popular medical remedies. Finally, Guerrini discusses Cheyne's religious faith and his late relationship with the Methodists.

Topics: *XI*, VII, VIII, XII Names: Cheyne, Newton, Pitcairne

0585 Guibbory, Achsah. *The Map of Time: Seventeenth-Century English Literature and Ideas of Pattern in History*. Urbana, Ill.: University of Illinois Press, 1986.

Guibbory relates views of history to the literary productions of selected writers and poets of the period. She uses a basic three-part scheme of: (1) history as decay; (2) history as cyclical; and (3) history as progress. In Chapter 2, Guibbory argues that Bacon's idea of history includes both cyclical and progressive views. She sees Bacon's views of the new science, of apocalyptic interpretation, and of history as all interrelated. The chapters on Donne, Milton, and Dryden do not include any mention of science. The introduction includes a short discussion of Boyle's view of history and a brief discussion of the "ancients vs. moderns" controversy.

Topics: *X*, IV Names: Bacon, Boyle, Donne, Dryden, Milton

0586 Guinsburg, Arlene Miller. "Henry More, Thomas Vaughan and the Late Renaissance Magical Tradition." *Ambix* 27(1980): 36-58.

Guinsburg analyzes the pamphlet war between More and Vaughan; she then traces the subsequent development of More's thought concerning the creation. She shows in detail how, despite More's atomism, the hermetic tradition acted as a catalyst to his ideas about space, matter, spirit and motion—as well as his view of authority.

Topics: *VII*, II Names: More, Vaughan, T.

0587 Gunther, Robert W. T. *Early Science in Oxford*. 14 vols. Oxford: Oxford University Press, 1923-1945.

This is a combination of primary materials and secondary commentary which is useful for issues having to do with science and social institutions. Gunther does not discuss any direct connection to religion. (The volumes have a more complicated publishing history than is shown above.)

Topic: *V*

0588 Guthke, Karl S. *The Last Frontier: Imagining Other Worlds from the Copernican Revolution to Modern Science Fiction*. Translated by Helen Atkins. Ithaca, N.Y.: Cornell University Press, 1990.

This is a fascinating survey of writings on the question of the "plurality of worlds" theme—especially after the observations of Galileo. The chapter on "The Baroque Period" is within the scope of this bibliography. Among the topics explored are: Donne's turmoil; Milton's referral to the inscrutability of God and his emphasis on human salvation; and More's vision of innumerable planetary worlds. Wilkins' belief in other inhabited worlds is thoroughly analyzed. Guthke decides that Wilkins saw the issue as a scientific one: the Bible does not contradict the book of nature and deals with a different subject. Further, the plurality of worlds increases a sense of the power and wisdom of God. Finally, Guthke categorizes Francis Godwin's *The Man in the Moon* as one of the first works of science fiction. The encounter between its hero and the people on the moon raised questions about human nature and the possibility that other beings may be of a higher, but still

Christian, order. Guthke's book was first published under the rather different title *Der Mythos der Neuzeit*.
Topic: *X* Names: Donne, Godwin, Milton, More, Wilkins

0589 Gysi, Lydia. *Platonism and Cartesianism in the Philosophy of Ralph Cudworth*. Bern: Herbert Lang, 1962.
Gysi explains the "dogmatic and critical tendencies" in Cudworth's thought by tracing the influences of Platonism and Cartesianism. Her chapters, in order, discuss Cudworth's theories of: substance; knowledge; truth; knowledge of God and ethics. Although these categories are treated separately, it becomes apparent that they are interrelated. For example, "plastic nature" (as Cudworth's solution to Cartesian dualism) is the intermediary between body and soul and also between God and matter. And the Trinity (a doctrine of God), for Cudworth, is the foundation of all knowledge.
Topic: *VII* Name: Cudworth

0590 Gysi, Lydia. *Ralph Cudworth: Mystical Thinker*. Newport Pagnell, England: Greek Orthodox Monastery of the Assumption, 1973.
This little book is a continuation of *Platonism and Cartesianism in the Philosophy of Ralph Cudworth*. It is written by Mother Maria (the name which Gysi took when she became a Greek Orthodox nun). In the chapter on "Participation in Reason," she briefly proposes how Cudworth understood participation in the Truth—God's mind—and how ideas in that mind are reproduced in humans. Gysi's books serve as a reminder of the attempts at a God-based unity of reason, will, and holiness in seventeenth-century thought.
Topic: *VII* Name: Cudworth

0591 Haber, Francis C. *The Age of the World: Moses to Darwin*. Baltimore: Johns Hopkins Press, 1959.
In this survey, Haber is concerned with the effects of modern science on biblical chronology and on concepts of historical time. The treatment is superficial (and a little outdated) but still useful for our period. See especially the chapter entitled "Historicism and the Scientific Revolution."
Topics: *X*, VII, XII Names: Burnet, T., Hooke

0592 Haber, Francis C. "Time, Technology, Religion, and Productivity Values in Early Modern Europe." In *Time, Science,*

Society in China and the West, edited by J. T. Fraser and others, 79-92. (The Study of Time, Vol. 5.) Amherst, Mass.: University of Massachusetts Press, 1986.

Haber proposes that there emerged "a tradition in the Christian West of justifying technology on religious grounds." This "technological idealism" is composed of the following tenants: God as maker; human inventions as imitations of God; and an eschatology emphasizing a gradual (progressive) recovery from the fall. Bacon and his disciples both culminate and consolidate this tradition.

Topic: *VI* Names: Bacon, Blith, Hartlib, Plattes

0593 Habicht, Hartwig. *Joseph Glanvill, ein spekulatuver Denker im England des XVII. Jahrhunderts: Eine Studie uber des fruhwissenschaftliche Weltbild*. Zurich: A.-G. Gebr. Leemann, 1936.

In this published doctoral dissertation in philosophy, Habicht treats Glanvill as a Latitudinarian within the tradition of Cambridge Platonism. He sees Glanvill's religion as the foundation of his natural science. Among other topics, Habicht discusses Glanvill's physics, physiology, and demonology.

Topics: *VII*, V Names: Cudworth, Glanvill, More

0594 Hacking, Ian. *The Emergence of Probability: A Philosophical Study of Early Ideas about Probability, Induction and Statistical Inference*. London: Cambridge University Press, 1975.

Although Hacking's emphasis is on continental thinkers, his book is good for background material as indicated in the subtitle. One short chapter focuses on the design argument among "Royal Society theologians" and Newtonians.

Topics: *VII*, VIII, XII Names: Bacon, Hobbes, Newton, Wilkins

0595 Haden, James C. "The Challenge of the History of Science." *Rev. Metaphysics* 7(1953): 74-88 and 262-281.

This is a review article of most of the important histories of science written before 1953. Haden wants the history of science to be influenced by a method more akin to the history of philosophy.

Topic: *I*

0596 Hall, A. Rupert. "Cultural, Intellectual and Social Foundations, 1600-1750." In *Technology in Western Civilization*, edited by

Melvin Kranzberg and Carroll W. Pursell, Jr., Vol. 1, 107-118. New York: Oxford University Press, 1967.

Hall does not find any significant linkages between Protestantism and technological progress. In fact, he finds that sermons have had no effect either on technology or on business at any time!

Topics: *VI*, III

0597 Hall, A. Rupert. *From Galileo to Newton, 1630-1720*. New York: Harper and Row, 1963.

In this survey, Hall shows little understanding of seventeenth-century religion.

Topics: *VII*, XII Name: Newton

0598 Hall, A. Rupert. "Henry More and the Scientific Revolution." In *Henry More (1614-1687) Tercentenary Studies*, edited by Sarah Hutton, 37-54. Dordrecht: Kluwer, 1990. See (0782).

This could be another chapter in (and somewhat overlaps various parts of) Hall's book on Henry More, (0600). Hall minimizes More's significance with respect to the scientific revolution in general and with respect to Newton in particular. "Whereas More was always a man looking backwards, and indeed proud of the fact, Newton's was the vision of the future." Hall barely mentions that either More or Newton had a theologically-based metaphysics.

Topics: *VII*, XII Names: More, Newton

0599 Hall, A. Rupert. *Henry More: Magic, Religion, and Experiment*. Oxford: Blackwell, 1990.

This is part of a series entitled "Blackwell Science Biographies" but it is not a biography of More. Rather, it is a series of essays on such topics as: Platonism and the scientific revolution; the Cambridge Platonists; More's philosophy; More on witchcraft and the "spirit world"; More and Descartes; More and the Royal Society; More and Newton (with the main focus on Newton). Hall generally argues that More should be treated only as a theologian and metaphysician—*not* as a natural philosopher. He argues that More is an important example of a seventeenth-century respondent to the new scientific movements—but that More had no major or lasting impact in intellectual history or in the history of science. Seemingly attracted to More as a man, repelled by some aspects of More's thought, retrospectively judging More and his contemporar-

ies, Hall appears ambivalent towards the subject of his book. The intended audience seems to vary from chapter to chapter, ranging from advanced undergraduates to the elite of Newtonian scholars.
Topics: *VII*, II, V, XII Names: Barrow, Boyle, Conway, Cudworth, Glanvill, Helmont, Hobbes, Hooke, More, Newton

0600 Hall, A. Rupert. "Infant Giants Are Not Pygmies: The 'Merton Thesis' and the Sociology of Science." In *Robert K. Merton: Consensus and Controversy*, edited by Jon Clark and others, 371-383. New York: Falmer Press, 1990. See (0216).
This is a rambling harangue attacking radical relativism in the history and sociology of science. Hall touches on various writers in anthropology, philosophy, sociology, and history—but barely discusses Merton and the Merton thesis.
Topics: *I*, III

0601 Hall, A. Rupert. "Magic, Metaphysics and Mysticism in the Scientific Revolution." In *Reason, Experiment, and Mysticism in the Scientific Revolution*, edited by M. L. Rhigini Bonelli and William R. Shea, 275-282. New York: Science History, 1975. See (1329).
Hall argues for the special character of historians of science. "The history of science is concerned with the rational discourse between men" and focuses on the chronological development of rational scientific theories. Hermeticism, alchemy, and (by implication) religion are—for Hall—irrational and have nothing to do with the history of science. Such "alternative modes of discourse" are of auxiliary interest only.
Topic: *I*

0602 Hall, A. Rupert. "Merton Revisited, or Science and Society in the Seventeenth Century." *Hist. Sci.* 2(1963): 1-16. Reprinted in *Science and Religious Belief*, edited by Colin A. Russell, 55-73. London: University of London Press, 1973. See (1377). Excerpted in *Puritanism and the Rise of Modern Science: The Merton Thesis*, edited by I. Bernard Cohen, 224-233. New Brunswick, N.J.: Rutgers University Press, 1990. See (0246).
This is a famous critical review of Merton's 1938 monograph (1058) from the position of internalist historiography of science.
Topics: *III*, I

0603 Hall, A. Rupert. "Newton and His Editors." *Notes Rec. Royal Soc. London* 29(1974): 29-52. Also published in *Proc. Royal Soc. Ser. A* 338(1974): 397-417.

This essay generally has nothing to do with religion. But Hall does address the significance and originality of Roger Cotes and his Preface to the second edition of the *Principia*.

Topics: *XII*, VII Names: Cotes, Newton

0604 Hall, A. Rupert. "On the Historical Singularity of the Scientific Revolution of the Seventeenth Century." In *The Diversity of History: Essays in Honour of Sir Herbert Butterfield*, edited by J. H. Elliott and H. G. Koenigsberger, 199-221. London: Routledge and Kegan Paul; Ithaca, N.Y.: Cornell University Press, 1970.

Here, Hall states the historiographic position of rationalist internalism.

Topic: *I*

0605 Hall, A. Rupert. *Philosophers at War: The Quarrel Between Newton and Leibniz*. Cambridge: Cambridge University Press, 1980.

This is a well-written book which, however, should be entitled "Mathematicians at War." Hall focuses almost entirely on the calculus priority dispute—and considers as rather irrelevant the other differences between Leibniz and Newton. The Leibniz-Clarke exchange receives five pages of coverage and the scientific and wider philosophic differences about thirty pages.

Topics: *XII*, VII Names: Clarke, Newton

0606 Hall, A. Rupert. "Science, Technology and Utopia in the Seventeenth Century." In *Science and Society 1600-1900*, edited by Peter Mathias, 33-53. Cambridge: Cambridge University Press, 1972. See (0998).

For Hall, the utopian idealism of persons such as Hartlib had "little or nothing to do with the development of science in mid-seventeenth-century England." Likewise, for him, science and technology were basically irrelevant to attaining an ideal society (which had to do with religion and virtue). Hall's anachronistic definitions of key words predetermine his conclusions. But this does represent a good example of "internalist" historiography.

Topics: *IV*, I, VI Names: Boyle, Hartlib

0607 Hall, A. Rupert, and Marie Boas Hall. "Clarke and Newton." *Isis* 52(1961): 583-585.

The authors argue that Newton was actively involved in the Leibniz-Clarke correspondence.

Topics: *XII*, VII Names: Clarke, Newton

0608 Hall, A. Rupert, and Marie Boas Hall. "Newton and the Theory of Matter." *Texas Q.* 3(1967): 54-68. Reprinted in *The Annus Mirabilis of Sir Isaac Newton*, edited by Robert Palter, 54-68. Cambridge, Mass.: MIT Press, 1970. See (1201).

The Halls argue that Newton's theory of matter, his exact understanding of mechanical principles, and his view of aetherial hypotheses are complex and even unsettled. They note that an important element in all of his attempts to understand and make statements on these topics is the role of God in the universe.

Topics: *XII*, VII Name: Newton

0609 Hall, A. Rupert, and Marie Boas Hall. "Newton's Theory of Matter." *Isis* 51(1960): 131-144.

Here the Halls discuss the difficulties and dilemmas of analyzing Newton's remarks on matter, particles, forces, and the aether. And they relate Newton's theory of matter to his understanding of God. They conclude that there were irreconcilable elements in Newton's philosophy and "forced to choose, Newton preferred God to Leibniz."

Topics: *XII*, VII Names: Clarke, Newton

0610 Hall, A. Rupert, and Marie Boas Hall. *Unpublished Scientific Papers of Isaac Newton: a Selection from the Portsmouth Collection in the University Library, Cambridge*. Cambridge: Cambridge University Press, 1962.

This publication includes relevant introductory commentaries—with some attention to Newton's understanding of the role of God in the mechanism of the universe.

Topics: *XII*, VII Name: Newton

0611 Hall, Basil. "Puritanism: the Problem of Definition." In *Studies in Church History*, edited by G. J. Cuming, Vol. 2, 283-296. London: Nelson, 1965.

Although not explicitly related to science, this discussion provides useful background for the Protestantism-and-the-rise-of-science debates.

Topic: *III*

0612 Hall, Marie Boas. "The Establishment of the Mechanical Philosophy." *Osiris* 10(1952): 412-541.

This is an important early work on Boyle, atomism, and the acceptance of both empiricism and the mechanical philosophy in England. However, Boas Hall does not explicate the significance of religion for these topics; indeed, she barely even mentions religion.

Topics: *VII*, XII Names: Boyle, More, Newton

0613 Hall, Marie Boas. Introduction to *Robert Boyle on Natural Philosophy: An Essay with Selections from His Writings*, edited by Marie Boas Hall, 1-115. Bloomington, Ind.: Indiana University Press, 1965.

In her introduction, Hall summarizes: Boyle's life; his position regarding the "new learning"; and his writings on the mechanical philosophy, chemistry, and pneumatics. She also touches on his theology and indicates that Boyle thought of mechanical philosophy "less as an hypothesis than as a theological doctrine." Rather oddly, she entitles the selections from his theology "Natural Religion."

Topics: *VII*, VIII Names: Boyle, Hartlib, Hooke, Ranelagh

0614 Hall, Marie Boas. "Matter in Seventeenth-Century Science." In *The Concept of Matter*, edited by Ernan McMullin, 344-367. Notre Dame, Ind.: University of Notre Dame Press, 1963. Reprinted in *The Concept of Matter in Modern Philosophy*, edited by Ernan McMullin, 76-99. Notre Dame, Ind.: University of Notre Dame Press, 1978. See (1039).

Boas Hall describes the fall of the Aristotelian view of matter, the rise and fall of ancient atomism, and the eventual dominance of the experimental corpuscularianism of Boyle and Newton. Theological assumptions and theological implications are noted but not emphasized.

Topics: *VII*, XII Names: Boyle, Newton

0615 Hall, Marie Boas. *The Mechanical Philosophy*. New York: Arno Press, 1981.

This is a reprint of her 1952 *Osiris* article (0612).

Topic: *VII*

0616 Hall, Marie Boas. "Newton's Voyage in the Strange Seas of Alchemy." In *Reason, Experiment, and Mysticism in the Scientific Revolution*, edited by M. L. Rhigini Bonelli and William R. Shea, 239-246. New York: Science History, 1975. See (1329).

In a commentary on Westfall's article in the same book, M. B. Hall plays down the significance of Newton's alchemy. He was a voyager and not really an alchemist. Boas Hall assumes the separation of science and religion.

Topics: *XII*, II Name: Newton

0617 Hall, Michael G. "The Introduction of Modern Science into Seventeenth-Century New England: Increase Mather." In *Proceedings of the Tenth International Congress of the History of Science*, Vol 1, 261-264. Paris: Hermann, 1964.

Michael Hall provides a chronological sketch of Mather's interest and reading in contemporary natural philosophy. The author notes that Mather's reading was not restricted by religious, political, or national prejudices. And he describes Mather as preeminent among New Englanders interested in natural philosophy during the last two decades of the seventeenth century.

Topic: *VII* Name: Mather, I.

0618 Hall, Michael G. *The Last American Puritan: The Life of Increase Mather, 1639-1723*. Middleton, Conn.: Wesleyan University Press, 1988.

This is the definitive biography of this Puritan who had universal interests. The chapter entitled "New Worlds of Science and Hope," describes Mather's activities as an observer and theologian of comets, his association with John Foster (printer and astronomer), and his attempt to found the Boston Philosophical Society. Hall believes that Mather's "dual vision" (his combination of astronomy and providentialist theology) results in contradictions.

Topic: *VII* Names: Foster, J., Mather, I.

0619 Hall, Thomas S. *Ideas of Life and Matter: Studies in the History of General Physiology, 600 B.C.-1900 A.D.* (Volume One: From Pre-Socratic Times to the Enlightenment.) Chicago: University of Chicago Press, 1969.

This book includes ten chapters (about 150 pages) relevant to our period. The subject matter, of course, is effectively at the crossroads of biology and religion. Hall describes the views of J. B. van Helmont, Bacon, Harvey, Descartes, Boyle, Hooke, Willis, and Mayow. He postulates natural science on the one hand and—grouped on the other—religion, magic, superstition, alchemy, occult interests, and nonsense. He says that it is difficult to abstract the sense (science) from the nonsense when studying the writings of these individuals. Despite such historiographic principles, the book is useful for introductory purposes. He at least admits that religion, spirit, and alchemy are important to those he describes.

Topics: *VII*, II, XI Names: Bacon, Boyle, Harvey, Hooke, Mayow, Willis

0620 Haller, William. *The Rise of Puritanism*. New York: Columbia University Press, 1938.

Haller does not discuss science but his book is good for background to the Puritanism-and-the-rise-of-science debates. It is especially important for anyone trying to define Puritanism.

Topics: *III*, V

0621 Hamilton, David. *The Healers: A History of Medicine in Scotland*. Edinburgh: Canongate, 1981.

Hamilton summarizes various types of healing and healers in a chapter on the seventeenth century. Included are not only physicians and apothecaries—but also "quacks," folk healers, healing wells and stones, and witchcraft.

Topics: *XI*, II

0622 Hamilton, Gertrude R. "Thomas Vaughan and the Divine Art of Alchemy." *Cauda Pavonis* 4(1985): 1-3.

This is an excellent summary of Thomas Vaughan's Christian Hermetic philosophy of nature.

Topics: *II*, X Name: Vaughan, T.

0623 Hanen, Marsha P., Margaret J. Osler, and Robert G. Weyant, eds. *Science, Pseudo-Science, and Society*. Waterloo, Ont.: Wilfrid Laurier University Press, 1980.
This anthology includes articles by J. R. Jacob (0794) and Westfall (1633) which are annotated elsewhere in this bibliography.
Topic: *II*

0624 Hankins, Thomas L. *Science and the Enlightenment*. Cambridge: Cambridge University Press, 1985.
This is a solid survey of the categories of natural philosophy during the Enlightenment. Where relevant, Hankins does recognize English roots and religious motives.
Topic: *VII*

0625 Hans, Nicolas. *New Trends in Education in the Eighteenth Century*. London: Routledge and Kegan Paul, 1951.
This is a survey which addresses some issues in our field. Hans corrects Irene Parker (1205) regarding dissenting academies; and he relates Puritans, along with Anglicans, to a supporting role in the emergence of modern scientific education. He argues that extreme Calvinist groups *opposed* the advancement of science. An appendix contains a list of schoolmasters.
Topics: *V*, III Name: Desaguliers

0626 Hargreaves-Mawdsley, W. N. *Oxford in the Age of John Locke*. Norman, Okla.: University of Oklahoma Press, 1973.
This is a semipopular book directed to the general reader. It discusses both the city and the university. The author gives some emphasis to science but deals with religion only in passing. The book is characterized by a rather old-fashioned historiography—part cheerleading and part retrospective evaluation. Functionally speaking, Anthony Wood is as central to the book as is John Locke.
Topic: *V* Names: Locke, Wood

0627 Harley, David. "Mental Illness, Magical Medicine, and the Devil in Northern England, 1650-1700." In *The Medical Revolution of the Seventeenth Century*, edited by Roger French and Andrew Wear, 114-144. Cambridge: Cambridge University Press, 1989. See (0510).

Harley uses an anthropological approach to the historiographical issue of "the decline of magic." Using a variety of particular cases and pamphlet debates, he shows differences: (1) between dissenters and Anglicans; and (2) between both of these groups (as educated groups) and the ordinary people. Each group had its theological and medical understanding of particular phenomena. Harley uses, as an example, the case of the "Surrey Demoniak."

Topics: *XI*, II, IV Names: Jolly, Taylor, Z.

0628 Harman, Peter M. *Metaphysics and Natural Philosophy: The Problem of Substance in Classical Physics*. Brighton, England: Harvester Press, 1982.

This is a philosophical analysis which emphasizes both the diversity of theories and the importance of metaphysics in eighteenth- and nineteenth-century physics. It includes a chapter on Newton in which the importance of God in Newtonian natural philosophy is noted.

Topics: *VII*, XII Name: Newton

0629 Harre, Romano. "Knowledge." In *The Ferment of Knowledge*, edited by George S. Rousseau and Roy Porter, 11-54. Cambridge: Cambridge University Press, 1980. See (1368).

This essay includes a section on "British empiricism" which stresses the importance both of "knowledge of the world" and "knowledge of God" in eighteenth-century epistemological discussions. Harre discusses this with respect to the changing historiographic tradition of recent years.

Topics: *I*, VII

0630 Harris, R. W. *Reason and Nature in the Eighteenth Century*. London: Blandford, 1968.

The new sciences, according to Harris, contributed to the disintegration of the humanist tradition (which had assumed that the universe was a single, rational, hierarchical creation). Harris is more interested in the implications of this disintegration for social theory, political theory, and literature than for religion. He sees Locke playing a major role, not only in a shift in political theory, but also in shifting relations between faith, philosophy, and science.

Topics: *IV*, VII, IX, X Names: Addison, Defoe, Locke, Pope, Swift

0631 Harris, Victor. *All Coherence Gone*. Chicago: University of Chicago Press, 1949.

This book deals with the effects of science on seventeenth-century culture generally and with respect to literature especially. It includes a discussion of the decay-of-nature controversy.

Topics: *X*, IV Names: Browne, T., Goodman, Hakewill

0632 Harrison, Charles. "Ancient Atomists and English Literature of the Seventeenth Century." *Harvard Studies in Classical Philology* 4(1934): 1-79.

See the next entry (0633).

Topics: *VII*, IX, X Names: Blackmore, Browne, T., Digby, Hobbes, Milton, More, Stanley, Wilkins

0633 Harrison, Charles. "Bacon, Hobbes, Boyle and the Ancient Atomists." *Harvard Studies and Notes in Philology and Literature* 15(1933): 191-213.

In his two valuable articles, Harrison discusses theological aspects of atomism within the contexts indicated by the titles. He describes both theistic and atheistic positions with respect to theories of matter.

Topics: *VII*, IX Names: Bacon, Boyle, Hobbes

0634 Harrison, John. *The Library of Isaac Newton*. Cambridge: Cambridge University Press, 1978.

Harrison shows Newton's wide-ranging interests—including the large number of titles in theology, biblical studies, and alchemy. In his introduction, Harrison notes that the most heavily annotated theological volume in Newton's collection is his English Bible.

Topics: *XII*, II, X Name: Newton

0635 Harrison, John, and Peter Laslett. *The Library of John Locke*. London: Oxford Bibliographical Society, 1965. 2nd ed., Oxford: Clarendon Press, 1971.

This work contains a list of books in Locke's library (prepared by Harrison) and an introductory essay by Laslett. Both the list and the essay show Locke's wide-ranging interests—including theology and natural philosophy.

Topics: *VII*, X Name: Locke

0636 Harrison, John L. "Bacon's View of Rhetoric, Poetry and the Imagination." *Hunt. Lib. Q.* 20(1957): 107-125. Reprinted in *Essential Articles for the Study of Francis Bacon*, edited by Brian Vickers, 107-125. Hamden, Conn.: Archon Books, 1968. See (1588).

Harrison discusses Bacon's views on the relations between poetry and religion, as well as between poetry and science. On the one hand, religion uses poetry. On the other hand, poetry and science have their sources in imagination and reason—two branches of Bacon's threefold division of knowledge. The imagination, expressed as poetry, is free to form complexes of images without reference to reason—and, therefore, without reference to external nature—for the benefit of human morality and dignity.

Topics: *X*, VII Name: Bacon

0637 Hartenstein, Gustav. *Locke's Lehre von Menschlichen Erkenntniss, in Vergleichung mit Leibniz's Kritik Derselben*. Leipzig: S. Hirzel, 1861.

This is a philosophical (rather than historical) analysis of Locke's metaphysical and epistemological views; it includes analyses of his views on innate ideas, substance, and space.

Topic: *VII* Name: Locke

0638 Harth, Phillip. *Contexts of Dryden's Thought*. Chicago: University of Chicago Press, 1968.

This is an argumentative and detailed exegesis of Dryden's religious poetry. Harth takes issue with Bredvold and others who assert Dryden's "Pyrrhonism" and fideism. His analysis of deism and freethinking places these movements in a political and broadly philosophical framework. Harth argues, in effect, that these movements had little or no connection to natural science. Harth's analyses of different kinds of Christian apologetics show that both deists and biblical Christians used natural theology in a free-standing way. (This also undermines the sometimes supposed connection between science and deism.) Harth sees Dryden as interested in the new science and as a sincere member of the Royal Society. But he somewhat downplays the religious significance of science for this poet because, he argues, Dryden consistently separated the province of reason from the province of revelation.

Harth is theologically sophisticated and explicates quite well the different theologies of the seventeenth century.

Topics: *X*, IV, VII, IX Names: Blount, Boyle, Charleton, Culverwel, Dryden, Herbert, E., Tillotson

0639 Harth, Phillip. "Empson's Interpretation of *Religio Laici.*" *Essays in Criticism* 20(1970): 446-450.

Here Harth defends his own interpretation (0638) while he argues that Empson (0432) misinterprets Dryden.

Topics: *X*, IX Name: Dryden

0640 Harth, Phillip. *Swift and Anglican Rationalism: The Religious Background of A Tale of a Tub*. Chicago: University of Chicago Press, 1961.

This book concerns Anglican theology of Swift's period and its relation to philosophy, natural religion, atheism, and deism. There is good discussion of the philosophical and social background—and especially of the role of apologetics. Harth sees the increasing emphasis on reason during the period as a development coming out of the polemical defense of Anglicanism against Puritanism, Catholicism, and atheism. He does not see science as a major factor in the increasing emphasis on reason. Harth does an excellent job of showing that, in its context, Anglican rationalism was quite different from deism.

Topics: *X*, IV, VII, VIII, IX Names: Burton, Cudworth, Glanvill, More, Parker, Stillingfleet, Swift, Tillotson

0641 Hartill, Isaac, and others. "The Faith of Newton." *Journal of the Transactions of the Victoria Institute or Philosophical Society of Great Britain* 78(1946): 75-90.

Hartill's outdated and amateurish lecture/paper (and a following discussion) reflect the struggles of the time over how to understand Newton the theologian. He views Newton as "a great Christian" and probably a Trinitarian who disliked Athanasius.

Topic: *XII* Name: Newton

0642 Hartley, Harold, ed. *The Royal Society: Its Origins and Founders*. London: Royal Society, 1960.

This is the tercentenary number of *Notes and Records of the Royal Society of London* reprinted in book form. It includes twenty-two short articles on various men—Ashmole, Boyle, Digby, Evelyn, Hooke, Oldenburg, Wallis, Wilkins, Wren, Charles II, and other less famous ones. It also includes a longer article by Douglas McKie on the origins of the Society. The essays are useful.

Topic: *V* Names: Ashmole, Boyle, Charles II, Digby, Evelyn, Hooke, Oldenburg, Wallis, Wilkins, Wren

0643 Harwood, John T. "Rhetoric and Graphics in *Micrographia*." In *Robert Hooke: New Studies*, edited by Michael Hunter and Simon Schaffer, 119-147. Woodbridge, England: Boydell Press, 1989. See (0765).

Harwood argues that Hooke's *Micrographia* was concerned with rhetorical or apologetic aims as well as purely scientific ones. He notes that Hooke's graphics (or pictures) were extremely effective in accomplishing these aims. One such aim was communication within the scientific community; another was justifying the new science and promoting a positive image of the Royal Society to the larger Restoration society. Harwood draws parallels between Hooke's book and Sprat's *History* in terms of their rhetorical aims. He recognizes and briefly discusses the religious dimension and the natural theological content of *Micrographia* in this context.

Topics: *X*, IV, V, VIII Names: Hooke, Sprat

0644 Harwood, John T. "Science Writing and Writing Science: Boyle and Rhetorical Theory." In *Robert Boyle Reconsidered*, edited by Michael Hunter, 37-56. Cambridge: Cambridge University Press, 1994. See (0759).

Harwood argues (among other things) that Boyle's literary style and rhetorical purposes show his natural philosophy and theology to be of a consistent whole and not separate from each other. He shows that Boyle used both microscopes and metaphors as instruments. Moreover, both Boyle's scientific and his theological writings are characterized by movement between literal and figurative styles. Finally, the metaphor of the "Two Books" (Scripture and nature—with nature itself understood as "God's epistle") is central to all of Boyle's writings.

Topics: *X*, I, VII Name: Boyle

0645 Hattaway, Michael. "Bacon and 'Knowledge Broken': Limits for Scientific Method." *J. Hist. Ideas* 39(1978): 183-197.

Hattaway analyzes "the interpenetration of the physical and metaphysical in Bacon's method." This he does in order to prove that Bacon was a Christian Aristotelian affected by a Renaissance amalgam of various philosophical traditions.
Topic: *VII* Name: Bacon

0646 Haydn, Hiram. *The Counter-Renaissance.* New York: Scribner, 1950.
This massive survey of 700 pages is written by a literary historian. It focuses mainly on the sixteenth century but it also covers the early seventeenth century and the whole book is good for seventeenth-century background. Haydn is trying to deal with the complexities and inconsistencies which face the historian of this period. (He is opposed to those historians who postulate a single worldview or temperament for the period.) Haydn traces three movements: the classicist, the romanticist, and the naturalist. The first of these represents the Renaissance and the latter two represent the "Counter-Renaissance." All of these movements can be traced in literature, religion, philosophy, science, politics, and other aspects of cultural history. Among other topics, Haydn treats: the scholastic traditions and "Christian Humanism"; scepticism and fideism; the new science; occult traditions and Neoplatonism; and the ancients vs. moderns controversy. He notes parallels between religion and the new science as well as their interaction. This is a good book and well-written.
Topics: *X*, II, IV, VII Names: Bacon, Donne, Shakespeare

0647 Hayes, Thomas W. "Alchemical Imagery in John Donne's 'A Nocturnal Upon S. Lucies Day'." *Ambix* 24(1977): 55-62.
Donne's poem, in Hayes' analysis, interweaves alchemical images: first of the world's decay and personal death; and then of renewal, resurrection, and union in love.
Topics: *X*, II Name: Donne

0648 Hayward, J. C. "New Directions in Studies of the Falkland Circle." *Seventeenth Century* 2(1987): 19-48.
Hayward argues that members of the Falkland Circle were influenced by Platonism and that, in turn, they influenced the Cambridge Platonists. They were united in their opposition to Hobbes and Descartes. This mainly concerns their moral philosophy—but

with some attention to natural philosophy. The article includes an appendix which describes the books found in the libraries of Clarendon and Morley.
Topic: *VII* Names: Clarendon, Falkland, Hobbes, Morley, Sheldon

0649 Hazard, Paul. *The European Mind (1680-1715)*. Translated by J. Lewis May. Cleveland, Ohio: World, 1963. Originally published as *La Crise de la Conscience Europeenne*. Paris: Boivin, 1935.
This now-classic volume has good insights concerning the rationalistic tendencies of the period—especially with respect to social ideals and religion. Hazard basically downplays the significance of the new science. The book mainly focuses on French culture and French writers, but the chapters dealing with the following topics are relevant to our field: deism and natural religion; the rationalists; "science and progress"; and Locke.
Topics: *VII*, IV, VIII, IX, X, XII Names: Hobbes, Locke, Newton, Swift, Toland

0650 Hazard, Paul. *John Locke (1632-1704) und sein Zeitalter*. Hamburg, Germany: Hoffman and Campe, 1947.
This is a short book on Locke's philosophy and its cultural context. It touches on materialism, deism, and the nature of the soul.
Topics: *VII*, IX Name: Locke

0651 Hefelbower, Samuel G. *The Relation of John Locke to English Deism*. Chicago: University of Chicago Press, 1918.
Hefelbower argues a set of clear-cut theses, the most important of which are that Locke was not a deist and that no causal connection exists between Locke and deism. Both Locke and deism are seen as parts of a broader rationalistic and critical movement. Although old, this book is still useful.
Topics: *IX*, VII Names: Collins, A., Locke, Tindal, Toland

0652 Heimann, Peter M. "'Nature is a Perpetual Worker': Newton's Aether and Eighteenth-Century Natural Philosophy." *Ambix* 20(1973): 1-25.
Heimann shows that Newton's changing views of the relations between active principles and aether was conflated by natural philosophers in the eighteenth century. They thereby "rejected

Newton's theory that the activity of nature was maintained by divine agency."

Topics: *VII*, XII Name: Newton

0653 Heimann, Peter M. "Newtonian Natural Philosophy and the Scientific Revolution." *Hist. Sci.* 11(1973): 1-7.

This is an historiographic discussion of "Newtonianism" and an essay review of several books related to eighteenth-century history of science. The focus is on the post-1720 period but Heimann's insights concerning the *complexity* of the terms ("Newtonian" and "Newtonianism") are useful for our period also.

Topics: *XII*, I, VII Name: Newton

0654 Heimann, Peter M. "Science and the English Enlightenment." *Hist. Sci.* 16(1978): 143-151.

This is a good essay review of M. C. Jacob's *The Newtonians and the English Revolution* (0817) and of Redwood's *Reason, Ridicule and Religion* (1313). Heimann briefly discusses Boyle, Thomas Burnet, and Newton as cases by which to discuss recent historiography in general and the books under review in particular.

Topics: *IV*, I, VII, X, XII Names: Boyle, Burnet, T., Newton

0655 Heimann, Peter M., and J. E. McGuire. "Newtonian Forces and Lockean Powers: Concepts of Matter in Eighteenth-Century Thought." *Hist. Stud. Phys. Sci.* 3(1971): 233-306.

Heimann and McGuire summarize and discuss the thought of selected eighteenth-century writers. They focus on the epistemological and ontological problems of matter theory. The topics covered include: forces, including the forces of attraction and repulsion; intrinsic and extrinsic powers of matter; causality; active principles; the paucity of solid matter in the universe; atomism; the void; the plenum; the aether; God and ongoing divine activity in the material universe. Theological issues are not emphasized but they are recognized. About thirty pages cover the period into the 1720s. Besides Newton and Locke, the authors discuss Robert Greene.

Topics: *VII*, XII Names: Greene, Locke, Newton

0656 Heinemann, F. H. "John Toland and the Age of Enlightenment." *Review of English Studies* 20(1944): 125-146.

Within Heinemann's grand scheme of western history, Toland is "the inaugurator of the second stage" of modern man's coming of age. Some attention is given to Toland's religious views; only marginal attention is given to his philosophy of nature.

Topics: *IX*, VII Name: Toland

0657 Heinemann, F. H. "Toland and Leibniz." *Philos. Rev.* 54(1945): 437-457.

Heinemann describes both the personal meetings and the philosophical interaction between Toland and Leibniz. He sees Toland's "dynamic materialism" as influenced by Leibniz yet leading to his pantheism. According to Heinemann, Leibniz strongly disagreed with Toland's physics and theology—yet Leibniz should be seen as Toland's mentor.

Topics: *VII*, IX Name: Toland

0658 Heninger, S. K., Jr. *Touches of Sweet Harmony: Pythagorean Cosmology and Renaissance Poetics*. San Marino, Calif.: Huntington Library, 1974.

Heninger presents a literary critical approach to renaissance literature. He argues that understanding Pythagorean cosmology is prerequisite to analyzing the work of sixteenth and seventeenth-century writers. He reconstructs this cosmology using convenient headings (number, cosmos, deity and time, occult sciences, moral philosophy) and notes their related aesthetic assumptions. He assumes a clear separation between Pythagorean universal harmony and the celestial mechanics of Newtonian natural science.

Topics: *X*, II Name: Browne, T., Fludd, Ralegh, Stanley

0659 Henry, John C. "Boyle and Cosmical Qualities." In *Robert Boyle Reconsidered*, edited by Michael Hunter, 119-138. Cambridge: Cambridge University Press, 1994. See (0759).

This is a study of Boyle's tracts about cosmical qualities and other related publications. Henry proposes that Boyle hinted at a new interpretation of the traditional term "quality:" i.e., "those unheeded relations and impressions" which "bodies owe to the determinate fabrick of the grand system" of which they are a part. Henry finds both the tradition of light metaphysics and a voluntarist theology to have played important roles in this conception of the world system. Thus, nature could not be confined to mathemat-

ical terms; and the hidden qualities could only be understood empirically through their effects. Yet Boyle was obscure on this issue—probably because he wished not to disrupt the arguments linking the new mechanical philosophy with Christian religion.
Topic: *VII*, X Names: Boyle, Worsley

0660 Henry, John C. "A Cambridge Platonist's Materialism: Henry More and the Concept of the Soul." *J. Warb. Court. Instit.* 49(1986): 172-195.
Henry discusses More's philosophical efforts to prove the immortality of the soul. He argues: (1) that More's eclectic blend of Platonism, Cartesianism, and other elements results in logical inconsistencies and philosophical confusion; and (2) that More's pneumatology is implicitly materialistic and only nominally dualistic. Henry compares More's views on spirit to those of Richard Baxter and other contemporaries. He relates More's philosophical confusion to a variety of nonphilosophical factors motivating More.
Topics: *VII*, VIII Names: Baxter, More

0661 Henry, John C. "Henry More and Newton's Gravity." *Hist. Sci.* 31(1993): 83-97.
Ostensibly, this is an essay review of A. Rupert Hall's *Henry More: Magic, Religion and Experiment*; actually, this is a critique of all historians who postulate that Newton was influenced by More. Henry argues that More and Newton had significantly different views on space, matter, and God. Theologically, Henry says, More was an intellectualist/necessitarian/rationalist while Newton was a voluntarist. Hall is especially weak, Henry notes, when it comes to the role of theology in the natural philosophies of More and Newton.
Topics: *VII*, XII Names: Bentley, More, Newton

0662 Henry, John C. "Henry More versus Robert Boyle: The Spirit of Nature and the Nature of Providence." In *Henry More (1614-1687) Tercentenary Studies*, edited by Sarah Hutton, 55-76. Dordrecht: Kluwer, 1990. See (0782).
Henry argues the following: In what appears to be a minor defense of his hydrostatic experiments, Boyle *actually* opposed More because of basic differences about God, providence, and the moral

order. As a "devout voluntarist," Boyle disagreed with More's intellectualist theology. Specifically, God could not be restricted by More's logically derived conclusions about the Spirit of Nature and utterly dead matter.
Topic: *VII* Names: Boyle, More

0663 Henry, John C. "The Matter of Souls: Medical Theory and Theology in Seventeenth-Century England." In *The Medical Revolution of the Seventeenth Century*, edited by Roger French and Andrew Wear, 87-113. Cambridge: Cambridge University Press, 1989. See (0510).
Henry begins this essay with an analysis of why most anti-atheistic writers ignored the obvious vitalistic and monistic theories of medical writers. He then considers two exceptions—the Cambridge Platonists, Cudworth and More. Whereas others could set aside problems regarding the immortality of the soul, they felt the need to demonstrate both the immortality and the essential immateriality of the soul. Because of their rationalist theology, they had to argue against medical theories that blurred the distinction between body and spirit. Thus, they saw medical theory "as heralding 'the rising sun of atheism'."
Topics: *XI*, VII, IX Names: Cudworth, More

0664 Henry, John C. "Newton, Matter, and Magic." In *Let Newton Be!*, edited by John Fauvel and others, 127-145. Oxford: Oxford University Press, 1988. See (0449).
This is a good introductory discussion of Newton's theories of matter and of forces in the context of the Aristotelian, mechanical, and natural magic traditions. The importance of natural magic is emphasized and explicated. Henry makes passing references to the role of God in Newton's theory.
Topics: *XII*, II Name: Newton

0665 Henry, John C. "Occult Qualities and the Experimental Philosophy: Active Principles in Pre-Newtonian Matter Theory." *Hist. Sci.* 24(1986): 335-381. Reprinted in *Seventeenth-Century Natural Scientists*, edited by Vere Chappell, 1-47. New York: Garland, 1992. See (0205).
Henry argues both historical and historiographical theses. Historically, he seeks to show that: (1) English mechanical philosophers

tried to overcome or circumvent problems of the mechanical philosophy by introducing unexplained active principles and occult qualities into matter theory; (2) the use of such active principles in matter theory was a clear and undeniable tradition before Newton; (3) active principles were used to avoid theological difficulties connected with theories of passive matter; and (4) an emphasis on the possible existence of occult active principles combined with theological voluntarism was an important aspect of the promotion of the experimental method. Historiographically, Henry argues that: (1) simplistic views of seventeenth-century English matter theory, in which matter is held to be passive and inert, need to be modified; (2) views which credit Newton with radical innovation as regards active principles need to be rejected; and (3) most Newton scholars should be criticized as too hagiographic in their approach.

Topics: *VII*, I, IV, VIII, XII Names: Boyle, Charleton, Glisson, Hale, Mayow, Newton, Petty, Warner (1570)

0666 Henry, John C. "'Pray Do Not Ascribe that Notion to Me': God and Newton's Gravity." In *The Books of Nature and Scripture*, edited by James E. Force and Richard H. Popkin, 123-147. Dordrecht: Kluwer Academic, 1994. See (0498).

Henry argues (contra Koyre, Hall, Cohen, Westfall, and Dobbs) that Newton believed gravity to be inherent in matter and believed in action-at-a-distance. What Newton rejected is that gravity is an *essential* property of matter; rather, according to Henry, Newton saw gravity as a property superadded to matter by God. Henry argues his point on the basis of relevant texts and describes in general terms the historiographic implications of his thesis.

Topics: *XII*, II, VII Names: Bentley, Newton

0667 Henry, John C. "Robert Hooke: The Incongruous Mechanist." In *Robert Hooke: New Studies*, edited by Michael Hunter and Simon Schaffer, 149-180. Woodbridge, England: Boydell, 1989. See (0765).

Henry shows that Hooke, like many other seventeenth-century English natural philosophers, incorporated various aspects of natural magic into his mechanical philosophy. This Hooke did both in theory (e.g., his use of an occult "spiritus") and in empirical

practice. Henry concludes that magic and science lie along a continuous spectrum and an interchange has always been possible.
Topics: *II*, VII Name: Hooke

0668 Henry, John C. "Thomas Harriot and Atomism: A Reappraisal." *Hist. Sci.* 20(1982): 267-296.
Henry finds Harriot to be unorthodox but a believer—and not a follower of Bruno. Harriot turned to atomism as a plausible explanation for problems (such as condensation and rarefaction) in physics.
Topics: *VII*, IX Name: Hariot

0669 Henry, Nathaniel H. "Milton and Hobbes: Mortalism and the Intermediate State." *Studies in Philology* 48(1951): 234-249.
The author argues against Williamson (1685) and others regarding Milton's relationship to philosophy generally and to Hobbes specifically. He sees Milton's theological views as fundamentally scriptural (based on revelation). For Henry, natural philosophy basically is irrelevant to Milton's views.
Topics: *X*, IX Names: Hobbes, Milton

0670 Herries Davies, Gordon L. *The Earth in Decay: A History of British Geomorphology, 1578-1878*. London: McDonald, 1969.
See (0319) by Gordon L. Davies.
Topic: *VII*

0671 Herrman, Rold-Dieter. "Newton's Positivism and the A Priori Constitution of the World." *Int. Philos. Q.* 15(1975): 205-214.
Herrman argues that Newton's scientific "positivism" is *not* independent of metaphysics or of his assumptions about the God of the Bible. Rather, Newton's scientific and historical inquiries are closely related to his understanding of God as creator. This can be seen especially when one analyzes Newton's views on the creation (the a priori constitution) of the world.
Topics: *XII*, VII Name: Newton

0672 Hesse, Mary B. "Francis Bacon." In *A Critical History of Western Philosophy*, edited by Daniel J. O'Connor, 141-152. New York: Free Press, 1964. See (1159).

This is a summary, explication, analysis, and evaluation of Bacon's philosophy of science and of his theory of matter. It includes a few references to God, theology, and metaphysics. Hesse presents theology as influencing Bacon's larger philosophy but as being separate from and outside the scope of his natural philosophy.
Topic: *VII* Name: Bacon

0673 Hesse, Mary B. "Francis Bacon's Philosophy of Science." In *Essential Articles for the Study of Francis Bacon*, edited by Brian Vickers, 114-139. Hamden, Conn.: Archon Books, 1968. See (1588).
This is an edited reprint of (0672). (Two biographical paragraphs are omitted and a different form is used for the references and endnotes.)
Topic: *VII* Name: Bacon

0674 Hesse, Mary B. "Hermeticism and Historiography: An Apology for the Internal History of Science." In *Historical and Philosophical Perspectives of Science*, edited by Roger H. Struewer, 134-160. Minneapolis: University of Minnesota Press, 1970. See (1514).
This is an example of extreme internalism in the history of science. Hesse argues against the study of Hermeticism in the history of science generally and in Newtonian studies in particular. Although Hesse does not address religion in this and other historiographic writings, the effect of her position is a philosophically-based position that science and religion have nothing to do with each other—even in seventeenth-century England.
Topics: *I*, II, XII Name: Newton

0675 Hesse, Mary B. "Reasons and Evaluation in the History of Science." In *Changing Perspectives in the History of Science*, edited by Mikulas Teich and Robert Young, 127-147. See (1535). London: Heinemann, 1973. Reprinted in *Revolutions and Reconstructions in the Philosophy of Science*, by Mary B. Hesse, 1-28. Bloomington, Ind.: Indiana University Press, 1980.
Hesse sees "rationality" as a legitimate criterion of internalist history of science. She does recognize the difficulties encountered

when the historian tries to define this criterion. A response by Piyo Rattansi (1298) follows Hesse's essay.
Topics: *I*, II, XII Name: Newton

0676 Hessen, Boris. "The Social and Economic Roots of Newton's 'Principia'." In *Science at the Cross Roads*, edited by N. I. Bukharin and others, 149-212. London: Kniga, 1931. See (0152).
This famous Marxist interpretation relates the new philosophy to the economic, military, and technical needs of the rising bourgeoisie. Hessen's essay has been influential as a historiographic symbol but its actual content today is not taken too seriously. The author's name sometimes is transliterated from Russian as Gessen.
Topics: *IV*, I, III, XII Name: Newton

0677 Hessen, Boris. *The Social and Economic Roots of Newton's "Principia."* Sydney, Australia: Current Book Distributors, 1946. New York: H. Fertig, 1971.
This is a reprint of (0676).
Topics: *IV*, XII Name: Newton

0678 Heyd, Michael. "The Emergence of Modern Science as an Autonomous World of Knowledge in the Protestant Tradition of the Seventeenth Century." *Knowledge and Society* 7(1988): 165-179.
Heyd argues for a dialectic between "negative" and "positive" autonomy in the emergence of modern science in the seventeenth century. By "negative" autonomy, he means independence from control by theologians (particularly by scholastics). By "positive" autonomy, he means that science gained legitimacy by "contributing to knowledge of ultimate concerns"—as, for example, in the "two books" metaphor, the new physico-theology, and the voluntarist doctrine of God. Overall, Heyd's approach is to examine "the relationship between soteriological concerns and the organization of knowledge" in England and Geneva.
Topic: *VII*

0679 Heyd, Michael. "Protestant Attitudes towards Science in the Seventeenth and Early Eighteenth Centuries." In *Les Eglises aux Sciences du Moyen Age au XXe Siecle. Actes du Colloque de la Commission Internationale d'Histoire Ecclesiastique Comparee tenu a Geneve en aout 1989*, edited by Oliver Fatio, 71-89. (*Histoire des Idees et Critique Litteraire*, 300.) Geneva: Libr. Droz, 1991.

Heyd ranges across Western Europe and England from the late-sixteenth through the early-eighteenth centuries as he examines Protestant attitudes towards science. He finds that a positive approach to experimental method was not an exclusively Puritan or Protestant attitude. Heyd also focuses on the nonsoteriological natural theology (aimed against deists and enthusiasts) of the late-seventeenth and early-eighteenth centuries. Historiographically, Heyd appeals to church historians to join in the examination of these issues.

Topics: *VIII*, I, III, VII Names: Derham, Wilkins

0680 Heyd, Michael. "The Reaction against Enthusiasm in the Seventeenth Century: Towards an Integrative Approach." *J. Mod. Hist.* 53(1981): 258-280.

This is a programmatic essay which both surveys the secondary literature and posits its own claims. Heyd's main thesis is that medicine and science were enlisted by the political and ecclesiastical establishment to diminish the epistemological importance of imagination and inspiration. Physiological and psychological theories, for example, were developed to explain ecstasies in natural terms. Heyd also notes that some elements of the program of the enthusiasts were adopted by the establishment elites.

Topics: *IV*, VII, XI Names: Boyle, Casaubon, M., More

0681 Heyd, Michael. "The Third Force in Seventeenth-Century Thought: A Comment." In *The Prism of Science*, edited by Edna Ullmann-Margalit, 51-56. Dordrecht: D. Reidel, 1986.

Heyd generally agrees with Popkin's interpretations but adds a few qualifications and suggestions. Heyd would: emphasize the complexity of the seventeenth-century intellectual situation; further study the *social* context of the "third-force" intellectuals; and seek an explanation for the transition to the materialism and positivism of the eighteenth century.

Topics: *VII*, IV, X

0682 Hicks, Louis E. *A Critique of Design Arguments*. New York: Charles Scribner's Sons, 1883.

This book contains both an historical review—with three chapters on seventeenth-century natural theology—and an analysis of Hicks' contemporary issues. (Hicks, a nineteenth-century professor of geology, preferred arguments from order as against arguments from ends.)
Topics: *VIII*, VII

0683 Hill, Christopher. *Change and Continuity in Seventeenth-Century England*. London: Weidenfeld and Nicolson, 1974. Cambridge, Mass.: Harvard University Press, 1975.
This is an anthology of twelve essays—four of which are relevant to this bibliography and annotated separately. There is also a seven-page "Conclusion," which summarizes the theme indicated by the title of the volume. See (0687), (0688), (0690), and (0691).
Topics: *IV*, III, V

0684 Hill, Christopher. *The Collected Essays of Christopher Hill*. 3 vols. Amherst, Mass.: University of Massacusetts Press, 1985.
These volumes contain several essays and articles annotated separately in this bibliography. See: (0685), (0692), and (0696).
Topic: *IV*

0685 Hill, Christopher. "Henry Vaughan (1621 or 1622?-1695)." In *The Collected Essays of Christopher Hill*, Vol. 1, 207-225. Amherst, Mass.: University of Massachusetts Press, 1985. See (0684).
This is an essay on Vaughan's poetry with a brief discussion of his life and thought. Hill finds Vaughan to be a Protestant Anglican. He suggests that Vaughan's interest in Hermeticism came *via* medicine. Vaughan's Hermeticism is seen in "the close feeling of identity with nature which shines in his best poetry."
Topics: *X* , II, XI   Name: Vaughan, H.

0686 Hill, Christopher. *The Intellectual Origins of the English Revolution*. Oxford: Clarendon Press, 1965.
Hill does not see Puritanism as a causal factor of the scientific revolution; rather, he sees Puritanism as part of a large set of political, social, religious, scientific, and economic aspects of one larger revolution. Thus, all of these movements facilitated each

other—and one could even say that the ideas of the scientists favored the Puritan and Parliamentary causes. Hill agrees with the Merton thesis and argues against its opponents. He gives his own, rather more elaborate, understanding of Puritanism—which now amounts to radical Protestantism. Hill is a master of primary sources and something of an aggressive, radical writer himself.

Topics: *IV*, III, V Names: Bacon, Briggs, Charles I, Comenius, Dury, Hakewill, Hartlib, Hobbes, James I, Milton, Ralegh

0687 Hill, Christopher. "The Medical Profession and Its Radical Critics." In *Change and Continuity in Seventeenth-Century England*, by Christopher Hill, 157-178. London: Weidenfeld and Nicolson, 1974. Cambridge, Mass.: Harvard University Press, 1975. See (0683).

Hill here describes parallels in the arguments against clergy, physicians, and lawyers among their mid-seventeenth-century English critics. He notes a social class pattern both in the professions and in the criticisms of the professions. He points out the sociopolitical connections between official medicine and the established church. And he notes the ideological connections between chemical medicine, public alchemy, religious heterodoxy, and political radicalism.

Topics: *XI,* IV, V Names: Chamberlen, Culpepper, Fox, Walwyn, Webster

0688 Hill, Christopher. "Newton and His Society." *Texas Q.* 10(1967): 30-50. Reprinted in *The Annus Mirabilis of Sir Isaac Newton*, edited by Robert Palter, 26-47. Cambridge, Mass., 1970. See (1201). Reprinted in *Change and Continuity in Seventeenth-Century England*, by Christopher Hill. London: Weidenfeld and Nicolson, 1974. Cambridge, Mass.: Harvard University Press, 1975. See (0683).

This is a suggestive social and psychological interpretation of Newton.

Topics: *XII*, III, IV Name: Newton

0689 Hill, Christopher. "Puritanism, Capitalism and the Scientific Revolution." *Past and Present* 29(1964): 88-97. Reprinted in *The Intellectual Revolution of the Seventeenth Century*, edited by Charles Webster, 243-253. London: Routledge and Kegan Paul, 1974. See (1618).

Hill argues against Kearney (0863) and for the Puritanism thesis. He relates both Puritanism and the new science to socioeconomic change.

Topic: *III*

0690 Hill, Christopher. "The Radical Critics of Oxford and Cambridge in the 1650's." In *Universities in Politics*, edited by J. W. Baldwin and R. Goldthwaite, 107-132. Baltimore: Johns Hopkins Press, 1972. Reprinted in *Change and Continuity in Seventeenth-Century England*, by Christopher Hill, 127-148. London: Weidenfeld and Nicolson, 1974. Cambridge, Mass.: Harvard University Press, 1975. See (0683).

The social and religious radicals (defined as those who rejected any state church) wanted to reform the universities—putting more mathematics and science into the curriculum. The social and religious establishment opposed the reforms. Hill discusses the ideological pattern of each side.

Topics: *V*, III, IV Names: Dell, Winstanley

0691 Hill, Christopher. "'Reason' and 'Reasonableness' in Seventeenth-Century England." *British Journal of Sociology* 20(1969): 235-252. Reprinted in *Change and Continuity in Seventeenth-Century England*, by Christopher Hill, 103-123. London: Weidenfeld and Nicolson, 1974. Cambridge, Mass.: Harvard University Press, 1975. See (0683).

Hill sees "old reason" (connected with logic, authority, tradition, correspondences, and analogies) as replaced by "new reason" (connected with common sense, the senses, evidence, experience, and experiment) in seventeenth-century England. Radical Protestantism and magical traditions in natural philosophy were instrumental, he argues, in the attack on the old reason. The new science and the Anglican establishment which eventually triumphed were part of yet another ideological structure of the reasonable.

Topics: *IV*, III, V, VII

0692 Hill, Christopher. "Science and Magic." In *Culture, Ideology and Politics: Essays for Eric Hobsbawm*, edited by Raphael Samuel and Gareth Stedman Jones, 176-193. London: Routledge and Kegan Paul, 1982. Reprinted in *The Collected Essays of Christopher Hill*, Vol. 3, 274-299. Amherst, Mass.: University of Massachusetts Press, 1986. See (0684).

In this public lecture, Hill indicates that while the revival of the hermetic or intellectual magic tradition had acted as a stimulus to the scientific imagination, by the Restoration it had been opposed and essentially defeated by orthodox Calvinists. He sees magical traditions linked to popular radical sects and revolutionary politics. Their defeat "created social conditions which favored the victory of the mechanical philosophy." For Hill, the mechanical philosophy was "secularized Calvinism."

Topics: *II*, IV

0693 Hill, Christopher. "Science, Religion and Society in the Sixteenth and Seventeenth Centuries." *Past and Present* 32(1965): 110-112. Reprinted in *The Intellectual Revolution of the Seventeenth Century*, edited by Charles Webster, 280-283. London: Routledge and Kegan Paul, 1974. See (1618).

Hill here defends his previous publications against Kearney (0862-0863) and Rabb (1279). He emphasizes that he is focusing on English society in a particular period and not on the "Scientific Revolution" as a general event in European history.

Topics: *III*, IV

0694 Hill, Christopher. *Society and Puritanism in Pre-Revolutionary England*. New York: Schocken Books, 1964.

In the chapter entitled "The Industrious Sort of People," Hill finds Puritan asceticism among "industrious artisans and aspiring peasants." While not directly discussing natural philosophy, Hill indicates that Puritan pamphlets assumed that the arts and sciences, as well as economic progress, would not be promoted as much by idle Catholics as by hardworking English Puritans.

Topic: *III*

0695 Hill, Christopher. *Some Intellectual Consequences of the English Revolution*. Madison, Wisc.: University of Wisconsin Press, 1980.

In a lecture series, Hill makes summary remarks concerning selected long-term effects of the revolutionary ferment of the 1640s and 1650s. About fifteen pages (out of ninety) deal with science

and religion—generally in terms of bourgeois winners and defeated radicals.
Topics: *IV*, V, IX

0696 Hill, Christopher. "Till the Conversion of the Jews." In *The Collected Essays of Christopher Hill*, Vol. 2, 269-303. Amherst, Mass.: University of Massacusetts Press, 1986. See (0684). An earlier version appeared in *Millenarianism and Messianism in English Literature and Thought, 1650-1800*, edited by Richard H. Popkin, 12-36. Leiden: Brill, 1988. See (1238).

Hill discusses how the idea of the conversion of the Jews and the related English policies of state were understood by interpreters of Biblical prophecies as a sign of the millennium. Natural philosophers were among those who believed and advocated these commonly held ideas (especially during the interregnum).
Topics: *X*, IV Names: Brightman, Hartlib, Mede

0697 Hill, Christopher. "William Harvey and the Idea of Monarchy." *Past and Present* 30(1964): 54-72. Reprinted in *The Intellectual Revolution of the Seventeenth Century*, edited by Charles Webster, 160-181. London: Routledge and Kegan Paul, 1974. See (1618).

Hill places within a single ideological context: (1) the acceptance of the circulation of the blood; (2) the heart displaced by the blood as the seat of the soul; (3) mechanical models such as the water pump; (4) Protestantism; (5) religious heresy; and, possibly (6) Parliamentarianism. He supports the Merton thesis as well—arguing that Protestantism was a liberating force for science.
Topics: *IV*, III, XI Name: Harvey

0698 Hill, Christopher. "William Harvey (No Parliamentarian, No Heretic) and the Idea of Monarchy." *Past and Present* 31(1965): 97-103. Reprinted in *The Intellectual Revolution of the Seventeenth Century*, edited by Charles Webster, 189-196. London: Routledge and Kegan Paul, 1974. See (1618).

Here, Hill defends his position (0697) against Whitteridge (1666) and clarifies his interpretation of Harvey. Hill argues that he is treating the possible implications of Harvey's publications (history

of ideas) and is not so concerned with what Harvey actually thought (biography)!
Topics: *IV*, III Name: Harvey

0699 Hill, Christopher. *The World Turned Upside Down: Radical Ideas During the English Revolution*. London: Temple Smith, 1972. New York: Viking Press, 1972.
Chapter 14 ("Mechanic Preachers and the Mechanical Philosophy") and parts of Chapters 6, 7, and 18 are relevant to our field. Hill argues for connections between radical religious views, radical social views, and the astrological, alchemical and Renaissance magic traditions. After the Restoration, according to Hill, Latitudinarianism and the new science were used to oppose fanaticism (and any radical views) and to support the hierarchical status quo.
Topics: *IV*, II, VI Names: Overton, Webster, Winstanley

0700 Hinman, Robert. *Abraham Cowley's World of Order*. Cambridge, Mass.: Harvard University Press, 1960.
Hinman is a literary historian and critic who is writing an apologia for Cowley and for the seventeenth-century effect of science on poetry. Over 300 pages are devoted to the harmony of science, religion, and poetry in the work of Cowley. "The poet of science is the poet of divine love." Hinman argues that Cowley was a follower both of Bacon and of Hobbes—even as he believed in the central importance of poetry and faith. He also sees Cowley as a master both of traditional natural philosophy and of the new science. In his enthusiasm, Hinman shows a tendency to read too much into the poetry of Cowley and the writings of Cowley's contemporaries.
Topics: *X*, VII Names: Bacon, Cowley, Hobbes

0701 Hirst, Desiree. *Hidden Riches: Traditional Symbolism from the Renaissance to Blake*. London: Eyre and Spottiswoode, 1964.
This is a study of the tradition of Christian Neoplatonic symbolism found in sources possibly used by William Blake. Concerning Blake's seventeenth-century predecessors, Hirst includes alchemists, Milton, and the Cambridge Platonists.
Topics: *X*, II Names: Conway, Donne, Fludd, Milton, More

0702 Hobhouse, Stephen. "Isaac Newton and Jacob Boehme: An Inquiry." In *Selected Mystical Writings of William Law*, edited by Stephen Hobhouse, 397-422. New York: Harper, 1948. An earlier version appeared as "Isaac Newton and Jacob Boehme." *Philosophia* (Belgrade) 2(1937): 25-54.

Hobhouse argues that Newton was not influenced by Boehme (the mystical German writer of the sixteenth century); and that the old tradition dating back to William Law (which assumes this influence) is wrong.

Topics: *XII*, II Name: Newton

0703 Hollander, John. *The Untuning of the Sky: Ideas of Music in English Poetry, 1500-1700*. Princeton, N.J.: Princeton University Press, 1961.

This is an extensive and well-written survey of the "music of the spheres" idea from classical times to the early eighteenth century. Hollander argues that this phrase, however, was empty of any substantial meaning for most poets by the end of the seventeenth century. Although Joseph Addison and the early eighteenth century are treated, Hollander makes no mention of Addison's paraphrase of Psalm 19, "The spacious firmament on high." (Perhaps, this is because Addison's poem would counter Hollander's thesis.)

Topic: *X* Names: Bacon, Browne, T., Cowley, Donne, Dryden, Mace, Vaughan, H.

0704 Holmes, Elizabeth. *Henry Vaughn and the Hermetic Philosophy*. Oxford: B. Blackwell, 1932.

Holmes shows how Hermetic themes shape the poetic images of Vaughn and other seventeenth-century poets.

Topics: *X*, II Name: Vaughn, H.

0705 Holmes, Geoffrey. "Science, Reason, and Religion in the Age of Newton." *Brit. J. Hist. Sci.* 11(1978): 164-171.

This is an essay review of M. C. Jacob's *The Newtonians and the English Revolution* (0817). Holmes praises Jacob's book for its provocative themes; but he effectively devastates some of Jacob's ecclesiastical terminology, some of her assumptions about the situation in England after 1689, and some of her omissions regarding factional strife in the Church of England.

Topics: *IV*, V, VIII, XII Name: Newton

0706 Holmyard, E. J. *Alchemy*. Harmondsworth, England: Penguin Books, 1957.

This is still a good introductory survey of the history of pre-seventeenth-century alchemy. It is dated, however, regarding its interpretations of and neglect of seventeenth-century figures.
Topic: *II* Name: Digby

0707 Holtgen, Karl Josef. "Richard Haydocke [c.1570-c.1642]: Translator, Engraver, Physician." *Library* 33(1978): 15-32.
Holtgen traces Haydocke's activities as an engraver of brasses as well as translator of an Italian treatise on painting. Haydocke's emblematical works indicate that a Puritan physician could produce religious images with multiple allegorical meanings.
Topics: *VI*, XI Name: Haydocke

0708 Holton, Gerald. "Frank Manuel's Isaac Newton." *The New Republic* 172, No.9(1975): 26-28. Reprinted in *The Scientific Imagination*, by Gerald Holton, 268-274. Cambridge: Cambridge University Press, 1978. See (0710).
This essay review popularized (in a general periodical) Manuel's view of Newton's psychic terror and Newton's invocation of God both through biblical studies and through science.
Topic: *XII* Name: Newton

0709 Holton, Gerald. "Presupposition in the Construction of Theories." In *Science as a Cultural Force*, edited by Harry Woolf, 77-108. Baltimore: Johns Hopkins Press, 1964.
Holton's original proposal for a thematic analysis of the history of scientific thought. He indicates clearly that theological presuppositions are to be included in discussions of themata.
Topics: *I*, VII, XII Name: Newton

0710 Holton, Gerald. *The Scientific Imagination: Case Studies*. London: Cambridge University Press, 1978.
This collection includes two relevant essays—"Frank Manuel's Isaac Newton" (0708) and "Themata in Scientific Thought" (0711).

0711 Holton, Gerald. "Themata in Scientific Thought." Reprinted in *The Scientific Imagination*, by Gerald Holton, 3-24. London: Cambridge University Press, 1978. See (0710). A previous, shorter version, appeared in *Science* 188(1974): 328-334.

Holton argues for a thematic analysis as an additional component of research in the history of science. Themata are concepts, methods of expression, or hypotheses to which scientists maintain loyalty over time. They can be explicit or implicit—and include theological as well as other philosophical elements.

Topic: *I*

0712 Holton, Gerald. “The Thematic Imagination in Science.” In *Thematic Origins of Scientific Thought: Kepler to Einstein*, by Gerald Horton, 47-68. Cambridge, Mass.: Harvard University Press, 1973.

This is a reprint of “Presupposition in the Construction of Theories,” (0709).

Topics: *I*, VII, XII Name: Newton

0713 Home, R. W. “Force, Electricity, and the Powers of Living Matter in Newton’s Mature Philosophy of Nature.” In *Religion, Science, and Worldview*, edited by Margaret J. Osler and Paul Lawrence Farber, 95-117. Cambridge: Cambridge University Press, 1985. See (1184). Reprinted in *Seventeenth-Century Natural Scientists*, edited by Vere Chappell, 245-267. New York: Garland, 1992. See (0205).

A historian of science who has specialized in theories of electricity argues here that Newton believed in mechanical explanations to a greater degree than is commonly supposed. He argues this by considering cases of electricity, magnetism, and the vegetative power in living matter. As regards God, Home says, Newton’s well-known view that God’s activity is the ultimate causal explanation is true only as a first cause. A major point for Home is that Newton was not as quick to invoke God as an explanatory factor as some historians have suggested.

Topics: *XII*, VII Name: Newton

0714 Hoopes, Robert. *Right Reason in the English Renaissance*. Cambridge, Mass.: Harvard University Press, 1962.

Hoopes surveys the change in the meaning of the words “reason” and “rational” during the sixteenth and seventeenth centuries. For the most part, science and religion are discussed only indirectly.

Topics: *X*, VII Names: Cudworth, Hobbes, Milton, More, Taylor, J., Whichcote

0715 Hooykaas, Reijer. "Answer to Dr. Bainton's Comment on 'Science and Reformation'." *J. World Hist.* 3(1957): 781-784.
Hooykaas defends the connection between Protestantism and the rise of science while distinguishing his position from Weber and Merton regarding economic activity. This short essay has been superseded by the Hooykaas book (0722).
Topic: *III*

0716 Hooykaas, Reijer. "L'historie des sciences, ses problemes, sa methode, son but" *Revista da Faculdade de Ciencias da Universidade de Coimbra* 22(1963): 5-35. Reprinted in *Selected Studies in History of Science*, by Reijer Hooykaas, 9-41. Coimbra, Portugal: Universidade, 1983. See (0724).
Among other issues, Hooykaas briefly outlines his position regarding science and religion. He: (1) distinguishes Christianity from clericalism; (2) points out the Biblical sense of the contingency of the world (necessitating an empirical method); and (3) emphasizes the Biblical theme of humans as collaborators with God.
Topic: *I*

0717 Hooykaas, Reijer. "Historiography of Science, Its Aims and Methods." *Organon* (Warsaw) 7(1970): 37-49.
This discussion of the historiography of science is wise and still applicable. Hooykaas says that the history of science provides "a critical self-examination of science" by making evident "that the scientists of the past were as adult, as human, and also as fallible as we are." Historians of science, then, need to be aware of the full humanity of past scientists and of their religious, philosophical, technological, economic, and social contexts.
Topic: *I*

0718 Hooykaas, Reijer. *Natural Law and Divine Miracle: The Principle of Uniformity in Geology, Biology and Theology.* Leiden: Brill, 1963.
Though primarily focusing on discussions in the nineteenth century, the chapter on theology (outlining four different metaphysical positions) is also relevant to earlier debates.
Topic: *VII*

0719 Hooykaas, Reijer. *The Principle of Uniformity in Geology, Biology, and Theology*. Leiden: Brill, 1959. 2nd ed. *Natural Law and Divine Miracle: The Principle of Uniformity in Geology, Biology and Theology*. Leiden: Brill, 1963.

See previous entry, (0718).

Topic: *VII*

0720 Hooykaas, Reijer. "The Reception of Copernicanism in England and the Netherlands." In *Selected Studies in History of Science*, by Reijer Hooykaas, 635-663. Coimbra, Portugal: Universidade, 1983. See (0724). Originally published in *The Anglo-Dutch Contribution to the Civilization of Early Modern Society*. London: Oxford University Press, 1976.

Primarily this is a discussion of Copernicanism in the Netherlands, with a few Englishmen introduced for comparative purposes. Hooykaas points out that some natural philosophers were selective Copernicans. He emphasizes the accommodation theory of scriptural exegesis as the main theological issue and sees this as derived mainly from Calvin. Hooykaas also intimates, almost as an aside, a distinction between Laudian High Churchmen and Copernicans.

Topics: *X*, VII Names: Gilbert, Wilkins, Wright

0721 Hooykaas, Reijer. *Religion and the Rise of Modern Science*. Edinburgh: Scottish Academic Press; Grand Rapids, Mich.: William B. Eerdmans, 1972.

This book is a systematic and articulate attempt to show the philosophical as well as sociological connections between science and Protestantism in the sixteenth and seventeenth centuries. Hooykaas tends to oversimplify when he categorizes "types" of Christianity and of philosophy. His own theological biases sometimes intrude. But the book remains important for anyone doing work in our field. It is excellent for an introductory discussion of the philosophical issues—and especially as regards the relation of the "voluntaristic doctrine of God" to early modern natural philosophy. Hooykaas examines continental as well as English Calvinists and considers why and how they believed science should be cultivated: (1) to the glory of God and to the benefit of humankind; (2) empirically, in spite of human authorities; and (3) by using our hands. The book is a veritable mine of relevant biblical texts.

Topics: *III*, IV, VII, X, XII Names: Bacon, Boyle, Newton, Wilkins

0722 Hooykaas, Reijer. *Robert Boyle: een Studie over Natuurwetenschap en Christendom.* N.p., 1941.

This work is important but in Dutch. It has been used as evidence by some scholars advancing the Protestantism-and-the-rise-of-science thesis. Hooykaas describes well Boyle's voluntaristic doctrine of God, his religious motivation and his justification for doing natural philosophy.

Topics: *VII*, III, IV Name: Boyle

0723 Hooykaas, Reijer. "Science and Reformation." *J. World Hist.* 3(1956): 109-139. Reprinted in *The Evolution of Science*, edited by Guy S. Metraux and Francois Crouzet, 258-290. New York: New American Library, 1963. See (1062).

In this once-important article defending the Protestantism thesis, Hooykaas shows "how the religious attitude of so-called 'ascetic' Protestantism, which more or less stood under Calvin's influence, furthered the development of science." This article is an acknowledged summary of (and thus has been superseded by) Hooykaas' book (0721).

Topics: *III*, IV

0724 Hooykaas, Reijer. *Selected Studies in History of Science.* Coimbra, Portugal: Universidade, 1983.

See the annotations of Hooykaas' studies (0716) and (0720).

0725 Hoppen, K. Theodore. *The Common Scientist in the Seventeenth Century: A Study of the Dublin Philosophical Society, 1683-1708.* Charlottesville, Va.: University Press of Virginia, 1970.

This is an excellent book about a little-studied group. Hoppen also discusses the Royal Society, the Oxford Philosophical Society, and other attempts to form scientific groups as well.

Topic: *V* Names: Ashe, Boyle, Flamsteed, Foley, Hooke, Keogh, King, Marsh, Molyneux, S., Molyneux, T., Molyneux, W., Mullin, Petty, Smyth, Swift

0726 Hopper, Jeffrey. *Understanding Modern Theology I.* Philadelphia: Fortress Press, 1987.

Chapter 1 briefly summarizes the challenge to Christian theology which the seventeenth-century scientific revolution presented.

Topic: *VII*

0727 Hornberger, Theodore. "The Date, the Source, and the Significance of Cotton Mather's Interest in Science." *American Literature* 6(1935): 413-420.

This brief article summarizes Mather's interest in science. Hornberger was one of the early proponents of the theses that Mather was unconsciously vacillating between science and religion, used science for a religious purpose, and tended toward deism(!).

Topics: *VII*, IX Name: Mather, C.

0728 Hornberger, Theodore. "Puritanism and Science: The Relationship Revealed in the Writings of John Cotton." *New Engl. Q.* 10(1937): 503-515.

From his analysis of Cotton's sermons and biblical commentaries, Hornberger finds that Cotton clearly was interested in comparing theories about physical causes and natural phenomena with biblical texts. Cotton's natural philosophy basically was that of the medieval scholastics—yet he also asserted that to study God's works is a duty imposed by God.

Topics: *III*, X Name: Cotton

0729 Hornberger, Theodore. "Samuel Lee (1625-1691), a Clerical Channel for the Flow of New Ideas to Seventeenth-Century New England." *Osiris* 1(1936): 341-355.

This is a short biographical article outlining Lee's interest in science (from the time of participation in Wilkins' circle at Oxford to his years in New England). Lee's books reflected a mixture of Cartesianism and of a Boyle-like natural theology; his books may have influenced Cotton Mather.

Topics: *VIII*, V, VII Names: Boyle, Lee, Mather, C., Wilkins

0730 Hornberger, Theodore. *Scientific Thought in the American Colleges, 1638-1800*. Austin, Texas, 1945.

Hornberger here discusses the curriculum, methods, the teachers, and their religious motives. He deals primarily with the eighteenth century but he does deal with Harvard in the seventeenth century.

Topic: *V* Names: Dunster, Greenwood

0731 Horne, Colin J. "The Phalaris Controversy: King versus Bentley." *Review of English Studies* 22(1946): 289-303.

Horne focuses on the late stages of the ancients-versus-moderns controversy. He shows the personal insults and some of the anti-scientific aspects of the defenders of the ancients.

Topics: *X*, IV Names: Bentley, King

0732 Horstmann, Ute. *Die Geschichte der Gedankenfreiheit in England. Am Beispiel von Anthony Collins: A Discourse of Free-Thinking*. Konigstein, Germany: Forum Academicum in der Verlagsgrupe Athenaum, Hain, Scriptor, Hanstein, 1980.

Although natural philosophy was a secondary factor in Collins' thought, it was part of the larger context of the development of "Free-Thinking" in early eighteenth-century England. Horstmann's book includes much historical background to Collins' work—including analyses of the writings of Chillingworth, Hobbes, Tillotson, Locke, Bentley, Toland, and the Boyle lectures of Benjamin Ibbot.

Topic: *IX* Names: Bentley, Chillingsworth, Collins, A., Hobbes, Locke, Tillotson, Toland

0733 Hoskin, Michael A. "Cosmology and Theology: Newton and the Paradoxes of an Infinite Universe of Stars." In *Science and Imagination in Eighteenth-Century British Culture*, edited by Sergio Rossi, 237-240. Milan: Edizioni Unicopli, 1987. See (1362).

Hoskin briefly examines the discussions among Newton, Bentley, Halley, and Stukeley concerning: (1) the possible uniformity of the distribution of the stars; and (2) the relation of the universe of stars to gravity and divine providence.

Topics: *VII*, XII Name: Newton

0734 Hoskin, Michael A. *The Mind of the Scientist*. New York: Taplinger, 1971.

This book contains popular scripts of BBC "dialogues" between Hoskin and famous scientists of the past. In the dialogue with Newton, they discuss Newton's views concerning God and the world as well as other religious questions. The script for Newton is done either in his own words or in accurate paraphrases.

Topics: *XII*, VII Name: Newton

0735 Hoskin, Michael A. "Mining All Within: Clarke's Notes to Rohault's *Traite de Physique.*" *The Thomist* 24(1962): 353-363.

Hoskin discusses the notes which Samuel Clarke added to Jacques Rohault's popularization of Cartesian science—a textbook used widely in England and America. Clarke's notes were Newtonian and contradicted Rohault. The only reference to theology is the important note that God imposed the law of gravity on matter.

Topics: *XII*, V, VII Name: Clarke

0736 Hoskin, Michael A. "Newton and the Beginnings of Stellar Astronomy." In *Newton and the New Direction in Science*, edited by G. V. Coyne, 55-63. Vatican City: Specola Vaticana, 1988. See (0281).

Hoskin describes Newton's attempts to reconcile universal gravity with the fixity of the stars and the consequences for cosmology. He notes the role of Providence in Newton's theory. There is nothing new here: this repeats in shorter form what is found in Hoskin's earlier essays.

Topics: *XII*, VII Name: Newton

0737 Hoskin, Michael A. "Newton, Providence and the Universe of Stars." *J. Hist. Astron.* 8(1977): 77-101. Reprinted in *Stellar Astronomy: Historical Studies*, edited by Michael A. Hoskin, 71-95. Chalfont St. Giles, England: Science History, 1982.

Basing his discussion on manuscript sources, Hoskin describes Newton's attempts to develop a consistent cosmology in which universal gravitation occurs but the universe of stars does not collapse into the center. God, both as creator and as the one who maintains the system, is an important factor in Newton's view.

Topics: *XII*, VII, VIII Names: Bentley, Newton

0738 Hoskin, Michael A. "Stukeley's Cosmology and the Newtonian Origins of Olbers's Paradox." *J. Hist. Astron.* 16(1985): 77-112.

This is a discussion and analysis of cosmological problems faced by Newton, Halley, David Gregory, and William Stukeley. Each was attempting to work out a consistent model which included infinite space, an infinity of stars, universal gravitation, physical stability, and a dark night sky. In addition, Stukeley wished to specifically locate heaven and hell within the universe. All included the providential activity of God. The sections on Newton repeat much of Hoskin's previous article on Newton, (0737).

Topics: *XII*, VII Names: Gregory, D., Halley, Newton, Stukeley

0739 Houghton, Walter E., Jr. "The English Virtuoso in the Seventeenth Century." *J. Hist. Ideas* 3(1942): 51-73 and 190-219.

Houghton discusses the motives and values of the men who called themselves "virtuosi." He argues that they were *not* interested in utilitarianism. He focuses on Bacon, Evelyn, and Henry Peacham; and he distinguishes them from "the real natural philosophers, men like Boyle and Hooke, Ray and Newton."

Topics: *IV*, III, X Names: Bacon, Evelyn, Peacham

0740 Houghton, Walter E., Jr. "The History of Trades: Its Relation to Seventeenth-Century Thought." *J. Hist. Ideas* 2(1941): 33-60.

This is a study of the *idea* of the history of trades. Houghton discusses science, technology, and utilitarianism from Bacon to the reformers of the Hartlib circle, to several men who were important in the rise of the Royal Society.

Topics: *IV*, III, V, VI Names: Bacon, Boyle, Evelyn, Hartlib, Petty

0741 Howell, Almonte C. "*Res et Verba*: Words and Things." *ELH* 13(1946): 131-142. Reprinted in *Essential Articles: For the Study of English Augustan Backgrounds*, edited by Bernard N. Schilling, 53-65. Hamden, Conn.: Archon Books, 1961. See (1411).

This is a discussion of seventeenth-century usage of the classical phrase indicated by the title. Howell traces the development, caused by the new science as well as other factors, of a plain prose style in place of a highly ornamental rhetoric. Preaching and religious prose are seen as affected by this development.

Topic: *X* Name: Bacon

0742 Howell, Almonte C. "Sir Thomas Browne and Seventeenth-Century Scientific Thought." *Studies in Philology* 22(1925): 61-80.

Howell describes Browne's interests as extremely wide. He should be seen not only as an important literary figure and religious writer but also as a scientist. Howell also describes "his most important scientific work, the *Vulgar Errors*." He sees Browne as both Baconian and Cartesian; and he explains Browne's defense of Ptolemaic astronomy.
Topics: *X*, VII Name: Browne, T.

0743 Howell, Wilbur Samuel. *Logic and Rhetoric in England, 1500-1700*. Princeton, N.J.: Princeton University Press, 1956.
The last chapter briefly outlines the rejection of tropes, figures, and exaggerations by members of the Royal Society as well as Glanvill's plain, practical homiletic.
Topics: *X*, V Name: Glanvill

0744 Hoyles, John. *The Waning of the Renaissance, 1640-1740: Studies in the Thought and Poetry of Henry More, John Norris and Isaac Watts*. The Hague: Nijhoff, 1971.
This book mainly concerns literary criticism and aesthetics. But it does include some passing remarks on the effects of science and various philosophies of nature on religion and on the general culture.
Topics: *X*, VII Names: Addison, More, Norris, Watts

0745 Huffman, William H. Introduction to *Robert Fludd: Essential Readings*, edited by William H. Huffman, 13-39. London: Aquarian Press, 1992.
Huffman writes a clear and helpful introduction to Fludd's writings (which are usefully excerpted in this volume). Drawing upon his book, Huffman finds Fludd to be the greatest summarizer and synthesizer of the Renaissance-Christian-Neoplatonist-alchemical tradition.
Topic: *II* Name: Fludd

0746 Huffman, William H. *Robert Fludd and the End of the Renaissance*. London: Routledge, 1988.
Huffman gives a very positive interpretation of Fludd which balances the prevailing view. He places Fludd at the end of the Platonic Renaissance—indeed, as its culmination. He shows that Fludd had many significant friends and patrons as well as oppo-

nents. Thus, Fludd was not an isolated figure. Huffman thoroughly discusses Fludd's *Mosaicall Philosophy*, showing how it combines Neoplatonic Hermeticism, Paracelsian alchemy, Genesis, and the Cabala. In so doing, Huffman indicates that Fludd believed that he was recovering Moses' ancient wisdom with the hope of purifying every aspect of human society. (But, the author says, Fludd was not a Rosicrucian.) After apologetic suggestions throughout the book, Huffman treats in his conclusion Fludd's subsequent reputation. Fludd's esteem declined not because of his "religious-Hermetic-Neoplatonist metaphysics," but "because of a change in the commonly accepted fashion about *sources* of knowledge and the consequent *methods* used to achieve it."

Topics: *II*, VII, X Names: Bacon, Fludd, Foster,W., Harvey, James I, Maier

0747 Humphreys, A. R. "Pope, God, and Man." In *Alexander Pope*, edited by Peter Dixon, 60-200. (Writers and their Background.) London: Bell, 1972.

Humphreys discusses the effects of the new science on the religious worldview of Pope and other literary figures in the early eighteenth century.

Topics: *X*, XII Names: Newton, Pope

0748 Hunter, Michael. "Alchemy, Magic and Moralism in the Thought of Robert Boyle." *Brit. J. Hist. Sci.* 23(1990): 387-410.

Hunter begins by examining the unpublished manuscript of a memorandum Boyle dictated to Gilbert Burnet in the early 1690s. He finds that in it, and in other writings, Boyle linked "the issue of the philosopher's stone with that of intercourse with spirits." Boyle's religion, however, led him to examine these issues through "his acutely developed sense of moral scruple," hence his reliance on Bishop Burnet for counsel. Hunter then thoroughly analyzes the complex question of Boyle's ambivalent attitude toward alchemy: on the one hand, he was an active alchemist; on the other hand, he was concerned about the morality of its power and its possible associations with evil magic.

Topic: *II* Names: Boyle, Burnet, G.

0749 Hunter, Michael. "Casuistry in Action: Robert Boyle's Confessional Interviews with Gilbert Burnet and Edward Stillingfleet, 1691." *J. Eccles. Hist.* 44(1993): 80-98.

This essay focuses on Boyle's conscience, his interviews with Burnet and Stillingfleet, and the related casuistry. The two bishops treated Boyle's almost obsessive concern with naturalistic responses. In this, Hunter sees "parallel traditions of thought" among religious men of the period (clergy and lay natural philosophers). Dictated notes of the interviews are included as appendices.

Topics: *X*, VII Names: Barlow, Boyle, Burnet,G., Stillingfleet

0750 Hunter, Michael. "The Conscience of Robert Boyle: Function alism, 'Dysfunctionalism' and the Task of Historical Understanding." In *Renaissance and Revolution*, edited by J. V. Field and Frank A. J. L. James, 147-159. Cambridge: Cambridge University Press, 1993. See (0466).

By considering the intensity of Boyle's religiosity, Hunter seeks to see him as a whole person and to add a third dimension to those historians of Boyle's thought and social functioning. He illustrates Boyle's scrupulosity and concern for casuistry by looking at his attitude towards his experiments and at his "great tenderness in point of oaths" (regarding his refusal to be President of the Royal Society).

Topics: *I*, V Name: Boyle

0751 Hunter, Michael. "The Early Royal Society and the Shape of Knowledge." In *The Shapes of Knowledge from the Renaissance to the Enlightenment*, edited by Donald R. Kelley and Richard H. Popkin, 189-202. Dordrecht: Kluwer Academic, 1991. See (0872).

Hunter notes several movements in the ideology of the Royal Society in the late seventeenth century. In its early years, there was a stress on utility and empiricism with a separation of science and religion. Later, the emphasis turned to "understanding over application," an increased role of science as a vindication of the design argument, and a shift to arguing that they only wanted to avoid religious controversies (not to totally avoid religion).

Topics: *V*, IV, VIII Names: Hooke, Moray, Sprat

0752 Hunter, Michael. *Establishing the New Science: The Experience of the Early Royal Society*. Woodbridge, England: Boydell Press, 1989.

This is a combination of some essays published only in this book and some essays also published elsewhere. All of those which deal significantly with science and religion are ones which also have been published elsewhere. Both these essays and the effect of the book generally support Hunter's view that science and religion were in complex interaction during our period and that no simple generalization describes this interaction. Hunter sees religion as one factor among many relevant to the study of the Royal Society; and he believes no single form of religion is dominant in relation to the Royal Society. The book ends with a thirteen-page bibliographic essay entitled "Recent Studies of the Early Royal Society and its Milieu." Also see (0756) and (0758).

Topic: *V* Names: Oldenburg, Sprat, Wilkins

0753 Hunter, Michael. Introduction to *Elias Ashmole 1617-1692: The Founder of the Ashmolean Museum and His World*, edited by Michael Hunter, 1-27. Oxford: Ashmolean Museum, 1983.

In his introduction to the exhibition of Ashmole's books, manuscripts, and artifacts, Hunter describes the subject's various interests and activities. Ashmole was a collector, a restorer of heraldic ceremony, an alchemist, and a numismatic expert. The museum built to house his collections and those of others was completed in 1683 and is the longest surviving public museum. Hunter finds the unbounded curiosity of Ashmole and others like him to be the backbone of the contemporary Baconian scientific movement. Religion is not addressed directly in this essay but is implicit in the person and worldview of Ashmole.

Topics: *V*, II Name: Ashmole

0754 Hunter, Michael. Introduction to *Robert Boyle Reconsidered*, edited by Michael Hunter, 1-18. Cambridge: Cambridge University Press, 1994. See (0759).

This is a very useful summary of the history of studies of Boyle and of the image of the "new" Boyle. Hunter points out that Boyle was taken for granted for many years. Few studied him because—with his obviously varied interests in theology, "strange reports," and alchemy—he did not fit into the intellectualistic historiography of the scientific revolution. But, Hunter goes on to suggest, Boyle also is more sophisticated and complex than how the recent social historians of science simplistically picture him.

Topic: *I* Name: Boyle

0755 Hunter, Michael. *John Aubrey and the Realm of Learning*. London: Duckworth, 1975.

After a chapter on Aubrey's life, Hunter elaborates on his primary purpose: to examine Aubrey's writings on natural philosophy and antiquities. Aubrey was a prolific writer on a wide range of subjects. Among those which Hunter highlights are natural history, technology (all knowledge should be useful), magic and astrology. Aubrey believed in deterministic explanations of human behavior in terms of environmental factors and heavenly bodies. And he did not see his occult interests as impious. Hunter also says that, for Aubrey, collecting—both new and old, both theories and phenomena—substituted for induction.

Topics: *VI*, II, VII, X Names: Ashmole, Aubrey, Bacon, Hobbes, Hooke, Plot, Ray, Wood, Wren

0756 Hunter, Michael. "Latitudinarianism and the 'Ideology' of the Early Royal Society: Thomas Sprat's *History of the Royal Society* (1667) Reconsidered." In *Philosophy, Science, and Religion in England 1640-1700*, edited by Richard Kroll and others, 199-229. Cambridge: Cambridge University Press, 1992. See (0917). Previously appeared in *Establishing the New Science*, by Michael Hunter, Chapter 2. Woodbridge, England: Boydell Press, 1989. See (0752).

This is a detailed examination of the context, production, and contemporary reactions to Sprat's *History*. Against Purver and J. R. Jacob, Hunter says that the book was not an official manifesto, written under the watchful care of the Society; but, against Webster, he argues that the *History* was not merely the private opinion of Sprat and Wilkins. Hunter emphasizes experimental natural philosophy as the dominant motive in the activities of the early Royal Society. As regards religion, the Royal Society and its members neither sought a Latitudinarian ecclesiastical comprehension (against J. R. Jacob) nor was the Society moving towards a complete indifference to religion (against Mulligan). Rather, it sought religious pluralism—allowing religious diversity while supporting the common cause of experimental science.

Topics: *V*, III, IV, X Names: Evelyn, Oldenburg, Sprat, Stubbe, Wilkins

0757 Hunter, Michael. "The Problem of 'Atheism' in Early Modern England." *Trans. Royal Hist. Soc.*, 5th series, 35(1985): 135-157.

Hunter argues that late-sixteenth and early-seventeenth century commentators on the threat of atheism encapsulated a complex range of phenomena that were perceived to be threats to religion. Included in this stereotype was naturalism—especially that drawn from classical pagan authors. Yet the anxiety (both over morality and over ideas) was exaggerated.

Topics: *IX*, VII Name: Hariot

0758 Hunter, Michael. "Promoting the New Science: Henry Oldenburg and the Early Royal Society." *Hist. Sci.* 26(1988): 165-181. Reprinted in *Establishing the New Science*, by Michael Hunter, 245-260. Woodbridge, England: Boydell Press, 1989. See (0752).

This essay includes a four-page Note entitled "On Oldenburg and Millenarianism." Here Hunter presents evidence that "Oldenburg may well have retained a commitment to the idea of an imminent apocalypse into the later years of his life."

Topics: *V*, X Name: Oldenburg

0759 Hunter, Michael, ed. *Robert Boyle Reconsidered.* Cambridge: Cambridge University Press, 1994.

The introduction by Hunter (0754) and the essays by E. B. Davis (0327), Harwood (0644), Henry (0659), MacIntosh (0971), Oster (1187), Principe (1265), Shanahan (1426), and Wojcik (1696) are annotated separately. This book also includes a good bibliography (not annotated) of writings about Boyle.

Topics: *VII*, I Name: Boyle

0760 Hunter, Michael. *The Royal Society and Its Fellows, 1660-1700: the Morphology of an Early Scientific Institution.* Chalfont St. Giles, England: British Society for the History of

Science, 1982. 2nd ed. Oxford: British Society for the History of Science, 1994.

This work includes: 50 pages of text; a 100-page "Catalogue of Fellows;" and 120 pages of notes, appendices, and indexes. It includes no direct treatment of religion and does not attempt to assign the Fellows to religious categories. But the Catalogue remains useful as a biographical reference source (especially in following the Protestantism-and-the-rise-of-science debates). The second edition includes updated details about the membership and Hunter's comments about other historians.

Topics: *V*, III Names: too many to index

0761 Hunter, Michael. "Science and Astrology in Seventeenth-Century England: An Unpublished Polemic by John Flamsteed." In *Astrology, Science and Society: Historical Essays*, edited by Patrick Curry, 261-300. Woodbridge, England: Boydell Press, 1987.

Hunter sets Flamsteed's manuscript in its historical context as one of the few attacks on astrology by a seventeenth-century proponent of the new science. That the new astronomy, with its empirical methods, played an important role in the critique is clear. Yet Hunter concludes that traditional religious objections may have had more significance for Flamsteed. A transcription of portions of the manuscript follows Hunter's essay.

Topic: *II* Name: Flamsteed

0762 Hunter, Michael. "Science and Heterodoxy: an Early Modern Problem Reconsidered." In *Reappraisals of the Scientific Revolution*, edited by David C. Lindberg and Robert S. Westman, 437-460. Cambridge: Cambridge University Press, 1990. See (0957).

Hunter finds that the new natural philosophers distanced themselves from real and imagined heterodox trends among fashionable leaders of society. Although particular writers (such as Hobbes and Toland) served their purpose as targets, it was the general disrespect by "scoffers" (and not political radicals) that the scientists saw as antithetical to science and Christianity. In addition, they felt it necessary to distance themselves lest they themselves be seen as tending towards atheism.

Topics: *IX*, VII Names: Boyle, Clarke, Glanvill

0763 Hunter, Michael. *Science and Society in Restoration England.* Cambridge: Cambridge University Press, 1981.

Hunter surveys science and society in a variety of ways. Topics include: the Baconian background; the Puritan Revolution and Interregnum as background; the Royal Society; science in relation to technological and utilitarian aims; politics and reform; science and the universities; science in relation to other ideals of learning; the ancients-versus-moderns debates; science, atheism, and the fear of atheism; and science as a support for traditional religion. (He does *not* pursue science and ideology in the sense favored by J. R. Jacob and M. C. Jacob.) For each topic, Hunter considers various theses proposed by previous scholars and then states his own independent conclusions. Usually, he takes a moderate position—emphasizing the complexity and diversity of the seventeenth-century circumstances. This is true especially with respect to religion: Hunter notes many interactions between science and religion but repeatedly concludes that there are no simple correlations which can be proven. Hunter also likes to compare actual results to the aims, schemes, and theories of both seventeenth-century writers and contemporary scholars. (Usually, he shows the gaps, disappointments, or exaggerations involved.) At the end, Hunter includes a twenty-one-page bibliographic essay. This is an excellent book—both for scholars and for general readers.

Topics: *IV*, III, V, VII, IX, XII Names: Boyle, Burnet, T., Casaubon, M., Evelyn, Glanvill, Hobbes, Keill, Newton, Ray, Sprat, Stubbe, Wilkins

0764 Hunter, Michael. "The Social Basis and Changing Fortunes of an Early Scientific Institution: An Analysis of the Membership of the Royal Society, 1660-1685." *Notes Rec. Royal Soc. London* 31(1976): 9-114.

This is an earlier and shorter version of *The Royal Society and Its Fellows 1660-1700* (0761). It is completely superseded by the later book.

Topic: *V*

0765 Hunter, Michael, and Simon Schaffer, eds. *Robert Hooke: New Studies*. Woodbridge, England: Boydell Press, 1989.

See the essays by Harwood (0643), Henry (0667), Oldroyd (1167), and Shapin (1432) which are annotated separately. The volume also

includes a bibliography of secondary sources on the life and work of Hooke.
Topic: *VII* Name: Hooke

0766 Hunter, William B., Jr. "Milton and Thrice Great Hermes." *J. Engl. Germ. Philol.* 45(1946): 326-336.
Hunter shows Milton's indebtedness to the Hermetic tradition. He notes the centrality of the sun, for Milton, in the cosmos and in ideas about the soul or life principle.
Topics: *X*, II Name: Milton

0767 Hunter, William B., Jr. "Satan as Comet: Paradise Lost, II, 708-11." *Engl. Lang. Notes* 5(1967): 17-21.
Hunter argues that Milton has Satan appear both as a comet and as a supernova in *Paradise Lost*; and Hunter analyzes the view of Milton's contemporaries that comets are omens for evil.
Topic: *X* Name: Milton

0768 Hunter, William B., Jr. "The Seventeenth-Century Doctrine of Plastic Nature." *Harv. Theo. Rev.* 43(1950): 197-213.
This is a brief survey of concepts of "plastic nature" during our period. It includes coverage of Boyle, Cudworth, Harvey, and More.
Topic: *VII* Names: Boyle, Cudworth, Harvey, More

0769 Huntley, Frank. *Sir Thomas Browne: A Biographical and Critical Study*. Ann Arbor, Mich.: University of Michigan Press, 1962.
Writing in a popular style, Huntley describes Browne's early life and discusses Browne's writings. Chapters on Browne and the world of science, his apprenticeship in medicine, the *Religio Medici*, and *Vulgar Errors* lead to Huntley's conclusion that Browne lived in what the author calls "two worlds." Huntley finds that Browne unified religion and science through poetic metaphor.
Topics: *XI*, X Names: Digby, Browne, T.

0770 Huntley, Frank L. "Sir Thomas Browne and the Metaphor of the Circle." *J. Hist. Ideas* 14(1953): 353-364.
Huntley shows that Browne used the metaphor of the circle to unify his natural philosophy and his doctrine of God. In so doing,

Browne drew from contemporary medicine and astronomy—but also from hermeticism.

Topics: *VII*, II, XI Name: Browne, T.

0771 Hurlbutt, Robert H., III. *Hume, Newton, and the Design Argument*. Lincoln, Neb.: University of Nebraska Press, 1965. Revised ed.: University of Nebraska Press, 1985.

Hurlbutt is a philosopher who dislikes religion and identifies with Hume. (He believes in "the logical wall, erected by Hume, which divides science and theology, reason and religion.") Hurlbutt presents an analysis of what he calls "scientific theism" in the eighteenth century. This theism relied especially on the Newtonian design argument and it was the main target of Hume's skeptical attack against natural theology. One part of the book (about 90 pages) is devoted to Newtonian natural theology and its "blind spots." A second part is devoted to "The Ancient and Medieval Context of the Design Argument." Despite the title of the second part, Hurlbutt uses a retrospective historiography (geared towards Hume and contemporary philosophy). The second edition is an exact reprint except for an additional chapter on Hume and a short additional preface.

Topics: *VIII*, VII, IX, XII Names: Bacon, Boyle, Clarke, More, Locke, Newton, Toland, Whiston

0772 Hutcheson, Harold R. *Lord Herbert of Cherbury's "De Religione Laici." With a Critical Discussion of His Life and Philosophy, and a Comprehensive Bibliography of His Works*. New Haven, Conn.: Yale University Press, 1944.

The eighty-page introduction is useful for understanding the deistic tradition in terms other than those of natural philosophy.

Topics: *IX*, VII Name: Herbert, E.

0773 Hutchinson, F. E. *Henry Vaughn: A Life and Interpretation*. Oxford, England: Clarendon Press, 1947.

This is a straightforward biography by a literary historian. The chapters regarding occult philosophy and Vaughan as physician downplay Vaughan as a serious natural philosopher. There is an emphasis on Vaughan as a poet with somewhat of a romantic philosophy of nature. Hutchinson presents Vaughan as an orthodox and royalist Anglican with deep religious sensibilities.

Topics: *X*, II, IV, XI Names: Vaughn H., Vaughan, T.

0774 Hutchison, Keith. "Dormitive Virtues, Scholastic Qualities, and the New Philosophies." *Hist. Sci.* 29(1991): 245-278.

Hutchison analyzes the nature of the disagreements between the scholastic philosophers and their opponents in the sixteenth and seventeenth centuries. He argues that this was "a dispute about the type of universe we inhabit, about the instruments God was believed to use to achieve his ends. It involved then, a clash of ontologies and aetiologies as well as methodologies." Hutchison's discussion is mainly continental but he does discuss Alexander Ross and Isaac Newton.

Topics: *VII*, XII Names: Newton, Ross

0775 Hutchison, Keith. "Supernaturalism and the Mechanical Philosophy." *Hist. Sci.* 21(1983): 297-333.

The main purpose of this paper is to "suggest that the mechanical philosophy must also be recognized as a significant interruption to the development of naturalism." Hutchison takes a long view—tracing his theme from Aristotle (through the Middle Ages, Luther, Calvin, and Pomponazzi) to Descartes, Boyle, Cudworth, and Clarke. He argues that Aristotelianism was fundamentally naturalistic and the mechanical philosophy (with its emphasis on barren matter) was a form of supernaturalism. At least in its seventeenth-century forms, he says, God was ontologically and epistemologically necessary to mechanical philosophers. Hutchison presents his theses and arguments in a clear, thorough, and convincing manner.

Topics: *VII*, III, IV, XII Names: Boyle, Clarke, Cudworth, Newton

0776 Hutchison, Keith. "What Happened to Occult Qualities in the Scientific Revolution." *Isis* 73(1982): 233-253.

This is an important detailed discussion of the changing meanings of the term "occult" in Christian Aristotelianism and in the new philosophies. "With the acceptance of insensible agencies into the scope of natural philosophy, the word occult lost its connotation of insensible and henceforth referred solely to unintelligibility." Hutchison argues that for Newton and Newtonians, however, unintelligibility was not a token of non-effectiveness—as the dispute over gravity indicates.

Topics: *II*, VII, XII Names: Boyle, Charleton, Newton

0777 Hutin, Serge. *Henry More, Essai sur les Doctrines Theosophiques chez les Platoniciens de Cambridge*. Hildesheim, Germany: Olms, 1966.

Hutin argues that More was a theosophist and that metaphysical intuition was more important to him than rational philosophy. He treats More as an immaterialist and Kabbalist. Hutin also argues that Newton's theory of space and time was influenced by More. These are extreme theses, weakly supported.

Topics: *VII*, II, XII Names: Conway, More, Helmont, Newton

0778 Hutin, Serge. *A History of Alchemy*. Translated by Tamara Alferoff. New York: Walker, 1962.

This is a brief introductory description of the history of alchemy. Hutin postulates five main patterns or understandings of alchemy.

Topic: *II* Names: Fludd, Starkey

0779 Hutin, Serge. "Leibniz a-t-il subi l'Influence d'Henry More?" *Studia Leibnitiana* 2(1970): 59-62.

Hutin argues that Leibniz, in his development of monadism, was influenced by More (via *The Immortality of the Soul* and through the agency of F. M. van Helmont). It is based on (and uses the same questionable assumptions as) pages 194-197 of Hutin's book on More (0777).

Topic: *VII* Name: More

0780 Hutin, Serge. "Rationalisme, Empiricisme, Theosophie: La Theorie de la Connaisance de Henry More." *Filosofia* 12(1962): 570-583.

This is a brief outline of More's epistemology with extensive quotation in the notes. While intending to propose a "reasonable Christianity," More's complex philosophy (in Hutin's view) moves from a rational philosophy to a theosophic illuminationism. The essay concludes with a discussion of More's belief in a secret "primitive philosophy" and the Cabbala.

Topics: *VII*, II Name: More

0781 Hutton, Sarah. "Edward Stillingfleet, Henry More, and the Decline of *Moses Atticus*: A Note on Seventeenth-Century Anglican Apologetics." In *Philosophy, Science, and Religion in England 1640-1700*, edited by Richard Kroll and others, 68-

84. Cambridge: Cambridge University Press, 1992. See (0917).

Hutton shows the shift from More's *prisca theologia* argument to Stillingfleet's *consensus gentium* argument for demonstrating the primacy of scripture and the necessary relationship between natural philosophy and natural theology.

Topics: *VII*, II, VIII, X Names: More, Stillingfleet

0782 Hutton, Sarah, ed. *Henry More (1614-1687) Tercentenary Studies*. Dordrecht: Kluwer, 1990.

A bibliography by Crocker (1751) and essays by S. Brown (0139), Coudert (0277), Crocker (0295 and 0296), Gabbey (0522), A. R. Hall (0598), Henry (0662), and Popkin (1247) are annotated separately.

Topic: *VII* Name: More

0783 Hutton, Sarah, ed. "Introduction to the Revised Edition." In *Conway Letters*, edited by Marjorie Hope Nicolson and Sarah Hutton, vii-xix. Oxford: Clarendon Press, 1992. See (1119).

This is an appreciation of Marjorie Hope Nicolson's original work and a brief explanation of the additions to the revised edition. It also includes a review of the related scholarship since 1930 and many footnotes with bibliographic references. As with Nicolson's book, science is rather tangential.

Topics: *VII*, II, X Names: Conway, More

0784 Hutton, Sarah. "More, Newton and the Language of Biblical Prophecy." In *The Books of Nature and Scripture*, edited by James E. Force and Richard H. Popkin, 39-53. Dordrecht: Kluwer Academic, 1994. See (0498).

On the language of biblical prophecy, Hutton compares More and Newton to each other and to Joseph Mede, "their point of departure." She finds that More had a greater sense of allegory and emphasized the spiritual significance of the visions; Newton, on the other hand, had "a constant drive towards the literal" (although sometimes obscure) and stressed their historical significance.

Topics: *X*, XII Names: Mede, More, Newton

0785 Hutton, Sarah, ed. *Of Mysticism and Mechanism: Tercentenary Studies of Henry More (1614-1687)*. Dordrecht: Kluwer, 1989.

This title appears in some bibliographies; but the work was actually published as *Henry More (1614-1687) Tercentenary Studies*, 1990. See (0783).

Topic: *VII* Name: More

0786 Hutton, Sarah. "Science, Philosophy, and Atheism: Edward Stillingfleet's Defence of Religion." In *Scepticism and Irreligion in the Seventeenth and Eighteenth Centuries*, edited by Richard H. Popkin and Arjo Vanderjagt, 102-120. Leiden: E. J. Brill, 1993. See (1250).

Hutton focuses on the two versions of Stillingfleet's *Origines Sacrae*, 1662 and 1702 (the latter version published three years after his death). Based on a comparison of these works, Hutton argues that Stillingfleet: (1) was motivated by a desire to keep Christian apologetics up-to-date with the latest trends in contemporary thought; and (2) was highly receptive to the new science in his apologetics. In the course of arguing her main points, Hutton takes issue with several theses about Newton and Newtonianism put forward by M. C. Jacob. The essay (although characterized by awkward sentence constructions and typographical errors) is well-argued.

Topics: *VIII*, XII Names: Bentley, Boyle, Harris, More, Newton, Ray, Stillingfleet

0787 Huxley, George L. "Two Newtonian Studies." *Harvard Library Bulletin* 13(1959): 348-361.

Part I is entitled "Newton's Boyhood Interests" and Part II is entitled "Newton and Greek Geometry." The studies include a few references to Newton's Bible study, his learning of Hebrew script, and his interest in theology. Huxley emphasizes Newton's deep "feeling for antiquity."

Topics: *XII*, VII, X Name: Newton

0788 Iliffe, Rob. "'Making a Shew': Apocalyptic Hermeneutics and Sociology of Christian Idolatry in the Work of Isaac Newton and Henry More." In *The Books of Nature and Scripture*, edited by James E. Force and Richard H. Popkin, 55-88. Dordrecht: Kluwer Academic, 1994. See (0498).

After presenting a short account of the work of Joseph Mede, Iliffe examines the expository works written by More from the 1660s into the 1680s. He then considers the unpublished manuscripts written by Newton during the same period. He analyzes in detail the dispute between Newton and More over the interpretation of the apocalyptic books. The sociological feature of their respective hermeneutics, he argues, is seen in the division over who can correctly understand the Apocalypse. For More, it was comprehensible to anyone who made an effort to learn the key. For Newton, only a remnant of natural philosophers could even provisionally understand the text and provide guides for others to follow. Each saw, in his own interpretation, the causes of the idolatry of the Catholic Church.

Topics: *X*, XII Names: Mede, More, Newton

0789 Inglis, Brian. *Natural and Supernatural: A History of the Paranormal from Earliest Times to 1914*. London: Hodder and Stoughton, 1977.

This is a serious popular survey which treats paranormal phenomena *as if* they may have occurred. Inglis carefully chooses incidents which had credible witnesses. Brief descriptions of occurrences in the seventeenth and early-eighteenth centuries are found in the chapters on "Miracles," "Witchcraft," and "Ghosts." The book, however, is mostly a history of nineteenth-century investigations.

Topic: *II*

0790 Jacob, Alexander. "Henry More's *Psychodia Platonica* and its Relationship to Marsilio Ficino's *Theologia Platonica*." *J. Hist. Ideas* 46(1985): 503-522.

Concerning Ficino's influence on More, Jacob argues against the strong claims made by C. A. Staudenbaur (1486). Jacob systematically notes differences between More's views and those of Ficino. He concludes that More's vision of the cosmos was informed by a variety of ancient and modern Neoplatonists.

Topic: *VII* Name: More

0791 Jacob, Alexander. Introduction and Notes to *Henry More: The Immortality of the Soul*, edited by Alexander Jacob, i-ciii and 328-446. (International Archives of the History of Ideas, 122) Dordrecht: Nijhoff, 1987.

An extensive bibliography brings the scholarly apparatus to over 240 pages. In his sections, Jacob focuses on More's philosophical

predecessors and the seventeenth-century context. He emphasizes More's Neoplatonism.
Topics: *VII*, XII Names: Hobbes, More, Newton

0792 Jacob, Alexander. "The Neoplatonic Conception of Nature in More, Cudworth, and Berkeley." In *The Uses of Antiquity*, edited by Stephen Gaukroger, 101-121. Dordrecht: Kluwer Academic, 1991. See (0530).
More, Cudworth, and Berkeley all adapted Neoplatonic metaphysics to the physics and physiology of their own time. They agreed on a vitalistic interpretation of nature and believed this to be a bulwark against the threat of atheism posed by materialistic and mechanistic philosophies. But they also had subtle differences in their understandings of God, the physical universe, and any soul or universal spirit intermediate between God and the sensible world. Jacob pursues these themes and explicates the Neoplatonic conceptions with detailed precision.
Topic: *VII* Names: Cudworth, More

0793 Jacob, James R. "Aristotle and the New Philosophy: Stubbe versus the Royal Society." In *Science, Pseudo-Science, and Society*, edited by Marsha P. Hanen, 217-236. Waterloo, Ont.: Wilfrid Laurier University Press, 1980. See (0623).
Jacob sees Stubbe as a political and theological (Arian) radical—yet as an Aristotelian! His opposition to the new philosophy was not that of an orthodox conservative. This essay is incorporated into Jacob's later book (0798).
Topics: *IV*, V, VII, IX Name: Stubbe

0794 Jacob, James R. "The Authorship of *An Account of the Rise and Progress of Mahometanism*." *Notes and Queries* 26(1979): 10-11.
Jacob argues that the author was Henry Stubbe and that Stubbe was one of the first Englishmen to appreciate Islam in an unbiased way. This essay is incorporated into and superseded by Chapter 4 of Jacob's later book on Stubbe (0798).
Topic: *IX* Name: Stubbe

0795 Jacob, James R. "Boyle's Atomism and the Restoration Assault on Pagan Naturalism." *Soc. Stud. Sci.* 8(1978): 211-233.

Jacob emphasizes the political and religious motives of Boyle's physico-theology, especially as expressed in his *Free Enquiry into the Vulgarly Received Notion of Nature* (written in 1666 but published in 1686). It is incorporated into Chapter 8 of Jacob's later book, (0799).

Topics: *IV*, VIII Name: Boyle

0796 Jacob, James R. "Boyle's Circle in the Protectorate: Revelation, Politics, and the Millennium." *J. Hist. Ideas* 38(1977): 131-140.

Jacob focuses on Boyle, his sister (Katherine, Lady Ranelagh), and their circle of friends. He argues that this group believed that the knowledge of both nature and scripture—and understanding their harmony—would produce an irenical effect. They believed that the millennium would be marked by the unity of religion and liberty of conscience as well as a moral, intellectual and spiritual reformation. (It would not be marked by a social or institutional revolution.) Jacob also argues that through Boyle's brother (Roger, Baron Broghill) the group was influential in the councils of the Protectorate. This article has been incorporated into Jacob's book on Boyle (0804).

Topics: *X*, IV, V Names: Boyle, Broghill, Ranelagh

0797 Jacob, James R. "'By an Orphean Charm': Science and the Two Cultures in Seventeenth-Century England." In *Politics and Culture in Early Modern Europe*, edited by Phyllis Mack and Margaret Jacob, 231-250. Cambridge: Cambridge University Press, 1987.

Jacob first sets this frame: during the seventeenth century, the social and intellectual elite became increasingly suspicious and hostile to popular culture. For religious and ideological reasons, they sought to control and direct the culture of the masses. Jacob then argues that the development of natural philosophy in England and Italy conforms (and contributed) to this pattern. Moreover, Henry Stubbe was right when "he saw that the Royal Society was attempting to make the conduct of science serve the pursuit of conservative religious and political goals." If some historians see the new science as a substitute for religion—such would be true only for the elite and certainly not for the common people.

Topics: *IV*, V Names: Bacon, Stubbe, Wren

0798 Jacob, James R. *Henry Stubbe, Radical Protestantism and the Early Enlightenment*. Cambridge: Cambridge University Press, 1983.

Jacob presents Stubbe as a consistent radical (theologically, philosophically, and ideologically) whose apparent shifts of position before and after 1660 simply reflect the changing political and rhetorical situation. Jacob sees Stubbe as a vitalistic materialist and paganizing naturist who also was a radical republican. This book is well-written and carefully articulated. See also the annotations for (0793) and (0794).

Topics: *IV*, IX Names: Beale, Blount, Boyle, Glanvill, Greatrakes, Hobbes, More, Sprat, Stubbe, Toland, Wallis, Willis

0799 Jacob, James R. "The Ideological Origins of Robert Boyle's Natural Philosophy." *J. Euro. Stud.* 2(1972): 1-21.

Jacob relates Boyle's science to theological and social factors and to his ethical views. This article overlaps with and is superseded by parts of Chapters 2 and 3 of Jacob's book (0804).

Topics: *IV*, III, VIII Name: Boyle

0800 Jacob, James R. "The New England Company, the Royal Society, and the Indians." *Soc. Stud. Sci.* 5(1975): 450-455.

Jacob argues that, in the 1660s and 1670s, an attempt was made to use the natural philosophy of the Royal Society as an instrument in civilizing and Christianizing the Indians of New England. The Protestant "work ethic" was central to this ideology. The essay overlaps with and has been superseded by part of Chapter 4 of Jacob's book (0804).

Topics: *IV*, III, V Names: Beale, Boyle, Sprat

0801 Jacob, James R. "Restoration Ideologies and the Royal Society." *Hist. Sci.* 18(1980): 25-38.

Here, Jacob explores the polemical use to which the science of the Royal Society was put by Tory pamphleteers during the decade of 1678-1688. He argues that, while the cause of Protestantism was central to Whig sociopolitico-natural-philosophic ideology, religion was basically ignored in the Tory use of natural philosophy for ideological purposes. This is an essay *not* reproduced in any of Jacob's other publications.

Topics: *IV*, V Names: Boyle, Houghton, Pett

0802 Jacob, James R. "Restoration, Reformation and the Origins of the Royal Society." *Hist. Sci.* 13(1975): 155-176.

Jacob argues that the views of the founders of the Royal Society were *not* (as some historians have claimed) apolitical, unrelated to religion, and without commitment to a specific social and economic ideology. He details "the aggressive, acquisitive, mercantilist ideology justified in the name of both Restoration and Reformation" for which these men understood themselves and their natural philosophy to stand. This article is reproduced (almost word for word) in Chapter 4 of Jacob's book (0804).

Topics: *IV*, V Names: Boyle, Pett, Sprat

0803 Jacob, James R. "Robert Boyle and Subversive Religion in the Early Restoration." *Albion* 6(1974): 275-293.

Jacob applies his ideological analysis to Boyle in the 1660s. He argues that Boyle and other fellows of the Royal Society attempted to use their natural philosophy to counter religious sectarianism. This essay has been superseded by Jacob's later works, especially by his book on Boyle (0804).

Topics: *IV*, V Names: Boyle, Sprat

0804 Jacob, James R. *Robert Boyle and the English Revolution: A Study in Social and Intellectual Change*. New York, 1977.

Jacob here analyzes Boyle's theological, ethical, social, political, and natural philosophical views as intertwining threads of a single whole. He relates all of these to Boyle's experience of the revolution in England in the 1640s and 1650s—as well as to his experience of the Restoration in the 1660s. For Jacob, Boyle's science and ideology are deeply affected by his presuppositions about God's creative power and providential purpose in both the natural world and in the political world. This is an important book in the history of science, religion, and ideology. See also the annotations for (0796), (0799), (0800), (0802), and (0803)—all of which articles have been incorporated into this book.

Topics: *IV*, III, V, VII, X Names: Beale, Boyle, Dury, Greatrakes, Hartlib, Oldenburg, Pett, Ranelagh, Sprat, Stubbe, Ussher, Wallis, Wilkins

0805 Jacob, James R., and Margaret C. Jacob. "The Anglican Origins of Modern Science: The Metaphysical Foundations of the Whig Constitution." *Isis* 71(1980): 251-267.

The Jacobs emphasize the importance of ideological factors in the development of science in seventeenth-century England. The key religious party after 1660, they argue, was latitudinarian Anglicanism—and not Puritanism. The Anglican ideology which supported the new science included some elements of a Puritan reforming vision but it rejected Puritan sectarianism and radical Puritan politics. The "Newtonian Enlightenment," (a form of Anglicanism but yet also a transformed Puritanism) was a vast holding action against materialism and its concomitant republicanism: it must be distinguished from the eighteenth-century "Radical Enlightenment." And the triumph of Newtonian science represents a victory for the Whig constitution.
Topics: *IV*, III, XII Names: Boyle, Newton, Stubbe

0806 Jacob, James R., and Margaret C. Jacob. "The Saints Embalmed: Scientists, Latitudinarians and Society: A Review Essay." *Albion* 24(1992): 435-442.
This is an essay review of *Philosophy, Science, and Religion in England, 1640-1700*, edited by Richard Kroll and others (0917). In a lively critique of Kroll's "Introduction" and of Hunter's essay in the volume, the Jacobs reiterate their view of the social, political, and ideological role of the Latitudinarians in developing Restoration natural philosophy. The Jacobs point to the other essays in the volume as justifying their own position.
Topics: *IV*, VII

0807 Jacob, James R., and Margaret C. Jacob. "Scientists and Society: The Saints Preserved." *J. Euro. Stud.* 1(1971): 87-92.
This is an essay review of Shapiro (1439) and Manuel (0988). The Jacobs note some value in each book. But they criticize each author for not enough awareness of social context and not enough emphasis on the ideological significance of science and religion.
Topics: *IV*, I, III, V, VII, XII Names: Newton, Wilkins

0808 Jacob, James R., and Margaret C. Jacob. "Seventeenth Century Science and Religion: the State of the Argument." *Hist. Sci.* 14(1976): 196-207.
This is an essay review of several books and articles specifically in terms of a historiography which emphasizes socioeconomic and political context. The Jacobs criticize the works not only for

underestimating the larger historical context but also for overworking historical labels and categories. The focus is especially on the *Past and Present* collection edited by Charles Webster (1618) and Manuel's *The Religion of Isaac Newton* (0989).

Topics: *I*, III, XII Name: Newton

0809 Jacob, Margaret C. "Christianity and the Newtonian World View." In *God and Nature*, edited by David C. Lindberg and Ronald Numbers, 238-255. Berkeley and Los Angeles: University of California Press, 1986. See (0954).

Jacob here summarizes her theses about a Newtonian worldview articulated by latitudinarian Anglican clerics. This, she says, offered a model of a stable, providentially-guided universe that justified the post-revolutionary socioeconomic order of commercial capitalism, political stability, a hierarchical society, and scientific progress. Jacob also comments on the relation of a second generation of Newtonians to deism—and the eighteenth-century Christian opposition to Newtonianism.

Topics: *IV*, VII, IX, X, XII Names: Bentley, Clarke, Newton

0810 Jacob, Margaret C. "The Church and the Formulation of the Newtonian Worldview." *J. Euro. Stud.* 1(1971): 128-148.

This was a groundbreaking article, which was later incorporated into *The Newtonians and the English Revolution* (0817). Jacob sees the "Newtonian ideology" as more the creation of Churchmen than of Newton. This ideology consists of the application of Newton's ideas on natural order to Anglican ideas of social order.

Topics: *IV*, V, VIII, XII Names: Evelyn, Lloyd, Newton, Tenison

0811 Jacob, Margaret C. "The Crisis of the European Mind: Hazard Revisited." In *Politics and Culture in Early Modern Europe: Essays in Honor of H. G. Koenigsberger*, edited by Phyllis Mack and Margaret C. Jacob, 251-271. Cambridge: Cambridge University Press, 1987.

Jacob uses a reconsideration of Hazard's theses to extend her own theories concerning politics, ideology, liberal Protestant theology, and the new science. Political developments and the new science (or the breakdown of the old science) provoked *la crise*; the creation of an elite and enlightened culture (liberal religion wedded to the new science) resolved it. Jacob argues that Hazard was incorrect in connecting *la crise* to the French Revolution and the development of democracy; actually, she tries to show, the Newtonian-Lockean resolution of the crisis served the interests of the social and intellectual elites.

Topics: *IV*, VII, XII Names: Locke, Newton

0812 Jacob, Margaret C. *The Cultural Meaning of Scientific Revolution*. New York: A. A. Knopf, 1988.

This volume is part of an advanced college textbook series. In general, Jacob seeks to explain the complex historical process by which scientific and technological knowledge became a dominant part of Western European culture in the seventeenth and eighteenth centuries. The "cultural meaning" of science and technology are described in social, political, and religious terms. This book is a good introduction to positions and theses expressed by Jacob in her other publications. She deals with natural philosophy and its interaction with religion: (1) in Bacon and Baconianism; (2) among Puritan moderates and radicals; (3) among Latitudinarians; (4) among deists; and (5) among Protestant dissenters. Although she emphasizes the elitism of natural philosophers, she also is concerned with the audiences and promoters of science and industrialization. The book includes a useful glossary of terms for the advanced undergraduate level.

Topics: *IV*, III, VI, VII, IX, XII Names: Bacon, Boyle, Clarke, Desaguliers, Hobbes, Newton

0813 Jacob, Margaret C. "Early Newtonianism." *Hist. Sci.* 12(1974): 142-146.

This is a call for more research directed toward the social and ideological aspects of early Newtonianism. (It now is somewhat outdated in that her call has since been somewhat answered!)

Topics: *IV*, I, V, VIII, XII Name: Newton

0814 Jacob, Margaret C. "John Toland and the Newtonian Ideology." *J. Warburg Cour. Inst.* 32(1969): 307-331.

Jacob analyzes the interaction between the philosophical and religious views of Toland and the natural philosophy of the Newtonians. She treats Toland's views as a transformation of the Hermetic tradition (and especially of Bruno's philosophy of nature). There is significant overlap, but by no means an identity,

between this essay and Chapter 6 of Jacob's book (0817). She includes references to the social and political context—but there is less explication of these matters than in Jacob's later writings.
Topics: *IX*, II, IV, XII Names: Clarke, Newton, Toland

0815 Jacob, Margaret C. "Millenarianism and Science in the Late Seventeenth Century." *J. Hist. Ideas* 37(1976): 335-341. Reprinted in *Philosophy, Religion and Science in the Seventeenth and Eighteenth Centuries*, edited by John W. Yolton, 493-499. Rochester, N.Y.: University of Rochester Press, 1990. See (1720).
Jacob here describes the understanding by Latitudinarian divines of an approaching earthly paradise characterized by a stable political and social environment. They applied a Newtonian form of natural philosophy as a proof by analogy of the providential approval of such order. Thus, they used natural philosophy for religo-ideological purposes.
Topics: *IV*, X, XII Names: Evelyn, Newton

0816 Jacob, Margaret C. "Newton and the French Prophets: New Evidence." *Hist. Sci.* 16(1978): 134-142.
Jacob here extends some of her themes found in *The Newtonians and the English Revolution* (0817). She argues that Newton knew of (and perhaps agreed with) Fatio de Duillier's millenarian interpretations of current events in the first decade of the eighteenth century. Other than Newton's person, the article has nothing to do with science.
Topics: *XII*, IV Names: Fatio, Newton

0817 Jacob, Margaret C. *The Newtonians and the English Revolution, 1689-1720*. Ithaca, N.Y.: Cornell University Press, 1976.
Jacob forcefully argues the ideological significance of science. She correlates sociopolitical and ecclesiastical positions with corresponding approaches to natural philosophy. And she delineates four basic ideological views in late-seventeenth and early-eighteenth-century England. These four positions are: (1) the Latitudinarians (represented by Wilkins, Boyle, the Cambridge Platonists, and the founders of the Royal Society); (2) the Newtonians (represented by Newton, Clarke, Bentley, and other Boyle lecturers); (3) the Freethinkers (represented by Hobbes, Stubbe, and Toland); and (4) the

Enthusiasts (represented by radical groups of the Interregnum and the French Prophets in the first decade of the eighteenth century). The Newtonians, for Jacob, are a later form of Latitudinarianism. She maintains that various social, political, and religious factors were: (a) interpreted in terms of the new science; (b) shaped by the new science; and (c) facilitated the acceptance of the new science. In other words, the early success of the new mechanical philosophy in England was *not* simply because it offered the most plausible explanation of nature. She says that the noted correlations are not logically necessary but are "linkages" which historically did occur. Stylistically, the book reads like a series of articles or essays organized around a given set of themes; it is not thorough or systematic. Her book is open to some criticism: she does not correlate any "ideology" or natural philosophical position to Roman Catholics or to "high-church" Anglicans; she uses her broad categories (with just *one* ideology for each) for extremely heterogeneous groups; in places, she seems unaware that political ideologies have always used analogies from philosophy of nature. But her work remains influential, provocative, and useful against noncontextual historiography of science.

Topics: *IV*, V, VII, VIII, IX, XII Names: Barrow, Bentley, Boyle, Burnet, T., Clarke, Collins, A., Cudworth, Derham, Evelyn, Harris, Hobbes, Moore, More, Newton, Sprat, Stubbe, Tenison, Tillotson, Tindal, Toland, Whiston, Wilkins, Wotton

0818 Jacob, Margaret C., and W. A. Lockwood. "Political Millenarianism and Burnet's *Sacred Theory*." *Science Studies* 2(1972): 265-279.

Jacob and Lockwood interpret Thomas Burnet's natural philosophy, biblical theology, and millenarianism by correlating two editions of his work to the differing political situations before and after the events of 1688-89. There is significant overlap but not an identity between this article and portions of Chapter 3 of Jacob's book (0817).

Topics: *IV*, X, XII Names: Burnet, T., Newton

0819 Jacquot, Jean. "Harriot, Hill, Warner and the New Philosophy." In *Thomas Harriot: Renaissance Scientist*, edited by John W. Shirley, 107-128. Oxford, England, 1974. See (1450).

Jacquot proposes that the transition from "Neoplatonic and Hermetic speculation to seventeenth-century mechanism" may be seen in the writings of the three men named in the title. He briefly touches on Harriot's atomism and the problem it posed for the doctrine of creation. Nicholas Hill's *Philosophia* and Walter Warner's papers receive more attention from Jacquot. He summarizes Hill's views on God as the source of energy in the infinite, eternal, and animated universe. And he analyzes Warner's understanding of cosmology and of the human organism.

Topic: *VII*, II Names: Harriot, Hill, Warner (1570)

0820 Jacquot, Jean. "Thos. Harriot's Reputation for Impiety." *Notes Rec. Royal Soc. London* 9(1952): 164-187.

Jacquot is responding to the question as to why Harriot, later in the seventeenth century, was perceived to be an unbeliever. He discusses issues related to whether Harriot actually rejected the book of Genesis. His atomism may have been the primary cause of his reputation for impiety because his trusted executor, Nathaniel Torporley, wrote a refutation of Harriot's reported atomistic views. Jacquot finds Harriot's infamous phrase *ex nihilo nihl fit* to have been taken out of context.

Topics: *IX*, VII, X Names: Harriot, Torporley

0821 Jaki, Stanley L. *The Origin of Science and the Science of its Origin*. South Bend, Ind.: Regnery/Gateway, 1979.

In these Fremantle public lectures, Jaki elaborates on his basic thesis that science originated in the Gospel (and particularly in the doctrine of creation). In five chapters and beginning with Bacon, Jaki evaluates other theories of origins—using various historical examples to illustrate his historiographic arguments.

Topics: *I*, VII, X Name: Bacon

0822 Jaki, Stanley L. *The Road of Science and the Ways of God*. Chicago: University of Chicago Press, 1978.

In these Gifford Lectures, Jaki's thesis is simple: "the road of science . . . is a logical access to the ways of God." His erudition is remarkable as he ranges through the history of science in the first half of the book (and treats twentieth-century science in the second half). Yet his single-mindedness makes the book have more value for apologetics than as a history. In Chapter 4, Jaki discusses

Bacon as empiricist-politician. In Chapter 6, he deals with Newton, whose "middle road" means having "an epistemological median, the common basis of the road of science" and natural theology.
Topics: *VIII*, I, VII, XII   Names: Bacon, Newton

0823 Jammer, Max. *Concepts of Force: A Study in the Foundations of Dynamics*. Cambridge, Mass.: Harvard University Press, 1957. New edition. New York: Harper Torchbooks, 1962.
See (0824).
Topics: *VII*, XII   Names: More, Newton

0824 Jammer, Max. *Concepts of Space*. Cambridge, Mass.: Harvard University Press, 1954. 2nd ed. Cambridge, Mass.: Harvard University Press, 1969.
Jammer's two books are broad surveys. For our period and topics, his discussions are not dependable in all details but he does show an awareness of the importance of the interplay between science and religion. He specifically discusses the issue of God as an explanatory factor in natural philosophy. His books are worth reading for survey background. (Jammer's *Concepts of Mass*, incidentally, is not relevant to the topics of this bibliography.)
Topics: *VII*, XII   Names: More, Newton

0825 Jardine, Lisa. *Francis Bacon: Discovery and the Art of Discourse*. London: Cambridge University Press, 1974.
Jardine basically assumes (rather than describes) Bacon's separation of theology from natural philosophy. There are only a few references in this book to religion, theology, God, or the Bible. Otherwise, there is significant attention given to Bacon's theory of knowledge, logic, and natural philosophy.
Topics: *VII*, X   Name: Bacon

0826 Jenkins, Alan C. *The Naturalists: Pioneers of Natural History*. London: Hamish Hamilton, 1978. New York: Mayflower Books, 1978.
Chapter 4 is a brief pictorial biography of John Ray with an acknowledgement of his religious views.
Topics: *VII*, VIII   Name: Ray

0827 Jeske, Jeffrey. "Cotton Mather: Physico-Theologian." *J. Hist. Ideas* 47(1986): 583-594.

Jeske emphasizes a contradictory rhetoric found in Mather's sermons and books on science. Jeske proposes an updated version of the thesis that there is a "fundamental split" in the Puritan mind between Calvinism and the new science. Likewise, he proposes that Mather's writings (unintentionally) show intellectual forces which were "inexorably drawing Puritanism toward deism."

Topics: *VIII*, III, IX, X Name: Mather, C.

0828 Jobe, Thomas H. "The Devil in Restoration Science: The Glanvill-Webster Witchcraft Debate." *Isis* 72(1981): 343-356.

Jobe contends that, behind this struggle between two kinds of science—a Paracelsian-Helmontian science and a mechanical corpuscularianism—lay a clash between a radical Protestant and an orthodox Anglican theology. He finds three essential issues in the debate: (1) the charge that the Devil plays a role in Paracelsian science; (2) a dispute over reliable testimony (human or biblical) regarding the spirit world; and (3) a controversy over the relation between spirit and matter, especially regarding specters.

Topics: *II*, VII, X Names: Glanvill, Webster

0829 Johannisson, Karin. "Magic, Science, and Institutionalization in the Seventeenth and Eighteenth Centuries." In *Progress in Science and Its Social Conditions: Nobel Symposium 58*, edited by Tord Ganelius, 51-59. Oxford: Pergamon Press, 1985. Reprinted in *Hermeticism and the Renaissance*, edited by Ingrid Merkel and Allen G. Debus, 251-262. Washington, D.C.: Folger Shakespeare Library; London: Associated Universities Presses, 1988. See (1051). Originally published as "Magi, Vetenskap och Institutionalisering under 1600-och 1700-talen." *Lychnos* (1984): 121-131.

Johannisson argues that Hermetic magic provided several ideological elements in the development of modern science. In particular, she concentrates on institutionalization (science must be propagated through institutions in order for it to be a transnational instrument of transformation and progress). This is a sociological follow-up to a Francis Yates thesis.

Topics: *V*, II

0830 Johanssen, Bertil. *Religion and Superstition in the Plays of Ben Jonson and Thomas Middleton*. Upsala, Sweden: Lundsquist; Cambridge, Mass.: Harvard University Press, 1950.

Johanssen deals with religion and astrology, alchemy, witchcraft and magic in the early seventeenth century.

Topics: *X*, II Names: Jonson, Middleton

0831 Johnson, Francis R. *Astronomical Thought in Renaissance England: A Study of the English Scientific Writings from 1500 to 1645*. Baltimore: Johns Hopkins Press, 1937.

Johnson only briefly touches on issues of science and religion in the seventeenth century. Samuel Purchas, for one, found no conflict between the Genesis account of creation and the new astronomy.

Topics: *X*, VII Names: Purchas, Tymme

0832 Johnson, Francis R. "Gresham College: Precursor of the Royal Society." *J. Hist. Ideas* 1(1940): 413-438. Reprinted in *Roots of Scientific Thought*, edited by Philip P. Wiener and Aaron Noland, 328-353. New York: Basic Books, 1957. See (1670).

Johnson proposes that there existed an informal club of natural philosophers, prominent officials, shipbuilders, and navy captains who met under the sponsorship of the Gresham professors of astronomy and geometry for fifty years prior to 1645. In the 1620s and 1630s, this group was led by men known to be Puritan. Thus, an acknowledged precursor had a history which pre-dated the memories of the original members of the Royal Society.

Topics: *V*, III Names: Briggs, Gellibrand, Wallis, Wells

0833 Jolley, Nicholas. "Leibniz on Locke and Socinianism." *J. Hist. Ideas* 39(1978): 233-250. Reprinted in *Philosophy, Religion and Science in the Seventeenth and Eighteenth Centuries*, edited by John W. Yolton, 170-187. Rochester, N.Y.: University of Rochester Press, 1990. See (1720).

Jolley discusses doctrines of matter and spirit within the metaphysical and theological contexts of Leibniz and Locke.

Topics: *VII*, IX Name: Locke

0834 Jones, Gordon W. Introduction to *The Angel of Bethesda by Cotton Mather*, edited by Gordon W. Jones, xi-xl. Barre,

Mass.: American Antiquarian Society and Barre Publishers, 1972.

The brief section on the unpublished manuscript, "The Angel of Bethesda," usefully describes Mather's style, sources, and library. Jones notes the "religious tone" of the text, which is "a fascinating summary of medical practice, beliefs, and theory of the seventeenth century."

Topics: *XI*, X Name: Mather, C.

0835 Jones, Richard F. *Ancients and Moderns: A Study of the Background of the Battle of the Books*. St. Louis: Washington University, 1936. 2nd ed., with new preface and minor revisions, *Ancients and Moderns: A Study of the Rise of the Scientific Movement in Seventeenth-Century England*. St. Louis: Washington University, 1961.

This was a pioneering work in the 1930s by a literary historian. It is an important book even for those who disagree with Jones' theses. He emphatically stresses "Baconianism" as the most important movement in the rise of science in seventeenth-century England. To Baconianism, he relates Puritanism, the advancement of learning, the ancients and moderns controversies, the idea of progress, and questions regarding the Royal Society and its precursors. It also represents an important (and independent) discovery of the "Merton thesis" as regards Puritanism and the rise of science. Absent from Jones' interpretation, of course, is the serious consideration of Paracelsian and other Renaissance magic traditions.

Topics: *X*, III, IV, V Names: Bacon, Boyle, Glanvill, Goodman, Hakewill, Hartlib, Petty, Stubbe, Wilkins

0836 Jones, Richard F. "The Attack on Pulpit Eloquence in the Restoration: an Episode in the Development of the Neo-Classical Standard for Prose." *J. Engl. Germ. Philol.* 30(1931): 188-217. Reprinted in *The Seventeenth Century*, edited by Richard F. Jones and others, 111-142. Stanford. Calif.: Stanford University Press, 1951. See (0845). Reprinted in *Essential Articles: For the Study of English Augustan Backgrounds*, edited by Bernard N. Schilling, 103-136. Hamden, Conn.: Archon Books, 1961. See (1411).

In three articles, including (0842) and (0843), Jones presents and defends his theses concerning the effects of science on literary style during the last quarter of the seventeenth century. The scientific ideal of style (plainness, directness, clearness) gained ascendancy over the elaborate, figurative, and rhetorical ideal characteristic of religious writers and preachers of the period before 1660.
Topics: *X*, IV Names: Glanvill, Parker

0837 Jones, Richard F. "The Background of the Attack on Science in the Age of Pope." In *Pope and His Contemporaries*, edited by James L. Clifford and Louis A. Landa, 96-113. Oxford: Clarendon Press, 1949. See (0224).
Jones here discusses the social sources, including the religious ones, of anti-scientific sentiment in restoration England. He also discusses the socially-oriented attempts to defend the new science.
Topics: *IV*, X Names: Pope, Swift

0838 Jones, Richard F. "The Background of *The Battle of the Books*." *Wash. Univ. Stud.*, 7, Humanistic Series 2(1920): 99-162. Reprinted in abridged form in *The Seventeenth Century*, edited by Richard F. Jones and others, 10-40. Stanford. Calif.: Stanford University Press, 1951. See (0845).
This is a shorter form of Jones' *Ancients and Moderns: A Study of the Background of the "Battle of the Books"* (0835). The full-length book has revised and less extreme forms of Jones' theses and thus supersedes this essay in either printing.
Topics: *X*, III, IV Names: Bacon, Glanvill, Wotton

0839 Jones, Richard F. "The Bacon of the Seventeenth Century." In *Essential Articles for the Study of Francis Bacon*, edited by Brian Vickers, 3-27. Hamden, Conn.: Archon Books, 1968. See (1588).
This is an edited reprint of pp. 41-61 from Jones' book, *Ancients and Moderns*, 1961. See (0835).
Topic: *X* Name: Bacon

0840 Jones, Richard F. "The Humanistic Defense of Learning in the Mid-Seventeenth Century." In *Reason and the Imagination*, edited by Joseph Anthony Mazzeo, 71-92. New York: Columbia University Press, 1962. See (1004).

Jones discusses debates carried on between Anglicans and Puritans concerning the proper functions and content of rhetoric, education and learning.

Topics: *IV*, III, X   Names: Gauden, Waterhouse

0841 Jones, Richard F. "Puritanism, Science and Christ Church." *Isis* 31(1939): 65-67.

Jones provides one more bit of evidence supporting the Puritanism thesis of Merton. It is also relevant to the issue of educational reform.

Topics: *III*, IV

0842 Jones, Richard F. "Science and English Prose Style in the Third Quarter of the Seventeenth Century." *PMLA* 45(1930): 977-1009. Reprinted in *The Seventeenth Century*, edited by Richard F. Jones and others, 75-110. Stanford, Calif.: Stanford University Press, 1951. See (0845). Reprinted in *Essential Articles: for the Study of English Augustan Backgrounds*, edited by Bernard N. Schilling, 66-102. Hamden, Conn., 1961. See (1411). Also reprinted as "Science and English Prose Style, 1650-1675" in *Seventeenth-Century Prose: Modern Essays in Criticism*, edited by Stanley E. Fish. New York: Oxford University Press, 1971.

See (0836).

Topics: *X*, IV   Names: Boyle, Glanvill, Tillotson, Wilkins

0843 Jones, Richard F. "Science and Language in England of the Mid-Seventeenth Century." *J. Engl. Germ. Philol.* 31(1932): 315-321. Reprinted in *The Seventeenth Century*, edited by Richard F. Jones and others, 143-160. Stanford, Calif.: Stanford University Press, 1951. See (0845). Also reprinted in *Seventeenth-Century Prose: Modern Essays in Criticism*, edited by Stanley E. Fish. New York: Oxford University Press, 1971.

See (0836).

Topics: *X*, IV   Names: Bacon, Boyle, Wilkins

0844 Jones, Richard F. "Science, Technology, and Society in Seventeenth Century England." *Isis* 21(1940): 438-441.

This is Jones' favorable review of Merton's controversial monograph (1057).

Topics: *III*, VI

0845 Jones, Richard F., and others. *The Seventeenth-Century: Studies in the History of English Thought and Literature from Bacon to Pope*. Stanford, Calif.: Stanford University Press, 1951.

This is an anthology of twenty essays and articles, seven of which are listed elsewhere in this bibliography. Three of the relevant ones are reprints of articles previously published by Jones. See the following entries: Fulton (0515); R. F. Jones (0836), (0842), and (0843); Nicolson (1136); Sherburn (1447); and Willey (1682).

Topic: *X*

0846 Jones, William Powell. *The Rhetoric of Science: A Study of Scientific Ideas and Imagery in Eighteenth-Century English Poetry*. London: Routledge and Kegan Paul, 1966.

For the period 1660-1720, this book provides an excellent discussion of the influence of science on the intellectual milieu. It does not break new ground but synthesizes most of the important work on this topic done by literary historians up to the mid-1960s.

Topics: *X*, IV, VIII, XII Names: Blackmore, Newton, Pope

0847 Jordonova, L. J., and Roy S. Porter, eds. *Images of the Earth: Essays in the History of the Environmental Sciences*. (BSHS Monographs, No. 1.) Chalfont St. Giles, England: British Society for the History of Science, 1979.

This anthology includes two relevant items: the introduction (0848) and the essay by Cantor (0179).

Topics: *I*, X

0848 Jordanova, L. J., and Roy S. Porter. "Introduction." In *Images of the Earth*, edited by L. J. Jordanova and Roy S. Porter, v-xx. Chalfont St. Giles, England: British Society for the History of Science, 1979. See (0847).

Jordanova and Porter describe how the historiographic patterns of the general history of science relate to studies of the history of geology. Attention is drawn to the intertwining of religion and theories of the earth.

Topic: *I*

0849 Josten, C. H. *Elias Ashmole (1617-1692): His Autobiographical and Historical Notes, His Correspondence, and Other Contemporary Sources Relating to His Life and Work*. Volume 1. Oxford: Clarendon Press, 1966.

This first volume (out of five devoted to Ashmole) is a lengthy and detailed biography. Arranged as a chronicle and "composed of many very small and many fragmentary pieces of information," Josten suggests that Mercury is a fitting symbol for his subject. Ashmole mediated between the public activities (of collecting antiquities, creating museums, and being a catalyst for the new sciences) and the private realms (of alchemy and hermeticism). He was, Josten points out, a loyal member of the Church of England, yet "was never worried by the unorthodoxy of his magical pursuits."

Topics: *II*, V Names: Ashmole, Lilly, Wharton (1617)

0850 Josten, C. H. "Robert Fludd's 'Philosophical Key' and His Alchemical Experiment on Wheat." *Ambix* 11(1963): 1-23.

Josten discusses Fludd's wheat experiment as described in the manuscript "Philosophical Key" without going into detail about its main purpose—the secret foundation of Fludd's macro- and microcosmical philosophy. Yet Josten does point out that Fludd distinguished between vulgar and celestial (or spiritual) alchemy.

Topic: *II* Name: Fludd

0851 Josten, C. H. "William Backhouse of Swallowfield." *Ambix* 4(1949): 1-33.

Josten does his best to reconstruct what was known (in 1949) about Backhouse's life and alchemical work. The sources for that work are the reproductions by Ashmole—and Josten analyzes their symbolism. What emerges is a portrait of a country squire who was a quiet and secretive scholar as well as a linguist and generous friend. He also was a man devoted to the hermetic doctrine and one who modestly claimed to possess the final secrets of alchemy.

Topic: *II* Names: Ashmole, Backhouse

0852 Jourdain, Philip E. B. "Newton's Hypotheses of Ether and Gravitation from 1672 to 1679." *Monist* 25(1915): 79-106.

See (0854). This first installment has no reference to God or religion.

Topics: *XII*, VII Name: Newton

0853 Jourdain, Philip E. B. "Newton's Hypotheses of Ether and Gravitation from 1679 to 1693." *Monist* 25(1915): 234-254.

See (0854).
Topics: *XII*, VII Name: Newton

0854 Jourdain, Philip E. B. "Newton's Hypotheses of Ether and Gravitation from 1693 to 1726." *Monist* 25(1915): 418-440.
These three essays form a unity. Among many and long quotations from primary sources, Jourdain makes interesting comments on Newton's views of an ether, gravity, and God. Considering Jourdain's era and his own apparent bias against religion, his recognition of Newton's piety and its relevance to Newton's natural philosophy is impressive.
Topics: *XII*, VII Name: Newton

0855 Kaiser, Christopher B. *Creation and the History of Science.* Grand Rapids, Mich.: W. B. Eerdmans, 1991.
This is a wide-ranging survey of the interaction of the doctrine of creation and physical science from the patristic writers of the early church to Einstein and Bohr. It is a useful volume in a series of introductory textbooks in the history of theology. His approach is similar to that of the history-of-ideas school. Kaiser argues that the basic theme in the "creationist tradition" is "that the entire universe is subject to a single code of law which was established along with the universe at the beginning of time." This theme manifested itself in four flexible ideas: (1) the comprehensibility of the world; (2) the unity of earth and heaven; (3) the relative autonomy of nature; and (4) the ministry of health care and reconciliation. A majority of the book deals with theologians and natural philosophers from the sixteenth through the eighteenth centuries. Kaiser discusses many of the topics important for this bibliography—from hermeticism to the Puritan thesis. For example, he has an interesting insight into the relation between Newton's Arianism and his views of space and time. Kaiser is good on the details of theological differences and theological rationale. In this, his work complements those historians of science who miss the nuances within theological debates. The usefulness of Kaiser's book is partially frustrated by a lack of footnotes, some correctable mistakes, and some overapplications of his thesis to particular figures. Overall, however, it is recommended as a replacement for Dillenberger's comparable survey.
Topics: *VII*, II, III, XII Names: Bacon, Boyle, Newton

0856 Kaplan, Barbara B. "Greatrakes the Stroker: The Interpretations of His Contemporaries." *Isis* 73(1982): 178-185.

Kaplan examines various seventeenth-century attempts to interpret Greatrakes' healings in rational terms. She suggests that such natural explanations were motivated both by Greatrakes' successes and by the preoccupation of science with problems involving nonobservables.

Topics: *XI*, VII Names: Boyle, Greatrakes, Lloyd, D., More, Oldenburg, Stubbe

0857 Kargon, Robert Hugh. *Atomism in England from Hariot to Newton*. Oxford: Clarendon Press, 1966.

Kargon discusses social, philosophical, and theological aspects of atomism and the mechanical philosophy in England. Among these are the possible connections between atomism and the mechanical philosophy on the one hand and atheism on the other. The book also includes annotated bibliographic lists on topics relevant to the study.

Topics: *VII*, V, VIII, IX, XII Names: Boyle, Cavendish, Charleton, Harriot, Hobbes, Newton, Percy

0858 Kargon, Robert Hugh. "Thomas Hariot, the Northumberland Circle and Early Atomism in England." *J. Hist. Ideas* 27(1966): 128-136.

Here Kargon summarizes the first few chapters of his book (0857). He discusses Harriot's science and the atomistic views of most of the Northumberland circle.

Topics: *VII*, IX Names: Harriot, Percy

0859 Kargon, Robert Hugh. "Walter Charleton, Robert Boyle and the Acceptance of Epicurean Atomism in England." *Isis* 55(1964): 184-192.

This work appears in Kargon's book noted above (0857).

Topic: *VII* Names: Boyle, Charleton

0860 Kassler, Jamie C. "The Paradox of Power: Hobbes and Stoic Naturalism." In *The Uses of Antiquity*, edited by Stephen Gaukroger, 53-78. Dordrecht: Kluwer Academic, 1991. See (0530).

Kassler argues that Hobbes believed in a Stoic naturalism and identified God's will with creative activity in the universe. His materialism thus does not entail complete determinism: for Hobbes, self-government in conformity with nature's power is freedom. In explicating Hobbes' views, Kassler also discusses those of William Harvey and Abraham Cowley.
Topics: *VII*, IX, X Names: Cowley, Harvey, Hobbes

0861 Katz, David S. "Isaac Vossius and the English Biblical Critics 1650-1689." In *Scepticism and Irreligion in the Seventeenth and Eighteenth Centuries*, edited by Richard H. Popkin and Arjo Vanderjagt, 142-184. Leiden: E. J. Brill, 1993. See (1250).
There is no direct connection to science in this essay. But Katz does deal with Biblical chronology; and he mentions, in passing, the Royal Society and several individuals relevant to our field.
Topic: *X* Names: Evelyn, Ussher, Vossius

0862 Kearney, Hugh F. "Puritanism and Science: Problems of Definition." *Past and Present* 31(1965): 104-110. Reprinted in *The Intellectual Revolution of the Seventeenth Century*, edited by Charles Webster, 254-261. London: Routledge and Kegan Paul, 1974. See (1618).
See (0863).
Topics: *III*, V

0863 Kearney, Hugh F. "Puritanism, Capitalism and the Scientific Revolution." *Past and Present* 28(1964): 81-101. Reprinted in *The Intellectual Revolution of the Seventeenth Century*, edited by Charles Webster, 218-242. London: Routledge and Kegan Paul, 1974. See (1618).
In his two articles, Kearney argues against the Puritanism-and-the-rise-of-science thesis generally and against Hill (0689) in particular. As he notes, issues of definition are central. He discusses Gresham College at length in his analyses.
Topics: *III*, V Names: Bacon, Briggs, Gellibrand

0864 Kearney, Hugh F. *Scholars and Gentlemen: Universities and Society in Pre-Industrial Britain, 1500-1700*. London: Faber, 1970. Ithaca, N.Y.: Cornell University Press, 1970.

Kearney presents the universities as socially and intellectually conservative—and as resisting the new science. He discusses the influence of scholasticism as well as Puritanism. He emphasizes the period from 1530 to 1660.

Topics: *V*, III, IV

0865 Kearney, Hugh F. *Science and Change, 1500-1700*. London: McGraw-Hill, 1971.

Kearney bases his treatment on three traditions—the organic, the magical, and the mechanistic. He does not emphasize the important religious factors but he does recognize them. This is a good introductory survey because Kearney sees clearly the importance of the Renaissance magic tradition even as he presents the important philosophical aspects of the new science. He includes a brief but articulate summary of his argument against the Puritanism-and-the-rise-of-science thesis.

Topics: *VII*, II, IV, XII Names: Bacon, Boyle, Hobbes, Hooke, Newton

0866 Kearney, Hugh F. "Scientists and Society." In *The English Revolution: 1600-1660*, edited by E. W. Ives, 101-114. London: Edward Arnold, 1971.

This essay consists of introductory generalizations directed at a general audience. It is not Kearney's best work.

Topics: *IV*, III

0867 Keller, Alex. "Technological Aspirations and the Motivation of Natural Philosophy in Seventeenth Century England." *Hist. Technol.* 15(1993): 76-92.

Keller explores the thesis that the original recruitment and legitimation for "the pursuit of science" among the founders of the Royal Society "came from technological enthusiasm rather than from Puritan inspiration." Thereby, he follows a portion of Merton's original discussion while appearing to turn him "on his head."

Topics: *VI*, III, V

0868 Keller, Evelyn Fox. "Baconian Science: A Hermaphroditic Birth." *Philosophical Forum* 11(1980): 299-307. Reprinted as "Baconian Science: The Arts of Mastery and Obedience," in

Reflections on Gender and Science, by Evelyn Fox Keller, 33-42. New Haven, Conn.: Yale University Press, 1985. See (0869).

Keller finds Bacon's imagery of the relationships between the scientist, nature and God to be more complex than either recent defenders or recent critics of Bacon's philosophy. Bacon's images combine aggression and responsiveness, a transformation of the mind, and submission to God's cleansing. Keller begins by discussing Bacon's equation of power and salvation, and she ends with an image of the bisexual nature of science.

Topics: *IV*, X Name: Bacon

0869 Keller, Evelyn Fox. *Reflections on Gender and Science*. New Haven, Conn.: Yale University Press, 1985.

This anthology includes two essays (0868) and (0871) which are annotated separately in this bibliography.

Topic: *IV*

0870 Keller, Evelyn Fox. "Secrets of God, Nature, and Life." *History of the Human Sciences* (3)1990: 229-242. Reprinted in *Secrets of Life/Secrets of Death*, by Evelyn Fox Keller, 52-72. New York: Routledge, 1992.

Keller examines the term "secrets" between the sixteenth and early-eighteenth centuries: emphasis shifts from the secrets of God to the secrets of nature. She proposes that a linguistic reading of the writings of Boyle (as an example) indicates that a wedge was rhetorically inserted between Nature and God. Women, previously identified with nature and life, were thereby left "to oscillate ambiguously between deconstructed nature and reconstructed man."

Topics: *IV*, VII, X Name: Boyle

0871 Keller, Evelyn Fox. "Spirit and Reason at the Birth of Modern Science." In *Reflections on Gender and Science*, by Evelyn Fox Keller, 43-65. New Haven, Conn.: Yale University Press, 1985. See (0869).

Keller here poses the question of the oppositions of the alchemical (and witchcraft) tradition to the mechanical tradition in terms of each's relation to gender. For alchemists, "God was immanent in the material world, in women and in sexuality." For the new natural philosophers, "chastity was a condition for Godliness." This opposition reflected a shift in the ideology of gender during the early modern period.

Topics: *IV*, II Names: Glanvill, More, Vaughan, T., Webster

0872 Kelley, Donald R., and Richard H. Popkin, eds. *The Shapes of Knowledge from the Renaissance to the Enlightenment*. Dordrecht: Kluwer Academic, 1991.

See the relevant essays by Coudert (0276), Hunter (0751), and Olivieri (1168), and an "Epilogue" by Popkin (1229)—all annotated separately.

Topics: *IV*, I, V, VII

0873 Kemsley, Douglas S. "Religious Influences in the Rise of Modern Science: a Review and Criticism, Particularly of the 'Protestant-Puritan Ethic' Theory." *Ann. Sci.* 24(1968): 199-226. Reprinted in *Science and Religious Belief*, edited by Colin A. Russell, 74-102. London: University of London Press, 1973. See (1377).

This is a good review of various theses. Kemsley concludes that "religious denominational labels are simply not sufficient to explain all known aspects of the phenomenon." He leans toward the view that some form of Anglican biblical tradition was closely connected with the new philosophy.

Topics: *III*, X Names: Haak, Sprat

0874 Kerszberg, Pierre. "The Cosmological Question in Newton's Science." *Osiris* (Second Series) 2(1986): 69-106.

This is a systematic reconstruction of Newton's views having to do with the structure, creation, and evolution of the universe. Kerszberg stresses Newton's questions, the mathematical and physical complications of his tentative answers, his changing conceptions over time, and the logical implications of his views. Theological considerations are prominent in Kerszberg's analysis. This is a lengthy essay in the tradition of Alexandre Koyre.

Topics: *XII*, VIII Names: Bentley, Clarke, Newton

0875 Keynes, John Maynard. "Newton, the Man." In The Royal Society. *Newton Tercentenary Celebrations*, 27-34. Cambridge: Cambridge University Press, 1947.

This famous short discussion calls for more study into religious and alchemical aspects of Newton's views. An often-quoted passage from this essay refers to Newton as "not the first of the age of reason. He was the last of the magicians. . . "
Topics: *XII*, II, VII Name: Newton

0876 Kilgour, Frederick G. "Thomas Robie (1689-1729): Colonial Scientist and Physician." *Isis* 30(1939): 473-490.
Kilgour outlines the life and work of Robie as a physician and natural philosopher at Harvard. He briefly mentions Robie's theological writings and piety.
Topic: *XI* Name: Robie

0877 King, Lester S. *The Philosophy of Medicine: The Early Eighteenth Century*. Cambridge, Mass.: Harvard University Press, 1978.
This book focuses on the medical theories of continental and British physicians from the mid-seventeenth into the early-eighteenth century. Arranged topically, King treats the following concepts: nature and form; Iatrochemistry and Iatromechanism; the relations between soul, mind, and body; the causes of disease; rationalism and empiricism. He recognizes the role played by beliefs in God. Stress is placed on the complexity of changes in the philosophy of medicine. The book is written in a readable, but sometimes informal style, with many explanations of terms and concepts.
Topics: *XI*, VII Names: Bacon, Boyle, Pitcairn, Sydenham, Willis

0878 King, Lester S. *The Road to Medical Enlightenment: 1650-1695*. London: Macdonald, 1970.
King likes retrospective history but he also does a good job of relating seventeenth-century medicine to its own wider social and intellectual milieu. Biographically, this book gives lengthy discussion to Boyle and less lengthy attention to other important medical thinkers.
Topics: *XI*, IV Names: Bacon, Boyle, Digby, Glanvill, Hooke, Nedham, Stubbe, Sydenham, Twysden

0879 King, M. D. "Reason, Tradition, and the Progressiveness of Science." *History and Theory* 10(1971): 3-32.

King here analyzes and criticizes Merton, Hall, and Kuhn in an attempt to give a theoretical model of the best way of handling the problem of scientific change in sociological terms.
Topics: *I*, III

0880 Kirsanov, Vladimir. "Non Mechanistic Ideas in Physics and Philosophy: From Newton to Kant." In *Nature Mathematized*, edited by William R. Shea, 269-276. Dordrecht: D. Reidel, 1982. See (1445).
Kirsanov argues that, for eighteenth-century natural philosophers, religion provided the freedom to go beyond existing science. And, despite attempts to separate mechanical laws from God, God's existence was a guarantee of the existence of reasons and laws which were quite different from mechanical ones.
Topics: *VII*, XII Name: Newton

0881 Kirsch, Irving. "Demonology and the Rise of Science: An Example of the Misperception of Historical Data." *Journal of the History of the Behavioral Sciences* 14(1978): 149-157.
Kirsch argues that a "whig" interpretation of the relationship between science and superstition has resulted in the misperception of the history of demonology. He argues, in particular, that witch mania is not found in the "dark" middle ages but rather in the mid-seventeenth century—along with the rise of science. The main target of Kirsch's critique is not historians but rather behavioral science textbooks.
Topics: *II*, I

0882 Kittredge, George L. "Cotton Mather's Election into the Royal Society." *Pub. Colonial Soc. Mass. Trans.* 14(1911-1913): 81-114.
Through the statutes of the Royal Society and a sequence of letters, Kittredge documents how and when Mather was elected without being present at a meeting. He also outlines the inoculation controversy which involved Mather. See also his continuation of this discussion in (0884).
Topics: *XI*, V Name: Mather, C.

0883 Kittredge, George L. "Dr. Robert Child the Remonstrant." *Pub. Colonial Soc. Mass. Trans.* 19(1919): 1-146.

This is a thorough biographical study of Child—emphasizing his role in the Remonstrance of 1646. While stressing Child's high Presbyterianism, Kittredge deals with Child as a physician and (at some length) as an alchemist. Also treated in this work are Child's interests in natural history, mining, and agricultural advances. Kittredge supports his arguments with long quotations from Child's letters and papers.

Topics: *II*, VI, XI Names: Child, Hartlib, Stirk, Winthrop

0884 Kittredge, George L. "Further Notes on Cotton Mather and the Royal Society." *Pub. Colonial Soc. Mass. Trans.* 14(1911-1913): 281-292.

Kittredge continues his discussion from (0882).

Topics: *XI*, V Name: Mather,C.

0885 Klaaren, Eugene M. *Religious Origins of Modern Science: Belief in Creation in Seventeenth-Century Thought*. Grand Rapids, Mich.: Eerdmans, 1977.

Klaaren aims at a more complete explanation of the thesis that the new science had its roots in religious presuppositions and, more specifically, in the voluntarist theology of creation. His book thus summarizes and extends the previous work of Whitehead, Collingwood, Merton, Oakley, and Hooykaas. He argues that theological/religious orientations in the seventeenth century were less determinative for natural philosophy than direct causes but more determinative than simple conditions. Klaaren postulates three pairs of theological-natural philosophical correlations: (1) scholastic theology with traditional natural philosophy; (2) spiritualist theology with spiritual natural philosophy; and (3) voluntarist theology with modern natural science. He explicates the theology of creation for each theological position and shows how the corresponding natural philosophy correlates to it—not only apparently but at the deepest levels of presupposition. Klaaren emphasizes the voluntarist theology using Boyle as his main exemplar (along with Bacon and Newton). Of these men, he says "Belief in the transcendent otherness of God, who was the Author of all law and design in creation, was at the moving center of their thought." Newton did not assume that the physical laws of nature were divinely originated simply to make a place for God in the world; he shared with Boyle a voluntarist presupposition of the lawful structure of creation grounded in its Author. Klaaren explicates how the divinely imposed laws of nature led to a divine basis for nature and inquiry. The legal-mechanistic world view supported a new natural philosophy with a different view of matter (atoms or corpuscles); it replaced the hierarchical worldview (with Aristotelian essences) and the organic worldview (with Helmontian seeds). The presupposition of the voluntarist theology of creation also was connected to a view of scientific inquiry oriented to God's law as a religious task—expressed by Boyle's image of the natural philosopher as a reformed priest. (Later, Deistic, presuppositions went with a later type of mechanistic view—not to be confused with either Boyle or Newton). Klaaren also postulates an epochally new scheme of intellectual "differentiation." For him, modernity is characterized by: (1) differentiation of natural philosophy from revealed theology and metaphysics; (2) differentiation of the natural sciences from natural philosophy; and (3) differentiation of empirical attention and careful observation as distinct from theory and framing of hypotheses. He connects such differentiation to a new view of knowledge related to the same voluntarist theological presuppositions. Klaaren says that this theology of creation also was formative in conceptions of ethical, political and social order; and it penetrated deeply into the personal lives of believers. All of these patterns gained unique strength in seventeenth-century English Protestantism. Klaaren also goes into some detail for spiritualist theology—using J. B. van Helmont as exemplar. (Here, images of the processes of life and light serve as the root metaphors of creation.) Klaaren believes that "spiritualist orientations" were influential as targets, catalysts, and partial bearers of the emerging new science. Klaaren's book is an important one. It is written in a scholarly and fairly dense style but is also accessible to non-specialists.

Topics: *VII*, III, XII Names: Boyle, Bacon, Newton

0886 Klein, Jurgen. "The Problem of Tolerance in a Baconian Context." In *State, Science and Modernization in England, from the Renaissance to the Modern Times*, edited by Jurgen Klein, 31-54. Hildesheim, Germany: G. Olms, 1994.

Klein proposes that in Bacon's doctrine of the separation of the two branches of truth—theology and natural philosophy—lay the foundation for a more tolerant public scene and for private and

individual freedom in religion. That is, "within the emerging complex structure of scientific and intellectual innovation in general, the idea of tolerance was extended." And, while acknowledging Bacon's political conservatism, Klein indicates that Bacon's proposal for scientific progress was "a necessary condition for the rise of tolerance."
Topic: *IV* Names: Bacon, Locke

0887 Knappen, Marshall M. *Tudor Puritanism: A Chapter in the History of Idealism.* Chicago: University of Chicago Press, 1939.
This focuses mainly on the sixteenth century and mainly on the political dimension of Puritanism. But the chapters "The Spirit of Puritanism" and "Learning and Education" provide relevant background. Knappen mentions Merton and Stimson and generally disagrees with them.
Topic: *III*

0888 Knight, David M. "Religion and the 'New Philosophy'." *Renaissance and Modern Studies* 26(1982): 147-166.
Knight's brief survey goes from Galileo to the decline of magic to Newton and the Boyle Lectures. Knight concludes that science and religion were so intertwined in the seventeenth century that we have to forgo many inherited assumptions (including the conflict thesis).
Topics: *VII*, I, II, IV, VIII Names: Boyle, Wilkins

0889 Knight, David M. "Science Fiction of the Seventeenth Century." *Seventeenth Century* 1(1986): 69-79.
Knight discusses popular presentations of the plurality-of-worlds theme. He notes that this contributed to the acceptance of the Copernican theory and led to various religious speculations. He considers works from Kepler to Fontenelle—including Godwin and Wilkins. It reads as if these are Knight's reading notes and includes no original analyses.
Topics: *X*, VII Names: Godwin, Wilkins

0890 Kochavi, Matania. "One Prophet Interprets Another: Sir Isaac Newton and Daniel." In *The Books of Nature and Scripture*, edited by James E. Force and Richard H. Popkin, 105-122. Dordrecht: Kluwer Academic, 1994. See (0498).

Kochavi examines the methods Newton developed to understand languages, in particular to interpret the text of Daniel. The prophecy of Daniel, for Newton, foresaw the corruption of Christianity. Kochavi posits, however, that Newton's sense of chosenness "emanated from his unique attainments in natural philosophy." Just as Daniel's authority seemed to derive from his ability to decode the secrets of nature, so did Newton's. Hence, Newton saw himself as a prophet.

Topics: *XII*, X Names: Mede, More, Newton

0891 Kocher, Paul H. "Francis Bacon on the Science of Jurisprudence." *J. Hist. Ideas* 18(1957): 3-26. Reprinted in *Essential Articles for the Study of Francis Bacon*, edited by Brian Vickers, 3-26. Hamden, Conn.: Archon Books, 1968. See (1588).

The last few pages deal with the harmony and interdependence of all fields of knowledge for Bacon. Kocher notes the importance of God and the Bible for Bacon; and he notes the respective roles of revelation and induction (for Bacon) both in natural science and in jurisprudence.

Topics: *VII*, IV, X Name: Bacon

0892 Kocher, Paul H. *Science and Religion in Elizabethan England*. San Mario, Calif.: Huntington Library, 1953.

While focused primarily on the last half of the sixteenth century, this book touches on the early seventeenth century as well. Additionally, it can function as good background for our period. Kocher organizes his chapters by theological and scientific topics, and emphasizes the complexity of the interaction between "divines and scientists."

Topics: *VII*, III, X, XI Names: Bacon, Donne, Ralegh

0893 Koestler, Arthur. *The Sleepwalkers*. New York: Macmillan, 1959.

This book is a dramatically-written survey of the history of astronomy in the early modern period. The last chapters touch on the Newtonian synthesis which, according to Koestler, resulted in "the fatal estrangement" between God and geometry.

Topics: *VII*, XII Name: Newton

0894 Korshin, Paul J. *Typologies in England, 1650-1820*. Princeton: Princeton University Press, 1982.

This book is a systematic and thorough treatment of religious and literary typologies during and beyond our period. Korshin provides no direct discussion of science but he makes several important indirect connections. He sees science as part of the larger context which affected figural change in English literature in the seventeenth and eighteenth centuries—but he does not emphasize science as an independent cause. He discusses (in passing) Newton, Cudworth, More, and the Boyle Lectures. He also mentions briefly the use of natural phenomena as divine emblems. Korshin addresses his work to audiences of several different levels: for specialists in our field, it is useful mainly as background.

Topics: *X*, XII Names: Butler, Cudworth, Dryden, Milton, More, Newton, Pope, Vaughan, T., Whiston

0895 Korshin, Paul J., and Robert R. Allen, eds. *Greene Centennial Studies: Essays Presented to Donald Greene in the Centennial Year of the University of Southern California*. Charlottesville, Va.: University Press of Virginia, 1984.

The two relevant essays by Popkin (1228) and Reedy (1314) are annotated separately in this bibliography.

Topic: *VII*

0896 Koslow, Arnold. "Ontological and Ideological Issues of the Classical Theory of Space and Time." In *Motion and Time, Space and Matter: Interrelations in the History of Philosophy and Science*, edited by Peter K. Machamer and Robert G. Turnbull, 224-263. Columbus, Ohio: Ohio State University Press, 1976.

This is a contemporary philosophical analysis of Newtonian space and time theory. It includes some discussion of the related Newtonian concept of God.

Topics: *XII*, VII Name: Newton

0897 Koyre, Alexandre. "Attraction, Newton and Cotes." In *Newtonian Studies*, by Alexandre Koyre, 273-282. Cambridge, Mass.: Harvard University Press, 1965. See (0909).

This essay focuses on an issue, gravitational attraction, which stands at the intersection of Newton's physics, his metaphysics, and his philosophy of science.

Topics: *XII*, VII Name: Newton

0898 Koyre, Alexandre. "Commentary" (on H. Guerlac). In *Scientific Change*, edited by Alistair C. Crombie, 847-857. London: Heinemann, 1963. See (0299).

Koyre stresses that science is (and always has been) essentially *theoria*—a search for truth. Consequently, it has had an internal and autonomous development; and the understanding of this autonomous development is that which constitutes historical understanding in the field.

Topic: *I*

0899 Koyre, Alexandre. "Concept and Experience in Newton's Scientific Thought." In *Newtonian Studies*, by Alexandre Koyre, 25-52. Cambridge, Mass.: Harvard University Press, 1965. See (0909).

Koyre discusses various understandings of the word "hypothesis" and places Newton's philosophy of science within that context. He argues (against E. W. Strong) that, for Newton, God is *not* a metaphysical hypothesis—but rather a certainty.

Topic: *XII*, VII Name: Newton

0900 Koyre, Alexandre. "Etudes Newtoniennes: Attraction, Newton, and Cotes." *Arch. Int. d'Hist. Sci.* 14(1961): 225-236. Translated and slightly revised as "Attraction, Newton and Cotes." In *Newtonian Studies*, by Alexandre Koyre, 273-282. Cambridge, Mass.: Harvard University Press, 1965. See (0909).

See (0897).

Topics: *XII*, VII Name: Newton

0901 Koyre, Alexandre. "Etudes Newtoniennes, 2. Les 'Queries' L'Optique." *Arch. Int. d'Hist. Sci.* 13(1960): 15-29.

Koyre analyzes the English and Latin editions of Newton's "Queries." The two editions, which conveniently are quoted by Koyre, have a stylistic difference regarding God and the creation of matter.

Topic: *XII* Name: Newton

0902 Koyre, Alexandre. "Etudes Newtoniennes, 1. Les 'Regulae Philosophandi'." *Arch. Int. d'Hist. Sci.* 13(1960): 3-14. Translated and slightly revised as "Newton's Regulae Philosophandi." In *Newtonian Studies*, by Alexandre Koyre, 261-272. Cambridge, Mass.: Harvard University Press, 1965. See (0909).

See (0910).

Topics: *XII*, VII Name: Newton

0903 Koyre, Alexandre. *From the Closed World to the Infinite Universe*. Baltimore: Johns Hopkins Press, 1957.

Koyre was one of the most influential scholars in the post-World War II history of science. His intellectualistic approach to the great men of the scientific revolution became a dominant interpretation. In this—probably his most accessible work—he announces that "it is science, philosophy and theology...that join and take part in the great debate." That debate is reflected in the title—which connotes "the destruction of the cosmos" and "the geometrization of infinite space." In chapters 5-12, Koyre discusses the ideas of seventeenth-century Englishmen about cosmology, space, the nature of matter, and God's relation to the world. Henry More is at the center of Chapters 5 and 6, first in opposition to Descartes regarding atomism, space and God. Then, Koyre proposes that despite his syncretistic approach to philosophy, More grasped "the fundamental principle," the infinity of space. And, Koyre indicates, More's God acts through the spirit of nature and supports extension. Thus space begins to have divine attributes. This theme is carried further in the remaining chapters which basically focus on Newton's thought on the relation of God to space. Koyre explores the issues of absolute and relative space, gravity, space as God's sensorium, and God's dominion and omnipresence. He concludes with a discussion of the Leibniz-Clarke debate, focusing on the different attributes of God (reason versus will) in relation to the infinite universe. Koyre's sense of a tragic end appears as he notes that the success of Newton's physics proved Leibniz's contention: while space and matter "interpreted all the ontological attributes of Divinity," Newton's personal God departed.

Topics: *VII*, I, XII Names: Bentley, Clarke, More, Newton, Raphson

0904 Koyre, Alexandre. "L'Hypothese et l'Experience chez Newton." *Bulletin de la Societe Francaise de Philosophie*

50(1956): 59-79. Translated and slightly revised as "Concept and Experience in Newton's Scientific Thought." In *Newtonian Studies*, by Alexandre Koyre, 25-52. Cambridge, Mass.: Harvard University Press, 1965. See (0909).

See (0899).

Topics: *XII*, VII Name: Newton

0905 Koyre, Alexandre. *De la Mystique a la Science: Cours, Conferences et Documents, 1922-1962*. Edited by Pietro Redondi. Paris: Editiones de l'Ecole des Hautes Etudes en Sciences Sociales, 1986.

This is a collection of lectures, conferences, and notes of Koyre as well as letters by and to him. The book includes good introductions by Redondi and a bibliography of Koyre's published writings. The arrangement illustrates the range and sequence of Koyre's interests in religion and science. Beginning with a section on "Mysticism and Cosmology," the book proceeds from the history of religious ideas in early modern Europe through several revolutions in the history of thought and to the history of religious ideas in relation to scientific thought. For Koyre, the history of science always is in reciprocal relation to the history of religious thought. Yet he is always clear with his distinctions: for example, he keeps the Platonism of the mystical distinct from the Pythagorean traditions. The book concludes with two lectures from 1947 conferences on theology (primarily medieval) and science.

Topics: *VII*, I, XII Names: Locke, Newton

0906 Koyre, Alexandre. "Newton and Descartes." In *Newtonian Studies*, by Alexandre Koyre, 53-200. Cambridge, Mass.: Harvard University Press, 1965. See (0909).

By comparing Newton to Descartes, Koyre discusses a wide range of physical and metaphysical issues in Newton's natural philosophy. These include: the concept of inertia and the laws of motion; gravitation and universal attraction; the vortex; matter and the void; space, time, and the infinite; God and God's relation to all of the above. Koyre concludes that "his belief in an omnipresent and omniactive God" enabled "Newton to transcend both the shallow empiricism of Boyle and Hooke and the narrow rationalism of Descartes." Newton believed in the importance of induction "because our world was created by the pure will of God; we have not, therefore, to prescribe his action for him; we have only to find out what he has done."

Topics: *XII*, VII Names: Clarke, Newton

0907 Koyre, Alexandre. "Newton, Galileo, and Plato." *Actes du IX Congres International d'Histoire des Sciences* (1960): 165-197. Reprinted, *Annales* 6(1960): 1041-1059. Reprinted with minor revisions in *Newtonian Studies*, by Alexandre Koyre, 201-220. Cambridge, Mass.: Harvard University Press, 1965. See (0909).

Koyre here analyzes Newtonian cosmological questions in terms of the Bentley letters and Bentley's Boyle Lectures. He also considers related issues having to do with the "Galileo-Plato" hypothesis of the formation of the solar system.

Topics: *XII*, VIII Names: Bentley, Newton

0908 Koyre, Alexandre. *Newtonian Essays*.

This prepublication title announced by Chapman and Hall, though it appears in some bibliographies, was not used for the actual publication. See (0909).

Topic: *XII* Name: Newton

0909 Koyre, Alexandre. *Newtonian Studies*. Cambridge, Mass.: Harvard University Press, 1965. London: Chapman and Hall, 1965.

This collection includes six relevant and important essays. Five were published previously, but three of these are in English translation for the first time and much the longest study in the book ("Newton and Descartes") did not appear previously. Koyre's emphasis is on what he calls "conceptual analysis"—he mainly studies philosophical aspects of Newton's thought. Although his approach is "internalist," he does include the philosophical aspects of theology as relevant to the history of science. Koyre's style includes an extensive scholarly apparatus and requires some effort to read. See (0897), (0899), (0900), (0902), (0904), (0906), (0907), (0910) and (0911).

Topics: *VII*, XII Names: Bentley, Clarke, Cotes, More, Newton

0910 Koyre, Alexandre. "Newton's 'Regulae Philosophandi'." In *Newtonian Studies*, by Alexandre Koyre, 261-272. Cambridge, Mass.: Harvard University Press, 1965. See (0909).

Koyre provides textual criticism and exegesis with respect to Newton's "Rules of Reasoning in Philosophy." This essay is only tangentially relevant to theology (but see footnote 4 on p. 271).

Topics: *XII*, VII Name: Newton

0911 Koyre, Alexandre. "The Significance of the Newtonian Synthesis." *Arch. Int. d'Hist. Sci.* 3(1950): 291-311. Reprinted, *Journal of General Education* 4(1950): 256-268. Reprinted with minor revisions in *Newtonian Studies*, by Alexandre Koyre, 3-24. Cambridge, Mass.: Harvard University Press, 1965. See (0909).

The "Newtonian synthesis" is seen as the successful combining of mathematical science with the corpuscular philosophy; it also is seen as the effective combining of mathematics and experiment. Understood either way, Newton's conception of God is seen as a metaphysical foundation of the Newtonian synthesis. Koyre's discussion of the later significance of Newtonianism also includes both scientific and theological developments.

Topics: *XII*, VII Name: Newton

0912 Koyre, Alexandre; and I. Bernard Cohen. "The Case of the Missing Tanquam: Leibniz, Newton and Clarke." *Isis* 52 (1961): 555-566.

This is a detailed discussion of the documents relating to the dispute over Newton's meaning of the phrase "sensorium of God."

Topics: *XII*, VII, VIII Names: Clarke, Newton

0913 Koyre, Alexandre, and I. Bernard Cohen. "Newton and the Leibniz-Clarke Correspondence; with Notes on Newton, Conti, and Des Maizeaux." *Arch. Int. d'Hist. Sci.* 15(1962): 63-126.

This work includes textual criticism and exegesis of various documents relating to Newton's theological and metaphysical quarrel with Leibniz. The authors argue that the Clarke-Leibniz correspondence and the Newton-Conti-Leibniz correspondence were intertwined and provide evidence of Newton's participation in the Clarke-Leibniz debate. Koyre and Cohen write in a highly technical style with long quotations from primary documents.

Topics: *XII*, VII, VIII Names: Clarke, More, Newton

0914 Krieger, Leonard. "The Autonomy of Intellectual History." *J. Hist. Ideas* 34(1973): 499-516.

Krieger tries to show the integrity and autonomy of intellectual history despite the plurality of conflicting methods used. His discussion of socio-intellectual history is particularly relevant.

Topic: *I*

0915 Kroll, Richard W. F. Introduction to *Philosophy, Science, and Religion in England 1640-1720*, edited by Richard Kroll and others, 1-28. Cambridge: Cambridge University Press, 1992. See (0917).

This anthology is drawn from a conference held at the Clark library. In addition to introducing the essays, Kroll outlines the historiography of latitudinarianism and its relation to the new science. He concludes with a discussion of the language of the latitudinarians—pointing out how, for them, religious and scientific signs and actions were "a reflection and reinforcement of social and cultural practices."

Topics: *VII*, I

0916 Kroll, Richard W. "The Question of Locke's Relation to Gassendi." *J. Hist. Ideas* 45(1984): 339-359.

Kroll traces a positive influence of Gassendi on Locke through Epicureanism. In the process, Kroll discusses the epistemological and (for natural philosophy) methodological implications of a Christianized Epicurean revival both in England and on the continent.

Topic: *VII* Names: Locke, Stanley

0917 Kroll, Richard, Richard Ashcraft, and Perez Zagorin, eds. *Philosophy, Science, and Religion in England, 1640-1700*. Cambridge: Cambridge University Press, 1992.

The introduction by Kroll (0915) and the essays by Coudert (0278), Gabbey (0520), Hunter (0756), Hutton (0781), Levine (0946), and Osler (1182) are annotated separately. See also the review of this volume by J. R. and M. C. Jacob (0806).

Topic: *VII*

0918 Krook, Dorothea. "Thomas Hobbes's Doctrine of Meaning and Truth." *Philosophy* 31(1956): 3-22.

Krook argues that Hobbes' radically nominalistic understanding of meaning and truth affects all aspects of his philosophy. This includes his epistemology, his natural philosophy, and the concept of God—and she briefly discusses each of these.
Topic: *VII* Name: Hobbes

0919 Kubrin, David. "Newton and the Cyclical Cosmos: Providence and the Mechanical Philosophy." *J. Hist. Ideas* 28(1967): 325-346. Reprinted in *Science and Religious Belief*, edited by Colin A. Russell, 147-169. London: University of London Press, 1973. See (1377).
Kubrin discusses Newton's natural theology in the wider context of his cosmogonic speculations. He also discusses the development of Newton's ideas on the sources of activity in nature. This essay includes still-valuable interpretations of some otherwise commonly misunderstood seventeenth-century philosophical and theological doctrines.
Topics: *XII*, VII, VIII, X Name: Newton

0920 Kubrin, David. "Newton's Inside Out! Magic, Class Struggle, and the Rise of Mechanism in the West." In *The Analytic Spirit: Essays in Honor of Henry Guerlac*, edited by Harry Woolf, 96-121. Ithaca, N.Y.: Cornell University Press, 1981.
Kubrin here trys to prove that: (1) magical and Hermetic thought was crucial to Newton's natural philosophy; but (2) Newton himself repressed or hid this in his public statements; and (3) this was because magical and Hermetic traditions (as opposed to Baconian and mechanistic ones) had dangerous social, political, economic, and religious implications in England during Newton's lifetime.
Topics: *XII*, II, IV Name: Newton

0921 Kubrin, David. "'Such an Impertinently Litigious Lady': Hooke's 'Great Pretending' vs. Newton's *Principia* and Newton's and Halley's Theory of Comets." In *Standing on the Shoulders of Giants*, edited by Norman J. W. Thrower, 55-90. Berkeley and Los Angeles: University of California Press, 1990. See (1545).
This is a rambling essay based on the natural philosophic and priority disputes between Hooke and Newton and between Hooke and Halley. Kubrin discusses: differing conceptions of the

mechanical philosophy; divergent understandings of Hermeticism; comets and cosmic catastrophes; the earth as an organism or as a passive body; and the resulting social and economic ideologies. Theology is an important tangent in Kubrin's various discussions.
Topics: *VII*, II, IV, V, XII Names: Halley, Hooke, Newton

0922 Kuhn, Thomas S. "History of Science." In *International Encyclopedia of the Social Sciences*, edited by David L. Sills, Vol. 14, 75-83. New York: Free Press, 1968.
Kuhn gives advice on what he sees as the proper use of internalist and externalist approaches. Concerning the issues surrounding the Merton thesis, he concludes that the internalists are correct when discussing the "classic" branches of science but wrong when discussing fields which were "novelties during the Scientific Revolution."
Topics: *I*, III

0923 Kuhn, Thomas S. "The Relations between History and History of Science." *Daedalus* 100(1971): 271-304. Reprinted in *Historical Studies Today*, edited by Felix Gilbert and Stephen R. Graubard, 159-192. New York: W. W. Norton, 1972.
Kuhn discusses the reasons for and the consequences of what he sees as the chasm which separates "history" (as then understood in traditional university history departments) and "history of science" (as understood by those who then practiced it as a professional discipline). While he does not mention religion specifically, he does point to the "two-culture" problem as a probable source for the difficulties of separation.
Topic: *I*

0924 Kuhn, Thomas S. *The Structure of Scientific Revolutions*. Chicago: University of Chicago Press, 1962. 2nd ed. with postscript. Chicago: University of Chicago Press, 1970.
The various editions of this book have been enormously significant and influential as well as controversial. Kuhn's approach is a mixture of historiographic criticism, philosophy of science, and historical examples. Key terms and concepts, including "paradigm," have entered the vocabularies of disciplines even beyond the history and philosophy of science. For Kuhn, the particular scientific community and the particular scientific tradition in which

a given scientist worked is the key issue for understanding scientific change. The period in England between 1600 and 1720 appears only incidentally in the book: but Kuhn's claims are so fundamental and far-reaching that any serious study touching on "scientific change" in the history of science is affected. There is still dispute as to whether his patterns are useful in understanding natural philosophy in the early modern period when research fields were taking shape under the guidance of particular communities.
Topic: *I*

0925 Laird, John. *Hobbes*. London: E. Benn, 1934. Reprinted, New York: Russell and Russell, 1968.
This is an older book-length discussion of Hobbes' philosophy. It includes good summaries of Hobbes' philosophy of nature (his materialism), of his physics and biology, and of his philosophy of religion. But Laird says little about any relationship between science and religion in Hobbes' thought. He presents Hobbes *not* as an atheist—but as a perhaps inconsistent and limited theist.
Topics: *VII*, IX Name: Hobbes

0926 Lamprecht, Sterling P. "The Role of Descartes in Seventeenth-Century England." In *Studies in the History of Ideas*, Vol. 3, 181-240. New York: Columbia University, 1935. Reprinted, New York: AMS Press, 1970.
Lamprecht proposes a three-phase scheme: Descartes writings were first eagerly embraced (1640's), critically appraised (1660's and following) and finally authoritatively rejected (1690's). Lamprecht emphasizes Descartes' natural philosophy as part of his discussion. This scheme is over-simple but still useful.
Topic: *VII* Names: Cudworth, Glanvill, Locke, More

0927 Landa, Louis A. "Swift, the Mysteries, and Deism." In *Studies in English, Department of English, The University of Texas, 1944*, 239-256. Austin, Texas: University of Texas Press, 1945.
This is a good analysis of eighteenth-century English debates concerning the epistemological status of Christian faith. It is only indirectly related to science.
Topics: *IX*, VI, X Name: Swift

0928 Lang, Jane. *Rebuilding St. Paul's after the Great Fire of London*. London: Oxford University Press, 1956.

This is a detailed narrative history of Christopher Wren and the rebuilding of St. Paul's Cathedral. It is written for the intelligent public and published by Oxford University Press.

Topic: *VI* Names: Compton, Jones, Tillotson, Wren

0929 Lange, Frederick Albert. *History of Materialism; and Criticism of Its Present Importance*. 3 vols. Translated by Ernest Chester Thomas. 2nd ed. London: Trubner, 1879-1881. Originally published as *Geschichte des Materialismus und Kritik seiner Bedeutung in der Gegenwart*. Leipzig: P. Reclam, 1873-1875.

Written by a philosopher, this is one of the better nineteenth-century works. It includes discussion of the natural philosophic views of Bacon, Hobbes, Boyle, Newton, Locke, and Toland. Lange notes the religious attitude and theological position of each philosopher in the context of "the peculiar combination of faith and materialism" which characterizes English thinkers. Judgmental but not polemical, Lange is interesting for *not* using the "warfare metaphor" in discussing science and religion.

Topics: *VII*, IX, XII Names: Bacon, Boyle, Hobbes, Locke, Newton, Toland

0930 Larsen, Robert E. "The Aristotelianism of Bacon's *New Organum*." *J. Hist. Ideas* 23(1972): 435-450.

Larsen argues against Anderson that Bacon was not a great innovator with respect to induction. Rather, Bacon used an Aristotelian/alchemical method—and hence was not the forefather of modern science. No direct connection to religion is made in this discussion.

Topics: *VII*, II Name: Bacon

0931 Lash, Nicholas. "Science and the Reformation." *Hist. Sci.* 11(1973): 145-148.

This is an insightful essay review of Hooykaas' *Religion and the Rise of Modern Science* (0721). Lash notes Hooykaas' strengths but also criticizes his oversimplifications and theological biases.

Topic: *III*

0932 Lasky, Melvin J. *Utopia and Revolution*. Chicago: University of Chicago Press, 1976.

This massive essay in the history of ideas is focused on "the fateful point in human history" at which the two concepts, utopia and revolution, meet. Thus, the thinkers and doers of the seventeenth century take up a large portion of the book. For Lasky, a new politics, mixed with "the ancient religiosity," and linked to the new science, included competing eschatologies. Lasky treats all of the major figures fairly but his own moderate position is never concealed.

Topic: *IV*, X Names: Bacon, Boyle, Burnet, G., Charles I, Comenius, Cromwell, Dryden, Dury, Hartlib, Hobbes, Howell, Locke, Nedham, Milton, Winstanley

0933 Lasswitz, Kurd. *Geschichte der Atomistik vom Mittelalter bis Newton*. 2 vols. Hamburg, Germany: L. Voss, 1890. Reprinted, Hildesheim, Germany: Olm, 1963.

This work includes chapters on Boyle, Hobbes, and Newton. The emphasis is on the physical theories as such—but with some reference to metaphysical and theological assumptions.

Topics: *VII*, XII Names: Boyle, Hobbes, Newton

0934 Laudan, Laurens. "The Clock Metaphor and Probabilism: The Impact of Descartes on English Methodological Thought, 1650-1665." *Ann. Sci.* 22(1966): 73-104.

Laudan finds Boyle and others echoing Descartes in using the clock metaphor to "buttress up a hypothetico-deductive methodology." God appears in Laudan's quotations of primary sources but not in Laudan's commentary.

Topic: *VII* Names: Boyle, Glanvill, Power

0935 Laudan, Laurens. "Comment." (On paper by A. Thackray) In *Historical and Philosophical Perspectives of Science*, edited by Roger H. Struewer, 127-132. Minneapolis: University of Minnesota Press, 1970. See (1514).

Laudan argues that debates about the general nature of the history of science are in vain. Each particular problem must be handled according to whatever techniques are most appropriate.

Topic: *I*

0936 Laudan, Laurens. "Comment." (On paper by G. Buchdahl). In *Historical and Philosophical Perspectives of Science*, edited by Roger H. Struewer, 230-238. Minneapolis: University of Minnesota Press, 1970. See (1514).

See the Buchdahl annotation (0143).

Topic: *I*

0937 Laudan, Laurens. "The Nature and Sources of Locke's Views on Hypotheses." *J. Hist. Ideas* 28(1967): 211-223. Reprinted in *Philosophy, Religion and Science in the Seventeenth and Eighteenth Centuries*, edited by John W. Yolton, 271-283. Rochester, N.Y.: University of Rochester Press, 1990. See (1720).

Laudan argues, against Yolton generally and Yost (1723) specifically, that Locke accepted both hypotheses and the corpuscularian philosophy. There is no direct discussion of theology or religion. (But Locke's methodological position regarding natural philosophy is related implicitly to his theological views and to questions about the separation of science and theology.)

Topic: *VII* Name: Locke

0938 Laudan, Laurens. "Theories of Scientific Method from Plato to Mach: A Bibliographical Review." *Hist. Sci.* 5(1967): 1-63.

This historiographical analysis and twenty-five-page bibliographical list surveys the history of theories of scientific method. Religion is ignored and our period is only part of the survey. But methodological issues are so important in the interplay between theology and natural philosophy that this article is useful (even if indirectly so).

Topics: *I*, VII, XII Name: Newton

0939 Lecky, W. E. H. *The History of the Rise and Influence of the Spirit of Rationalism in Europe*. 2 vols. London: Longman, Roberts, and Green, 1865.

This survey once was a classic of the "warfare" school. It is now outdated; but Lecky's chapter on witchcraft, magic, and their decline under "the spirit of rationalism" is interesting even today.

Topics: *VII*, II, IX Name: Glanvill

0940 LeClerc, Ivan. "Concepts of Space." In *Probability, Time, and Space in Eighteenth-Century Literature*, edited by Paula

Backscheider, 209-216. New York: AMS Press, 1979. See (0052).

This is an etymological and philosophical analysis of the concept of space in the eighteenth century. LeClerc argues that the so-called "Newtonian" understanding of absolute space was not Isaac Newton's view.

Topics: *VII*, XII Name: Newton

0941 Lemmi, Charles W. *The Classical Deities in Bacon: A Study in Mythological Symbolism*. Baltimore: Johns Hopkins Press, 1933.

This is an important summary of Bacon's interpretations of Greek myths. Many of Bacon's interpretations were adaptations from other writers, especially the Italian Noel Conti (Comes) and the alchemists. With Comes, Bacon believed that the purpose of myths was "to conceal philosophic arcana from the multitude." Lemmi concludes that Bacon looked to the future yet also listened to the medieval past.

Topics: *X*, II, VII Name: Bacon

0942 Lemmi, Charles W. "Mythology and Alchemy in *The Wisdom of the Ancients*." In *Essential Articles for the Study of Francis Bacon*, edited by Brian Vickers, 51-92. Hamden, Conn.: Archon Books, 1968. See (1588).

This is a long excerpt from Lemmi's book (0941).

Topics: *X*, II Name: Bacon

0943 Lennon, Thomas M., John M. Nicholas, and John W. Davis, eds. *Problems of Cartesianism*. Kingston, Ontario: McGill-Queens University Press, 1982.

The relevant essays by Gabbey (0523), Popkin (1226), and Roger (1337) are annotated separately.

Topics: *VII*, IX

0944 Leroy, O. *Le Chevalier Thomas Browne (1605-1682): Medicin, Styliste et Metaphysicien*. Paris: J. Gamber, 1931.

Leroy's large study of Browne is divided into four parts: life; thought; art; and critics. Besides portions in the section on Browne's life, the most relevant chapters are those on his beliefs and his understanding of science and medicine. For Leroy, Browne never ceased to be (simultaneously) a moralist, a philosopher, and a theologian.

Topics: *XI*, VII, X Names: Browne, T., Digby

0945 Levack, Brian P., ed. *Articles on Witchcraft, Magic and Demonology*. 12 Vols. N.Y.: Garland, 1992.

This is a collection of photocopied reprints. Vol. 11 (*Renaissance Magic*) includes reprints of articles by Rossi (1357), Westfall (1644) and Yates (1707).

Topic: *II*

0946 Levine, Joseph M. "Latitudinarians, Neoplatonists, and the Ancient Wisdom." In *Philosophy, Science, and Religion in England 1640-1700*, edited by Richard Kroll and others, 85-109. Cambridge: Cambridge University Press, 1992. See (0917).

Levine traces the tradition of "ancient wisdom" in John Sherman, Benjamin Whichcote, and Henry More during the interregnum. He argues that they used it in more than one way. On the one hand, it helped them to defend the pagan classics against their Puritan colleagues. But, on the other hand, it assisted them in their refutation of Hobbes. With the aid of this tradition, they could harmonize reason with revelation as well as the new (or recovered) natural philosophy with Neoplatonic Christianity.

Topics: *II*, VII Names: Conway, More, Sherman, Whichcote

0947 Levy, Ron. "A Clash of Wills: Voluntarism in the thought of Robert Boyle." In *Science, Technology, and Religious Ideas*, edited by Mark H. Shale and George W. Shield, 157-176. Lanham, Maryland: University Press of America, 1994.

Levy qualifies the received understanding of Boyle's voluntarism. In considering Boyle's epistemology and empirical approach to particular phenomena, Boyle adopted what Levy calls a "disjunctive voluntarism" (in which God's transcendent will is emphasized). But, in considering general natural laws for the whole cosmos, Levy says that Boyle assumes a "conjunctive voluntarism" (in which "the goodness, wisdom, and power of God" are bound together). Thus, for Boyle, all species of law are moral and accessible to human reason.

Topic: *VII* Name: Boyle

0948 Leyden, Wolfgang von. "Locke and Nicole: Their Proofs of the Existence of God and Their Attitudes towards Descartes." *Sophia* (1948): 41-55.

The author explores the possible use Locke made of the writings of Pierre Nicole regarding his natural theology and his responses to Descartes. Leyden also points to Cudworth as a lesser influence. But Leyden notes that issues of sources in Locke's writings are difficult to sort out.

Topics: *VIII*, VII Names: Cudworth, Locke

0949 Leyden, Wolfgang von. "What is a Nominal Essence the Essence of?" In *John Locke: Problems and Perspectives*, edited by John W. Yolton, 224-233. London: Cambridge University Press, 1969. See (1715).

This is a philosophical analysis of Locke's usage of the terms "substance" and "essence." It is relevant to Locke's metaphysics and to understanding Bishop Stillingfleet's criticism of Locke's doctrines.

Topic: *VII* Names: Locke, Stillingfleet

0950 Lichtenstein, Aharon. *Henry More, the Rational Theology of a Cambridge Platonist*. Cambridge, Mass.: Harvard University Press, 1962.

Lichtenstein focuses on the More's theory and practice concerning the role of the intellect in true religion. There are only occasional passages directly regarding science or natural philosophy. (But there is a rather major section on More's psychology.) Lichtenstein writes as an intellectual historian and occasionally as something of a philosophical preacher. He argues for the decline of religion in the late seventeenth and early eighteenth centuries: in this decline, science was a factor but not the main culprit.

Topics: *VII*, X Name: More

0951 Lilley, S. "Social Aspects of the History of Science." *Arch. Int. d'Hist. Sci.* 28(1949): 376-443.

Lilley criticizes and revises de Candolle. He praises Merton and emphasizes the importance of specific historical context in discussing these matters.

Topics: *III*, I

0952 Lindberg, David C. "Conceptions of the Scientific Revolution from Bacon to Butterfield: A Preliminary Sketch." In *Reappraisals of the Scientific Revolution*, edited by David C. Lindberg and Robert S. Westman, 1-26. Cambridge: Cambridge University Press, 1990. See (0957).

Lindberg's historical sketch usefully summarizes a series of historiographic understandings of what came to be called "the scientific revolution." For the interaction of science and religion, see especially his discussions of Duhem and "the rehabilitation of medieval science" and Koyre and "the evolution of *idees trans-scientifiques*."

Topic: *I* Name: Bacon

0953 Lindberg, David C., and Ronald L. Numbers. "Beyond War and Peace: A Reappraisal of the Encounter between Christianity and Science." *Church History* 55(1986): 338-354. Reprinted in *Perspectives on Science and Christian Faith* 39(1987): 140-149.

This is a direct refutation of A. D. White's thesis that there has been "simple bipolar warfare" between science and theology (1659). Lindberg and Numbers respond by using some of White's historical examples (none of which occurred in English-speaking areas in the seventeenth century). The authors also note that apologetic historiographies are unacceptable—whether scientistic celebration of the rise of science or that which sees Christianity and science as "perennial allies." They propose a more subtle research program and stress the complexity of the mutual encounter between Christianity and science. Portions of this essay originally appeared in the introduction (0956) to the same authors' anthology, *God and Nature*.

Topic: *I*

0954 Lindberg, David C., and Ronald Numbers, eds. *God and Nature: Historical Essays on the Encounter between Christianity and Science*. Berkeley and Los Angeles: University of California Press, 1986.

This is a good collection of essays on the encounter from the early church to the present. The essays are written by leading scholars. Written for the general reading public, they summarize the latest conclusions rather than break new ground. See the Introduction

(0955) and the essays by Ashworth (0036), Deason (0333), and Webster (1623).
Topic: *I*

0955 Lindberg, David C., and Ronald L. Numbers. Introduction to *God and Nature: Historical Essays on the Encounter between Christianity and Science*, edited by David C. Lindberg and Ronald Numbers, 1-18. Berkeley and Los Angeles: University of California Press, 1986. See (0954).
This essay includes a review and analysis of the warfare metaphor in historical interpretations of the encounter between Christianity and science. Lindberg and Numbers describe why the essays included in their book portray a complex and diverse interaction in the history of science and Christianity—a complexity and diversity which defies reduction to simple conflict.
Topic: *I*

0956 Lindberg, David C., and Robert S. Westman. Introduction to *Reappraisals of the Scientific Revolution*, edited by David C. Lindberg and Robert S. Westman, xvii-xxvii. Cambridge: Cambridge University Press, 1990. See (0957).
The editors go so far as to question the possibility of an historical unit named "the scientific revolution." They find a tendency toward "a new and more thoroughgoing historicism" in the articles included in their book.
Topic: *I*

0957 Lindberg, David C., and Robert S. Westman, eds. *Reappraisals of the Scientific Revolution*. Cambridge: Cambridge University Press, 1990.
The papers in this volume cover a wide range of topics and reflect the historiographic developments of recent scholarship. The "Introduction" (0956) and papers by Copenhaver (0269), Eamon (0411), Hunter (0762), and Lindberg (0952) are annotated separately in this bibliography.
Topics: *III*, I

0958 Lindberg, David C., and others. "Science and Religion." *Metascience* 1(1992): 31-52.

This symposium contains reviews of John Hedley Brooke's *Science and Religion* and his response. The theme of Brooke's work and of the reviewers is the complexity of the interactions between Western Christianity and some of the sciences. All agree that this theme should serve as the foundation for future research. And Brooke adds that he hopes that this theme will serve "as a corrective to the view that the latest debates are necessarily the most sophisticated."
Topic: *I*

0959 Lindeboom, Gerrit Arie. *Boerhaave and Great Britain: Three Lectures on Boerhaave with Particular Reference to His Relations with Great Britain*. Leiden: Brill, 1974.
Boerhaave admired Robert Boyle's religious convictions, his scientific method, and their separation. Lindeboom discusses this point and emphasizes the influence of Boerhaave as teacher on British students of medicine and chemistry.
Topics: *XI*, VII Name: Boyle

0960 Linden, Stanton J. "Francis Bacon and Alchemy: The Reformation of Vulcan." *J. Hist. Ideas* 35(1974): 547-560.
From his background as a scholar of the literary satires of alchemy, Linden examines the ambiguous complexity of Bacon's attitude toward alchemy. He finds that Bacon thought that a truly restored alchemy was possible if based in natural philosophy, theoretical and applied, as distinct from imagination and belief. Linden also shows that Bacon used traditional categories (spirit, for example) when detailing alchemical methods. He concludes that Bacon held to "a complicated mixture of doubt and confidence" regarding alchemy.
Topics: *II*, VII, X Name: Bacon

0961 Llasera, Margaret. "Concepts of Light in the Poetry of Henry Vaughan." *Seventeenth Cent*. 3(1988): 47-61.
Llasera explicates light as an element in the poetry of Vaughan both in religious/metaphysical and in scientific terms. She sees Vaughan's philosophy of light as a blend of Greek, Christian, Neoplatonic, Hermetic, and Keplerian concepts. She explains the difference between immaterial *lux* and the material *lumen* for Vaughan in relation to his received tradition.
Topics: *X*, II Names: Vaughan, H.

0962 Locke, Louis G. *Tillotson: A Study in Seventeenth-Century Literature*. Copenhagen: Rosenkilde and Bagger, 1954.

This is a rather old-fashioned biography and appreciation of John Tillotson (1630-1694). Tillotson was a famous preacher, F.R.S., and (at the very end of his life) Archbishop of Canterbury. As the title indicates, the emphasis is on Tillotson as a figure in English literary history. Locke makes only occasional references to science or its influence.

Topics: *X*, V Names: Tillotson, Wilkins

0963 Loptson, Peter J. Introduction to Anne Conway, *The Principles of the Most Ancient and Modern Philosophy*, edited Peter J. Loptson, 1-60. The Hague: Martinus Nijhoff, 1982.

This is a summary of and philosophical commentary on the thought of Anne Conway. Her unorthodox theology, metaphysics, and vitalistic philosophy of nature are compared to Descartes, Hobbes, and Leibniz. Her discussions of God, infinite time, Christ, and creation are unorthodox and show the influence of F. M. van Helmont. For Conway, God is perfect, rational and immutable; Christ is the mutable intervening agent; and everything in creation is of one substance (monads)—with gradations from the material to the spiritual. Loptson is a philosopher who emphasizes the contemporary relevance of Conway's philosophy.

Topics: *VII*, IX Names: Conway, Helmont, Hobbes, More

0964 Lovejoy, Arthur. *The Great Chain of Being*. Cambridge, Mass.: Harvard University Press, 1936.

In this classic book, Lovejoy focuses on a complex of ideas which formed what he calls "the great chain of being." He examines its history from the ancient Greeks to the romantics of the nineteenth century. He includes a few direct references to England during our period: these have to do with God, the principle of plentitude, the infinity of the physical world, the plurality of worlds, and the place of humankind in nature. The book contains many suggestive ideas and theses. The first chapter is a relevant historiographic essay which outlines Lovejoy's views on the study of the history of ideas.

Topics: *VII*, I, X Names: Milton, More, Pope

0965 Lovejoy, Arthur O. "The Historiography of Ideas." *Proceed. Am. Phil. Soc.* 78(1938): 529-563. Reprinted in *Essays in the*

History of Ideas, Arthur O. Lovejoy, 1-13. Baltimore: Johns Hopkins Press, 1948.

Lovejoy begins by pointing out that the divisions between various temporarily isolated disciplines that deal with ideas are breaking down. He then makes three suggestions: (1) begin with the history of philosophy; (2) study "unit-ideas;" and (3) work collaboratively. Lovejoy uses Milton's poetry and theology in connection with the new astronomy as an example in this sixty-year-old, but still interesting, proposal.

Topic: *I*

0966 Lovejoy, Arthur O. "Milton's Dialogue on Astronomy." In *Reason and the Imagination*, edited by Joseph Anthony Mazzeo, 129-142. New York: Columbia University Press, 1962. See (1004).

Lovejoy exegetes Adam's dialogue with Raphael in *Paradise Lost*. This could be one more chapter in *The Great Chain of Being*.

Topics: *X*, VII Name: Milton

0967 Lovejoy, Arthur O. "Reflections on the History of Ideas." *J. Hist. Ideas* 1(1940): 3-23. Reprinted in *Ideas in Cultural Perspective*, edited by Philip P. Wiener and Aaron Noland, 3-23. New Brunswick, N.J.: Rutgers University Press, 1962.

Lovejoy here outlines the purpose of the *Journal of the History of Ideas* and argues for the history of ideas as having an independent value. The journal, he hopes, will contribute toward a liaison between specialists. He specifically calls for studies of the influence of science on culture (including religion).

Topic: *I*

0968 Lyons, Henry. *The Royal Society, 1660-1940: A History of Its Administration under Its Charters*. Cambridge: Cambridge University Press, 1944.

As its title indicates, this is an administrative history. It is useful for basic facts and statistics, but is not directly relevant to issues of science and religion.

Topic: *V*

0969 MacDonald, Michael. *Mystical Bedlam: Madness, Anxiety, and Healing in Seventeenth-Century England*. Cambridge: Cambridge University Press, 1981.

McDonald analyzes some 2000 cases of Richard Napier, the astrological physician and Anglican cleric. From these cases, MacDonald formulates an historical picture of mental illnesses of ordinary people and their treatments in the early seventeenth century. He finds that Napier's various treatments (astrological charts, herbs, amulets, spiritual counsel, social advice) situated the person in the cosmos. Napier's eclectic treatments aimed at restoring harmony to the person simultaneously on the corporeal, social, and cosmic levels.
Topics: *XI*, II Name: Napier

0970 MacDonald, Michael. "Religion, Social Change, and Healing in England, 1600-1800." In *The Church and Healing*, edited by W. J. Sheils, 101-125. Oxford: B. Blackwell, 1982.
MacDonald focuses on the Puritan tradition of healing fasts and prayer as one example of the eclectic religious therapies that appealed to ordinary people. This tradition, he shows, continued into the Restoration and—among dissenters and Methodists—into the eighteenth century. But it was rejected by the Anglican elite.
Topics: *XI*, IV

0971 MacIntosh, J. J. "Locke and Boyle on Miracles and God's Existence." In *Robert Boyle Reconsidered*, edited by Michael Hunter, 193-214. Cambridge: Cambridge University Press, 1994. See (0759).
MacIntosh provides an analysis of Locke and Boyle on miracles and God's existence—giving historical interpretations and philosophical evaluations. For Boyle, God is necessary to explain laws of nature as well as the initial creation. Miracles are a subset of "naturalistically inexplicable features of the world." A Christian virtuoso will be aware of these supramechanical phenomena but, as a naturalist, will not be concerned with them. Boyle, according to MacIntosh, was interested both in the regularities and in all interventions in nature; such would lead both to belief in God and to an awareness that God's will may well be different from ours.
Topics: *VII*, VIII Names: Boyle, Locke

0972 MacIntosh, J. J. "Robert Boyle on Epicurean Atheism and Atomism." In *Atoms, Pneuma, and Tranquility*, edited by

Margaret J. Osler, 197-219. Cambridge: Cambridge University Press, 1991. See (1179).

This is a detailed discussion of Boyle's unpublished arguments against ancient Epicurean atomism. Macintosh recreates Boyle's sophisticated three-stage argument for God and Christianity. In addition, he proposes a possible reason for Boyle's failure to publish his arguments in any comprehensive form.

Topics: *IX*, VIII, VII Name: Boyle

0973 MacIntosh, J. J. "Robert Boyle's Epistemology: The Interaction-Between Scientific and Religious Knowledge." *Int. Stud. Philos. Sci.* 6(1992): 91-121.

MacIntosh presents a solid systematic analysis of Boyle's epistemology. At the same time, he argues that Boyle's epistemology reflects his personality and that, for Boyle, scientific and religious knowledge overlap seamlessly. MacIntosh discusses Boyle's views on the "limitations of human knowing," arguments for the existence of God, and the epistemological standing of scientific claims. He also examines Boyle's use of the terminology of hypotheses, theories, and facts—Boyle's Law being taken as an example.

Topics: *VII*, VIII Name: Boyle

0974 Mackinnon, Flora I. Introduction and Notes to *The Philosophical Writings of Henry More*, edited by Flora I. Mackinnon, ix-xxvii and 233-333. London: Oxford University Press, 1925.

Mackinnon's thirteen-page "Outline Summary of More's Philosophical Theory" is still useful. Her extensive notes focus on such topics as More's arguments on: the nature of space; the Spirit of Nature; witchcraft; and the reality of spirit from the phenomena of nature.

Topic: *VII* Name: More

0975 Macklem, Michael. *The Anatomy of the World: Relations between Natural and Moral Law from Donne to Pope*. Minneapolis: University of Minnesota Press, 1958.

Macklem traces what he considers to be the transformation from Donne's sense of disorder in nature and humanity to Pope's sense of order in both. Involved in these images is the question of moral and material evil: is it a consequence of sin or a condition of existence? Central to his discussion is Burnet's theory of the earth and the responses which it stimulated. Also playing roles in this transformation, according to Macklem, were: Newton's theory of gravity and its relation to the original and present providence of God; and Clarke's application of this continuous view of providence to morality.

Topics: *X*, VII, XII Names: Burnet, T., Clarke, Donne, Newton, Pope

0976 Maddison, R. E. W. *The Life of the Honourable Robert Boyle, F.R.S.* London: Taylor and Francis, 1969.

This detailed biography is organized by the periods of Boyle's residence: his early life; his life at Stalbridge; at Oxford; at London. Maddison then adds chapters on: Boyle's death; his will and the resulting Boyle Lectures; and miscellanea and appendixes. Maddison incorporates autobiographical texts and letters into the text. Maddison notes Boyle's religious outlook, his theological and scientific publications, his chemistry and the air pump. But no mention is made of alchemy.

Topics: *VII*, V, X Names: Boyle, Hartlib, Hooke, Oldenburg, Ranelagh

0977 Mamiani, Maurizio. "The Rhetoric of Certainty: Newton's Method in Science and in the Interpretation of the Apocalypse." In *Persuading Science: The Art of Scientific Rhetoric*, edited by Marcello Pera and William R. Shea, 157-172. Canton, Mass.: Science History, 1991.

Mamiani proposes that Newton used the same rhetorical device in his interpretation of the Apocalypse as in his *Principia*. They both reflect a common "reference to certainty, to which they come by means of devices that can eliminate hypotheses."

Topics: *XII*, VII, X Name: Newton

0978 Mandelbaum, Maurice. *Philosophy, Science, and Sense Perception: Historical and Critical Studies*. Baltimore: Johns Hopkins, 1964.

This is a well-written, influential, and (in some quarters) controversial work in the history of philosophy. Two of the four studies are entitled "Locke's Realism" and "Newton and Boyle and the Problem of 'Transdiction'." As part of the larger agenda of supporting his own philosophical position of critical realism,

Mandelbaum addresses issues of epistemology, scientific method, seventeenth-century atomism, and what he calls "transdiction." This last term refers to the inductive inference from observable phenomena to conclusions about unobservable reality. Boyle, Newton, and Locke, argues Mandelbaum, all believed in the legitimacy of transdiction. Only indirect references to God or theology are included in these studies, but the significance is nonetheless great for the overlap between seventeenth-century science and theology. (Mandelbaum does not seem to realize it, but virtually all natural theological claims of an empirical kind are dependent upon "transdiction." Especially to be noted here is that form of transdiction which Mandelbaum calls "the use of analogical reasoning for the translation of explanatory principles from the observed to the unobserved.") And one could easily demonstrate with natural theological examples that all three of Mandelbaum's subjects—but particularly Boyle and Newton—clearly believed in transdiction.

Topics: *VII*, VIII, XII Names: Boyle, Locke, Newton

0979 Mandelbrote, Scott. "'A Duty of the Greatest Moment': Isaac Newton and the Writing of Biblical Criticism." *Brit. J. Hist. Sci.* 26(1993): 281-302.

Although Mandelbrote does not much discuss "biblical criticism" in the usual sense of the phrase, this is a good summary of Newton's biblical and theological views. Mandelbrote notes that, for Newton, studying the Bible was an important religious duty. And, although Newton's biblical study was done in a rather solitary manner, there is a unity to Newton's theological, natural philosophical, and mathematical thought.

Topics: *XII*, VII, X Names: Newton

0980 Mandelbrote, Scott. "Isaac Newton and Thomas Burnet: Biblical Criticism and the Crisis of Late Seventeenth-Century England." In *The Books of Nature and Scripture*, edited by James E. Force and Richard H. Popkin, 149-178. Dordrecht: Kluwer Academic, 1994. See (0498).

Mandelbrote first sets the discussions between Newton and Burnet over the interpretation of Genesis in the context of the history of Augustinian exegesis. He then develops the thesis that key differences between the two lay in their understandings of the use of

the interpretive technique of accommodation and in Burnet's appeal to the established church. Ironically, Newton (the secret heretic) found royal favor while Burnet retired in disgrace.
Topics: *X*, V, XII Names: Burnet, T., Newton

0981 Manuel, Frank E. *The Eighteenth Century Confronts the Gods*. Cambridge, Mass.: Harvard University Press, 1959.
The parts on Newton are superseded by *Isaac Newton, Historian* (0985).
Topics: *XII*, VII, X Name: Newton

0982 Manuel, Frank E. *Freedom from History and Other Essays*. New York: New York University Press, 1972.
The author's essays on Newton (0986), Newton and Locke (0983), and pansophia (0987) are annotated separately in this bibliography.
Topics: *XII*, IV

0983 Manuel, Frank E. "The Intellectual in Politics: Locke, Newton, and the Establishment." In *Freedom from History*, by Frank E. Manuel, 189-203. New York: New York University Press, 1971. See (0982).
Manuel presents Locke and Newton as ideological representatives of early eighteenth-century English society.
Topics: *XII*, I, V Names: Locke, Newton

0984 Manuel, Frank E. *Isaac Newton, Historian*. Cambridge, Mass.: Harvard University Press, 1963.
Here Manuel deals with Newton's chronological and biblical studies. Particular chapters deal with astronomical dating, euhemerism, the accommodation theory of biblical interpretation, and Newton's connecting of sacred and profane history. This is a good (early) study of Newton and of the historical context in which he did his thinking.
Topics: *XII*, X Names: Locke, Newton, Stukeley, Whiston

0985 Manuel, Frank E. "The Lad from Lincolnshire." In *The Annus Mirabilis of Sir Isaac Newton*, edited by Robert Palter, 2-21. Cambridge, Mass.: MIT Press, 1970. See (1201).
This essay represents one chapter from Manuel's *A Portrait of Isaac Newton* (0988).
Topic: *XII* Name: Newton

0986 Manuel, Frank E. "Newton as Autocrat of Science." *Daedalus* 97(1968): 969-1001. Reprinted in *Freedom from History*, by Frank E. Manuel, 151-187. New York: New York University Press, 1971. See (0982).

Manual argues that Newton became "the grand administrator of science" in the years following his appointments as Master of the Mint (1696) and President of the Royal Society (1702). Newton was "responsible for weaving a sacred and aristocratic aura around science." He did this through his disciples, clerical and lay, and by encouraging a Newtonian natural theology as well as demanding a Newtonian natural philosophy. Manual describes the sessions of the Royal Society as ceremonies officiated by "the first high priest of modern science"—yet, in Newton's mind, for the glory of God.

Topics: *XII*, V Name: Newton

0987 Manuel, Frank E. "Pansophia, A Seventeenth-Century Dream of Science." In *Freedom from History*, by Frank E. Manuel, 89-113. New York: New York University Press, 1971. See (0982).

Manuel shows that, for an international community of thinkers, the new science was to be used as a tool for discovering universal wisdom (pansophia) and for establishing a utopian Christian society. This program also was intimately related to religious views.

Topics: *IV*, II, III, IV Names: Bacon, Comenius, Hartlib

0988 Manuel, Frank E. *A Portrait of Isaac Newton*. Cambridge, Mass.: Harvard University Press, 1968.

Manuel here presents an extensive psychological (neo-Freudian) interpretation of Newton. The book has come under heavy criticism, some of it deserved. However, it does have some good insights and has been influential. It directly confronts Newton's eccentric, obsessive, and unattractive personality traits. Manuel relates Newton's science and religion both to each other and to the emotional traumas of his childhood.

Topics: *XII,* II, V Name: Newton

0989 Manuel, Frank E. *The Religion of Isaac Newton*. Cambridge: Clarendon Press, 1974.

Manuel emphasizes Newton's historical and scriptural studies. He does note parallels in Newton's method in these studies to that of his scientific work. As in *A Portrait of Isaac Newton*, Manuel at times seems to reduce Newton's ideas and work to his psychological traumas. This book also reproduces a number of important and previously unpublished Newtonian manuscripts.
Topics: *XII*, II, III, X Name: Newton

0990 Markley, Robert. "Isaac Newton's Theological Writings: Problems and Prospects." *Restoration* 13(1989): 35-48.
Markley discusses the recent recovery of Newton's theological manuscripts and assesses their significance for current and future Newtonian studies. He argues that the works of Manuel, Westfall, Dobbs, and Castillejo represent only preliminary accounts. Markley proposes a postmodern approach to Newton's writings and suggests that "it is only through a sophisticated theoretical awareness of the problems of language, narrativity, and referentiality that we can begin to explore the relationships that may exist among Newton's works in science, theology, and history."
Topics: *XII*, I, VII, X Name: Newton

0991 Marks, Carol L. "Thomas Traherne and Cambridge Platonism." *PMLA* 81(1966): 521-534.
This is a good summary of Cambridge Platonism and of the poet Traherne's relation to the movement. As a manifestation of divinity, in Traherne's Neoplatonist view, the physical world merits both praise and study. Here is the juncture between the new philosophy and Cambridge Platonism. Marks discusses Traherne's views on infinity, space, and God. She sees him as a Christian Platonist but yet as one who, in some ways, defies such labels.
Topics: *X*, VII Names: More, Traherne

0992 Marks, Geoffrey, and William K. Beatty. *The Story of Medicine in America*. New York: Scribner, 1973.
This is a popular historical overview—parts of which are relevant and useful.
Topic: *XI* Names: Boylston, Firmin, Mather, C.

0993 Mason, Stephen F. *A History of the Sciences*. New York: Collier Books, 1962.

See (0994).
Topic: *VII*

0994 Mason, Stephen F. *Main Currents of Scientific Thought*. New York: H. Schuman, 1956. Revised Ed., *A History of the Sciences*. New York:Collier Books, 1962.
This is a well-written standard survey. Mason includes external factors in the history of science. He gives a summary of his view that the scientific revolution was encouraged by the Protestant Reformation.
Topics: *VII*, III

0995 Mason, Stephen F. "Religion and the Rise of Modern Science." In *Science and Religion/Wissenschaft und Religion*, edited by Anne Baumer and Manfred Buttner, 2-13. Bochum, Germany: Universitatsverlag N. Brockmeyer, 1989. See (0066).
In this summary of his larger pieces, Mason argues that Protestant thought tended to promote the development of science "without much overt intent." It did so by promoting common patterns of thought which replaced the medieval cosmic and social hierarchy. The new pattern included, first, an absolute monarchical God and, later, a mechanistic universe and polis with the ruler limited by natural and constitutional laws.
Topics: *III*, IV

0996 Mason, Stephen F. "Science and Religion in Seventeenth Century England." *Past and Present* 3(1953): 28-44. Reprinted with corrections in *The Intellectual Revolution of the Seventeenth Century*, edited by Charles Webster, 197-217. London: Routledge and Kegan Paul, 1974. See (1618).
Mason argues in favor of the connection between Protestantism and science. He stresses the Calvinist doctrines concerning God's governance of the universe.
Topics: *III*, VII Name: Wilkins

0997 Mason, Stephen F. "The Scientific Revolution and the Protestant Reformation." *Ann. Sci.* 9(1953): 64-87 and 154-175.
This has to do with the sixteenth century on the continent. But it has been used in the rise-of-science debate in favor of Merton's position.
Topic: *III*

0998 Mathias, Peter, ed. *Science and Society 1600-1900*. Cambridge: Cambridge University Press, 1972.

The two relevant essays by A. R. Hall (0606) and by Rattansi (1297) are annotated separately.

Topic: *IV*

0999 May, Henry F. *The Enlightenment in America*. New York: Oxford University Press, 1976.

May proposes (and discusses in a clear and useful way) four categories of the Enlightenment. He also discusses the impact of the Enlightenment on America. Most of the book is concerned with the post-1730 period: Section I, on the moderate or rational Enlightenment, is most relevant to our bibliography.

Topics: *VII*, IX Names: Clarke, Locke, Tillotson

1000 Mayo, Thomas Franklin. *Epicurus in England (1650-1725)*. Dallas: Southwest Press, 1934.

Mayo argues that the Epicurean revival in England began with Charleton and Evelyn in 1656 and came from France. He ties its success to socioeconomic and political forces—and its demise to an alliance between Anglican Christianity and the new science combined with the moral and intellectual conservatism of the rising middle class. Mayo suggests provocative theses but they are not very well-supported. This book has been superseded but some scholars still find it to be useful.

Topics: *VII*, IV, X Names: Charleton, Evelyn

1001 Mazzeo, Joseph Anthony. "Cosmos and Meaning: The Theme of Cosmic Order in Donne's *Anniversary* Poems and Pope's *Essay on Man*." In *Nature and the Cosmos: Essays in the History of Ideas*, 59-108. Oceanside, N.Y.: Dabor Science, 1977. See (1002).

This essay focuses on the history of the question: does the cosmos offer a home for human values and aspirations? Mazzeo argues: (1) that there was a crisis in the tradition of cosmic piety in the sixteenth and seventeenth centuries; and (2) that Newtonian science and natural theology reestablished cosmic order and piety (at least for the poets of the eighteenth century).

Topics: *VIII*, VII, X, XII Names: Donne, Newton, Pope

1002 Mazzeo, Joseph Anthony. *Nature and the Cosmos: Essays in the History of Ideas*. Oceanside, N.Y.: Dabor Science, 1977.

Two articles (1001) and (1007) are annotated separately in this bibliography.

Topics: *I*, X

1003 Mazzeo, Joseph Anthony. "Notes on John Donne's Alchemical Imagery." *Isis* 48(1957): 103-123.

Mazzeo shows that Donne understood the Hermetic tradition and the "principle of universal analogy." He finds in Donne's poetry imagery derived from the purifying tincture—the quintessence—which is sometimes likened to Christ's blood.

Topics: *X*, II Name: Donne

1004 Mazzeo, Joseph Anthony, ed. *Reason and the Imagination: Studies in the History of Ideas 1600-1800*. New York: Columbia University Press, 1962.

The relevant essays by Colie (0252), R. F. Jones (0840), and Lovejoy (0966) are annotated separately.

Topic: *X*

1005 Mazzeo, Joseph Anthony. *Renaissance and Revolution: The Remaking of European Thought*. New York: Pantheon Books, 1965.

This is a somewhat dated but well-written survey of the intellectual history of the period and its implications for the present. It is aimed at a general audience and written by an historian of literature and culture. Three chapters are relevant: "Bacon: The New Philosophy;" "Hobbes: The Scientific Secularization of the World;" and "The Idea of Progress: Science and Poetry." Mazzeo sees the seventeenth century as a "dividing point in the nature of the influence of science in literature." But he disagrees with those who believe that science caused a degeneration of poetry. Mazzeo's discussions of various interpretations of Christian thought and their interplay with scientific and political thought are insightful (even though, as he acknowledges, he bases his survey on the work of others).

Topics: *X*, VII Names: Bacon, Goodman, Hakewill, Hobbes

1006 Mazzeo, Joseph Anthony. "Some Interpretations of the History of Ideas." *J. Hist. Ideas* 33(1972): 379-394. Reprinted as

"The History of Ideas and the Study of Literature." In *Nature and the Cosmos: Essays in the History of Ideas*, 1-17. Oceanside, N.Y.: Dabor Science, 1977. See (1002).

Mazzeo discusses the history of the methods of historians of ideas and of the ways in which conceptual thought has entered the literary imagination. Special attention is given to Hegel, Burckhardt, Dilthey, Whitehead, Lovejoy, and Spitzer. "They have helped us," Mazzeo concludes, "to grasp the subtle relations which obtain between mythic, religious, and poetic activities of the mind, on the one hand, and conceptual thinking of a philosophical and scientific kind, on the other."

Topics: *I*, X

1007 Mazzeo, Joseph Anthony. "The World's Meaning: The Theme of Cosmic Piety in Western Thought." *Centennial Review* 21(1977): 355-373.

This is a shorter version of (1001).

Topics: *VIII*, VII, X, XII Names: Donne, Newton

1008 McAdoo, Henry R. *The Spirit of Anglicanism: A Survey of Anglican Theological Method in the Seventeenth Century*. New York: Scribner; London: A. and C. Black, 1965.

McAdoo is an Anglican bishop writing a history of theology. He emphasizes Scripture, reason, and the appeal to antiquity (i.e., the early Church Fathers) as the three important methodological criteria in seventeenth-century Anglican theology. He explicates the evolution and usage of these criteria by briefly summarizing the individual publications of dozens of theologians from the period. (With the help of the index, the book can be used as an encyclopedia for these theologians.) McAdoo sees the use of all three of his criteria as characteristic of liberal (and true) theological method and tends to see them in all of the writers he studies. He thus emphasizes what the various writers had in common more than their differences. His ten chapters include two on the Cambridge Platonists, two on the Latitudinarians, and two on "The New Philosophy and Theological Method." He emphasizes the compatibility of science and religion during the period. Because McAdoo understands Anglican theology so well, his book functions as a good counter to works in cultural history and the history of science which press Enlightenment and deistic themes backwards into the seventeenth century.

Topics: *VII*, VIII, X, XII Names: Barrow, Bentley, Boyle, Browne, T., Chillingworth, Cudworth, Glanvill, More, Newton, Patrick, Ray, Stillingfleet, Tillotson, Wilkins

1009 McCaffery, Ellen. *Astrology: Its History and Influence in the Western World*. New York: C. Scribner's and Sons, 1942. Reprinted several times, including New York: S. Weiser, 1970.

McCaffery's book is a popular overview of the history of astrology. Chapters 20 and 21 deal with the seventeenth and early-eighteenth centuries—touching on relations between astrology and religion. This book is dated but readable.

Topics: *II*, X Names: Ashmole, Lilly, Partridge

1010 McColley, Grant. "The Astronomy of Paradise Lost." *Studies in Philology* 34(1937): 209-247.

McColley here summarizes the differing astronomical systems contemporary with Milton and describes Milton's own reactions.

Topic: *X* Name: Milton

1011 McColley, Grant. "A Facet from the Life of Newton." *Isis* 28(1938): 94-95.

Newton, Barrow, and More all subsidized a book in 1676—one of the aims of which was to improve the singing of Psalms in the churches.

Topics: *XII*, X Names: Barrow, More, Newton

1012 McColley, Grant. "John Wilkins—a Precursor of Locke." *Philos. Rev.* 47(1938): 642-643.

McColley argues that Wilkins had some similar ideas to those which Locke made famous (though he did not necessarily influence Locke).

Topic: *VII* Names: Locke, Wilkins

1013 McColley, Grant. "Milton's Dialogue on Astronomy: The Principal Immediate Sources." *PMLA* 52(1937): 728-762.

McColley here relates *Paradise Lost* to the Ross-Wilkins debate.

Topics: *X*, VII Names: Milton, Ross, Wilkins

1014 McColley, Grant. "Nicholas Hill and the *Philosophia Epicurea*." *Ann. Sci.* 4(1939): 390-405.

This is a summary of Hill's book (1601). McColley outlines Hill's atomism, his affirmations of the Copernican theory and the infinity of the cosmos, his nonecclesiastical theology, and his understanding of human nature.

Topic: *VII* Name: Hill

1015 McColley, Grant. "The Ross-Wilkins Controversy." *Ann. Sci.* 3(1938): 153-189.

This is a good discussion of Aristotelianism and biblical conservatism in the last aggressive stand taken against the new astronomy.

Topics: *VII*, IV, X Names: Ross, Wilkins

1016 McColley, Grant. "The Theory of a Plurality of Worlds as a Factor in Milton's Attitude toward the Copernican Hypothesis." *Mod. Lang. Notes* 47(1932): 319-325.

McColley here argues that, in early-seventeenth-century England, the Copernican (heliocentric) theory was associated with the view that there are a plurality (or even an infinity) of inhabited planets. He discusses several writers, including Milton, who seem to have distrusted the Copernican theory for this very reason.

Topics: *X*, VII Names: Milton, More

1017 McCutcheon, Elizabeth. "Bacon and the Cherubim: An Iconographical Reading of the *New Atlantis*." *English Language Renaissance* 2(1972): 334-355.

McCutcheon considers the image of the cherubim on the title page of *Sylva Sylvarum* to be a key to the thought and structure of the *New Atlantis* and to be "a symbol of the singular promise of Bensalem." She explores the verbal metaphors in Bacon's work and finds that the people of Bensalem are "extraordinarily like the cherubim" as they are traditionally visualized. For Bacon, she says, cherubim symbolize human potential by joining a dedication to true knowledge with its loving and life-giving use.

Topics: *X*, VII Name: Bacon

1018 McGiffert, Arthur Cushman. *Protestant Thought Before Kant*. New York: C. Scribner's Sons, 1911. Reprinted, New York: Harper Torchbooks, 1962.

This is an old classic. It is still worth looking at for its treatment of "Rationalism" and English religious thought during our period. McGiffert makes little reference to science—but his work indirectly supports the position that deism and heretical forms of Christianity developed independently of science.

Topics: *VII*, IX Names: Clarke, Locke, More, Tillotson, Toland

1019 McGrath, Alister E. "Anglicanism." In *The History of Christian Theology*, edited by Paul Avis, Vol. 1, 179-205. Grand Rapids, Mich.: Eerdmans, 1986.

This is a useful introduction to seventeenth-century English theology—including its relation to natural philosophy. McGrath emphasizes natural theology, Locke's universal method (with a critique of it), and deism.

Topics: *VII*, VIII, IX Name: Locke

1020 McGuire, J. E. "Atoms and the 'Analogy of Nature': Newton's Third Rule of Philosophizing." *Stud. Hist. Philos. Sci.* 1(1970): 3-58. Reprinted in *Tradition and Innovation*, by J. E. McGuire, 52-102. Dordrecht: Kluwer Academic, 1996. See (1032).

The main focus of this long article is on the overlap between Newton's theory of matter and the epistemological problem of "transduction" (i.e., making claims about the unobservable from analogies with the observable). The basic framework, according to McGuire, is formed from the ancient metaphysics of the macrocosm and microcosm. "Thus the theological chain of being is transformed by Newton to provide a general framework with which to comprehend material phenomena and in so doing he combines it with the primordials of the atomic hypothesis." And, behind the third rule "which seems merely to be a methodological statement lies a complete ontology based on the notion of the great chain of being and the Christian doctrine of the perfection of God." McGuire briefly discusses George Cheyne's elaborate metaphysics of analogy and (in footnote 126) presents an early thesis about the ideological applications (political and social) of Newtonian natural philosophy—all rooted in the archetypal patterns in the mind of God.

Topics: *XII*, IV, VII Names: Cheyne, Locke, Newton

1021 McGuire, J. E. "Body and Void in Newton's *De Mundi Systemate*: Some New Sources." *Arch. Hist. Exact Sci.* 3(1966). 206-248. Reprinted in *Tradition and Innovation*, by J. E. McGuire, 103-150. Dordrecht: Kluwer Academic, 1996. See (1032).

McGuire focuses on Newton's doctrines of matter and void spaces, the related theological and philosophical sources, and Newton's differences with Leibniz. McGuire was one of the first historians to see just how deeply Newton's doctrine of God was integrated with his natural philosophy. As with most of his essays, this one is highly technical but well-written.

Topics: *XII*, VII Name: Newton

1022 McGuire, J. E. "Boyle's Conception of Nature." *J. Hist. Ideas* 33(1972): 523-542.

McGuire here discusses the philosophical implications of voluntarist theology with respect to Boyle and to other seventeenth-century natural philosophers. This includes: the theological implications of mechanism and materialism; Boyle's understanding of space; and the relevance of Boyle's doctrine of Providence to his natural philosophy. McGuire tries to keep the picture in our period undistorted by retrospective theses about science "changing" traditional Christianity. This is an excellent article.

Topics: *VII*, II Name: Boyle

1023 McGuire, J. E. "Existence, Actuality and Necessity: Newton on Space and Time." *Ann. Sci.* 35(1978): 463-508. Reprinted in *Seventeenth-Century Natural Scientists*, edited by Vere Chappell, 269-314. New York: Garland, 1992. See (0205). Reprinted in *Tradition and Innovation*, by J. E. McGuire, 1-51. Dordrecht: Kluwer Academic, 1996. See (1032).

McGuire here presents Newton's metaphysical and epistemological views regarding God, space, time, existence, and necessity. He does this in a systematic and historical way—considering also selected positions of Gassendi, Charleton, More, Descartes, and Spinoza. McGuire concludes that Newton's natural philosophy is grounded in his metaphysics and theology. Moreover, "Newton's aim goes beyond explaining the phenomena of nature, important though this is. Ultimately, he wished to render the Creator and the whole of creation intelligible to human understanding."

Topics: *XII*, VII, VIII Names: Charleton, More, Newton

1024 McGuire, J. E. "Force, Active Principles, and Newton's Invisible Realm." *Ambix* 15(1968): 154-208. Reprinted in *Tradition and Innovation*, by J. E. McGuire, 190-258. Dordrecht: Kluwer Academic, 1996. See (1032).

McGuire here argues that some basic concepts in Newton's natural philosophy—matter, the void, causation, force and others—can be ultimately clarified only in terms of the theological framework. In explicating this framework, McGuire emphasizes Newton's voluntaristic doctrine of God and the traditional Judeo-Christian doctrines of creation and providence.

Topics: *XII*, II, VII Names: More, Newton

1025 McGuire, J. E. "Neoplatonism and Active Principles: Newton and the *Corpus Hermeticum*." In *Hermeticism and the Scientific Revolution*, edited by Robert S. Westman and J. E. McGuire, 93-142. Los Angeles: William Andrews Clark Library, University of California, 1977. See (1655).

McGuire argues that the Neoplatonism of the Cambridge Platonists formed Newton's basic intellectual orientation. The role of Hermeticism in this tradition, however, is very limited (for McGuire) in that Newton rejected hermetic magic. While grudgingly acknowledging Newton's alchemy, McGuire points out the differences between the hermetic corpus and Newton's writings regarding the relation of God to nature, active forces, and the phenomena of light. Further, he argues that Hermeticism itself was never "a separate intellectual tradition" but almost always was "disseminated through the revival of Neoplatonism."

Topics: *VII*, II, XII Names: Cudworth, More, Newton

1026 McGuire, J. E. "Newton and the Demonic Furies: Some Current Problems and Approaches in History of Science." *Hist. Sci.* 11(1973): 21-48.

This is a wide-ranging historiographic discussion with one section devoted to a critique of Frank Manuel's *Portrait of Isaac Newton* (0988).

Topics: *XII*, I, II Name: Newton

1027 McGuire, J. E. "Newton on Place, Time, and God: An Unpublished Source." *Brit. J. Hist. Sci.* 11(1978): 114-129.

Most of this article is given over to a manuscript from the Portsmouth Collection at Cambridge. In five pages of commentary, McGuire relates the manuscript to *De Gravitatione* on the one hand and to the General Scholium of the *Principia* on the other. McGuire basically reinforces the interpretations of Newton which appear in his other writings.

Topics: *XII*, VII Name: Newton

1028 McGuire, J. E. "Newton's 'Principles of Philosophy': An Intended Preface for the 1704 *Opticks* and a Related Draft Fragment." *British Journal for the History of Science* 5(1970): 178-186.

In his brief introduction to the text of Newton's manuscripts, McGuire indicates that God is included among Newton's "principles." (The main manuscript itself includes an argument for God as the cause of "the frame of nature" and "the contrivance of ye bodies of living creatures.")

Topics: *XII*, VII Name: Newton

1029 McGuire, J. E. "Predicates of Pure Existence: Newton on God's Space and Time." In *Philosophical Perspectives on Newtonian Science*, edited by Philip Bricker and R. I. G. Hughes, 91-108. Cambridge, Mass.: MIT Press, 1990. See (0123).

McGuire here defends against John Carriero (0190) the positions taken by McGuire in articles published in 1978. He modifies his language and admits a philosophical difficulty ("that the unity of God's essence and existence is threatened") in Newton's view. McGuire continues to insist that, regarding absolute space and time, Newton's theology was fundamentally voluntarist.

Topics: *XII*, VII Name: Newton

1030 McGuire, J. E. "Space, Geometrical Objects and Infinity: Newton and Descartes on Extension." In *Nature Mathematized*, edited by William R. Shea, 69-112. Dordrecht: D. Reidel, 1982. See (1445). Reprinted in *Tradition and Innovation*, by J. E. McGuire, 151-189. Dordrecht: Kluwer Academic, 1996. See (1032).

McGuire gives a historico-philosophical analysis of selected passages from Newton's *De Gravitatione et aequipondio fluidorum*

and from Descartes' *Meditations*. God is not a main focus here but does come into McGuire's discussion at several points.
Topics: *XII*, VII Name: Newton

1031 McGuire, J. E. "Space, Infinity, and Indivisibility: Newton on the Creation of Matter." In *Contemporary Newtonian Research*, edited by Zev Bechler, 145-190. Dordrecht: D. Reidel, 1982. See (0071).
This is a high-powered philosophical and historical analysis of Newton's early (1660s) manuscript treatise *De Gravitatione et aequipondio fluidorum*. As he does in other essays, McGuire focuses on Newton's metaphysical views and their epistemological and other implications. He relates Newton's views to the philosophical context of the time, especially to Newton's opposition to Descartes. Newton's "voluntarist" doctrine of God is basic to his metaphysics of space and matter and McGuire effectively analyzes this as well.
Topics: *XII*, VII Name: Newton

1032 McGuire, J. E. *Tradition and the Innovation: Newton's Metaphysics of Nature*. Dordrecht: Kluwer Academic, 1996.
This is a collection of seven reprinted essays—six of which are included in this bibliography. The book includes a six-page introduction and updated bibliographic references in the endnotes of each reprinted article. See: (1020), (1021), (1023), (1024), (1030), and (1033).
Topics: *XII*, VII Name: Newton

1033 McGuire, J. E. "Transmutation and Immutability: Newton's Doctrine of Physical Qualities." *Ambix* 14(1967): 69-95. Reprinted in *Tradition and Innovation*, by J. E. McGuire, 262-286. Dordrecht: Kluwer Academic, 1996. See (1032).
In one of his early essays, McGuire traces the development of Newton's views concerning transmutation and the immutable qualities of matter. He does not directly discuss religion, but God and theology are addressed indirectly in connection with atomism and the *prisca sapientia*—in which terms McGuire frames his analysis.
Topics: *XII*, II, VII Name: Newton

1034 McGuire, J. E., and P. M. Rattansi. "Newton and the 'Pipes of Pan'." *Notes Rec. Royal Soc. London* 21(1966): 109-143.

In this early work connecting Newton to ancient wisdom, McGuire and Rattansi argue strongly for the positive influence of a *prisca theologia* in Newton's natural philosophy. In so doing, they also discuss: (1) theology and doctrines of matter; and (2) Newton's natural theology. This was a groundbreaking article which has since become something of a classic.

Topics: *XII*, II, IV, VII, VIII Names: Cudworth, More, Newton

1035 McGuire, J. E., and Martin Tamny. "Commentary." In *Certain Philosophical Questions: Newton's Trinity Notebook*, edited by J. E. McGuire and Martin Tamny, 1-325. Cambridge: Cambridge University Press, 1983.

This is a long and thorough commentary on the young Newton's notebook (probably from the years 1664 and 1665). "Theological entries are few in number" in the manuscript itself and, consequently, Newton's religious views are not a major part of the commentary. But about thirty pages of the commentary relate to religious topics: God, infinity, and space (pp. 118-125); the Cartesian influence (127-145); whether God created time (307-309); of souls (316-317); and Newton's heterodox theology (309 and 317). McGuire and Tamny also discuss the early Newton's reactions to Charleton, Hobbes, and More.

Topics: *XII*, VII Names: Charleton, Hobbes, More, Newton

1036 McKnight, Stephen A. "The Renaissance Magus and the Modern Messiah." *Rel. Stud. Rev.* 5(1980): 81-89.

This is an historiographical essay covering major works in the field (Yates, Walker, and others) as of 1979. McKnight perceives a shift from seeing the Hermetic tradition as a form of gnosticism to seeing it as a basis for the eschatological dreams of modernity.

Topics: *II*, I

1037 McLachlan, Herbert. Introduction to *Isaac Newton: Theological Manuscripts*, edited by Herbert McLachlan, 1-25. Liverpool: Liverpool University Press 1950.

McLachlan does not relate Newton's religion to his science. He stresses Newton's rationalism and his unitarianism. McLachlan's

work on Newton has been superseded—especially by the work of Westfall and Manuel.
Topics: *XII*, IX, X Name: Newton

1038 McLachlan, Herbert. *The Religious Opinions of Milton, Locke and Newton*. Manchester, England: Manchester University Press, England, 1941. Reprinted, New York: Russell and Russell, 1972.
This work mainly deals with the anti-trinitarianism of the subjects of the book; there is very little consideration of seventeenth-century science.
Topics: *IX*, III, X, XII Names: Locke, Milton, Newton

1039 McMullin, Ernan, ed. *The Concept of Matter in Modern Philosophy*. Notre Dame, Ind.: University of Notre Dame Press, 1978.
The essays by McMullin (1042) and M. B. Hall (0614) are annotated separately.
Topics: *VII*, XII Name: Newton

1040 McMullin, Ernan, ed. *Galileo: Man of Science*. New York: Basic Books, 1967.
This anthology includes two relevant essays: see I. B. Cohen (0232) and Stillman Drake (0400).
Topic: *VII*

1041 McMullin, Ernan. "The History and Philosophy of Science: A Taxonomy." In *Historical and Philosophical Perspectives of Science*, edited by Roger H. Struewer, 12-67. Minneapolis: University of Minnesota Press, 1970. See (1514).
McMullin primarily is interested in establishing the value of the history of science for the philosophy of science. He includes some insightful statements (which apply to studying the history of science and religion) concerning historiographic fallacies to which philosophers often are vulnerable in their doing of history.
Topic: *I*

1042 McMullin, Ernan. "Introduction: The Concept of Matter in Transition." In *The Concept of Matter in Modern Philosophy*, edited by Ernan McMullin, 1-55. Notre Dame, Ind.: University of Notre Dame Press, 1978. See (1039).

McMullin surveys the philosophical presuppositions, implications, and controversies connected with the concept of matter from "the Greek and medieval inheritance" to the present. A major focus is the seventeenth century with special attention given to Newton. There is some discussion of God and spirit in relation to matter and materialism.

Topics: *VII*, XII Name: Newton

1043 McMullin, Ernan. "Natural Science and Belief in a Creator: Historical Notes." In *Physics, Philosophy, and Theology*, edited by Robert John Russell, 49-79. Vatican City: Vatican Observatory, 1988. See (1381). An earlier version appears in *Religion, Science and the Search for Wisdom*, edited by David M. Byers, 13-41. Washington, D.C.: United States Catholic Conference, 1987.

This is an historical survey from the tension between Jerusalem and Athens (between the Biblical God and Aristotle's "links of causation") to contemporary theological issues. In the few pages on the seventeenth and eighteenth centuries, McMullin presents an insightful analysis of the possible meanings of the "gaps" type of argument in physico-theology.

Topics: *VIII*, X, XII Names: Boyle, Newton

1044 McMullin, Ernan. *Newton on Matter and Activity*. Notre Dame, Ind.: University of Notre Dame Press, 1978.

McMullin approaches Newton's texts from the perspective of the history and philosophy of science. He presents and analyzes Newton's views (both published and manuscript) on matter, motion, force, gravity, active principles, and the mechanical philosophy. With respect to each of these subjects, McMullin consistently (though sometimes tangentially) relates Newton's natural philosophy to Newton's understanding of God. McMullin treats Newton's God as an important and legitimate metaphysical foundation of Newton's science. This is a fairly short but rather intense book.

Topics: *XII*, VII Name: Newton

1045 McRae, Robert. "Unity of Sciences: Bacon, Descartes, Leibniz." *J. Hist. Ideas* 18(1957): 27-48.

McRae includes a good discussion of Bacon's understanding of the relations between science and religion as well as the separation of science from religion.
Topic: *VII* Name: Bacon

1046 *Melanges Alexandre Koyre*. 2 vols. Paris: Hermann, 1964.
Relevant items from this festschrift include an "Hommage" to Koyre by Cohen and Tate (0247), a bibliography of Koyre's principal publications (1790), and an article by Fleming (0482).
Topic: *VII*

1047 Mendelsohn, J. Andrew. "Alchemy and Politics in England, 1649-1665." *Past and Present* 135(1992): 30-78.
Mendelsohn shows at great length that alchemy was not associated simply with or only practiced by radicals during the Interregnum. In addition to the variety of correlations during the Interregnum, he points to Nicaise LeFevre, the alchemist/apothecary whom Charles II brought to England at the Restoration. Thus, the author proposes that "across the discontinuity at 1660" there was a continuity in the meaning of "chymistry"—"a continuity of reform ideology." At the same time, he notes the fluidity of alchemy's ideological alignments (especially during the Interregnum). Mendelsohn recognizes the historiographical complexity of this topic and period.
Topics: *II*, I, IV, XI Names: Charles II, Charleton, LeFevre, Overton, Starkey, Vaughan, T.

1048 Merchant, Carolyn. *The Death of Nature*. San Francisco: Harper and Row, 1980.
Merchant's book is a fascinating, influential, and controversial study combining ecological and feminist concerns. From this perspective, she reexamines the scientific, economic, cultural and technological transformation of Europe and England in the sixteenth and seventeenth centuries. She highlights the changes in the religious imagery of nature and of women. Especially relevant is her presentation of how an organic, ecological worldview—with nature as a nurturing mother—became a minority opinion expounded by a few men and women philosophers (particularly by Anne Conway). The dominant view, she says, became one of mechanical order, with machines as the symbols of the new ordering of life.

Merchant concludes with a discussion of Newton's attempt to incorporate fermentation as an antidote to the "death of nature."

Topics: *IV*, I, VI, VII, XII Names: Bacon, Conway, Cudworth, Helmont, Hobbes, Newton

1049 Merchant, Carolyn. "The Vitalism of Anne Conway: Its Impact on Leibniz's Concept of the Monad." *J. Hist. Philos.* 7(1979): 255-269. Reprinted with modifications in *The Death of Nature*, Chapter 11. San Francisco: Harper and Row, 1980. See (1048). Reprinted in *Guilford Review* 23 (1986): 2-13

This is a good introduction to Conway's philosophical vitalism, Quakerism, and interest in Cabala. Merchant indicates that Conway was a monist (body and spirit being the same substance). She also held to the interconnectedness of all spirits and the great chain of being. Merchant (at each point) compares Conway to Leibniz—who had read her book and probably derived the term "monad" from it.

Topic: *VII* Names: Conway, Helmont

1050 Merchant, Carolyn. "The Vitalism of Francis Mercury van Helmont: Its Influence on Leibniz." *Ambix* 26(1979): 170-183. Reprinted with modifications in *The Death of Nature*. San Francisco: Harper and Row, 1980. See (1048).

Merchant summarizes van Helmont's vitalistic philosophy of nature. For him, spirit and matter emanated from God's perfection—hence matter is like a sleeping spirit rather than "dead" as in mechanistic philosophy. Further, Merchant points out that the earth is then seen as a living spiritual being, a "nurturing mother."

Topics: *VII*, II Name: Helmont

1051 Merkel, Ingrid, and Allen G. Debus, eds. *Hermeticism and the Renaissance: Intellectual History and the Occult in Early Modern Europe*. Washington, D.C.: Folger Shakespeare Library, 1988. London: Associated Universities Presses, 1988.

The articles by Dobbs (0389), Johannisson (0829), Ormsby-Lennon (1176), Rousseau (1364), and Van Pelt (1584) are annotated separately.

Topic: *II*

1052 Merton, Egon Stephen. *Science and Imagination in Sir Thomas Browne*. New York: Kings Crown Press, 1949.

Merton is a literary historian whose theme is that "the circumference of Browne's being is represented by his science; the center, by his imagination." He emphasizes Browne's eclectic natural philosophy—with Neoplatonic, Aristotelian, Baconian, and Christian elements. Browne had "an animistic view of matter and a vitalistic view of life which have significant implications for his philosophy and, ultimately, for his art." The author presents Browne's views on: scientific method; cosmology; ontology; teleology; body and soul; matter and spirit; generation; angels, the Devil, and witches; effluvia; and the great chain of being. He says that "the great impulse animating all [Browne's] scientific research is the desire to interpret this 'stenography' of the world and so to arrive at an understanding of the mind of God." Merton emphasizes the complexity of Browne's attitudes—especially as regards science, magic, and the Devil. He gives a running commentary on Browne's incoherence and inconsistencies as well as the interconnections and unity of Browne's thought. The influence of science on Browne's artistic imagination was basically indirect, according to Merton, providing some metaphors and images—but mainly effecting a certain detachment and objectivity in Browne's style. In terms of the effects of science on Browne's philosophy, Merton emphasizes the importance of his biology. Browne's science served as a catalyst for revitalizing traditional philosophical and theological concepts—not for evolving original or new ones.

Topics: *X*, VII, VIII, XI Name: Browne, T.

1053 Merton, Robert K. "Bibliographical Postscript to 'Puritanism, Pietism and Science'." In *Social Theory and Social Structure*, by Robert K. Merton, 595-606. 2nd ed. New York: Free Press, 1957. Reprinted in *Social Theory and Social Structure* 649-660. 3rd ed. New York: Free Press, 1968. See (1059). Reprinted in *Puritanism and the Rise of Modern Science: The Merton Thesis*, edited by I. Bernard Cohen, 322-333. New Brunswick, N.J.: Rutgers University Press, 1990. See (0246).

This is a review essay of selected works relevant to the Puritanism-and-the-rise-of-science debate. Merton answers a couple of critics

but mostly reviews works which support the theses found in his early work (1057).
Topics: *III*, IV

1054 Merton, Robert K. "The Fallacy of the Latest Word: The Case of 'Pietism and Science' (by G. Becker)." *Am. J. Socio.* 89(1984): 1091-1121.
Merton defends his specific claims concerning science and German pietism against the specific critique of George Becker (0075); and he defends his general socio-historical claims about science and ascetic Protestantism against its critics in general. He indicates how he would improve his original monograph after half a century: interestingly, he now sees "the element of rationality" as rather irrelevant in the religious ethos. He would emphasize even more that the support of science was an unintended consequence of the religious ethic and *not* the result of direct and deliberate support of science by religious leaders.
Topics: *III*, I, IV

1055 Merton, Robert K. "Preface: 1970." In *Science, Technology, and Society in Seventeenth-Century England*, by Robert K. Merton, vii-xxix. New York: Harper Torchbooks, 1970. See (1058). Reprinted in *The Sociology of Science*, Chapter 7. Chicago: University of Chicago Press, 1973. See (1060). Reprinted in *Puritanism and the Rise of Modern Science: The Merton Thesis*, edited by I. Bernard Cohen, 303-321. New Brunswick, N.J.: Rutgers University Press, 1990. See (0246).
Merton defends his 1938 monograph against critics of the following three decades. He points out that the focus is science, society, and culture in general—not just religion and science. As regards religion and science, Merton emphasizes the cautious yet significant claim made in the monograph: that Puritanism inadvertently made a contribution to the rise of modern science.
Topics: *III*, IV

1056 Merton, Robert K. "Puritanism, Pietism, and Science." *Sociological Review* 28(1936): 1-30. Reprinted in the various editions of *Social Theory and Social Structure*, New York: Free Press, 1949, 1957, and 1968. See (1059). Reprinted in

Science and Religious Belief, edited by Colin A. Russell, 20-54. London: University of London Press, 1973. See (1377).

This essay is incorporated into and superseded by Chapter 6 of "Science, Technology and Society." It can be used as a precis of the Puritanism-and-the-rise-of-science thesis.

Topics: *III*, IV

1057 Merton, Robert K. "Science, Technology and Society in Seventeenth-Century England." *Osiris* 4(1938): 360-632. Reprinted, with a new preface by the author and an additional bibliography, as *Science, Technology, and Society in Seventeenth-Century England*. New York: Harper Torchbooks, 1970. See (1058).

This monograph was a pioneering work in the 1930s and has been a controversial piece of sociohistorical scholarship ever since. Merton was a doctoral student in sociology at Harvard University and went on to become an influential sociologist of science. At a time when the warfare and separation metaphors were dominant in the study of science and religion, Merton produced provocative theses and evidence in support of the interaction between religion and science. Moreover, he chose seventeenth-century England for his key examples. In the tradition of Max Weber (1610), Merton studied ascetic Protestantism and related it to the rise of science. The larger framework was a search for the social processes and cultural conditions which played a role: (1) in the increased interest in science and technology; and (2) in scientific and technological progress. Merton argued that the social and cultural soil of seventeenth-century England was fertile for the growth and spread of science. In this monograph of 260 pages: about eighty pages directly focus on religion and science (Chapters 4, 5, and 7); and about ninety pages are important for background or for methodological discussion (Chapters 1, 2, 3, 10, and 11). Merton assumes that Protestant religion in general (Puritanism specifically) was a dominating expression of cultural values and orientation in seventeenth-century England. The so-called "Merton thesis" is that, sociologically speaking, this Protestant/Puritan religious ethos led men towards scientific pursuits and provided justification for natural science at a time when it needed such social support. Some of Merton's subtheses are the following. (1) Puritans/Protestants valued nature as God's creation and the study of nature, for them, was a way of glorifying God. (2) The understanding of "vocation" and the stress on disciplined labor in so-called ascetic Protestantism indirectly supported scientific activity. (3) Puritans and the Baconian scientists of the time had shared ideals concerning the importance of "reason," good works, and utility for the welfare of society. (4) Both Puritans and followers of the new philosophy were characterized by antiauthoritarianism and by experiential or empirical orientation. (5) The ideas of reform and of progress were inherent in both the Puritan and the scientific movements. (6) An analysis of the members of the Royal Society in its early years shows that a disproportionately high number were from a Puritan background. (7) A study of educational institutions during the period shows a Puritan connection in support of natural philosophy generally and the new philosophy specifically. (8) Demographically, a statistical analysis of where scientists were active—even beyond seventeenth-century England—shows a great preponderance in Protestant areas. Merton takes Richard Baxter as a representative Puritan writer and Robert Boyle as a representative natural philosopher in his analysis. He also puts forward Hartlib, Petty, Ray, Sprat, and Wilkins as significant examples in support of his conclusions. He contrasts the Puritan ethos to the medieval Catholic ethos. It is important to note the essentially sociological nature of Merton's argument. His emphases are on: religious ethos more than doctrine; correlation of social factors more than direct causality; and social patterns more than given individual cases.

Topics: *III*, IV, V, VI, VII Names: Baxter, Boyle, Hartlib, Petty, Ray, Sprat, Wilkins

1058 Merton, Robert K. *Science, Technology and Society in Seventeenth-Century England*. New York, 1970.

This is a reprint of the famous *Osiris* monograph (1057) combined with a twenty-three page "Preface: 1970," (1055), a seven-page "Selected Bibliography: 1970," and an index.

Topic: *III*

1059 Merton, Robert K. *Social Theory and Social Structure*. New York: Free Press, 1949. 2nd ed. New York: Free Press, 1957. 3rd ed. New York: Free Press, 1968.

This includes a revised version of Merton's 1936 article from the *Sociological Review* (1056). The second and third editions also

include a "Bibliographic Postscript" (1053) regarding Puritanism and science.
Topic: *III*

1060 Merton, Robert K. *The Sociology of Science*. Chicago: University of Chicago Press, 1973.
Chapter 7, "Social and Cultural Contexts of Science," is a reprint of "Preface: 1970" (1055); Chapter 8, "Changing Foci of Interests on the Sciences and Technology," is a reprint of "Foci and Shifting Interest in the Sciences and Technology," from the *Osiris* monograph (1057); and Chapter 11, "The Puritan Spur to Science," is a reprint of "Motive Forces of the New Science" from the same work (1057).
Topic: *III*

1061 Merton, Robert K. "STS: Foreshadowings of an Evolving Research Program in the Sociology of Science." In *Puritanism and the Rise of Modern Science: The Merton Thesis*, edited by I. Bernard Cohen, 334-371. New Brunswick, N.J.: Rutgers University Press, 1990. See (0246).
This autobiographical essay mainly concerns the sociology of science outside of our period. It includes a few pages on the Puritanism-and-rise-of-science discovery and debate.
Topic: *III*

1062 Metraux, Guy S., and Francois Crouzet, eds. *The Evolution of Science*. New York: New American Library, 1963.
The articles by Hooykaas (0723) and Russo (1382) are annotated separately.
Topic: *III*

1063 Metz, Rudolf. "Bacon's Part in the Intellectual Movement of his Time." In *Seventeenth Century Studies Presented to Sir Herbert Grierson*, 21-32. Oxford: Clarendon Press, 1938. See (1424).
Metz sets himself the task of judging Bacon's place in the history of Western philosophy. He concludes that Bacon stands Janus-like between the Renaissance past and the modern age to come. Metz praises Bacon as a religious agnostic who strengthened science by

secularizing it—helping to set "learning free from the shackles of faith."
Topics: *X*, VII Names: Bacon

1064 Metzger, Helene. *Attraction universelle et religion naturelle chez quelques commentateurs anglais de Newton. Premiere partie: Introduction philosophique; deuxieme partie: Newton-Bentley-Whiston-Toland; troisieme partie: Clarke-Cheyne-Baxter-Priestly*. Nos. 621-623 of *Actualites scientifiques et industrielles*. Paris: Hermann, 1938.
Metzger provides an excellent commentary on Newton and various English Newtonians of the eighteenth century. She gives significant and specific study to pre-1720 writings as well as to those of later periods. Relevant topics include: theological aspects of atomism and doctrines of matter; God as an explanatory factor in natural philosophy; cosmological and teleological arguments; and deism.
Topics: *VIII*, VII, IX, XII Names: Bentley, Cheyne, Clarke, Newton, Toland, Whiston

1065 Middlekauff, Robert. *The Mathers: Three Generations of Puritan Intellectuals, 1596-1728*. New York: Oxford University Press, 1971.
Chapter 16, "The Experimental Philosophy," is especially relevant. Middlekauff finds Cotton Mather to be well read in English natural philosophy and a natural theologian who yet emphasized the Spirit (as against science) as the main source of religion.
Topics: *VII*, VIII, XI Names: Mather, C., Mather, I.

1066 Mijuskovic, Ben Lazare. *The Achilles of Rationalist Arguments: The Simplicity, Unity, and Identity of Thought and Soul from the Cambridge Platonists to Kant. A Study in the History of an Argument*. The Hague: Martinus Nijhoff, 1974.
This revised history-of-philosophy dissertation covers both the continent and the British Isles during our period. It includes a good chapter on the immortality of the soul question.
Topic: *VII* Names: Cudworth, Locke, More, Norris, Smith

1067 Miles, Roger B. *Science, Religion, and Belief: The Clerical Virtuosi of the Royal Society of London, 1663-1687*. New York: P. Lang, 1992.

Miles begins with a useful prosopographic study of the 53 clerics who were elected to the Royal Society between 1663 and 1687. He follows this with chapters on their level of activity in the Society, on "The Problem of 'Belief'," and on "The Mysteries of Christianity." The latter two chapters focus on how the clerical virtuosi dealt with the issue of certainty and on their defense of Christian doctrines. Miles concludes that both the clerical virtuosi and their opponents "were rationalists to the core."

Topics: *V*, VII, VIII, X Names: Ardeme, Barrow, Beale, Burnet, Flamsteed, Gale, Glanvill, Laney, Pearson, Pell, Ray, Sheldon, Sprat, Wallis, Ward, Wilkins

1068 Milhac, Francois. *Essai sur les Idees Religieuses de Locke.* Geneva: Rivera et Dubois, 1886.

This short book of 115 pages is hard to obtain. Milhac, a Protestant evangelical student, presents a systematic theology gathered from various writings of Locke. The author concludes that Locke, although heretical with respect to some traditional doctrines, was a true disciple of Christ. There is no discussion of science in the book.

Topics: *VII*, IX Name: Locke

1069 Milic, Louis T. "The Metaphor of Time as Space." In *Probability, Time, and Space in Eighteenth-Century Literature*, edited by Paula R. Backscheider, 249-258. New York: AMS Press, 1979. See (0052).

Milic makes provocative suggestions regarding the use of time and space as metaphors in the seventeenth and eighteenth centuries.

Topic: *X* Name: Addison

1070 Miller, Genevieve. "A Seventeenth-Century Astrological Diagnosis." In *Science, Medicine and History*, Vol. 2, 26-33. Edited by E. Ashworth Underwood. Oxford: Oxford University Press, 1953. See (1580).

While not directly touching on religion, Miller's brief essay usefully describes one example of a seventeenth-century astrological medical diagnosis. This case is found in a collection of John Winthrop the Younger.

Topics: *XI*, II Name: Winthrop

1071 Miller, Perry. "Bentley and Newton." In *Isaac Newton's Papers & Letters On Natural Philosophy*, edited by I. Bernard Cohen, 271-278. Cambridge, Mass.: Harvard University Press, 1958 and 1978. See (0245).

This is a patronizing and unsympathetic commentary on the Bentley-Newton correspondence with harsh words for Bentley and his Boyle Lectures.

Topics: *XII*, VII, VIII Names: Bentley, Newton

1072 Miller, Perry. "The Marrow of Puritan Divinity." *Pub. Colonial Soc. Mass.* 32(1937): 247-300. Reprinted in *Errand into the Wilderness*, by Perry Miller, 48-98. Cambridge, Mass.: Harvard University Press, 1956.

Miller's classic description of the covenant theology includes discussions of the relation of God to natural law and why natural philosophy was an important part of theology for the Puritans.

Topics: *VII*, III, X Names: Ames, Cotton, Preston

1073 Miller, Perry. *The New England Mind: The Seventeenth Century*. New York: Macmillan, 1939.

The part labeled "cosmology" deals with reason, knowledge, and nature. For our purposes, this book is valuable because it can serve as a detailed starting point for a discussion of the Puritan understanding of the interactions of science and religion. The coherence of Ramist logic, natural theology, biblical interpretation, and the doctrine of providence is demonstrated through the textbooks which the Puritans read and the sermons which they preached. For Miller, a Puritan was "required to maintain a unified theory that would meet the requirements of both his piety and his logic." But it was more important to employ nature as a symbol of providence than to decide "which particular system of physics was used to explain nature." Miller's work is based on a vast knowledge of the sources.

Topics: *VII*, VIII, X Names: Ames, Cotton, Mather, C., Mather, I., Morton, Perkins, Preston, Richardson, Willard

1074 Miner, Earl. "Dryden and the Issue of Human Progress." *Philol. Q.* 40(1961): 120-129.

The sources of Dryden's views on progress are seen as other than either religion or the new philosophy.

Topics: *X*, IV Name: Dryden

1075 Miner, Earl. "The Poets and Science in Seventeenth-Century England." In *The Uses of Science in the Age of Newton*, edited by John G. Burke, 1-19. Berkeley and Los Angeles: University of California Press, 1983. See (0157).

Miner summarizes attitudes towards science (and religion) amongst selected poets of the period. He includes some good quotations from primary sources.

Topic: *X* Names: Burton, Butler, Donne, Dryden

1076 Mintz, Samuel I. *The Hunting of Leviathan: Seventeenth-Century Reactions to the Materialism and Moral Philosophy of Thomas Hobbes*. Cambridge: Cambridge University Press, 1962.

This well-written book includes a discussion of the various conceptions of atheism found in seventeenth-century England. Beyond stating the general point that mechanistic philosophy supports materialism, Mintz does not argue theses concerning the new science; but the book has the effect of minimizing the importance of natural philosophy in the controversies related to Hobbes and Hobbism.

Topics: *IX*, V, VII Names: Bramhall, Cudworth, Glanvill, Hobbes, More

1077 Mitchell, W. Fraser. *English Pulpit Oratory from Andrewes to Tillotson*. New York: Russell and Russell, 1932.

Mitchell provides a detailed and thorough study of the rhetoric of seventeenth-century English sermons. One of his emphases is the influential role played by members of the Royal Society in the simplification of preaching style.

Topics: *X*, V Names: Andrewes, Barrow, Baxter, Burnet,G., Donne, Glanvill, More, Sprat, Taylor, J., Tillotson, Wilkins

1078 More, Louis Trenchard. "Boyle as Alchemist." *J. Hist. Ideas* 2(1941): 61-76.

This is an early study: of Boyle's published works on alchemy and transmutation, and of Newton's letters to Boyle and Locke on these matters. More also considers the questions of secrecy and plagiarism. As regards religion, he simply notes the importance of the will of God for Boyle.

Topics: *II*, IV, XII Name: Boyle, Newton

1079 More, Louis Trenchard. *Isaac Newton: A Biography*. New York: C. Scribner's Sons, 1934. Reprinted, New York: Dover, 1962.

At one time, this was a standard work. Now, it has been made fundamentally obsolete by the greater availability of Newtonian manuscripts, improved historiography in Newtonian studies, and the works of Westfall, Manuel, and Christianson.

Topics: *XII*, II, V, X, VII Names: Bentley, Boyle, Cotes, Flamsteed, Locke, Newton

1080 More, Louis Trenchard. *The Life and Works of the Honourable Robert Boyle*. London: Oxford University Press, 1944.

This generally has been superseded, but it does include some good insights. More discusses Boyle's religion as a source of scientific motivation.

Topics: *VII*, II, IV, V, VIII, X Name: Boyle

1081 Morgan, John L. *Godly Learning: Puritan Attitudes towards Reason, Learning, and Education, 1560-1640*. Cambridge: Cambridge University Press, 1986.

This is a detailed reexamination of Puritan intellectual attitudes. While not directly attending to the question of Puritanism and science, Morgan implies that such a relation could only be indirect. He emphasizes the dialectical relationship between enthusiasm and reason (with enthusiasm being primary).

Topics: *IV*, III, X Names: Ames, Perkins, Preston

1082 Morgan, John L. "Puritanism and Science: A Reinterpretation." *Hist. J.* 22(1979): 535-560.

Morgan argues against the Puritanism-and-the-rise-of-science thesis by considering Puritan theology before 1640. Three of his subpoints are that these theologians: (1) were not Baconian; (2) warned against confidence in human reason; and (3) believed that natural philosophy is irrelevant to salvation. A fourth point is that many Puritan theologians did not explicitly approve of the study of nature. Morgan misses (or ignores) the sociological dimension of the Merton thesis. His argument generally is not convincing.

Topics: *III*, VII Name: Bacon

1083 Morison, Samuel Eliot. *Harvard College in the Seventeenth Century*. 2 vols. Cambridge, Mass.: Harvard University Press, 1936.

Morison's narrative history of Harvard from approximately 1650 to 1708 is unmatched in its combination of detail and readability. He is concerned not only with the subjects, materials, and methods of the curriculum. He also considers the "food and drink, and play and prayer" of the students as well as the financial and political manipulations of the administrators. The chapters on the curriculum are particularly relevant to this bibliography. The introduction of the new astronomy is described—as is the later revision in teaching physics through Charles Morton's *Compendium Physicae*. This period, Morison shows, saw a gradual transition away from a scholastic curriculum in natural philosophical subjects. Morison emphasizes the common education of clergy and lay-educated gentlemen in the undergraduate college. (The professional degree in theology came after the initial degree and not all clergy read for it.) The Puritan background of Harvard is clear throughout both volumes. But Morison also shows that, while Harvard was primarily modeled after Puritan Cambridge, its students read and discussed books that were used all over Western Europe.

Topics: *V*, X Names: Ames, Boyle, Brattle, T., Chauncy, Dunster, Mather, C., Mather, I., Morton, Wigglesworth

1084 Morison, Samuel Eliot. "The Harvard School of Astronomy in the Seventeenth Century." *New Eng. Q.* 7(1934): 3-24.

Using seventeenth-century almanacs and letters, Morison expounds the thesis that "the Puritan clergy . . . were the chief patrons and promoters of the new astronomy, and of other scientific discoveries in New England." The almanacs and helpful astronomical observations were institutionally sponsored by Harvard—where the new astronomy was considered to be consistent with the Bible.

Topics: *V*, III, VII, X Names: Brattle, T., Brigden

1085 Morison, Samuel Eliot. *The Intellectual Life of Colonial New England*. New York: New York University Press, 1956.

This is a revised edition of *The Puritan Pronoas* (1086). With Perry Miller, Morison attempted to counteract the contemporary disparaging accounts of Puritanism. Rather than hostility towards science and culture, Morison argues, the Puritans' ubiquitous religion stimulated science and culture.

Topics: *III*, VII, X Names: Dunster, Mather, C., Mather, I., Winthrop

1086 Morison, Samuel Eliot. *The Puritan Pronoas*. New York: New York University Press, 1936.

This book has been superseded by *The Intellectual Life of Colonial New England* (1085).

Topic: *III*

1087 Morrison, James C. "Philosophy and History in Bacon." *J. Hist. Ideas* 38(1977): 585-606.

In this analysis of Bacon's understanding of the kinds of history and their relation to philosophy and religion, Morrison points out the means and ends of knowledge for Bacon. Knowledge is the ability to make—or, rather, to remake—what has already been made in the creation of nature. Hence there is a coincidence of knowledge and human power in the praxis of experimentation. The end or purpose of this practical knowledge is charity—the removal of human pain and suffering—however ambiguous that end seems to us today.

Topic: *VII*, IV, X Name: Bacon

1088 Mosse, George L. "Puritan Radicalism and the Enlightenment." *Church History* 29(1960): 424-439.

Mosse emphasizes certain forms of philosophy and Puritan radicalism, acting independently of each other, as most important in the rise of deistic ideas. (He does not see the new science or natural theology as significant factors in the rise of deism.)

Topic: *IX* Names: Overton, Webbe

1089 Mother Maria. *Ralph Cudworth: Mystical Thinker*. Newport Pagnell, England: Greek Orthodox Monastery of the Assumption, 1973.

See (0590), under the name of Lydia Gysi.

Topic: *VII* Name: Cudworth

1090 Motzo Dentice de Accadia, Cecilia. *Preilluminismo e Deismo in Inghilterra*. Napoli: Libreria Scientifica Editrice, 1970.

The author draws a line of development from several Italians (Campanella, Bruno, Galileo) through English philosophers (Bacon, Hobbes, Cambridge Platonists, Locke) to English natural religion, English deists (Toland, Collins, Chubb, Tindal, Bolingbroke) and the early Enlightenment. This is based on work first published in the 1930s.
Topics: *IX*, VIII Names: Bacon, Collins, A., Herbert, E., Hobbes, Locke, Toland

1091 Mouton, Johann. "Reformation and Restoration in Francis Bacon's Early Philosophy." *Modern Schoolman* 60(1983): 101-112.
Mouton outlines Bacon's views concerning the Reformation and the authority of the Bible. He then shows that Bacon drew an analogy between the reformation of religious thought and the restoration of the sciences: in both, there is a return to origins—to the two books, God's Word and God's Works.
Topics: *VII*, X Name: Bacon

1092 Muirhead, John H. *The Platonic Tradition in Anglo-Saxon Philosophy*. London: G. Allen and Unwin, 1931.
Muirhead discusses Cambridge Platonists and especially Cudworth with passing reference to science and its relation to religion.
Topic: *VII* Name: Cudworth

1093 Mulder, David. *The Alchemy of Revolution: Gerrard Winstanley's Occultism and Seventeenth-Century English Communism*. New York: P. Lang, 1990.
Against both Marxist and purely Christian interpretations, Mulder argues that the ideology of Winstanley and the Digger movement can be called "revolution by magic." He relates their Christocentric and Hermetic ideology to the agricultural practices and economic conditions of the members of the movement. Thus the activity of the Diggers is understood as a form of divine alchemy.
Topics: *IV*, II, VI Name: Winstanley

1094 Mulligan, Lotte. "Anglicanism, Latitudinarianism and Scienc in Seventeenth Century England." *Ann. Sci.* 30(1973): 213-219.
Generally, Mulligan argues that "interest in science was correlated with no specific religious or political creeds." Specifically, she

argues against Kemsley (0873) and Shapiro (1439)—that is, against attempts to link science with Anglicanism or with Latitudinarianism.
Topics: *III*, IV

1095 Mulligan, Lotte. "Civil War Politics, Religion and the Royal Society." *Past and Present* 59(1973): 92-116. Reprinted in *The Intellectual Revolution of the Seventeenth Century*, edited by Charles Webster, 317-346. London: Routledge and Kegan Paul, 1974. See (1618).
Mulligan uses a quantitative method to argue against any connections between religion and the rise of science. She correlates interest in science with Royalists, gentlemen, and the "waning role of religion." She argues especially against Merton and Hill.
Topics: *III*, V

1096 Mulligan, Lotte. "Puritanism and English Science: A Critique of Webster." *Isis* 71(1980): 456-469.
This is an extended critique of Charles Webster's theses in *The Great Instauration* (1616) concerning Puritanism and the rise of science. Mulligan argues, with examples, that: (1) Webster's usage of the term "Puritanism" is too broad; (2) the religious and intellectual characteristics which Webster attributes to Puritans were common to many English Protestants; and (3) Webster's argument for Puritan influence on the formation of the Royal Society is somewhat contrived.
Topics: *III*, V

1097 Mulligan, Lotte. "'Reason', 'Right Reason', and 'Revelation' in Mid-Seventeenth Century England." In *Occult and Scientific Mentalities in the Renaissance*, edited by Brian Vickers, 351-374. Cambridge: Cambridge University Press, 1984. See (1591).
Mulligan argues that many of the men in the seventeenth century (whom we take to have sharply differing theological and philosophical positions) shared a view of "right reason" and revelation. Thereby, they could communicate with each other in their debates. "Right reason" was the path to God's external truths, both novel and natural. She further points out that the same person could hold

"two or more—to us incomprehensible—models" of the natural world.

Topics: *VII*, II Names: Charleton, Hobbes, More, Taylor, J., Vaughan, T., Winstanley

1098 Mulligan, Lotte. "A Rejoinder." *Past and Present* 66(1975): 139-142.

Mulligan here defends her theses and methods against Shapiro's critique (1435) in a formal debate. She does modify her position a bit by saying that she "has not attempted to 'disprove' the relation between science and religion in the seventeenth century."

Topic: *III*

1099 Mulligan, Lotte. "Robert Boyle, 'Right Reason,' and the Meaning of Metaphor." *J. Hist. Ideas* 55(1994): 235-257.

Mulligan argues that Boyle used "right reason" in its traditional sense as "natural reason seasoned by revelation." Through his rhetorical use of common metaphors in both natural philosophy and theology, Boyle interwove the two disciplines. And because of "right reason," Mulligan continues, Boyle believed that only the Christian virtuoso was able to be a good natural philosopher. She also argues that Boyle defended not just Christianity against deist and atheist opponents, but the Church of England against its Christian opponents. Although she does cite Klaaren, she does not acknowledge other scholars who interpret Boyle as more irenical and diffident than she does.

Topic: *VII*, X Name: Boyle

1100 Murdin, Leslie. *Under Newton's Shadow: Astronomical Practices in the Seventeenth Century*. Bristol, England: Avon, 1985. Boston: A. Hilger, 1985.

Murdin discusses the variety of astronomers active in the time of Newton. He includes not only professional astronomers and university men, but also amateurs—including clergy, tradesmen, and paid workers. He describes their personal lives and religious thought and activities as well as their astronomical practices and professional relations.

Topics: *VII*, V, VII, XII Names: Derham, Flamsteed, Gregory, Halley, Hooke, Newton, Pound, Wallis, Willis, Wren

1101 Nagy, Doreen Evenden. *Popular Medicine in Seventeenth-Century England*. Bowling Green, Ohio: Bowling Green State University Popular Press, 1988.

Nagy argues that the professional physicians and popular practitioners used the same methods. But geography, economics, and religion were major factors in the reliance of the majority of the English upon popular practitioners. In a major chapter, she examines the role of women in this latter category. Nagy's short book usefully supplements standard histories of physicians and medical theory.

Topics: *XI*, V Names: Boyle, Culpepper, Hall (1575), Hoby, Josselin, Symcotts

1102 Nakayama, Shigeru. "Galileo and Newton's Problem of World-Formation." *Japanese Studies in the History of Science* 1(1962): 76-82.

Nakayama focuses on the "Galileo-Plato" problem of accounting for the formation of the Copernican solar system. He briefly describes why, in Newton's cosmogony, God *is* a necessary hypothesis.

Topics: *XII*, VII

1103 Nakayama, Shigeru. "Galileo-Newton's Problem of Cosmogony." *Kagagusi Kenkyu* 56(1960): 1-9.

This is basically the same as (1102).

Topics: *XII*, VII Name: Newton

1104 Nathanson, Leonard. *The Strategy of Truth: A Study of Sir Thomas Browne*. Chicago: University of Chicago Press, 1967.

This is a discerning analysis by a literary historian of *Religio Medici*, *Urn Burial*, and *The Garden of Cyrus*. Nathanson emphasizes a particular form of coherence in Browne's "strategy of truth" which is based on Christian Platonism combined with a certain form of empiricism. His discussion of the Platonic tradition and of Browne's implied epistemology provides an insightful approach to the relationship between science and religion in Browne's writings.

Topics: *X*, VII, XI Names: Bacon, Browne, T.

1105 Needham, Joseph. "Review of Robert K. Merton, *Science, Technology, and Society in Seventeenth Century England.*" *Science and Society* 2(1938): 566-571.

This is a rambling discussion of science and culture in England in the seventeenth-century. Mixed in are positive remarks about Merton's book (1057) and support of the Puritanism-and-the-rise-of-science thesis.

Topic: *III*

1106 Nelson, Benjamin. "The Early Modern Revolution in Science and Philosophy." *Boston Stud. Philos. Science* 3(1967): 1-40.

Nelson discusses in broad terms (with bibliographic references) the complexity of relating various forms of Christianity to the rise of modern science. He argues against any unique significance of Protestantism. He also discusses the voluntaristic concept of God and opposes the theories of Oakley (1155), Popkin (1232), and Van Leeuwen (1583).

Topics: *VII*, I, III

1107 Nelson, Benjamin. "Review of Robert K. Merton, *Science, Technology, and Society in Seventeenth Century England.*" *Am. J. Sociol.* 78(1972): 223-231.

This is an essay review of the 1970 edition of Merton's book (1058). (It follows the editor's apology for a thirty-four-year delay of Nelson's critique which originally was caused by World War II!) Nelson raises several questions or riddles. He asks why have Merton's critics concentrated on the "Puritan thesis" and not the "military thesis"? And why have American sociologists in general and Merton himself not followed up his pathbreaking analysis of comparative military sociology?

Topic: *III*

1108 Nelson, Nicolas H. "Astrology, *Hudibras*, and the Puritans." *J. Hist. Ideas* 37(1976): 521-536.

Nelson describes the debate over astrology among the Puritans during the Interregnum and shows that it forms the background for Samuel Butler's satire *Hudibras*. A useful bibliographical appendix chronologically lists and annotates the works produced in the debate.

Topics: *II*, X   Names: Butler, Lilly

1109 Newell, Lyman C. "Newton's Work in Alchemy and Chemistry." In *Sir Isaac Newton: A Bicentenary Evaluation of His Work*, edited by Frederick E. Brasch, 203-255. Baltimore: Williams and Wilkins, 1927. See (0113).
This is an outdated survey.
Topics: *XII*, II Name: Newton

1110 Newman, William R. "Arabo-Latin Forgeries: The Case of the *Summa Perfectionis*." In *The 'Arabick' Interest of the Natural Philosophers in Seventeenth-Century England*, edited by G. A. Russell, 109-127. Leiden: E. J. Brill, 1994. See (1378).
Newman describes an alchemical manuscript from the thirteenth century and its usage in seventeenth-century England.
Topic: *II* Name: Vaughan, T.

1111 Newman, William R. *Gehennical Fire: The Lives of George Starkey, an American Alchemist in the Scientific Revolution*. Cambridge, Mass.: Harvard University Press, 1994.
The term "Lives" in the subtitle takes on importance because, as Newman shows, Starkey not only wrote under the pseudonym Eirenaeus Philalethes, but also created an alter ego as an alchemical adept and friend. In this excellent study, Newman traces Starkey's life, sets his thought within its context, compares him to his contemporaries, discusses Starkey's sources, and shows his influence. The relationships between alchemy, prophecy, and millennialism are demonstrated both for Starkey and for the alchemical tradition. This book is indispensable for an understanding of seventeenth-century alchemy or chemistry. (Newman also points out that these terms, based on Latin, were interchangable in that period.)
Topics: *II*, X, XI, XII Names: Boyle, Child, Hartlib, Lockyer, Newton, Richardson, Starkey, Vaughan, T., Wigglesworth, Winthrop, Worsley

1112 Newman, William R. "George Starkey and the Selling of Secrets." In *Samuel Hartlib and Universal Reformation*, edited by Mark Greengrass, 193-210. Cambridge: Cambridge University Press, 1994. See (0573).
Central to Starkey's concealment of his alchemical and chemical secrets was his belief that such secrets (sometimes received in

dreams) were divinely sanctioned revelations. Newman shows how this belief both affected his relationship to his patron Robert Boyle and led to his refusal to sell his "keys" to the vulgar. This article amounts to an extract from Newman's book on Starkey (1113).
Topic: *II* Name: Starkey

1113 Newman, William R. "Newton's *Clavis* as Starkey's Key." *Isis* 78(1987): 564-574.
A letter from William Starkey to Robert Boyle in 1651 (recently discovered by Newman) appears to settle a dispute over the authorship of the important alchemical text, *Clavis*. Attributed by some to Newton, the letter shows both that Starkey wrote the *Clavis* and that he probably was the anonymous author who wrote as Eirenaeus Philalethes. The letter is included and contains an English version of *Clavis*.
Topics: *II*, XII Names: Newton, Starkey

1114 Newman, William R. "Prophecy and Alchemy: The Origin of Eirenaeus Philalethes." *Ambix* 37(1990): 97-115.
Newman examines the background, content, and context of the alchemical text *Introitus apertus ad occlusum regis palatium* by the pseudonymous Eirenaeus Philalethes. He identifies the author as William Starkey and places him in the Hartlib circle with its combination of millennialist vision, utilitarianism, and alchemy.
Topics: *II*, IV Names: Hartlib, Starkey

1115 Newman, William R. "Thomas Vaughan as an Interpreter of Agrippa von Nettesheim." *Ambix* 29(1982): 125-140.
Newman argues that Vaughan's alchemical philosophy of nature is "an exegesis of Agrippa's *De occulta*." He traces themes through various writers but notes especially Vaughan's continual veneration of Agrippa. In particular, Vaughan adopts Agrippa's threefold system of worlds and elements—and sees them as corresponding to the Trinity.
Topics: *II*, VII Name: Vaughan, T.

1116 Nicholl, H. F. "John Toland: Religion Without Mystery." *Hermathena* 100(1965): 54-65.
This is a brief article on Toland and his most famous book. There is nothing original here.
Topic: *IX* Name: Toland

1117 Nicolson, Marjorie Hope. *The Breaking of the Circle. Studies in the Effect of the "New Science" Upon Seventeenth-Century Poetry*. Evanston, Ill.: Northwestern University Press, 1950. Revised ed., New York: Columbia University Press, 1960.

In discussing poetry, Nicolson indirectly deals with the effects of the new science upon religious views. She appreciates poetry which assumes the "Circle of Perfection" and the circle metaphor. She describes not only the breaking of this circle but also the breakdown of the microcosm-macrocosm analogy and the language of poetry being displaced by the new science and Baconian common sense.

Topics: *X*, IV, VII Name: Donne

1118 Nicolson, Marjorie Hope. "Christ College and the Latitude Men." *Modern Philology* 27(1929): 35-53.

Nicolson here describes some of the struggles of the Cambridge Platonists—centering her discussion on Christ Church College of Cambridge University. She eulogizes this group and this "brief space of time" when "science was truly religious, religion scientific." Science, however, is rather marginal in this particular essay.

Topics: *V*, VII Names: Cudworth, More

1119 Nicolson, Marjorie Hope, ed. *Conway Letters, The Correspondence of Anne, Viscountess Conway, Henry More, and their friends, 1642-1684*. New Haven, Conn.: Yale University Press, 1930. Revised edition, edited by Sarah Hutton. Oxford: Clarendon Press, 1992.

Primary sources (the correspondence) and Nicolson's biographical and literary commentary alternate in this edition. (Nicolson's part amounts to about 150 pages.) Religion is fairly prominent; science is tangential. There are good sections on Henry More, F. M. van Helmont, Valentine Greatrakes, and other related figures. For Hutton's introduction, see (0783).

Topics: *VII*, II, X Names: Conway, Digby, Finch, J., Greatrakes, More, Helmont

1120 Nicolson, Marjorie Hope. "Cosmic Voyages." *ELH* 7(1940): 83-107.

This article is superseded by *Voyages to the Moon* (1138).

Topic: *X*

1121 Nicolson, Marjorie Hope. "The Discovery of Space." In *Medieval and Renaissance Studies*, edited by O. B. Hardison, Jr., 40-59. (Proceedings of the Southeastern Institute of Medieval and Renaissance Studies, Summer, 1965.) Chapel Hill, N.C.: University of North Carolina Press, 1966.

This is an interesting lecture. It summarizes the effects of the popularization of the discovery of the vastness of space on the imaginations of seventeenth-century literary figures and philosophers.

Topics: *X*, VII Names: Godwin, Milton, More, Wilkins

1122 Nicolson, Marjorie Hope. "The Early Stage of Cartesianism in England." *Studies in Philology* 26(1929): 356-374.

Nicolson here traces English Cartesianism in the 1640s and 1650s. There is some discussion of science and religion but mainly the emphasis is on the psychological and literary effects of the early stage of the movement.

Topics: *X*, VII Name: More

1123 Nicolson, Marjorie Hope. "The History of Literature and the History of Thought." In *English Institute Annual, 1939*, 56-89. New York, 1940.

This is an early exposition and defense of the "History of Ideas" approach to intellectual history—applied specifically to English literature and to our chronological period.

Topics: *I*, X

1124 Nicolson, Marjorie Hope. "Kepler, the *Somnium*, and John Donne." *J. Hist. Ideas* 1(1940): 259-280.

Nicolson argues that Donne was the first English writer to be influenced by Kepler's fictional supernatural voyage to the moon. This work is repeated in Nicolson's *Voyages to the Moon* (1138).

Topic: *X* Name: Donne

1125 Nicolson, Marjorie Hope. "The Microscope and English Imagination." *Smith Coll. Stud. Mod. Lang.* 16, No.4(1935): 1-92.

Here Nicolson describes how microscopic observation influenced concepts of God and natural theology in the late-seventeenth and early-eighteenth centuries. It provided proof for the great chain of being and for seeing God as a Creator of lavish, prolific, almost unrestrained variety. The contrivance of God could be seen in the smallest insect or plant as well as in large-scale ways. The microscope was also relevant to the ancients vs. moderns controversy, the idea of progress, optimism, and discussion regarding human nature.

Topics: *VI*, IV, VII, VIII, X Names: Addison, Pope, Swift

1126 Nicolson, Marjorie Hope. "Milton and Hobbes." *Stud. Philol.* 26(1929): 356-374.

After discussing Hobbes and the negative responses he provoked, Nicolson answers the rhetorical question "Was not the most magnificent of all replies to Hobbes Milton's *Paradise Lost*?" Milton believed in a platonic "Spirit of Nature" against Hobbes' mechanical universe. Nicolson also discusses Milton on natural law, on the immortality of the soul, and as emphasizing God's reason and goodness (rather than God's will).

Topics: *X*, VII, IX Names: Hobbes, Milton

1127 Nicolson, Marjorie Hope. "Milton and the *Conjectura Cabbalis tica*." *Philol. Q.* 6(1927): 1-18.

Nicolson gives a good description of the seventeenth-century Christian understanding of the Jewish Cabbala and then shows its relevance to Milton. One of her theses is that the cabbalistic doctrines were popular in the second quarter of the century because they offered "*a means of reconciling religion and the new science*" (Nicolson's italics). She shows how both Milton and More applied cabbalistic doctrines to the creation of the world and to the first three chapters of Genesis. This is a surprisingly early article recognizing the importance of what later came to be called the *prisca* tradition.

Topics: *X*, II, VII Names: Milton, More

1128 Nicolson, Marjorie Hope. "Milton and the Telescope." *ELH* 2(1935): 1-32.

In her third article on the effect of the telescope on the literary and religious imagination in the seventeenth century, Nicolson proposes that Milton's imagination was deeply affected by his "actual sense experience of celestial observation." It "made *Paradise Lost* the first modern cosmic poem."

Topics: *X*, VI Name: Milton

1129 Nicolson, Marjorie Hope, ed. *Mountain Gloom and Mountain Glory: The Development of the Aesthetics of the Infinite*. New Haven, Conn.: Yale University Press, 1930.

Nicolson discusses the interplay between science and religion not only in the area of aesthetics but also in that of "Genesis and geology." She tends to interpret texts in terms of "the warfare between science and theology"—and she tends to be unsympathetic towards theology.

Topics: *X*, VII Names: Addison, Burnet, T., Donne, More, Pope

1130 Nicolson, Marjorie Hope. "The 'New Astronomy' and English Literary Imagination." *Stud. Philol.* 32(1935): 428-462.

The last ten pages of this essay focus on John Donne and the religious implications of the telescopic discoveries of the early seventeenth century.

Topics: *X*, VI Name: Donne

1131 Nicolson, Marjorie Hope. *Newton Demands the Muse. Newton's "Opticks" and Eighteenth Century*. Princeton, N.J.: Princeton University Press, 1946.

Nicolson deals mainly with the period after 1720. She does not say much about religion but she does include one whole chapter on metaphysical implications.

Topics: *X*, VII, XII Names: Addison, Locke, Newton, Pope

1132 Nicolson, Marjorie Hope. *Pepys Diary and the New Science*. Charlottesville, Va.: University Press of Virginia, 1965.

This book deals very little with religion or theology—but it is good for the social milieu of the Royal Society in the 1660s.

Topics: *X*, IV, V Name: Pepys

1133 Nicolson, Marjorie Hope. "The Real Scholar Gypsy." *Yale Review* 18(1929): 347-363.

This is an essay on Francis Mercury van Helmont and some of his adventures. It is partially repeated in *Voyages to the Moon* (1138).

Topic: *II* Name: Helmont

1134 Nicolson, Marjorie Hope. "Richard Foster Jones." In *The Seventeenth Century: Studies in the History of English Thought and*

Literature from Bacon to Pope, edited by Richard F. Jones and others, 1-9. Stanford, Calif., 1951. See (0846).

This is a tribute to Jones—emphasizing his role as a pioneer in showing the scientific background of seventeenth-century literature. (Jones was doing this as early as the 1920s.)

Topics: *I*, X

1135 Nicolson, Marjorie Hope. "The Spirit World of Milton and More." *Stud. Philol.* 22(1925): 433-452.

Nicolson compares the views of Henry More and John Milton regarding spirits—both human souls and angels. She discusses their physical qualities, their abstract qualities, and their place of abode. She concludes that the spirit worlds of More and Milton were identical, and that Milton might have been influenced by More.

Topic: *X* Names: Milton, More

1136 Nicolson, Marjorie Hope. "The Telescope and Imagination." *Modern Philology*: 32(1935): 233-260.

The "new astronomy" and the telescope jointly had a revolutionary impact upon human imagination and especially upon poetry; they disclosed new heavens and a new earth. The examples are mainly continental but this essay still is good for background.

Topics: *VI*, X

1137 Nicolson, Marjorie Hope. *Voyages to the Moon*. New York: Macmillan, 1948.

This book is aimed at the general reader. It concerns the themes of cosmic voyages and human flight in English books of the seventeenth and eighteenth centuries. Thus, it treats one aspect of the impact of science (and to some extent, technology) on the literature of the period. Religion is treated rather tangentially by Nicolson—mainly because of her own assumptions. Religious opposition to human flight, angelic and supernatural voyages, the theological implications of plurality of worlds—all are mentioned but then downplayed by her. (Nicolson treats "religion and superstition" as being displaced by science during the period.) The book generally is well-written and includes an annotated bibliography of primary sources.

Topics: *X*, VI Names: Addison, Burton, Cavendish, Donne, Godwin, Swift, Wilkins

1138 Nicolson, Marjorie Hope. "A World in the Moon: A Study of the Changing Attitude toward the Moon in the Seventeenth and Eighteenth Centuries." *Smith Coll. Stud. Mod. Lang.* 17, No.2 (1936): 1-72.

Nicolson here discusses ideas about the moon in relation to doctrines of God, to the widening of the boundaries of the universe, and to the plurality of worlds controversy. She emphasizes the changes in traditional Christianity caused by the new science.

Topics: *X*, IV, VII, XII Names: Godwin, Newton, Wilkins

1139 Nicolson, Marjorie Hope, and Nora M. Mohler. "The Scientific Background of Swift's *Voyage to Laputa*." *Ann. Sci.* 2(1937): 299-334.

This essay focuses on the social sources of antiscientific attitudes.

Topics: *X*, IV Name: Swift

1140 Nicolson, Marjorie Hope, and Nora Mohler. "Swift's 'Flying Island' in the *Voyage to Laputa*." *Ann. Sci.* 2(1937): 299-334.

Mohler and Nicolson discuss the scientific background and the literary symbolism of Swift's "Flying Island"—a fantasy which combines principles of flight with a cosmic voyage. There is no direct connection to religion. This work is repeated in Nicolson's *Voyage to the Moon* (1138).

Topics: *X*, VI Name: Swift

1141 Nicolson, Marjorie Hope, and George S. Rousseau. *This Long Disease, My Life: Alexander Pope and the Sciences*. Princeton, N.J.: Princeton University Press, 1968.

Researched by Rousseau and written by Nicolson, this book examines Pope's case history, his attitudes towards several topics in medicine, and his interest in astronomy and other sciences. In the long chapter on astronomy, the authors examine Pope's responses to Whiston's popular lectures on astronomy and theology. They discuss Pope's concern with the place of humans in an awesomely infinite universe. And they note how Newton's *Opticks* influenced Pope's poetic use of color and, especially, light and darkness.

Topics: *X*, IV, XI, XII Names: Arbuthnot, Gay, Newton, Pope, Swift, Whiston

1142 Noble, David F. *A World Without Women: The Christian Clerical Culture of Western Science*. New York: Knopf, 1992.

The title and subtitle indicate Noble's provocative thesis: the institutions of western science historically have been set within those of Latin Christianity. Hence the scientific community, like the clergy, has excluded women. In both, however, a world without women was neither necessary nor complete. In science as well as in Christianity, Noble postulates an "orthodoxy" (in which women were excluded) and a "heterodoxy" (in which women were accepted). The institutionalized scientific revolution of the seventeenth century is understood as an aspect of the restoration of male clerical authority. But over against this "orthodoxy" Noble sets "revivals of anticlerical religious heterodoxy"—which include humanism, the Hermetic tradition, and popular magic in our period. The scope of this book encompasses two thousand years and Noble relies almost entirely on secondary sources. (He does acknowledge this potential problem, however, and mostly uses the best recent literature.) His schematic division between orthodox and revivalist is somewhat oversimplified—as is sometimes implicit in the sources which he quotes.

Topics: *V*, I, II, IV

1143 North, J. D. "Finite and Otherwise: Aristotle and Some Seventeenth Century Views." In *Nature Mathematized*, edited by William R. Shea, 113-148. Dordrecht: D. Reidel, 1982. See (1445).

This is a highly sophisticated analysis of seventeenth-century doctrines of the infinite. North discusses Descartes, Locke and Newton—noting both the differences and the similarities found in their treatments of infinity, space, time, and extension. He emphasizes a common, Aristotelian inheritance which, however, was changed in various ways. North includes related metaphysical and theological issues and argues for their importance.

Topics: *VII*, XII Names: Locke, Newton

1144 North, J. D. "Thomas Harriot's Papers on the Calendar." In *The Light of Nature*, edited by J. D. North and J. J. Roche, 145-174. Dordrecht: M. Nijhoff, 1985. See (1145).

Two rather interesting areas of seventeenth-century science and mathematics used in the service of Christianity were: (1) computing the right day for celebrating Easter; and (2) determining ancient chronology (including the date of the Creation of the world). Harriot, as shown by North, expended serious efforts on both of these projects. His work shows him to have been a competent mathematician and chronologist—and sheds light on the nature of his faith as well.

Topics: *X*, VI Name: Harriot

1145 North, J. D., and J. J. Roche, eds. *The Light of Nature: Essays in the History and Philosophy of Science presented to A. C. Crombie*. Dordrecht: M. Nijhoff, 1985.

This anthology includes four essays annotated elsewhere in this bibliography. See: Brooks (0136), Feingold (0453), North (1144), and Rogers (1341).

Topics: *X*, VII

1146 Northrop, F. S. C. "Leibniz's Theory of Space." *J. Hist. Ideas* 7(1946): 422-446.

Northrop analyzes Leibniz's theories (physical, metaphysical, and epistemological) over against Newton and Locke. The importance of the differing concepts of God is recognized but not emphasized.

Topics: *VII*, XII Names: Locke, Newton

1147 Notestein, Wallace. *A History of Witchcraft in England from 1558 to 1718*. Washington, D.C.: American Historical Association, 1911.

This work is old, but it still is useful for its appendixes and for Notestein's hesitation in jumping to prejudged conclusions regarding the opinions of natural philosophers.

Topics: *II*, X Names: Casaubon, M., Glanvill, James I, More, Webster

1148 Nourrisson, Jean Felix. *Philosophies de la Nature: Bacon, Boyle, Toland, Buffon*. Paris: Perrin, 1887.

This nineteenth-century work includes individual chapters on Bacon, Boyle, and Toland, with special attention given to theories of matter and spirit and to Toland's pantheism.

Topics: *VII*, IX Names: Bacon, Boyle, Toland

1149 Nugent, Donald. "The Renaissance and/of Witchcraft." *Church History* 40(1971): 69-78.

Nugent draws parallels between Renaissance witchcraft and the counterculture of the late 1960s. Both developed out of despair during a time of transition—when the occult compensated for the success of scientific and bureaucratic rationality.

Topic: *II*

1150 Numbers, Ronald L. "Science and Religion." *Osiris* 1(1985): 59-80.

A few pages at the beginning of this historical survey of science and religion in America deal with the colonial period. Numbers argues against the "warfare" metaphor.

Topic: *VII*

1151 Numbers, Ronald L., and Darel W. Amundsen, eds. *Caring and Curing: Health and Medicine in the Western Religious Traditions*. New York: Macmillan, 1986.

This is a useful historical encyclopedia consisting of signed articles on topics indicated by the title. Most of the religions and most of the Christian denominations in Europe and America are covered. See, in particular, the articles on the Reformed and Anglican traditions which briefly discuss relations between religion, health and medicine in England and New England during our period.

Topic: *XI*

1152 Nuttall, Geoffrey F. *The Holy Spirit in Puritan Faith and Experience*. Oxford: B. Blackwell, 1946.

In this book, Nuttall does not mention science at all. But his discussion is helpful in regard to the problem of defining "Puritanism" and for distinguishing among Puritan parties and other separatists.

Topic: *III* Names: Baxter, Cromwell, Fox, Owen

1153 Nuttall, Geoffrey F. "'Unity with the Creation': George Fox and the Hermetic Philosophy." *Friends Quarterly* 1(1947): 134-143. Reprinted in *The Puritan Spirit*, by Geoffrey F. Nuttall, 194-203. London: Epworth Press, 1967.

Nuttall shows the connections between early Quakerism and the "Egyptian Learning." The associations are found both personally among the physicians and friends of Anne Conway and in the similar images of creation in the writings of Fox and Henry Vaughan.

Topics: *II*, VII, X Names: Conway, Fox, Helmont, Vaughan, H.

1154 Oakeshott, Michael. Introduction to *Leviathon*, by Thomas Hobbes, vii-lxvi. Oxford: Blackwell, 1946.

In his detailed introduction, Oakeshott argues that, for Hobbes, philosophy is reasoning. Thus, Hobbes distinguished philosophy from science (though imperfectly, the author suggests). Oakeshott also elaborates on his thesis that Hobbes was a "civil theologian." Overall, he contends that "the system of Hobbes' philosophy lies in the conception of the nature of philosophical knowledge, not in any doctrine about the world." For example, Hobbes is "a scholastic, not a 'scientific' mechanist": "The rational world is analogous to a machine." (But Hobbes does not say that it is a machine.)

Topic: *VII* Name: Hobbes

1155 Oakley, Francis. "Christian Theology and the Newtonian Science: The Rise of the Concept of the Laws of Nature." *Church History* 30(1961): 433-457. Reprinted in *Creation: The Impact of an Idea*, edited by Daniel O'Connor and Francis Oakley, 54-83. New York: Scribner, 1969. See (1161). Reprinted (with a confirming epilogue) in *Creation, Nature, and Political Order in the Philosophy of Michael Foster*, edited by Cameron Wybrow, 179-211. Lewiston, N.Y.: E. Mellen Press, 1992. See (1703).

Oakley traces the idea of the laws of nature through several centuries of the voluntarist-nominalist tradition. He argues that this concept played a major part in the overthrow of scholastic physics. This was a groundbreaking article in 1960.

Topics: *VII*, III, XII Names: Boyle, Charleton, Mather, I., Newton

1156 Oakley, Francis. *Omnipotence, Covenant, and Order: An Excursion in the History of Ideas from Abelard to Leibniz*. Ithaca, N.Y.: Cornell University Press, 1984.

Oakley has two purposes in this short monograph: (1) he presents an apologia for Lovejoy's approach to the history of ideas; and (2) he uses this method to trace the distinction between the absolute and ordained powers of God from the twelfth century into the seventeenth century. This theme contrasts with Lovejoy's "great chain of being" but Oakley shows that its trajectory is parallel to that of Lovejoy's. And, similarly, it underlies thought about the natural order as well as the social, political, and salvational orders. Chapters 1 and 3 are relevant to our field (Boyle's theology is the primary example in Chapter 3). Historiographically, Oakley indicates that historians of seventeenth-century science sometimes frame too narrowly the context of a particular text and thereby misunderstand its historical meaning in traditions of thought.

Topics: *VII*, I, IV Name: Boyle

1157 Oakley, Francis, and E. W. Urdang. "Locke, Natural Law, and God." *Natural Law Forum* 11(1966): 92-109.

Here, Oakley deals with Locke's understanding of the relationship between God and natural law in human nature and morality. He finds Locke to be an ambiguous voluntarist. This useful discussion also has implications for understanding seventeenth-century views of the relation between the laws of physical nature and God.

Topic: *VII* Name: Locke

1158 O'Brien, J. J. "Samuel Hartlib's Influence on Robert Boyle's Scientific Development." *Ann. Sci.* 21(1965): 1-14; 257-276.

Like the work of G. H. Turnbull, this two-part article is filled with details concerning Boyle's relations to Hartlib, the Hartlib circle, and various other individuals in the 1640s and 1650s. It includes no direct discussion of religion (but the indirect references are obvious and fairly beg to be developed in ways that Charles Webster later does so develop).

Topics: *VII*, II, V Names: Boyle, Hartlib

1159 O'Connor, Daniel J., ed. *A Critical History of Western Philosophy*. New York: Free Press, 1964.

This anthology includes three relevant essays annotated separately: Flew on Hobbes (0482); Hesse on Bacon (0672); and O'Connor on Locke (1160). The large double-columned pages make the essays of substantial length.

Topic: *VII*

1160 O'Connor, Daniel J. "Locke." In *A Critical History of Western Philosophy*, edited by Daniel J. O'Connor, 204-219. New York: Free Press, 1964. See (1159).

This is a rather ahistorical explication and evaluation of Locke's *Essay concerning Human Understanding*. It includes no direct discussion of Locke on science and religion but a number of O'Connor's passages are indirectly relevant to the topic.

Topic: *VII* Name: Locke

1161 O'Connor, Daniel, and Francis Oakley, eds. *Creation: The Impact of an Idea*. New York: Scribner, 1969.

Included in this anthology are Foster (0501) and Oakley (1155).

Topic: *III*

1162 Odom, Herbert H. "The Estrangement of Celestial Mechanics and Religion." *J. Hist. Ideas* 27(1966): 533-548.

In this retrospective view, Odom proposes that the "marriage" between Newtonian mechanics and natural theology was based on a "false bias" and on an irrelevant application of science. Although he discusses Newton, Clarke, and Maclaurin, the article basically presents the somewhat later views of d'Alembert and Laplace.

Topics: *VIII*, IX, XII Names: Clarke, Maclaurin, Newton

1163 Ogden, H. V. S. "Thomas Burnet's *Telluris Theoria Sacra* and Mountain Scenery." *ELH* 14(1947): 139-150.

Ogden explores the responses to Burnet's view of mountains and the mechanics of the biblical flood. While Burnet was ambivalent about mountains, the responses extolled their beauty and usefulness. There followed an increase in writings about mountain scenery.

Topic: *X* Names: Burnet, T., Ray, Warren

1164 Ogonowski, Zbigniew. "Le 'Christianisme sans Mysteres' selon John Toland et les Sociniens." *Archiwum Historii Filozifii i Mysli* Spolecsnej 12(1966): 205-223.

Ogonowski argues that Toland was influenced not only by Locke but also by continental Socinians in his views of Christian mysteries and miracles. He gives a good exposition of Toland's (and Locke's) position regarding the sense in which miracles can be supernatural.

Topics: *IX*, VII, X Names: Locke, Toland

1165 O'Higgins, James. *Anthony Collins: The Man and His Works*. The Hague: Martinus Nijhoff, 1970.

O'Higgins sees science as a minor causal factor in the rise of deism through a common emphasis on human reason. But generally he downplays the role of science in the thought of Collins. He explicates Collins' views by analyzing his writings within the context of Anglican liberal theology and of European radical religious views of Collins' day.

Topic: *IX* Names: Bentley, Clarke, Collins, A., Locke, Tillotson, Toland, Whiston

1166 O'Higgins, James. *Determinism and Freewill. Anthony Collins's A Philosophical Inquiry Concerning Human Liberty*. The Hague: Nijhoff, 1976.

The forty-five page introduction is similar to and represents an extension of O'Higgins' book on Collins.

Topic: *IX* Name: Collins, A.

1167 Oldroyd, David R. "Geological Controversy in the Seventeenth Century: 'Hooke vs. Wallis' and its Aftermath." In *Robert Hooke: New Studies*, edited by Michael Hunter and Simon Schaffer, 207-233. Woodbridge, England: Boydell, 1989. See (0765).

Oldroyd provides commentary on transcriptions of the texts of the debate. Hooke appealed to possible experiments, to classical writings, and to Genesis in his scriptural-geological understanding of the creation and modifications of the earth.

Topics: *X*, VII Names: Hooke, Wallis

1168 Olivieri, Grazia Tonelli. "Galen and Francis Bacon: Faculties of the Soul and the Classification of Knowledge." In *The Shapes of Knowledge from the Renaissance to the Enlightenment*, edited by Donald R. Kelley and Richard H. Popkin, 61-81. Dordrecht: Kluwer Academic, 1991. See (0872).

Olivieri describes Bacon's selection of three faculties of the soul (memory, imagination, and reason), their correlation to three ventricles of the brain, and their relation to Bacon's classification of knowledge. She relates the topic to various medical and intellectual traditions which preceded Bacon. Although Olivieri does

not discuss explicitly the interrelationships of science and religion in Bacon's views, the topic is implicit in her discussion.
Topics: *VII*, XI Name: Bacon

1169 Ollard, Richard. *Pepys: A Biography*. London: Hodder and Stoughton, 1974.
This excellent biography touches on Pepys' rational religion and emphasizes Pepys' sense that knowledge, like God, is indivisible. Ollard also discusses Pepys' participation in the Royal Society, as Fellow and as President (but not in much depth).
Topics: *X*, V Names: Evelyn, Pepys

1170 Olson, Richard. "On the Nature of God's Existence, Wisdom and Power: The Interplay between Organic and Mechanistic Imagery in Anglican Natural Theology, 1640-1720." In *Approaches to Organic Form*, edited by Frederick Burwick, 1-48. Dordrecht: D. Reidel, 1987.
This is a wide-ranging article directed against Westfall's generally accepted thesis regarding the conflict between mechanistic imagery of the world and Christian doctrine. Olson first argues that many of the misunderstandings are due to a lack of sensitivity to the long-standing tradition of natural theology. After outlining the major elements in this tradition, he proposes that Richard Hooker (not natural philosophy) was the chief spur to seventeenth-century Anglican natural theology. Second, Olson argues that those natural philosophers who adopted mechanistic imagery did so in order to maintain a divine conserving activity. They did not deny providence, but distinguished between ordinary and special providence. Finally, natural theology affected both mechanistic and organic images in natural philosophy. Olson, by noting these images and emphasizing natural theology, hopes to refocus discussion of religion and natural philosophy in the seventeenth and eighteenth centuries.
Topics: *VIII*, I, VII Names: Boyle, Charleton, Cudworth, Ray

1171 Olson, Richard. *Science Deified and Science Defied: The Historical Significance of Science in Western Culture*. 2 vols. Berkeley and Los Angeles: University of California Press, 1982 and 1990.

The combination of the title and subtitle accurately describes both the ambivalence and the scope of this survey. Olson summarizes the activities and attitudes of those who have studied natural phenomena from ancient times to the early nineteenth century. For our topic, Chapter 9 in Vol. 1 ("From the Renaissance to Modern Scientism in the Works of Johann Andreae and Francis Bacon") and Chapter 3 in Vol. 2 ("The Religious Implications of Newtonian Science") are most relevant. The former discusses the significance of *The New Atlantis*. The latter emphasizes natural theology, Newton, and Newtonianism. A portion of this chapter amounts to an edited version of Olson's article (1170). In both volumes, Olson shows that he is aware of the complex relationships between Christian theology and natural philosophy. As a set, these could be used as supplementary readings in a "Western Civilization" course.
Topics: *VII*, VIII, XII Names: Bacon, Bentley, Clarke, Newton

1172 Olson, Richard. "Spirits, Witches, and Science: Why the Rise of Science Encouraged Belief in the Supernatural in Seventeenth Century England." *Skeptic* 1, No.4(1992): 34-43.
In this semi-popular essay, the author shows that belief in ghosts and witches increased in conjunction with the empirical methods of the new philosophy after the Restoration. (Earlier in the century, such belief had declined and naturalistic interpretations had increased.) Olson sees a reaction to Hobbes' rationalistic materialism (and his perceived atheism) as the turning point.
Topics: *II*, VII, IX Names: Glanvill, Hobbes

1173 Ong, Walter J. *The Presence of the Word*. New Haven, Conn.: Yale University Press, 1967.
This is a study of the religious implications of the spoken word in the context of our technological culture. It includes the thesis that in the seventeenth century there was a shift "from preoccupation with sound to preoccupation with space" (p.73). But men at the time were unaware that their visual model for theological language was being shaped by their media.
Topics: *X*, VI

1174 Orchard, Thomas N. *The Astronomy of Paradise Lost*. London: Longmans, Green, 1896.

This work is included in and superseded by *Milton's Astronomy* (1175).
Topic: *X* Name: Milton

1175 Orchard, Thomas N. *Milton's Astronomy: The Astronomy of Paradise Lost*. London: Longmans, Green, 1913.
Orchard describes Milton's knowledge of astronomy and his poetic allusions to astronomical phenomena. He emphasizes that Milton knew both the Ptolemaic and Copernican systems and personally met Galileo. From about the age of thirty, Orchard argues, Milton was a Copernican. For poetic reasons, however, Milton based the cosmology of *Paradise Lost* on ancient, classical, Ptolemaic sources (as well as Biblical and even original ideas). Orchard describes the allusions to light, the sun, the moon, Venus, meteors and falling stars, comets, and the four seasons in Milton's various poetic works. All of this is set within the religious context of the Creator God and salvation history. (Orchard was a physician and Fellow of the Royal Astronomical Society.) He addresses the work to a general audience and mixes in his own love for poetry and his own appreciation of the wisdom of the Creator.
Topic: *X* Names: Horrox, Milton

1176 Ormsby-Lennon, Hugh. "Rosicrucian Linguistics: Twilight of a Renaissance Tradition." In *Hermeticism and the Renaissance*, edited by Ingrid Merkel and Allen G. Debus, 311-341. Washington, D.C.: Folger Shakespeare Library, 1988. London: Associated Universities Presses, 1988. See (1051).
Ormsby-Lennon analyzes the controversies over linguistics during the Puritan revolution and the Restoration. He shows that they reflect very complex interrelations between styles of language within the religious community—even in the writings of members of the Royal Society who attempted reforms. He argues, finally, that this period saw the apparent twilight of the linguistic Platonism of the Hermetic tradition. Both the Rosicrucian linguists and the "language planners" (such as Wilkins) finally lost out to mathematics as the universal, yet specialized, language of science.
Topics: *X*, II Names: Glanvill, Parker, Sprat, Vaughan, T., Webster, Wilkins

1177 Ornstein, Martha. *The Role of Scientific Societies in the Seventeenth Century*. Chicago: University of Chicago Press, 1928.

Originally submitted as a Ph.D. dissertation to Columbia University in 1913, this work was republished by her friends twelve years after Ornstein's death. It then became the standard general history of scientific societies throughout Europe. Ornstein concluded that these societies replaced universities as centers of support for the new science. There is little mention of religion—except negatively. Ornstein's work, of course, has been superseded.

Topics: *V*, III

1178 Orr, John. *English Deism: Its Roots and Fruits*. Grand Rapids, Mich.: Eerdmans, 1934.

Orr does not emphasize the importance of science in the rise of deism. His book has been superseded by more critical and more historically descriptive works.

Topic: *IX* Names: Blount, Herbert, E., Toland

1179 Osler, Margaret J., ed. *Atoms, Pneuma, and Tranquility: Epicurean and Stoic Themes in European Thought*. Cambridge: Cambridge University Press, 1991.

The essays by Dobbs (0392), MacIntosh (0972), and Wright (1701) are annotated separately in this bibliography.

Topic: *VII*

1180 Osler, Margaret J. "Certainty, Scepticism, and Scientific Optimism: The Roots of Eighteenth-Century Attitudes Toward Scientific Knowledge." In *Probability, Time, and Space in Eighteenth-Century Literature*, edited by Paula R. Backscheider, 3-28. New York, 1979. See (0052).

Osler traces two different epistemological traditions which converged on the mechanical philosophy and which produced two contradictory attitudes toward scientific knowledge. One of the traditions is the probabilism rooted in the skeptical crisis connected to post-Reformation theological debates. The article is mainly post-1720 in its thrust, but much of it is devoted to the seventeenth-century "roots."

Topic: *VII* Name: Locke

1181 Osler, Margaret J. "Descartes and Charleton on Nature and God." *J. Hist. Ideas* 40(1979): 445-456.

Osler shows that empiricist and rationalist "theories of knowledge were rooted in the theological traditions from which the systems emerged" These traditions emphasized, respectively, God's will and God's intellect. Osler compares the epistemology of Walter Charleton (working within the first tradition) with that of Descartes (working within the second).

Topic: *VII* Name: Charleton

1182 Osler, Margaret J. "The Intellectual Sources of Robert Boyle's Philosophy of Nature: Gassendi's Voluntarism and Boyle's Physico-Theological Project." In *Philosophy, Science, and Religion in England 1640-1700*, edited by Richard Kroll and others, 178-198. Cambridge: Cambridge University Press, 1992. See (0917).

Osler argues that Boyle drew on Gassendi's writings as he developed his Christianized corpuscular natural philosophy. She shows that their positions were almost identical—especially those regarding a voluntarist doctrine of God, nominalism, empiricism, and natural theology. Thus, she indicates that (contra James Jacob) this pattern of ideas preceded Boyle's particular political and social situation.

Topic: *VII* Name: Boyle

1183 Osler, Margaret J. "Locke and the Changing Ideal of Scientific Knowledge." *J. Hist. Ideas* 31(1970): 3-16. Reprinted in *Philosophy, Religion and Science in the Seventeenth and Eighteenth Centuries*, edited by John W. Yolton, 325-338. Rochester, N.Y.: University of Rochester Press, 1990. See (1720).

Osler here argues that Locke's epistemology reflected the changing conceptions of scientific method in the seventeenth century. In rejecting the Aristotelian and Cartesian ideal of the certain knowledge of real essences, Locke provided epistemological underpinnings for the empirical but uncertain science of Boyle and Newton. There are only tangential references to theology—but the whole point is implicitly important for the philosophical relations between science and theology. The essay is not original but it is nonetheless a useful summary.

Topics: *VII*, XII Names: Boyle, Locke, Newton

1184 Osler, Margaret J., and Paul Lawrence Farber, eds. *Religion, Science, and Worldview: Essays in Honor of Richard S. Westfall*. Cambridge: Cambridge University Press, 1985.

The essays by Home (0713) and Straker (1508) are annotated separately.

Topic: *VII*

1185 Osmond, Percy H. *Isaac Barrow: His Life and Times*. London: Society for Promoting Christian Knowledge, 1944.

In this somewhat flowery biography, Osmond includes many useful quotations (both from Barrow's writings and from those of his contemporaries). The book is divided into topical chapters. Those on "The First Lucasian Professor of Mathematics" and "The Preacher" are the most relevant to our field of study.

Topics: *VII*, X, XII Names: Barrow, Collins, J., Newton, Tillotson

1186 Oster, Malcolm. "Millenarianism and the New Science: The Case of Robert Boyle." In *Samuel Hartlib and Universal Reformation*, edited by Mark Greengrass, 137-148. Cambridge: Cambridge University Press, 1994. See (0573).

Oster briefly discusses the complexities and ambiguities in Boyle's writings on millenarianism. He sees Boyle not as a millenarian—but as one who perhaps saw himself as an instrument of God "in bringing about a new 'Revolution' in both science and divinity" before the eschaton.

Topics: *X*, VII Name: Boyle

1187 Oster, Malcolm. "Virtue, Providence and Political Neutralism: Boyle and Interregnum Politics." In *Robert Boyle Reconsidered*, edited by Michael Hunter, 19-36. Cambridge: Cambridge University Press, 1994. See (0759).

Oster here provides an analysis of Boyle's political experiences and developing sociopolitical views from the 1640s through the 1660s. He concludes that to ask whether Boyle was a Royalist or a Parliamentarian is to pursue the wrong question. Boyle's understanding of social stability and personal virtue, according to Oster, is both less ideological and less representative of others than is portrayed by Shapin, Shaffer, and the Jacobs. Moreover, while Boyle's natural philosophy previously has been distorted by internalist historiography, one should not minimize the epistemological, philosophical, and theological roots of Boyle's philosophy of nature by an unwarranted and oversimplified recourse to socio-political explanation.

Topics: *IV*, I, V Name: Boyle

1188 O'Toole, Frederick J. "Qualities and Powers in the Corpuscular Philosophy of Robert Boyle." *J. Hist. Philos.* 12(1974): 295-315. Reprinted in *Seventeenth-Century Natural Scientists*, edited by Vere Chappell, 101-121. New York: Garland, 1992. See (0205).

O'Toole shows in great detail how Boyle distinguished between "non-inherent relational and non-relational inherent properties of corporeal objects" (O'Toole's terms). He begins, as does Boyle, by indicating that God both created matter and governs it through natural laws.

Topic: *VII* Name: Boyle

1189 Otten, Charlotte F. *Environ'd with Eternity: God, Poems, and Plants in Sixteenth and Seventeenth Century England*. Lawrence, Kansas: Coronado Press, 1985.

Otten studies the literature of the "terraculturalists" (a comprehensive term that includes botanists, herbalists, poets who used nature images, and writers of garden books and agricultural manuals). These writers begin with God, she shows. In their view of the creation, Adam was a gardener and, in their eschatology, they present a paradisal garden. God is the Gardener-Architect of Eden who has commissioned them to follow his model. Likewise, Christ is portrayed as both redeemer of the earth and himself a "rose amongst thorns." Otten also explores the "Hortulan Saints" and the "ontology of vegetable smells."

Topics: *X*, VI, VIII Names: Austen, Blith, Coles, Donne, Evelyn, Gerard, Herbert, G., Milton, Parkinson, Traherne, Vaughan, H.

1190 Pacchi, Arrigo. *Cartesio in Inghilterra: Da More a Boyle*. Rome: Laterza, 1973.

This is a thorough study of Cartesianism in England. Pacchi adopts the common pattern of: (1) an early positive response in the 1640s and 1650s; and then (2) an ever more negative reaction in the 1660s through the 1680s. He includes many English writers—both

major and minor ones. Pacchi does not put particular emphasis on either science or religion but he does include both.
Topic: *VII* Names: Boyle, Charleton, Fludd, More, Stillingfleet

1191 Pagel, Walter. *New Light on William Harvey*. Basel: S. Karger, 1976.
In this little book, Pagel continues his close reading of Harvey's writings that he began in *William Harvey's Biological Ideas*. For Pagel, Harvey was a thinker whose thought, although appearing divided to us, was a unity for Harvey. Pagel again emphasizes his subject's Aristotelianism with its view of final causes, a monistic body and soul, and macro-microcosmic circular symbolism. He compares Harvey to his medieval and Renaissance predecessors. And he pictures Harvey as a friend of Robert Fludd and a listener to John Donne's sermons.
Topics: *XI*, VII Names: Donne, Fludd, Harvey

1192 Pagel, Walter. *Religion and Neoplatonism in Renaissance Medicine*. Edited by Marianne Winder. London: Variorum Reprints, 1985.
This is a volume published by Variorum Reprints. It includes the following essays annotated separately in this bibliography: "Religious Motives in the Medical Biology of the XVIIth Century;" (1193) and "The Vindication of 'Rubbish'" (1196).
Topic: *XI* Name: Harvey

1193 Pagel, Walter. "Religious Motives in the Medical Biology of the XVIIth Century." *Bulletin of the Institute of the History of Medicine* 3(1935): 97-128, 213-231, 265-312. Reprinted in *Religion and Neoplatonism in Renaissance Medicine*, by Walter Pagel. London: Variorum Reprints, 1985. See (1192).
This is an important early work on sixteenth- and seventeenth-century men who believed that medical research was divine service inspired by divine grace. Pagel examines the *Medicina Catholica* of Robert Fludd as well as Paracelsus and other continental writers. Fludd, he finds, assumed "the absolute unity of the world in God" and hence pursued "the method of analogy" in all of his medical and biological thought. Pagel concludes by comparing these medico-theologies to nineteenth-century naturphilosophie.
Topics: *XI*, II, VII Name: Fludd

1194 Pagel, Walter. "Review of Francis Yates' *Giordano Bruno and the Hermetic Tradition.*" *Ambix* 12(1964): 72-76.

Pagel approves of Yates theses—reminding his audience and her critics that religious mission did not exclude empirical or "proto-scientific" insight.

Topic: *II*

1195 Pagel, Walter. *The Smiling Spleen: Paracelsianism in Storm and Stress*. Basel: S. Karger, 1984.

Pagel outlines the support and critique of the Paracelsian tradition in the seventeenth century. Although Pagel is concerned primarily with Europeans, Boyle appears as an important late figure. As a generalization, the Paracelsians were motivated by "the mixture of mysticism with empiricism and naturalism, the product of religious scepticism towards scholastic and argumentative reasoning." Pagel finds this mixture expressed by Boyle in agreement with van Helmont: knowledge of God's purposes in creation are found by empirically studying specific individual units of that creation (and particularly in medical cases).

Topics: *XI*, II, VII Name: Boyle

1196 Pagel, Walter. "The Vindication of 'Rubbish'." *Middlesex Hospital Journal* 45(1945): 42-45. Reprinted in *Religion and Neoplatonism in Renaissance Medicine*, by Walter Pagel, 1-14. London: Variorum Reprints, 1985. See (1192).

Pagel proposes a historiography that takes "rubbish" (the religious and philosophical views of seventeenth-century figures in the history of medicine) seriously. He uses the writings of Harvey and van Helmont to support his position.

Topics: *I*, VII, XI Name: Harvey

1197 Pagel, Walter. "William Harvey and the Purpose of Circulation." *Isis* 42(1951): 22-38.

Pagel does not directly relate Harvey's views about circulation to his theology. But it is clear in the essay both that Harvey was an Aristotelian and that he assumed the circulation of blood to be a microcosmic image of the macrocosm (with the heart analogous to the sun). Pagel also points out that the perfection of the circle was a widely-held image by "mystics" who supported Harvey.

Topics: *XI*, VII Name: Harvey

1198 Pagel, Walter. "William Harvey Revisited." *Hist. Sci.* 8(1969): 1-31 and 9(1970): 1-41.

See Pagel's book, *New Light on William Harvey* (1191), which includes and expands upon this two-part essay.

Topic: *XI* Name: Harvey

1199 Pagel, Walter. "William Harvey: Some Neglected Aspects of Medical History." *J. Warburg Cour. Inst.* 7(1944): 144-153.

This early article has been superseded by Pagel's later work.

Topic: *XI* Name: Harvey

1200 Pagel, Walter. *William Harvey's Biological Ideas: Selected Aspects and Historical Background*. Basel: Karger, 1967.

Pagel's "selected aspects" picture Harvey as a whole person. Continuing to argue against both Whig and heroic historiography, the author discusses Harvey in his philosophical, cosmological, and religious setting. In much of the book, therefore, he describes the thought of Harvey's contemporaries and predecessors. He examines in detail Harvey's writings on the circulation of the blood and on generation. Pagel shows both Harvey's critical adherence to Aristotle and his use of microscopic symbolism. Hence, he argues, the view that modern science arose simply in opposition to Aristotle is misleading. Pagel also finds that, although Harvey was a Christian and a Royalist, neither of these elements played a major role in his biological thought or in the controversies which it raised.

Topics: *XI*, IV, VII Name: Harvey

1201 Palter, Robert, ed. *The Annus Mirabilis of Sir Isaac Newton, 1666-1966*. Cambridge, Mass.: MIT Press, 1970.

This is a reprint of the Autumn, 1967, issue of *Texas Quarterly* with additional commentaries. The articles focus on the background, development, and significance of Newton's work. Newton's religion is not emphasized. See: A. R. and M. B. Hall (0608); Hill (0688); Manuel (0985); Randall (1287); and Stein (1492).

Topics: *XII*, VII Name: Newton

1202 Palter, Robert. "The Newton Myths: Some Reflections on Westfall's *Never at Rest*." *Arch. Int. d'Hist. Sci.* 33(1983): 344-353.

In this essay review, Palter praises Westfall's biography of Newton (1640) but devotes about half of the essay to challenging him. Palter wishes that Westfall had devoted less attention to the development of Newton's thought and had engaged more in the philosophical analysis of Newton's final published concepts concerning force, motion, and absolute space. Palter suggests that Westfall overemphasizes the significance of Newton's theology and alchemy.

Topics: *XII*, I, VII Name: Newton

1203 Park, Catherine, and Lorraine J. Daston. "Unnatural Conceptions: The Study of Monsters in Sixteenth- and Seventeenth-Century France and England." *Past and Present* (92)1981: 20-54.

Park and Daston trace the evolution of the treatment of monsters (animal aberrations in nature) during the period from the 1500s to the 1700s. They do this on two fronts, theoretical and social. Theoretically and methodologically, they argue, there was a movement away from religious interpretations having to do with God's providence. Socially, they interpret the changing views and treatments of monsters as part of "the withdrawal of high from popular culture." Monsters came to be viewed, at least by the literate elite, first in a more secular and then in a more biological-medical way. The authors treat Francis Bacon's reflections and programme as a key intermediate stage in the evolution which they describe. They also discuss the Royal Society in their analysis.

Topics: *VII*, IV, V, X, XI Name: Bacon

1204 Parker, Derek. *Familiar to All: William Lilly and Astrology in the Seventeenth Century*. London: J. Cape, 1975.

This is a biography of one of the leading astrologers and almanac writers of seventeenth-century England. (He also is one of the few astrologers who left enough material to allow research into his life.)

Topic: *II* Name: Lilly

1205 Parker, Irene. *Dissenting Academies in England*. Cambridge: Cambridge University Press, 1914.

Parker argues that educational reform was an important part of Puritanism and that part of this reform was to introduce the new science into curricula. She stresses the continuity between "Puritan" plans for educational reform and the dissenting academies. She sees the Restoration and the Act of Uniformity as generally disastrous for English education (although the indirect causes of the formation of the dissenting academies). By later standards, Parker's theses are not very well supported, but her book remains provocative.

Topics: *V*, III Names: Hartlib, Morton

1206 Parry, G. J. R. "Puritanism, Science and Capitalism: William Harrison and the Rejection of Hermes Trismegistus." *Hist. Sci.* 22(1984): 245-270.

This deals with the sixteenth century, but it still is relevant to the argument against the Puritanism thesis.

Topics: *III*, II

1207 Passmore, John. *The Perfectibility of Man*. New York: Scribner Sons, 1970.

In this survey of the topic throughout Western Civilization, seventeenth-century philosophers play a pivotal role. The first half of the book focuses on the metaphysical perfection of the individual in relation to God. Later chapters focus on public moral improvement by social action, education and scientific progress. Passmore shows that different forms of Christianity have been operative in correlation to the various views toward human nature and its potential.

Topics: *IV*, VII Name: Locke

1208 Paterson, Antoinette Mann. *Francis Bacon and Socialized Science*. Springfield, Ill.: Thomas, 1973.

Paterson argues that "Bacon used Bruno's materialized model of the metaphysics of Cusanus for his scientific method." And he "cloaked his social materialism with Christianity." Further, for Bacon, the physical and civil sciences were in dynamic interplay. The result was a reformed tradition involving religion, science, and the state. Paterson writes in an argumentative style but with some pithy comments, e.g., "written science and philosophy is external the minute it is written down."

Topics: *VII*, IV Name: Bacon

1209 Patrick, James. "The Place of Michael Foster's Protest in the Controversial Historiography of Modern Science." In *Creation, Nature and Political Order in the Philosophy of Michael Foster*, edited by Cameron Wybrow, 245-254. Lewiston, N.Y.: E. Mellen Press, 1992. See (1703).

Patrick sets Foster and his articles in *Mind* clearly in the midst of debates between competing schools of thought in the 1930s. He argues that Foster's basic thesis was directed against Etienne Gilson and "the rising tide of historical interpretation which tended to rehabilitate Scholasticism." In contrast to the then generally accepted view that the presuppositions of high-medieval theology underlay the development of modern science, Foster emphasized the Reformation (i.e., Christianity liberated from its Greek aspects).

Topics: *III*, I, VII

1210 Patrides, C. A. *Milton and the Christian Tradition*. Oxford: Clarendon Press, 1966.

This is something of a systematic handbook of traditional and Renaissance Christian theology. Patrides argues that Milton can and should be presented in such terms. Natural philosophy, in this system, is basically a subheading of "The Doctrine of Creation" and the new science is simply irrelevant. This book includes approximately 1200 names and several thousand source-and-page references in its presentation of traditional Christianity.

Topic: *X* Names: Donne, Milton

1211 Paul, Leslie. *The English Philosophers*. London: Faber and Faber, 1953.

This is a popular survey of philosophers in the English-speaking world. Chapters 2-5 discuss those in the seventeenth century. Paul's book is outdated but very readable.

Topics: *VII*, XII Names: Bacon, Boyle, Hobbes, Locke, Newton

1212 Pauli, Wolfgang. "The Influence of Archetypal Ideas on the Scientific Theories of Kepler." In *Interpretation of Nature and the Psyche*, edited by C. G. Jung and W. Pauli, 147-240. New York: Pantheon Books, 1955.

Pauli, one of the leading physicists of the early twentieth century, gives an important interpretation of the controversy between Kepler

and Fludd. Each refers psychologically to a "completeness of experience" of the human soul. Yet "Fludd's symbolical *picturae* and Kepler's geometrical diagrams present an irreconcilable contradiction." Pauli includes long quotations from the sources which enhance the usefulness of this essay.
Topics: *II*, X Name: Fludd

1213 Pearson, Samuel C. "Science and the Theological Enterprise in Early Modern Europe." *Encounter* 47(1986): 27-39.
The aim of this survey article is to remember the positive, cordial relationship between science and theology in England in the seventeenth and early eighteenth centuries as an antidote to present-day "creation science." Pearson essentially equates scientific developments with "the mechanization of the world picture." He relates this to Locke's theory of religious toleration and expounds it in relation to the dissenting academies.
Topics: *VII*, IV, V Name: Locke

1214 Pedersen, Olaf. *The Book of Nature*. Vatican City and Notre Dame: Vatican Observatory, 1992.
This is a survey (in three lectures) of the history of the metaphor "the book of nature." In the third lecture, Pedersen attempts "to locate some of the historical prerequisites of a possible answer" whether the profound harmony between belief in the rationality of the universe and faith in Christ as the external logos of the world ended in the seventeenth century with a "victory" for science. He briefly discusses Newton's views on space and God. He judges the strengths and weaknesses of apologetics and natural theology in our period.
Topics: *VII*, I, X, XII Names: Boyle, Newton, Ray

1215 Pelseneer, Jean. "Les Influences Protestantes dans l'Histoire des Sciences." *Arch. Int. d'hist. Sci.* 1(1948): 348-353.
Pelseneer's three articles have to do with sixteenth-century continental thinkers. But his work has been drawn into the discussion on seventeenth-century England by both sides of the Protestantism-and-the-rise-of-science debate. Pelseneer himself argues in favor of the connection.
Topic: *III*

1216 Pelseneer, Jean. "L'Origine Protestante de la Science Moderne." *Lychnos* (1946-1947): 246-248.
See the annotation above (1215).
Topic: *III*

1217 Pelseneer, Jean. "La Reforme et L'Origine de la Science Moderne." *Revue de l'Universite de Bruxelles* 6(1954): 400-418.
See the annotation above (1215).
Topic: *III*

1218 Peuckert, W. E. *Pansophie*. Vol. 3, *Das Rosenkreutz*. 2nd ed. Berlin: E. Schmidt, 1973.
Though almost entirely a discussion of continental thought, Peuckert does set Fludd within the context of European Rosicrucianism.
Topic: *II* Name: Fludd

1219 Pennington, D., and K. Thomas, eds. *Puritans and Revolutionaries: Essays in Seventeenth-Century History Presented to Christopher Hill*. Oxford: Clarendon Press, 1978.
The essays by Aylmer (0051) and Tyake (1578) are annotated separately.
Topic: *VII*

1220 Perl, Margula R. "Physics and Metaphysics in Newton, Leibniz, and Clarke." *J. Hist. Ideas* 30(1969): 507-526. Reprinted in *Philosophy, Religion and Science in the Seventeenth and Eighteenth Centuries*, edited by John W. Yolton, 500-519. Rochester, N.Y: University of Rochester Press,1990. See (1720).
Perl argues that Leibniz and Newton did not differ on fundamental scientific questions. The similarities in their natural philosophies, in fact, were lost to sight because of their personal and metaphysical feuding. Clarke is represented as misunderstanding Newton's natural philosophy (although he can be taken as a fair representative of Newton's theological position).
Topics: *VII*, XII Names: Clarke, Newton

1221 Perry, Henry Ten Eyck. *The First Duchess of Newcastle and Her Husband as Figures in Literary History*. Boston: Ginn, 1918.

This book still is useful for biographical information and even for Perry's biased summaries of Margaret Cavendish's writings. Perry finds her writings on science to be fanciful—he agrees with Wilkins' quip that her writings are "castles in the air." But Perry does appreciate and respect the Duchess.

Topic: *X* Name: Cavendish

1222 Petersson, R. T. *Sir Kenelm Digby: The Ornament of England, 1603-1665*. London: J. Cape, 1956.

This flowery biography covers Digby's personality, activities and interests. These include: his religious conversion to Protestantism and his reconversion back to Catholicism; his philosophical writings, particularly the *Two Treatises*; his role in Gresham College and the Royal Society; his alchemy (though it is not emphasized); his politics; and his interaction with an amazing number and variety of persons.

Topics: *VII*, II, IV, V Names: Charles I, Charles II, Cromwell, Digby, Hobbes, Van Dyck, Villiers, White

1223 Phillips, Patricia. *The Scientific Lady: A Social History of Women's Scientific Interests, 1520-1918*. New York: St. Martin's Press, 1990.

Phillips investigates the British-leisured-class women's interest in science. This predilection, Phillips argues, resulted partly from the general exclusion of women from classical studies. She describes the arguments based on Genesis—both for and against this kind of education for women. She also points out the importance of Comenius' educational proposals during our period. Margaret Cavendish and Bathsua Makin are included among her influential models.

Topics: *V*, IV, X Names: Cavendish, Comenius, Makin

1224 Pilkington, Roger. *Robert Boyle: Father of Chemistry*. London: J. Murray, 1959.

Pilkington's biography, addressed to a popular audience, concludes with a chapter on Boyle's piety and theology.

Topic: *VII* Names: Boyle, Ranelagh

1225 Plumb, J. H. "Reason and Unreason in the Eighteenth Century: The English Experience." In *Some Aspects of Eighteenth-*

Century England, by J. H. Plumb and Vinton A. Dearing, 1-26. Los Angeles: William Andrews Clark Memorial Library, University of California, 1971.

This is a rambling essay of broad generalizations. It is marginally relevant to our field. Plumb argues that during the period from 1680 to 1720 reason and empiricism characterized the social and political establishment—but not the lower classes. (In the period from 1760 to 1800, by contrast, reason and empiricism characterized the dissenters and lower classes but not the establishment elites.)

Topics: *IV*, X

1226 Popkin, Richard H. "Cartesianism and Biblical Criticism." In *Problems of Cartesianism*, edited by Thomas M. Lennon and others, 61-81. Kingston, Ontario: McGill-Queens University Press, 1982. See (0943).

Popkin argues that Cartesianism, the new science, and modern biblical criticism were all interacting parts of a common intellectual drama in the seventeenth century. A major factor in the development of modern irreligion, says Popkin, was the application of Cartesian and scientific methodology to the evaluation of religious knowledge. This essay is mainly continental in its focus but it applies also, of course, to England. It includes brief discussions of Toland and Stillingfleet.

Topics: *X*, IX Names: Stillingfleet, Toland

1227 Popkin, Richard H. "The Crisis of Polytheism and the Answers of Vossius, Cudworth, and Newton." In *Essays on the Context, Nature, and Influence of Isaac Newton's Theology*, by James E. Force and Richard H. Popkin, 9-25. Dordrecht: Kluwer Academic, 1990. See (0499).

Popkin here discusses Cudworth's philosophy of religion and that of Newton in terms of their understanding of ancient pagan theology. Both relied on Gerard Vossius for their information about ancient religions; and both believed in the *prisca theologia* (the original, true revelation) which was corrupted by ancient pagans. In Newton's use of these materials, Popkin says, one can see his strong biblicism—and *not* deism.

Topics: *XII*, II, IX, X Names: Blount, Cudworth, Newton, Vossius

1228 Popkin, Richard H. "Divine Causality: Newton, the Newtonians, and Hume." In *Greene Centennial Studies*, edited by Paul J. Korshin and Robert R. Allen, 40-56. Charlottesville, Va.: University Press of Virginia, 1984. See (0895).

Popkin here argues that great mechanistic scientists in Britain, including Newton, saw the natural world functioning within the divine world. And creation was not just a general design for them: natural history, from the creation to the millennium and final destruction of the world, was part of the sacred history disclosed in the Biblical prophecies. The separation of science and religion is a later development: it does *not* characterize seventeenth- and eighteenth-century science in Britain.

Topics: *VII*, I, X, XII Name: Newton

1229 Popkin, Richard H. "Epilogue." In *The Shapes of Knowledge from the Renaissance to the Enlightenment*, edited by Donald R. Kelley and Richard H. Popkin, 215-220. Dordrecht: Kluwer Academic, 1991. See (0872).

Popkin briefly surveys the issues involved in the topic indicated by the title of the book. He states that our current views regarding the organization of knowledge (including our current views of the respective places of science and religion) are themselves an Enlightenment development. And he calls for revisionist historical research—particularly on this subject of the shapes of knowledge.

Topic: *I*

1230 Popkin, Richard H. *The High Road to Pyrrhonism*. Edited by Richard A. Watson and James E. Force. San Diego, Calif.: Austin Hill Press, 1980.

This is a collection of Popkin's articles dealing with scepticism in the early modern period. Two of the articles (1231) and (1237) are relevant to our field—especially with respect to the epistemological issues facing Glanvill and Locke.

Topic: *VII* Names: Glanvill, Locke

1231 Popkin, Richard H. "The High Road to Pyrrhonism." *Am. Philos. Q.* 2(1965): 1-15. Reprinted in *The High Road to Pyrrhonism*, by Richard H. Popkin, 11-37. San Diego, Calif.: Austin Hill Press, 1980. See (1230).

The focus in this essay is continental but is good for the background of the sceptical and epistemological issues regarding religion and the new philosophy.
Topic: *VII*

1232 Popkin, Richard H. *The History of Scepticism from Erasmus to Descartes*. Assen, the Netherlands: Van Gorcum, 1960.
This early edition is superseded by *The History of Scepticism from Erasmus to Spinoza* (1233).
Topic: *VII*

1233 Popkin, Richard H. *The History of Scepticism from Erasmus to Spinoza*. 3rd revised and augmented ed. Berkeley and Los Angeles: University of California Press, 1979.
Popkin's treatment is primarily continental in focus. But it is directly significant for Lord Herbert of Cherbury; and it provides some good background for Chillingworth, Hobbes, Glanvill, and other English writers. Popkin shows that the epistemological doubts generated by this intellectual movement were important both for science and for theology.
Topics: *VII*, VIII, IX Names: Chillingworth, Glanvill, Herbert, E., Hobbes

1234 Popkin, Richard H. "Hobbes and Scepticism." In *History of Philosophy in the Making: A Symposium of Essays to Honor James D. Collins on His 65th Birthday*, edited by Linus Thro, 133-148. Washington, D.C.: University Press of America, 1982. Reprinted as "Hobbes and Scepticism I," in *The Third Force in Seventeenth-Century Thought*, by Richard H. Popkin, 9-26. Leiden: E. J. Brill, 1992. See (1248).
Popkin asks in what sense can the label "sceptic" be applied to Hobbes; and he discusses this with respect to religion, science, and political theory. Specifically, he addresses biblical criticism and the questioning of the Mosaic authorship of the Pentateuch. Popkin concludes that Hobbes basically ignored epistemological scepticism but adopted an early form of political scepticism. This later evolved into a demonic scepticism of the modern state.
Topics: *VII*, IV, X Name: Hobbes

1235 Popkin, Richard H. "Hobbes and Scepticism II." In *The Third Force in Seventeenth-Century Thought*, by Richard H. Popkin, 27-49. Leiden: E. J. Brill, 1992. See (1248).

Popkin here elaborates on but repeats the same points found in "Hobbes and Scepticism" (1234).

Topics: *VII*, IV, X Name: Hobbes

1236 Popkin, Richard H. "Introduction." In *Joseph Glanvill's Essays on Several Important Subjects in Philosophy and Religion*, edited by Richard H. Popkin, v-xxxv. New York: Johnson Reprint, 1970.

Popkin summarizes each of the six essays included in Glanvill's book and provides a general introduction. His thesis is that Glanvill tried to show how a mitigated scepticism could "encourage scientific research of the sort then being initiated by the Royal Society" and how it "would harmonize with a liberal, reasonable, and tolerant religion." Popkin also explicates Glanvill's claim that scientific research provides evidence for witches and evidence against atheism. Glanvill's "rational fideism" is described as a rational faith and a faith in reason and his views on witchcraft are described as eminently sane in their context. This essay is an excellent introduction not only to Glanvill's essays but more generally to the intellectual issues of the time.

Topics: *VII*, VIII, X Name: Glanvill

1237 Popkin, Richard H. "Joseph Glanvill: Precursor of Hume." *J. Hist. Ideas* 14(1953): 292-303. Reprinted in *The High Road to Pyrrhonism*, by Richard H. Popkin, 181-195. San Diego, Calif.: Austin Hill Press, 1980. See (1230).

Popkin here discusses the role of scepticism in Glanvill's views of science and religion as well as Hume's relationship to Glanvill.

Topic: *VII* Name: Glanvill

1238 Popkin, Richard H., ed. *Millenarianism and Messianism in English Literature and Thought, 1650-1800: Clark Library Lectures, 1981-1982*. Leiden: Brill, 1988.

This anthology includes four relevant articles: Funkenstein (0516), Hill (0696), Quinn (1273), and Rousseau (1364), all of which are annotated separately in this bibliography.

Topic: *X*

1239 Popkin, Richard H. "Newton and Maimonides." In *A Straight Path: Studies in Medieval Philosophy and Culture. Essays in Honor of Arthur Hyman*, edited by Ruth Link-Salinger et al., 216-229. Washington, D.C.: Catholic University of America Press, 1988. Reprinted in *The Third Force in Seventeenth-Century Thought*, by Richard H. Popkin, 189-202. Leiden: E. J. Brill, 1992. See (1248).

Here, Popkin argues that Newton was influenced by the medieval Jewish theologian Moses Maimonides. This can be seen both in Newton's biblical/historical studies and in his concept of God as expressed in the General Scholium of Book III of the *Principia*. Popkin provides powerful counterevidence to Westfall's view of Newton as a proto-deist.

Topics: *XII*, IX, X Name: Newton

1240 Popkin, Richard H. "Newton and the Origins of Fundamentalism." In *The Scientific Enterprise*, edited by Edna Ullman-Margalit, 241-259. Dordrecht: Kluwer Academic, 1993. See (1579).

Popkin argues that Newton influenced those men who, from the revolutions of the late-eighteenth century onward, have read history in terms of providence and the millennium. He discusses both Newton's views on prophecy and on the millennium as well as the thought of those who can be seen as originating the millennial tradition in fundamentalism.

Topics: *X*, XII Name: Newton

1241 Popkin, Richard H. "Newton as a Bible Scholar." In *Essays on the Context, Nature, and Influence of Isaac Newton's Theology*, edited by James E. Force and Richard H. Popkin, 103-118. Dordrecht: Kluwer Academic, 1990. See (0499).

This is an explication of Newton's views towards, methods with, and applications of biblical texts. Although he used radically critical as well as scientific methods in his biblical studies, Newton was not a religious sceptic (as compared to Spinoza). Newton's millenarianism and Arianism are central to his understanding of and approach to the Bible.

Topics: *XII*, X Name: Newton

1242 Popkin, Richard H. "Newton's Biblical Theology and his Theological Physics." In *Newton's Scientific and Theological*

Legacy, edited by P. B. Scheurer and G. Debrock, 81-97. Dordrecht: Kluwer Academic, 1988. See (1409). Reprinted in *The Third Force in Seventeenth-Century Thought*, by Richard H. Popkin, 172-188. Leiden: E. J. Brill, 1992. See (1248).

Rather than viewing Newton as a great scientist who spent much time on religious matters, Popkin suggests that we might see him as an active theologian and outstanding biblical critic who took time off to write works on natural science. Popkin then summarizes Newton's biblical and theological writings; and, briefly, he indicates the connection between Newton's biblical theology and the metaphysics of his natural philosophy. This is a good summary (although it basically repeats what is found, collectively, in other of Popkin's articles).

Topics: *XII*, X Name: Newton

1243 Popkin, Richard H. "Polytheism, Deism, and Newton." In *Essays on the Context, Nature, and Influence of Isaac Newton's Theology*, edited by James E. Force and Richard H. Popkin, 27-42. Dordrecht: Kluwer Academic, 1990. See (0499).

Popkin argues that (in seventeenth-century England) "the vast taxonomic researches concerning the varieties of polytheistic religion, ancient and modern" produced in response both the deism of Herbert and Blount and a pro-Christian response from Cudworth and Newton. Newton's studies of ancient chronology and of "gentile theology" are interpreted by Popkin in a strongly non-deist way.

Topics: *XII*, II, IX Names: Blount, Cudworth, Herbert, E., Newton

1244 Popkin, Richard H. "Predicting, Prophesying, Divining and Foretelling from Nostradamus to Hume." *Hist. Euro. Ideas* 5(1984): 117-135. Reprinted in *The Third Force in Seventeenth-Century Thought*, by Richard H. Popkin, 285-307. Leiden: E. J. Brill, 1992. See (1248).

This focuses on the religious and epistemological issues connected with foretelling the future in the early modern period. Indirectly, it relates the topic to the science of the time; and it includes references to Newton, Whiston, and the Latitudinarians.

Topics: *VII*, II, X, XII Names: Newton, Tillotson, Whiston

1245 Popkin, Richard H. "The Religious Background of Seventeenth-Century Philosophy." *J. Hist. Philos.* 25(1987): 35-50. Reprinted in *The Third Force in Seventeenth-Century Thought*, by Richard H. Popkin, 268-284. Leiden: E. J. Brill, 1992. See (1248).

Presenting arguments for contextualist history of philosophy, Popkin uses the analogy of externalist history of science. He notes that, for most seventeenth-century thinkers, "knowledge involved the natural and revealed truth found in the Book of Nature and the Word of God: science and scripture." Furthermore, we must understand the variety of religious orientations and study how they interacted with the various philosophical positions.

Topics: *I*, VII, X

1246 Popkin, Richard H. "Some Further Comments on Newton and Maimonides." In *Essays on the Context, Nature, and Influence of Isaac Newton's Theology*, edited by James E. Force and Richard H. Popkin, 1-7. Dordrecht: Kluwer Academic, 1990. See (0499).

This is merely a summary of Popkin's other article (1239), "Newton and Maimonides," minus the attack on Westfall's view of Newton as a proto-deist.

Topics: *XII*, X Name: Newton

1247 Popkin, Richard H. "The Spiritualistic Cosmologies of Henry More and Anne Conway." In *Henry More (1614-1687) Tercentenary Studies*, edited by Sarah Hutton, 97-114. Dordrecht: Kluwer, 1990. See (0782).

Popkin finds that the cosmologies of More and Conway offered a genuine and important alternative to materialism. What he calls their "spiritology" explained why things happened and made the new science intelligible. Popkin says that their spiritualistic cosmologies "put the physical world . . . into the theodicy of the Cabbala, and into the Divine historical drama of the Bible and its prophecied outcome." Popkin sees Newton as privately being in this same tradition.

Topics: *VII*, X, XII Names: Conway, Mede, More, Newton

1248 Popkin, Richard H. *The Third Force in Seventeenth-Century Thought*. Leiden: E. J. Brill, 1992.

This is a collection of twenty-two essays: seven of them are relevant to our field; and two of these cannot be found in other of Popkin's writings. See the annotations for the following essays: (1234), (1235), (1239), (1242), (1244), (1245), and (1249).
Topic: *VII*

1249 Popkin, Richard H. "The Third Force in Seventeenth-Century Thought: Scepticism, Science and Millenarianism." In *The Prism of Science*, edited by Edna Ullmann-Margalit, 21-50. Dordrecht: Kluwer Academic, 1986. Reprinted in *The Third Force in Seventeenth-Century Thought*, by Richard H. Popkin, 90-119. Leiden: E. J. Brill, 1992. See (1248).
Cartesian rationalism and British empiricism, according to Popkin, are the first and second forces. The "third force," refers to theosophic speculations and Millenarian interpretation of Scripture: it influenced the group of thinkers whom Webster calls "the spiritual brotherhood." Among these intellectuals, Popkin includes Mede, Twisse, Comenius, Dury, Hartlib, the Cambridge Platonists, Lady Anne Conway—and possibly Newton. Popkin argues that this movement was one more reaction to the sceptical crisis that engulfed European thought in the sixteenth and seventeenth centuries. He relates the development of modern science both to the sceptical crisis and to this "third force." As in other of his writings, Popkin presents his findings in a story-telling style without attempting strict demonstration of his theses.
Topics: *VII*, X, XII Names: Conway, Dury, Mede, More, Newton, Twisse

1250 Popkin, Richard H., and Arjo Vanderjagt, eds. *Scepticism and Irreligion in the Seventeenth and Eighteenth Centuries*. Leiden: E. J. Brill, 1993.
The essays by Force (0485), Hutton (0786), and Katz (0861) are annotated separately in the present bibliography.
Topics: *IX*, VII

1251 Porter, Roy. "Creation and Credence: The Career of Theories of the Earth in Britain c. 1660-1820." In *Natural Order: Historical Studies of Scientific Culture*, edited by Barry Barnes and Steven Shapin, 97-124. Beverly Hills, Calif: Sage, 1979.

In discussing the genre of theories of the earth, Porter's basic question is: why did they have such a checkered career? Why did these theories fail? Unlike Newtonian natural theology, theories of the earth had to account for a unique sequence of events—what was essentially the drama of human salvation in the cosmos. And, Porter argues, the heterogeneous group of writers had no common loyalties (theological or political).

Topics: *X*, V Name: Burnet, T.

1252 Porter, Roy. "Introduction." In William Hobbs, *The Earth Generated and Anatomized by William Hobbs: An Early Eighteenth Century Theory of the Earth*, edited by Roy Porter, 5-46. Ithaca, N.Y.: Cornell University Press, 1981.

William Hobbs was a little-known Dorset man whose early-eighteenth century manuscript was found in 1973. Porter's introduction describes what is known about Hobbs and summarizes his theory of the earth. For Hobbs, the earth was an animated organic whole whose geological structures (e.g., fossilized rocks) were the result of alchemical-like processes. Rejecting literalism, he interpreted the days of Genesis to be metaphors for essentially simultaneous aspects of God's creative activity.

Topics: *VII*, II, X Name: Hobbs

1253 Porter, Roy. *The Making of Geology: Earth Science in Britain, 1660-1815*. Cambridge: Cambridge University Press, 1977.

This book is an adaptation of Porter's dissertation. The first three chapters provide a good summary of the developments of what would become the science of geology. They deal with the institutional background, the men involved, their field work, and their theories. References to natural theology and interpretations of scripture also appear in these discussions. Despite an anachronistic usage of the term "fundamentalist," Porter's book is useful. Historiographically, it is a good example of treating science as constructed—as made "by human choice and work."

Topics: *VII*, I, VIII, X Names: Burnet, T., Hooke, Plot, Ray, Whiston, Woodward

1254 Porter, Roy. "The Scientific Revolution: A Spoke in the Wheel?" In *Revolution in History*, edited by Roy Porter and Mikulas

Teich, 290-316. Cambridge: Cambridge University Press, 1986.

Porter first critiques the standard historiographic understanding of "The Scientific Revolution" as portrayed by Butterfield and others. He then finds a more subtle set of criteria: a conscious challenge by protagonists; a struggle; and conquest of the "new." In this transition, he admits, "religion, metaphysics, and ideology continued to play key roles *within* science." Yet he returns to emphasize the "extraordinary internal power" and "deep attractions of the New Science" which have played such important roles in its dominance in contemporary culture.

Topic: *I*

1255 Porter, Roy. "William Hobbs of Weymouth and His *The Earth Generated and Anatomized* (1715?)." *J. Soc. Bibliog. Natur. Hist.* 7(1976): 333-341.

This paper is a preliminary version of the introduction to Porter's edition of Hobbs' manuscript (1252). As in the introduction, Porter points out Hobbs' observations of fossils and his interesting interpretation of Genesis as well as his "animistic philosophy."

Topics: *VII*, II, X   Name: Hobbs

1256 Powell, Anthony. *John Aubrey and His Friends*. New York: C. Scribner's Sons,1948. Revised ed., New York: Barnes and Noble, 1963.

Although older, this still is a good biography of Aubrey.

Topic: *X*   Name: Aubrey

1257 Power, J. E. "Henry More and Isaac Newton on Absolute Space." *J. Hist. Ideas* 31(1970): 289-296.

Power sees absolute space as merely part of More's theological program; for Newton, however, he sees it as a metaphysical foundation of his natural philosophy.

Topics: *VII*, XII   Names: More, Newton

1258 Poynter, F. N. L. "Nicholas Culpepper and the Paracelsians." In *Science, Medicine and Society in the Renaissance*, edited by Allen G. Debus, Vol. 1, 201-220. New York: Science History, 1972. See (0357).

After a brief introduction to Culpepper as physician and Puritan, Poynter comments on two texts written or translated by Culpepper. In these works, his views of the relation between medicine and alchemy (as well as his piety and his Paracelsian hermeticism) are clearly shown.
Topics: *XI*, II Name: Culpepper

1259 Pratt, John P. "Newton's Date for the Crucifixion." *Quarterly Journal of the Royal Astronomical Society* 32(1991): 301-304
Pratt proposes that Newton estimated the date of the crucifixion by calculating the visibility of the crescent of the new moon and then correlating the Judean and Julian calendars.
Topics: *XII*, VI, X Name: Newton

1260 Preus, J. Samuel. *Explaining Religion: Criticism and Theory from Bodin to Freud*. New Haven, Conn.: Yale University Press, 1987.
Chapter 2 (of seventeen pages) deals with Herbert of Cherbury—deemphasizing the importance of innate ideas and presenting evidence that a design argument was central to Herbert's position.
Topics: *IX*, VIII Name: Herbert, E.

1261 Preus, J. Samuel. "Religion and Bacon's New Learning." In *Continuity and Discontinuity in Church History*, edited by Forrester F. Church and Timothy George, 267-284. Leiden: Brill, 1979.
Preus examines the "religious dimensions of Bacon's thought from three perspectives": (1) the theological framework for science in *The New Atlantis*; (2) the distinction between theology (regarding redemption) and the new learning (regrading creation); and (3) the extent to which Bacon allowed for a possible scientific study of religion. Preus argues that Bacon used religion to underwrite and provide the context for an entirely secular parallel to theology. But he had not yet thought of religion itself as a possible object of analysis.
Topics: *VII*, X Name: Bacon

1262 Price, John Vladimir. "Religion and Ideas." In *The Eighteenth Century*, edited by Pat Rogers, 120-152. New York: Holmes and Meier, 1978. See (1346).

With other of his related essays, Price aims at providing the context of English literature during the period. He is surprisingly unsympathetic, even mocking, towards the religion which he discusses. This essay only tangentially touches on the relationships between science and religion (which Price describes as "two irreconcilable disciplines").

Topic: *X*

1263 Priestley, F. E. L. "The Clarke-Leibniz Controversy." In *The Methodological Heritage of Newton*, edited by Robert E. Butts and John W. Davis, 34-56. Toronto: University of Toronto Press, 1969. See (0172).

Priestley makes the point that each antagonist (Leibniz and Clarke) is arguing from his own set of premises, from his own philosophical context, and seems incapable of even understanding the other. Thus, to the end, Leibniz believed that Newton's philosophy was destructive of natural religion while Clarke and Newton believed it to be a main support of natural religion. Priestley then explicates the opposing philosophical and theological positions, alternately, in a way sympathetic to each.

Topics: *VII*, VIII, XII Names: Clarke, Newton

1264 Priestley, F. E. L. "Newton and the Romantic Concept of Nature." *Univ. Toronto Q.* 17(1948): 323-336.

In a not very helpful way, Priestley relates Newton's views on God and nature to other seventeenth- and eighteenth-century currents.

Topics: *XII*, X Name: Newton

1265 Principe, Lawrence M. "Boyle' Alchemical Pursuits." In *Robert Boyle Reconsidered*, edited by Michael Hunter, 91-105. Cambridge: Cambridge University Press, 1994. See (0759).

Principe argues: (1) that traditional alchemy was a deep and long-term interest and commitment in Boyle's career; (2) that Boyle adopted alchemical principles and theories as well as mechanical and corpuscularian ones; and (3) that theological factors were central to his interest in alchemy. For Boyle, alchemy is the interface between the rational, mechanical functioning of the corporeal world and the suprarational, miraculous workings of the spiritual world. Against the received historiography and especially against Marie Boas Hall, Principe is arguing for fuller study of

Boyle's alchemy and for greater recognition of its place in Boyle's career and thought. This essay is written in a concise form but provides extensive documentation for Principe's claims.
Topics: *II,* I, VII Name: Boyle

1266 Principe, Lawrence M. "Robert Boyle's Alchemical Secrecy: Codes, Ciphers, and Concealments." *Ambix* 39(1992): 63-74.
Principe here argues that Boyle practiced open communication in all branches of natural philosophy except alchemy. He shows that, in his private papers and published writings, Boyle used various codes, ciphers, and methods of concealment. He was committed to secrecy in alchemy for two traditional reasons: (1) alchemical knowledge was divinely inspired and privileged; and (2) possible evil consequences would follow if it fell into the wrong hands.
Topics: *II*, IV Name: Boyle

1267 Prior, Moody E. "Bacon's Man of Science." *J. Hist. Ideas* 15(1954): 348-370. Reprinted in *Essential Articles for the Study of Francis Bacon*, edited by Brian Vickers, 140-163. Hamden, Conn.: Archon Books, 1968. See (1588).
Prior discusses "the intellectual, psychological, and ethical qualities which Bacon demanded of his new scientist." He then concludes that Bacon was one of the first writers to provide a secular and humanist stereotype of the scientist ("it seems less correct to say that they were religious men than that they constituted a religious cult in themselves"). But Prior's three dozen references to and quotations about God, the Bible, Adam, Abel, angels, and Christianity seem to counter his own conclusion.
Topics: *X*, IV Name: Bacon

1268 Prior, Moody E. "Joseph Glanvill, Witchcraft, and Seventeenth-Century Science." *Mod. Philol.* 30(1932): 167-193.
Prior shows why Glanvill's defense of the existence of witches was consistent with his scientific and philosophical views. Along with More and Boyle, Glanvill believed that the philosophical defense (and even experimental study) of spirits would help defend true religion against dogmatic (especially Hobbesian) materialism. In the eyes of these men, skeptical philosophy and experimental science were not at all inconsistent with a belief in witchcraft. This is a well-written article that now seems decades ahead of its time.
Topics: *VII*, II, IX Names: Boyle, Glanvill, More

1269 Pumfrey, Stephen. "'These 2 hundred years not the like published as Gellibrand has done de Magnete': The Hartlib Circle and Magnetic Philosophy." In *Samuel Hartlib and Universal Reformation*, edited by Mark Greengrass, 247-267. Cambridge: Cambridge University Press, 1994. See (0573).

Pumfrey begins with Hartlib's apparently exaggerated praise of a tract on magnetism. He then shows how Gellibrand's work on magnetism fits into the natural philosophy and the Puritanism of the Hartlib circle. Pumfrey concludes by noting that, in Hartlib's world, the compass was "a divine token of millennialist providential history."

Topics: *VI*, III, VII, X Names: Gellibrand, Hartlib

1270 Purver, Margery. *The Royal Society: Concept and Creation*. London: Routledge and K. Paul, 1967.

This is an aggressive defense: (1) of Sprat's *History of the Royal Society of London for the Improving of Natural Knowledge* (London, 1667) as the only authoritative source for the Society's early history and philosophy; (2) of the Oxford group as the only legitimate precursor of the Society; and (3) of the "true" Baconianism of the early Society. Purver argues against the Merton thesis in various ways. She notes the complexity of religious outlook of the Society (although she seemingly can not see a similar complexity with respect to natural philosophy).

Topics: *V*, III, VIII Name: Sprat

1271 Purver, Margery, and E. J. Bowen. *The Beginnings of the Royal Society*. Oxford: Clarendon Press, 1960.

This work is similar to and superseded by Purver's later book (1270).

Topic: *V*

1272 Quinn, Arthur. *The Confidence of British Philosophers: An Essay in Historical Narrative*. Leiden: E. J. Brill, 1977.

Quinn writes not only of the confidence of philosophers but also of their failure. He sees the history of British philosophy—typical of all philosophy—as characterized by proud and pretentious claims to the discovery of the first principles of truth. But, in hindsight, each philosophical school becomes obsolete and unconvincing. About seventy-five pages of this book are given over to the

Cambridge Platonists, Isaac Newton, and several Newtonians. Science and religion are presented as closely interrelated in the writings of these philosophers.

Topics: *VII*, VIII, XII Names: Cudworth, Desaguliers, Milton, More, Newton

1273 Quinn, Arthur. "On Reading Newton Apocalyptically." In *Millenarianism and Messianism in English Literature and Thought, 1650-1800*, edited by Richard H. Popkin, 176-192. Leiden: Brill, 1988. See (1238).

Quinn argues that prophecy was a way to resolve the crisis of skepticism for Newton and that the rediscovery of the *prisca theologia* was a crucial sign of the beginning of the millennium. He relates these beliefs to the secretive nature of Newton's rhetoric. And, in historiographic contrast to Westfall and others, Quinn argues that the staunchest advocates of modernity (positivist philosophers of the twentieth century) rhetorically have been millennialists.

Topics: *XII*, II, X Name: Newton

1274 Quintana, Ricardo. *Two Augustans: John Locke and Jonathan Swift*. Madison, Wisc.: University of Wisconsin Press, 1978.

Here are medium-length studies of Locke and Swift addressed to the general reader. Quintana approaches the Augustan age (1689 and following) as a period in which there were strains and ambiguities of thought and culture; and he believes that Locke and Swift well represent the age. He treats the differing religious preoccupations of the two men and their differing attitudes towards the new science as parts of a larger whole. He uses an eclectic historiographic approach but mainly this is literary history.

Topic: *X* Names: Locke, Swift

1275 Quinton, Anthony. *Francis Bacon*. New York: Hill and Wang, 1980.

This is a useful brief introduction to Bacon's philosophy of science. Quinton concludes that Bacon's main claim to importance rests on his role—not as a philosopher—but as a prophet. Regarding religion and science, Quinton only recognizes and appreciates Bacon's "firm separation of science from religion."

Topic: *VII* Name: Bacon

1276 Rabb, Theodore K. "Francis Bacon and the Reform of Society." In *Action and Conviction in Early Modern Europe: Essays in Memory of E. H. Harbison*, edited by Theodore K. Rabb and Jerrold E. Seigel, 169-193. Princeton, N.J.: Princeton University Press, 1969.

Rabb argues that Bacon emphasized the need for central authority in society, politics, church, and science. Bacon, he says, approved of gradual reforms within existing institutions and opposed wholesale innovations. Thus, Rabb challenges the thesis that Bacon's views corresponded to those of later Puritan revolutionaries.

Topics: *IV*, III, V Name: Bacon

1277 Rabb, Theodore K. "Puritan Science?" *Brit. J. Hist. Sci.* 11(1978): 156-159.

This is an essay review of Charles Webster's *The Great Instauration* (1616). Rabb praises Webster for his massive scholarship but takes issue with him on several points of form and content.

Topics: *IV*, III

1278 Rabb, Theodore K. "Puritanism and the Rise of Experimental Science in England." *J. World Hist.* 7(1962): 46-67.

Rabb argues that Merton's statistics are questionable and that Merton (1057) has found no inherently Puritan qualities which promoted scientific inquiry. In his articles, Rabb uses phrases such as "when Puritanism became more than a set of religious beliefs" and "genuine scientific inquiry" with which he can prove almost any thesis he cares to put forward.

Topic: *III*

1279 Rabb, Theodore K. "Religion and the Rise of Modern Science." *Past and Present* 31(1965): 111-126. A revised version appears in *The Intellectual Revolution of the Seventeenth Century*, edited by Charles Webster, 262-279. London: Routledge and Kegan Paul, 1974. See (1618).

In arguing against Hill (0689) and (0694), Rabb emphasizes a distinction between pre-1640 (the "rise of science") and post-1640 (when science had already risen). Rabb says that Puritanism was amenable to science after 1640 but had nothing to do with the *rise* of science.

Topics: *III*, I

1280 Rabb, Theodore K. "Science, Religion and Society in the Sixteenth and Seventeenth Centuries." *Past and Present* 32(1966): 148. Reprinted in *The Intellectual Revolution of the Seventeenth Century*, edited by Charles Webster, 284-285. London: Routledge and Kegan Paul, 1974. See (1618).

Here, Rabb continues his debate with Hill (0693).

Topic: *III*

1281 Raistrick, Arthur. *Quakers in Science and Industry*. New York: Philosophical Library, 1950. Reprinted, Newton Abbot, England: David and Charles, 1968.

This is mostly post-1720 and industrial in its content. But the book does include sections which are relevant to science and religion in our period. As an apologist for the Society of Friends, Raistrick argues for a positive correlation between Quakerism and support for science, technology, and medicine. He also describes a Quaker worldview (human equality, Christian fraternity, and social justice) which could be applied to some of the later debates concerning science, religion, and ideology during our period.

Topics: *VI*, III, IV, XI Names: Fox, Graham, Lawson, Logan, Quare, Tompion

1282 Ramsay, Mary Paton. *Les Doctrines Medievales chez Donne*. London: Oxford University Press, 1917.

A chapter on "Les Sciences" discusses Donne in relation to medicine, alchemy, astronomy, and astrology. Donne is presented as highly traditional, indeed medieval, in his views. Religion is treated separately.

Topics: *X*, II Name: Donne

1283 Ramsay, Mary Paton. "Donne's Relations to Philosophy." In *A Garland for John Donne*, edited by Theodore Spencer, 99-120. Cambridge, Mass.: Harvard University Press, 1931.

Donne is presented in the same manner as in Ramsay's book (1282).

Topic: *X* Name: Donne

1284 Ramsey, I. T. Editor's Introduction to *The Reasonableness of Christianity with A Discourse of Miracles and Part of A Third*

Letter Concerning Toleration, by John Locke, 9-20. Stanford, Calif.: Stanford University Press, 1958.

This is a philosophical (rather than historical) discussion with no explicit reference to natural science. But Ramsey does include a useful introductory account of Locke's distinction between faith and reason.

Topic: *VII* Name: Locke

1285 Randall, John Herman, Jr. *The Making of the Modern Mind: A Survey of the Intellectual Background of the Present Age*. Boston: Houghton Mifflin, 1926. Revised ed., Boston: Houghton Mifflin, 1940.

For our period, this is similar to Burtt (0160). Randall emphasizes the logical and historical consequences of the new science (including rationalism in Christianity, natural religion, deism, and atheism). He tends to read later eighteenth-century developments backwards into our period and to make judgments on the negative effects of science on religion. This work now is outdated.

Topics: *VII*, VIII, IX, XII Name: Newton

1286 Randall, John Herman, Jr. "Newton's Natural Philosophy: Its Problems and Consequences." In *Philosophical Essays in Honor of Edgar Arthur Singer, Jr.*, edited by F. P. Clarke and M. G. Nahm, 335-357. Philadelphia: University of Pennsylvania Press, 1942.

This is a retrospective analysis of the philosophical assumptions and presuppositions in Newtonian natural philosophy. Randall argues that the "theological foundation of Newton's natural philosophy needs no extraneous religious reasons to account for its presence." Rather, God is a logical necessity within the Newtonian system.

Topics: *XII*, VII Name: Newton

1287 Randall, John Herman, Jr. "The Religious Consequences of Newton's Thought." *Texas Q.* 10(1967): 275-285. Reprinted in *The Annus Mirabilis of Sir Isaac Newton, 1666-1696*, edited by Robert Palter, 333-343. Cambridge, Mass.: MIT Press, 1970. See (1201).

Randall here gives a retrospective evaluation of various elements of Newton's thought and of "Newton the symbol." Much in this

essay has to do with philosophy of religion in the nineteenth and twentieth centuries.
Topics: *XII*, VII, IX, X Name: Newton

1288 Randall, John Herman, Jr. "What Isaac Newton Started." In *Newton's Philosophy of Nature: Selections from His Writings*, edited by J. S. Thayer, ix-xvi. New York: Hafner Press, 1953.
Randall provides broad, retrospective generalizations about the ramifications of Newtonian thought for philosophy of science, religion, and social ideology.
Topics: *XII*, IV, VII Name: Newton

1289 Rattansi, Piyo M. "Alchemy and Natural Magic in Raleigh's 'History of the World'." *Ambix* 13(1966): 122-138.
Rattansi examines Raleigh's *History* (published in 1614) for its discussions of magic and alchemy. He also compares Raleigh's views with those of Bacon and James I. Natural magic, for Raleigh, was a practical part of as well as the God-given perfection of natural philosophy. His Neoplatonism is shown by his views concerning the role of light in God's creation.
Topics: *II*, X Names: Bacon, James I, Ralegh

1290 Rattansi, Piyo M. "The Helmontian-Galenist Controversy in Restoration England." *Ambix* 12(1964): 1-23.
While dealing primarily with the controversies between the College of Physicians (on the one hand) and a temporary alliance of Helmontian physicians and apothecaries (on the other), Rattansi points out the relation of anti-college attitudes to Puritan ascendancy. With the Great Plague, however, the influence of both sets of physicians diminished, despite the bravery shown by the Helmontians.
Topics: *XI*, III, IV

1291 Rattansi, Piyo M. "The Intellectual Origins of the Royal Society." *Notes Rec. Royal Soc. London* 23(1968): 129-143. Reprinted in *Seventeenth-Century Natural Scientists*, edited by Vere Chappell, 49-63. New York: Garland, 1992. See (0205).
Rattansi here discusses the influences of Renaissance and magical traditions on the early Royal Society. He shows that a number of

different traditions—Hermeticism, neo-Aristotelianism, and the mechanical philosophy—should be taken into account. By so doing, he downplays Puritan motives.
Topics: *IV*, II, III, V

1292 Rattansi, Piyo M. "Newton and the Wisdom of the Ancients." In *Let Newton Be!*, edited by John Fauvel and others, 185-201. Oxford: Oxford University Press, 1988. See (0449).
Rattansi gives an introductory description of: Newton's view of history; his understanding of his own philosophy of nature as a rediscovery of ancient truth; and his view that ancient truth was Mosaic in origin. He summarizes (for the general reader) and integrates the specialized studies of Manuel, Dobbs, Casini, McGuire, and Rattansi.
Topics: *XII*, II, X Name: Newton

1293 Rattansi, Piyo M. "Newton's Alchemical Studies." In *Science, Medicine and Society in the Renaissance*, edited by Allen G. Debus, Vol. 2, 167-182. New York: Science History, 1972. See (0357).
Rattansi provides strong arguments connecting Newton to alchemical and *prisca* traditions. He also discusses the historiographic implications of this.
Topics: *XII*, I, II Name: Newton

1294 Rattansi, Piyo M. "Paracelsus and the Puritan Revolution." *Ambix* 11(1963): 24-32.
Rattansi suggests that the rejection of Aristotelian natural philosophy and traditional religion by the Paracelsians helped give their mystical and reforming system appeal to Puritans in the 1640s and 1650s. He indicates that an attack on Paracelsian teachings began to be mounted by advocates of Baconianism in the 1650s. This is an early essay supporting the position that the acceptance of natural philosophic ideas can be strongly influenced by social and political factors.
Topics: *III*, II, IV Names: Charleton, Ward, Webster, Wilkins

1295 Rattansi, Piyo M. "Puritanism and Science: The 'Merton Thesis' after Fifty Years." In *Robert K. Merton: Consensus and*

Controversy, edited by Jon Clark and others, 351-369. New York: Falmer Press, 1990. See (0216).

Rattansi discusses Merton's thesis (1057) as an extension of Max Weber's *The Protestant Ethic and the Spirit of Capitalism* (1610). He summarizes: Rupert Hall's works which oppose the Merton thesis; and the works of Christopher Hill and Charles Webster which have supported and extended Merton's theory. He then argues for a more complex approach to the issues—an approach which would include Neoplatonic, Neo-Pythagorean, Hermetic, and Paracelsian traditions. He also argues that, in the seventeenth century, a variety of religious positions supported a variety of scientific movements. Merton's thesis is vulnerable, says Rattansi, in that it focuses specifically on "Puritanism" and a specific form of science; but Merton made a major contribution in bringing our attention at all to the interrelationships between science, society and religion.

Topics: *III*, II, IV Names: Bacon, Comenius

1296 Rattansi, Piyo M. "Science and Religion in the Seventeenth Century." In *The Emergence of Science in Western Europe*, edited by Maurice P. Crosland, 79-87. New York: Science History, 1976. See (0300).

Rattansi here argues that the complex relations between religion and science in England should be seen as an example of the more general "problem of defining the relation . . . in particular European societies." Rattansi emphasizes the variety and complexity of traditions and notes how the "mythical element in the ideal image of science" points to aspects of the value system of any given society.

Topics: *I*, IV, VII

1297 Rattansi, Piyo M. "The Social Interpretation of Science in the Seventeenth Century." In *Science and Society 1600-1900*, edited by Peter Mathias, 1-32. Cambridge: Cambridge University Press, 1972. See (0998).

This is a historiographical critique of the manner in which the social interpretation of seventeenth-century science has been approached both by externalists and by internalists. Rattansi modifies Merton's thesis (1057) by expanding it (both theologically and geographically) and by bringing to the discussion his knowledge of

Renaissance Hermeticism. For example, Rattansi suggests that Bacon transformed Hermetic ideas and that Boyle broke with Hermetic interests because of their association with extreme sectarianism. A more moderate and true religious view, Boyle decided, could be derived from a mechanical conception of nature.
Topics: *IV*, I, II, III Names: Bacon, Baxter, Boyle

1298 Rattansi, Piyo M. "Some Evaluations of Reason in Sixteenth- and Seventeenth-Century Natural Philosophy." In *Changing Perspectives in the History of Science*, edited by Nicholas Teich and Robert Young, 148-166. London: Heinemann, 1973. See (1535).

This is an answer to Mary Hesse's essay in the same volume (0675). Rattansi emphasizes the importance of Christian Renaissance Neoplatonism and Hermeticism for Newtonian studies.
Topics: *I*, II, XII Names: More, Newton

1299 Rattansi, Piyo M., and Antonio Clericuzio, eds. *Alchemy and Chemistry in the Sixteenth and Seventeenth Centuries*. Dordrecht: Kluwer Academic, 1994.

The essays by Clericuzio (0222) and Emerton (0429) are annotated separately.
Topic: *II*

1300 Raven, Charles E. *English Naturalists from Neckam to Ray: A Study of the Making of the Modern World*. Cambridge: Cambridge University Press, 1947.

In this study of medieval and early modern English naturalists, about 150 pages are devoted to men of the early seventeenth century. Raven's primary concern is to analyze the lives and the writings on natural history of his subjects. This concern is reflected in extensive indexes on flora and fauna. In two cases, Raven emphasizes religious beliefs. Edward Topsell wrote biblical commentaries as well as natural history. And Raven concludes with an epilogue on Thomas Browne and *Religio Medici*—which he entitles "The Coming of Modern Man." Raven understands this "modern man" to be those seventeenth-century natural historians and philosophers who attained a "syncretism" of the two books of nature and scripture.

Topics: *VII*, X Names: Browne, T., How, Johnson, Merret, Parkinson, Topsell

1301 Raven, Charles E. *John Ray, Naturalist: His Life and Works*. Cambridge: Cambridge University Press, 1942. 2nd ed., with minor addenda, Cambridge: Cambridge University Press, 1950. Reissued, with an introduction by S. M. Walters, 1986.

This is the most thorough study of the life and writings of John Ray. Raven pays detailed attention to Ray's botanical writings and supplies indexes of flora and fauna. He concludes by emphasizing the consequences of Ray's work—especially of *The Wisdom of God Manifested in the Works of Creation*. (Not only were many editions published, but it was widely imitated and plagiarized.) For one hundred and fifty years, Raven believes, this resulted both in British scientific progress and in a type of natural theology "capable of giving appropriate expression to the Christian faith in a scientific age." Thus, as Walters indicates in his introduction, Raven found in Ray a pioneer of his own liberal theology.

Topics: *VIII*, III, VII Names: Derham, Lister, Ray, Robinson, Sloane, Willughby

1302 Raven, Charles E. *Natural Religion and Christian Theology*. 2 vols. Cambridge: Cambridge University Press, 1953.

Raven's first series of Gifford Lectures (Vol. 1) surveys the history of Christian theologies of nature. It is a defense of natural theology at a time when neo-orthodoxy dominated the theological scene; and it is an affirmation of the historical importance of the biological sciences at a time when physics and astronomy dominated the discipline of the history of science. The two chapters on English thought during our period emphasize the view of nature as a coherent organic system—and the change to that of a more self-contained mechanical system. The second series (Vol. 2) contains Raven's present theology.

Topics: *VIII*, VII Names: Cudworth, More, Ray

1303 Raven, Charles E. *Organic Design: A study of Scientific Thought from Ray to Paley*. Oxford: Oxford University Press, 1954.

Raven here compares John Ray's (scholarly) organic design argument with William Paley's (amateurish) mechanistic natural theology. In this lecture, Raven declares that two mistakes were

made in the development of natural theology during the period under review. First, there was an emphasis on a mechanistic concept of teleology; second, there was a failure to face evidence of error, pain, and sin.
Topics: *VIII*, VII Name: Ray

1304 Raven, Charles E. *Synthetic Philosophy in the Seventeenth Century*. Oxford: B. Blackwell, 1945.
Raven is a theologian who stands against any form of division between science and Christianity. Here he draws attention to the development of the biological sciences in the holistic philosophy—Christian and scientific—of the seventeenth century. This is based upon a public Herbert Spencer lecture (hence the title); it provides a brief preview of his first series of Gifford lectures.
Topics: *VII*, VIII

1305 Ravetz, Jerome R. "Francis Bacon and the Reform of Philosophy." In *Science, Medicine and Society in the Renaissance*, edited by Allen G. Debus, Vol. 2, 97-119. New York: Science History, 1972. See (0357).
Ravetz points out that Bacon's strategy for the reform of natural philosophy was the means for the religious redemption of humankind. As the fall originated from vanity and a lust for knowledge, redemption would come with humility and charity. And part of that redemption will be dominion over the natural world. Ravetz also deals with Bacon's distinction between the "true sons of science" and the vulgar. Nature is a "holy" subject to be handled successfully only by the few "cleansed" humans. This reform is also a sign of the millennium. Ravetz concludes that, while we can no longer view scientific progress as leading to moral reform, the issue of the morality of science is still with us.
Topics: *VII*, IV, X Name: Bacon

1306 Ravetz, Jerome R. "The Symbol of Science in European Thought." In *Human Implications of Scientific Advance*, edited by Eric G. Forbes, 28-34. Edinburgh: Edinburgh University Press, 1978. See (0483).
Ravetz uses Bacon both to indicate the morality of the optimism of modern science and as one of few to realize its potential evils. The

article is directed at the present-day crisis and against those who would separate "pure" scientific knowledge from application.
Topics: *VII*, IV, VI Name: Bacon

1307 Raylor, Timothy. "Providence and Technology in the English Civil War: Edmond Felton and His Engine." *Renaiss. Stud.* 7(1993): 398-413.
Raylor sets Fulton and his engine of war in the context both of the Felton family misfortunes and of "Puritan attitudes towards technology and warfare." Felton promoted it to Parliament by referring to it as a "means" by which the providence of God could become effective and by arguing that it was an instrument of defense rather than of aggression. (Felton did not succeed with Parliament.)
Topics: *VI*, III, IV Names: Felton, Hartlib

1308 Read, John. *Prelude to Chemistry: An Outline of Alchemy.* London: G. Bell and Sons, 1936.
Read's survey history provides an outline of alchemy, its literature, and other selected topics. It includes many black-and-white plates. The only seventeenth-century figure discussed at length is Elias Ashmole.
Topic: *II* Name: Ashmole

1309 Redgrove, Herbert Stanley. *Bygone Beliefs.* London: W. Rider, 1920.
This older work contains two chapters on alchemy and one with brief biographies of the Cambridge Platonists. Redgrove believes that the origins and continuing power of alchemy is understood best in terms of mystical theology and the phallic element in its symbolism.
Topics: *II*, VII Name: More

1310 Redondi, Pietro. "Dieu et Mon Droit: Rivoluzioni Inglesi e Rivoluzione Scientifica." *Intersezioni: Rivista di Storia delle Idee* 14(1994): 261-277.
This is a review essay of some major writings on the social and political history of natural philosophy. Redondi discusses Merton, Webster, the Jacobs, Shapin, and Schaffer; he also raises the question of participation by women in the scientific revolution of the

seventeenth century. Redondi concludes by pointing to the importance of the urban setting for the Royal Society.
Topics: *IV*, I, III, V

1311 Redondi, Pietro, ed. *De la Mystique a la Science: Cours, Conferences et Documents, 1922-1962*. Paris: Editiones de l'Ecole des Hautes Etudes en Sciences Sociales, 1986.
See Alexandre Koyre (0905).
Topics: *VII*, I, XII

1312 Redwood, John A. "Charles Blount (1654-93), Deism, and English Free Thought." *J. Hist. Ideas* 35(1974): 490-498.
This is a good brief article which discusses Blount's deism, his contemporaries' conceptions of atheism, and the rather minor role of natural philosophy in his program.
Topic: *IX* Name: Blount

1313 Redwood, John A. *Reason, Ridicule, and Religion: The Age of Enlightenment in England, 1660-1750*. London: Thames and Hudson, 1976. Cambridge, Mass.: Harvard University Press, 1976.
The author appears to have read hundreds of books, pamphlets, and sermons from the period, and this book appears to be the partially organized reading notes from such research. Redwood proposes in shotgun fashion various (and sometimes contradictory) theses about atheism, deism, politics, natural philosophy, atomism, materialism, the origin of the earth, witches, antitrinitarianism, reason, nature, the church, the Bible, satire and ridicule. He does this amidst a barrage of detail and footnotes. (In one run of 195 pages there are 1,877 footnotes!) If there is a main theme, it is that of atheism and "atheist-hunters" in a "God-ridden" century. Almost all the topics and authors are treated in "negative" terms: either in terms of what a given author most opposed, criticized, or feared; or in terms of a given author's opponents and critics; or in terms of those who most feared any real or imagined movement. In some ways, this makes for an interesting approach—but it becomes wearisome. Although reading this book straight through could give one indigestion, it is useful when read selectively.
Topics: *IX*, IV, VII, VIII, X, XII Names: Bentley, Blackall, Blount, Burnet, T., Chubb, Clarke, Cudworth, Gastrell, Hobbes, Locke, Newton, Tindal, Toland, Whiston, Wise, Woodward

1314 Reedy, Gerard. "Barrow, Stillingfleet, and Tillotson on the Truth of Scripture." In *Greene Centennial Studies*, edited by, 22-39. Charlottesville, Va.: University Press of Virginia, 1984. See (0895).

Reedy describes the arguments and something of the context of these defenders of the truth of the Bible. Hobbes, radical dissenters, and Roman Catholics are seen as the opponents; new understandings of rationality, new epistemological questions, and the new science are seen as the positive context for understanding these men. If we understand the context, argues Reedy, we will see that they were traditional Christians; they were not proto-deists and they were not without emotion in their faith. This somewhat overlaps with portions of Reedy's book (1315).

Topics: *X*, VII, IX Names: Barrow, Stillingfleet, Tillotson

1315 Reedy, Gerard. *The Bible and Reason: Anglicans and Scripture in Late Seventeenth-Century England*. Philadelphia: University of Pennsylvania Press, 1985.

Reedy here focuses on various Latitudinarian theologians of the period 1660-1700. He makes good distinctions between various meanings of the word "reason" during the period and describes various understandings of the relationship between reason and revelation. He does not directly discuss science and religion—but he implicitly suggests that science largely was *irrelevant* to: (1) these understandings of reason; (2) contemporary controversies over principles of Biblical interpretation; and (3) the various deist and unitarian heterodoxies. This essay is good for the religious background of our field.

Topics: *X*, VII, IX Names: Barrow, Locke, Stillingfleet, Tillotson, Toland

1316 Reedy, Gerard. "Interpreting Tillotson." *Harv. Theo. Rev.* 86(1993): 81-103.

Reedy argues for a new model for interpreting Tillotson. This model is based on Tillotson's sermons and focuses on the central Christian mysteries. This model shows, among other things, that reason and revelation are mutually interdependent for Tillotson. Reedy suggests that reason for Tillotson is used in a traditional Christian way and not to be understood in a merely mathematical

way. The sermons present "Tillotson's strongest case against the reason of contemporary philosophers."
Topics: *X*, VII Name: Tillotson

1317 Reedy, Gerard. "Socinians, John Toland, and the Anglican Rationalists." *Harv. Theo. Rev.* 70(1977): 287-289.
This is a note *against* the view which sees a continuity between liberal Anglicans (Anglican rationalists) and deists (Toland) based on their common interest in nature. Rather there is discontinuity between Toland, Socinians, and liberal Anglicans based on biblical and theological differences.
Topics: *IX*, X Names: Nye, Stillingfleet, Toland

1318 Rees, Graham. "Bacon's Philosophy: Some New Sources with Special Reference to the 'Abecedariaum Novum Naturae'." In *Spiritus*, edited by M. Fattori and M. Bianchi, 223-241. Rome: Edizioni dell' Ateneo, 1984. See (0446).
An analysis of the terminology in a heretofore undiscovered manuscript of the "Abecedarium" substantiates Rees' claim that Bacon spent as much effort on his speculative system as on methodology. That system, Rees argues, is based on the distinction between spirit (a tenuous material substance and the source of activity in the universe) and tangible matter.
Topic: *VII* Name: Bacon

1319 Rees, Graham. "The Fate of Bacon's Cosmology in the Seventeenth Century." *Ambix* 24(1977): 27-38.
Rees finds that Bacon's speculative cosmology was ignored by almost everyone in the seventeenth century and asks "Why?" He suggests that the established Baconian tradition was set over against both enthusiasm and hermetic-scriptural cosmologies.
Topics: *VII*, II Names: Bacon, Ward, Webster

1320 Rees, Graham. "Francis Bacon's Biological Ideas: A New Manuscript Source." In *Occult and Scientific Mentalities in the Renaissance*, edited by Brian Vickers, 297-314. Cambridge: Cambridge University Press, 1984. See (1591).
This "preliminary report" on a recently discovered manuscript outlines Bacon's views on aging and the ways of slowing down the aging processes. Rees indicates that Bacon adopted an alchemical

aim—"prolongation of life as a protosalvation"—but proposed non-alchemical means. This portion of Rees' reconstruction of Bacon's speculative cosmology deals mainly with the roles of various "spirits."

Topics: *II*, VII Name: Bacon

1321 Rees, Graham. "Francis Bacon's Semi-Paracelsian Cosmology." *Ambix* 22(1975): 83-101.

Rees proposes "that while Bacon was working on his plans for the instauration of the sciences, he invested considerable energy" in the construction of a speculative system of the world. This included a revision of the Paracelsian chemico-physical cosmology. But Bacon rejected completely the Paracelsian exegesis of Genesis as a sanction for that cosmology.

Topics: *II*, VII, X Name: Bacon

1322 Reese, Ronald Lane, Steven M. Everett, and Edwin D. Craun. "'In the Beginning': The Ussher Chronology and Other Renaissance Ideas Dating the Creation." *Archaeoastronomy: The Bulletin of the Center of Archaeoastronomy* 5(1982): 20-23.

The authors propose that Ussher was not alone in his "pythagorean fixation with numbers, cosmic harmonies and precision"—as well as in his hermeneutic of the chronologies and of the creation story in Genesis.

Topic: *X* Name: Ussher

1323 Reif, Patricia. "The Textbook Tradition in Natural Philosophy, 1600-1650." *J. Hist. Ideas* 30(1969): 17-32.

Reif discusses scholastic authoritarianism and tradition-oriented conservatism as a broad and pervasive academic phenomenon. It affected the approach to natural philosophy in the schools and was not merely religious in origin.

Topics: *V*, IV

1324 Reimann, Hugo. *Henry Mores Bedeutung fur die Gegenwart.* Basel: R. Geering, 1941.

This is a published doctoral dissertation which focuses on More's theory of the soul and its immortality.

Topic: *VII* Name: More

1325 Renaldo, John J. "Bacon's Empiricism, Boyle's Science, and the Jesuit Response in Italy." *J. Hist. Ideas* 37(1976): 689-695.

Renaldo describes the Jesuit opposition to Bacon's motto, "Nulliam in verbo," and to Boyle's atomism. This article mainly deals with continental affairs but it does indicate some of the implications of this form of natural philosophy for certain Roman Catholics.

Topic: *VII* Names: Bacon, Boyle

1326 Reventlow, Henning Graf. *The Authority of the Bible and the Rise of the Modern World*. Translated by John Bowden. Philadelphia: Fortress Press, 1984.

In this thick and detailed work, Reventlow presents a history of biblical interpretation in its English setting from the late sixteenth into the eighteenth century. After pointing out "preparatory developments" (such as Humanism), Reventlow analyzes how different groups—Puritans, Latitudinarians, Deists—and particular individuals used the Bible and understood its authority. He sees this period as important in the development both of what would come to be called biblical criticism and of the emphasis on morality over revelation in Christianity. He fits the views of natural philosophers (including Newton and Locke) into the general, continuous picture which he portrays. This is an important book for understanding the biblical aspect of the general picture of religion and science in our period.

Topics: *X*, III, IX, XII Names: Blount, Chillingworth, Collins, A., Herbert, E., Hobbes, Locke, Newton, Stillingfleet, Swift, Tillotson, Tindal, Toland

1327 Reventlow, Henning Graf. *Bibelautoritat und Geist der Modern: Die Bedeutung des Bibelverstandnisses fur die Geistesgeschichtliche und Politische Entwicklung in England von der Reformation biz zur Aufklarung*. Gottingen: Vandenhoeck und Ruprecht, 1980.

This German original has been superseded by the English, second edition (1326).

Topic: *X*

1328 Rhys, Hedley Howell, ed. *Seventeenth Century Sciences and the Arts*. Princeton: Princeton University Press, 1961.

See the annotations for essays by Bush (0166) and Toulmin (1555).
Topic: *IV*

1329 Righini Bonelli, M. L., and William R. Shea, eds. *Reason, Experiment, and Mysticism in the Scientific Revolution*. New York: Science History, 1975.

Annotated separately are Shea's introduction (1444) and essays by Casini (0194), Debus (0335), A. R. Hall (0601), M. B. Hall (0616), Rossi (1357), and Westfall (1649).

Topics: *I*, II

1330 Rivers, Isabel. *Reason, Grace, and Sentiment: Vol. 1, Whichcote to Wesley*. Cambridge: Cambridge University Press, 1991.

Rivers explores the shift in the languages of grace and reason in the thought of late seventeenth-century Anglican moral theologians and reactions to it by dissenters. Although it does not deal directly with the latitudinarian interest in science, the chapter on "The Religion of the Reason: The Latitude Men" provides an excellent analysis of their intellectual and social milieu, their publications, and their ideas about religion and human nature.

Topics: *X*, VII Names: Barrow, Cudworth, Stillingfleet, Tillotson, Whichcote, Wilkins

1331 Robbins, Caroline. "Faith and Freedom (c.1677-1729)." *J. Hist. Ideas* 36(1975): 47-62.

Robbins sees the attempts of intellectuals in this period to interpret truth "according to the canons of contemporary scientific and philosophical thought" as a positive event. Further, Robbins suggests that this may be seen as the beginning of modern liberalism in theology.

Topic: *VII*

1332 Roberts, Gareth. *The Mirror of Alchemy: Alchemical Ideas and Images in Manuscripts and Books from Antiquity to the Seventeenth Century*. Toronto: University of Toronto Press, 1994.

This is an excellent introduction to alchemy using manuscripts in the British Library. It is nicely produced and has especially useful explanations and illustrations of alchemical symbolism. While

primarily concerned with pre-seventeenth-century manuscripts, Roberts does indicate the persistence of the images and metaphors in the seventeenth century (despite the apparent fragmentation of the religious and scientific unity of the alchemical tradition.)
Topic: *II*

1333 Roberts, J. Deotis. *From Puritanism to Platonism in Seventeenth-Century England*. The Hague: Martinus Nijhoff, 1968.
This book deals mainly with Benjamin Whichcote—whom the author claims is the father/founder of Cambridge Platonism. Roberts discusses various philosophical issues but touches natural philosophy only superficially.
Topic: *VII* Name: Whichcote

1334 Rodney, Joel M. "A Godly Atomist in 17th-Century England: Ralph Cudworth." *Historian* 32(1970): 243-249.
Rodney discusses how Cudworth tried to reconcile his theology with the atomist mechanism of Gassendi and Charleton.
Topics: *VII*, VIII, IX Names: Charleton, Cudworth

1335 Rodney, Joel M. "Newton Revisited: Foes, Friends, and Thoughts." *Research Studies* [Washington State University] 36(1968): 351-360.
In what is misleadingly labeled a review article, Rodney discusses selected topics in Newtonian philosophy and theology. Foremost among these is the Clarke-Collins debate concerning the nature of the soul. Rodney aims to vindicate Clarke.
Topics: *XII*, VII, IX Names: Clarke, Collins, A., Newton

1336 Rodney, Joel M. "Ralph Cudworth: The Legitimation of the Atomic Theory of Matter." *Research Studies* [Washington State University] 32(1964): 21-27.
This is a brief summary of selected parts of Book I of Cudworth's *The True Intellectual System of the Universe*. To it, Rodney attaches questionable theses concerning Cudworth's influence on Newton.
Topics: *VII*, XII Names: Cudworth, Newton

1337 Roger, Jacques. "The Cartesian Model and its Role in Eighteenth-Century 'Theories of the Earth'." In *Problems of Car-*

tesianism, edited by Thomas M. Lennon and others, 95-112. Kingston, Ontario: McGill-Queens University Press, 1982. See (0943).

Roger traces cosmogony from Descartes to Hutton. He includes brief discussions of Burnet, Newton, and Whiston. An interesting thesis regarding Newtonian epistemology is that "observation and induction allow us to establish the *laws* of nature but not to know the *genesis* of the *structure* of the world." Moreover, argues Roger, it is Newton's theological views which account for this aspect of his epistemological and cosmological positions.

Topics: *VII*, XII Names: Burnet, T., Newton, Whiston

1338 Roger, Jacques. "La Theorie de la Terre au XVIIe Siecle." *Rev. d'Hist. Sci.* 26(1973): 23-48.

Roger proposes that, from the seventeenth century to the beginning of the nineteenth century, theories of the earth constituted the "intellectual frame" of the earth sciences. This theoretical understanding, he finds, originated in Descartes' discussion. Much of Roger's article then describes Burnet's conjunction of Cartesianism and scripture.

Topics: *X*, VII Name: Burnet, T.

1339 Rogers, G. A. J. "The Basis of Belief: Philosophy, Science and Religion in Seventeenth-Century England." *Hist. Euro. Ideas* 6(1985): 19-40.

Rogers presents the *vita activa* as the dominant theme of seventeenth-century English philosophy, theology, and natural science. This approach, he indicates, recognized the problematic and tentative nature of all sorts of knowledge—while also allowing preferences (e.g., "justified belief"). It replaced the *vita contemplativa* as men of affairs replaced those of monasteries and universities.

Topic: *VII* Names: Bacon, Locke, More

1340 Rogers, G. A. J. "Boyle, Locke, and Reason." *J. Hist. Ideas* 27(1966): 205-216. Reprinted in *Philosophy, Religion and Science in the Seventeenth and Eighteenth Centuries*, edited by John W. Yolton, 339-350. Rochester, N.Y.: University of Rochester Press, 1990. See (1720).

This essay concerns Boyle's influence on Locke's theory of knowledge.

Topic: *VII* Names: Boyle, Locke

1341 Rogers, G. A. J. "Descartes and the English." In *The Light of Nature*, edited by J. D. North and J. J. Roche, 281-302. Dordrecht: M. Nijhoff, 1985. See (1145).

Rogers proposes that Descartes' impact should be measured "not by converts but by its challenge." In other words, Descartes made the English rethink their scientific method and their doctrine of God.

Topic: *VII* Name: Hobbes

1342 Rogers, G. A. J. "Locke, Newton, and the Cambridge Platonists on Innate Ideas." *J. Hist. Ideas* 40(1979): 191-206. Reprinted in *Philosophy, Religion and Science in the Seventeenth and Eighteenth Centuries*, edited by John W. Yolton, 351-365. Rochester, N.Y.: University of Rochester Press, 1990. See (1720).

Here, Rogers compares Locke and Newton as empiricists and in their rejection of innate ideas. He argues that they probably arrived at their views independently—but if there was any "influence" it was that Locke influenced Newton.

Topics: *VII*, XII Names: Locke, More, Newton, Parker

1343 Rogers, G. A. J. "Locke's *Essay* and Newton's *Principia*." *J. Hist. Ideas* 39(1978): 217-232. Reprinted in *Philosophy, Religion and Science in the Seventeenth and Eighteenth Centuries*, edited by John W. Yolton, 366-381. Rochester, N.Y.: University of Rochester Press, 1990. See (1720).

Rogers argues that Newton's *Principia* had no major influence on Locke's *Essay*. Some tangential remarks are made about religion and Locke's view of God.

Topics: *VII*, XII Names: Locke, Newton

1344 Rogers, G. A. J. "The System of Locke and Newton." In *Contemporary Newtonian Research*, edited by Zev Bechler, 215-238. Dordrecht: D. Reidel, 1982. See (0071). Reprinted in *Seventeenth-Century Natural Scientists*, edited by Vere Chappell, 315-338. New York: Garland, 1992. See (0205).

Rogers argues that Locke and Newton shared a set of epistemological, metaphysical, and natural philosophic views which was extremely influential in the eighteenth century. He summarizes the

key elements in the set of views and clearly notes the place of God within this philosophic system.
Topics: *XII*, VII Names: Locke, Newton

1345 Rogers, Jack B., and Donald K. McKin. *The Authority and Interpretation of the Bible: An Historical Approach.* San Francisco: Harper and Row, 1979.
Chapter 4 contains an introductory summary of understandings of the Bible in relation to reason in seventeenth-century Great Britain. Included are discussions of the Westminster Confession, John Owen and the Independents, and advocates of the new science. The general tendency of the last-named group is said to be away from the authority of faith and the Bible and toward that of reason.
Topics: *X*, VII, XII Names: Locke, Newton, Owen, Wilkins

1346 Rogers, Pat, ed. *The Eighteenth Century.* (The Context of English Literature Series.) New York: Holmes and Meier, 1978.
This includes chapters on "Religion and Ideas" by Price (1262) and "Science" by Rousseau (1366). The latter is worth reading.
Topic: *X*

1347 Romanell, Patrick. "Some Medico-Philosophical Excerpts from the Mellon Collection of Locke's Papers." *J. Hist. Ideas* 25(1964): 107-116.
In the medical manuscripts in the Mellon Collection, Romanell has found attacks on the "Rosycrucians" and on a German Paracelsian alchemist. He also finds that Locke was greatly influenced by Thomas Sydenham.
Topics: *II*, XI Names: Locke, Sydenham

1348 Ronan, Colin A. *Edmund Halley: Genius in Eclipse.* Garden City, N.Y.: Doubleday, 1969.
This biography is addressed to (and reasonably well-written for) a general audience. Ronan generally is retrospective and internalist in his historiography of science. He is thorough and enthusiastic with respect to Halley's scientific work. But his discussion of Halley's religious views are few and weak; and his discussions of the theological implications of Halley's cosmological and geological theories are dutiful and unenthusiastic. Ronan's treatment of

Newton, Bentley, Flamsteed, and Hooke are not very good (particularly with respect to religion).
Topics: *VII*, XII Names: Bentley, Flamsteed, Halley, Newton

1349 Rose, Harold Wickliffe. *The Colonial Houses of Worship in America*. New York: Hastings House, 1963.
This book includes analyses by denomination and by material. It has maps and photographs of churches still standing (whether still in use or not).
Topic: *VI*

1350 Rosen, George. "Left-wing Puritanism and Science." *Bulletin of the Institute of the History of Medicine* 15(1944): 375-380.
Rosen argues in favor of the connection between left-wing Puritanism and the new science. He claims that it was the radical sectarians who did the most to increase interest in science.
Topics: *III*, IV Name: Winstanley

1351 Rosenfeld, Leon. "Newton and the Law of Gravitation." *Arch. Hist. Exact Sci.* 2(1965): 365-386.
See following annotation (1352).
Topics: *XII*, VII Name: Newton

1352 Rosenfeld, Leon. "Newton's Views on Aether and Gravitation." *Arch. Hist. Exact Sci.* 2(1965): 29-37.
In his two articles, Rosenfeld discusses the philosophical and theological context of Newton's ideas on mechanical explanations of the cause of gravity. He shows the relevance of Newton's conceptions of God's omnipotence and God's will when considering Newton's views of gravitation.
Topics: *XII*, VII Name: Newton

1353 Ross, G. MacDonald. "Occultism and Philosophy in the Seventeenth Century." In *Philosophy, Its History and Historiography*, edited by A. J. Holland, 95-115 and 145-147. Dordrecht: D. Reidel, 1985.
In this study of seventeenth-century philosophers, Ross shows that. "there was no clear line of demarcation between occultism, philosophy, religion, and science." That is, there was no philosophical orthodoxy; and hence, historiographically, occultism as a

defiant form of thought should be of little use to historians of seventeenth-century thought. There follows a long response by Simon Schaffer; compare (1405).
Topics: *VII*, II Name: Locke

1354 Rossi, Mario M. *Alle Fonti del Deismo e del Materialismo Moderno*. Florence: Nuovo Italia, 1942.
This is a two-hundred-page monograph: the first half deals with Herbert of Cherbury and the origin of deism; the second half deals with deism and the evolution of the thought of Hobbes.
Topics: *IX*, VII Names: Herbert, E., Hobbes

1355 Rossi, Paolo. *Aspetti della Rivoluzione Scientifica*. Naples: Morano, 1971.
This is a collection which includes several essays dealing with Bacon and the Bible, Burnet's sacred theory of the earth, and Wilkins and the plurality of worlds theme. This last-mentioned essay has been translated into English (see 1359).
Topics: *VII*, X Names: Bacon, Burnet, T., Wilkins

1356 Rossi, Paolo. *Francis Bacon: from Magic to Science*. Translated by S. Rabinovitch. London: Routledge and K. Paul; Chicago: University of Chicago Press, 1968.
Rossi argues that Bacon's main objective was a dream of cooperative science and scientific institutions. He does this in six chapters: on the mechanical arts, magic, and science; on the refutation of philosophers; on the classic fables and myths; on logic; on language; and on the rhetorical tradition and the method of science. Rossi clearly sees Bacon as reacting to the Renaissance magical and hermetic tradition. On the one hand, its experimental nature was quite compatible with his vision of method. But, on the other hand, its elitist, individualistic, and secretive characteristics were not compatible with his vision of a cooperative science. Furthermore, the purpose of this science was "the improvement of the conditions of human existence." Bacon, Rossi indicates, believed that a new era in human history was at hand with the break with classical and medieval philosophies. These philosophies, among other features, mixed religion and science. Rossi argues that an important part of Bacon's reform was a distinction between "matters of science from matters of faith." Yet he also shows that

Bacon interpreted classical myths (e.g. Prometheus) allegorically to show his distinction was part of the most ancient wisdom. Originally written in 1957, Rossi's book has become a standard interpretation of Bacon.

Topics: *VII*, II, IV, X Name: Bacon

1357 Rossi, Paolo. "Hermeticism, Rationality and the Scientific Revolution." In *Reason, Experiment, and Mysticism in the Scientific Revolution*, edited by M. L. Righini Bonelli and William R. Shea, 247-273 and 317-318. New York: Science History, 1975. See (1329). Reprinted in Vol. 11 of *Articles on Witchcraft, Magic, and Demonology*, edited by Brian P. Levack, 133-159. New York: Garland, 1992. See (0945).

In discussing the Scientific Revolution in Part I, Rossi emphasizes "the discontinuity that separates the new science from the old" and lists fourteen essential and decisive factors of modern thought. About half of these have a theological dimension. In Parts II and III, Rossi discusses the study of the hermetic tradition as useful in some ways but then criticizes the new historiography most influenced by it as "interested only in the elements of continuity and the influence of traditional ideas." He sees Bacon as standing for a total separation of science and religion and implies that we too should adopt this stand. Parts IV and V constitute a miniature sermon on the dangers for our own time of taking magic and the occult too seriously.

Topics: *I*, II, VII Name: Bacon

1358 Rossi, Paolo. "Magic, Science, and Equality of Men." In *Human Implications of Scientific Advance*, edited by Eric G. Forbes, 64-69. Edinburgh: Edinburgh University Press, 1978. See (0483).

Rossi argues for a difference between a magician and a natural philosopher: the magician has reached a level of secret knowledge which sets him apart from the rest of humankind. Equality, or at least the possibility of equal knowledge, is the mark and goal of natural philosophy. Rossi briefly relates this distinction both to political ideology and to the desire to universalize control of all the human race over nature.

Topics: *II*, III, IV Name: Hobbes

1359 Rossi, Paolo. "Nobility of Man and Plurality of Worlds." In *Science, Medicine and Society in the Renaissance*, edited by Allen G. Debus, Vol. 2, 131-162. New York: Science History, 1972. See (0357). The Italian-language version appears *Aspetti della Rivoluzione Scientifica*. Naples: Morano, 1971. See (1355).

Rossi discusses the debate over the plurality of worlds and an infinite universe in sixteenth- and seventeenth-century Europe. Both exultation and bewilderment resulted from the new visions of the universe. Rossi notes Kepler and Galileo as among those who denied the plurality of worlds while Wilkins and Thomas Burnet were among the proponents. At issue were the centrality (and thereby the nobility) of humans in nature.

Topics: *VII*, I, X Names: Burnet, T., Donne, Wilkins

1360 Rossi, Paolo. *Philosophy, Technology, and the Arts in the Early Modern Era*. Translated by Salvator Attanasio. New York: Harper and Row, 1970. Originally published as *I Filosofi e le Macchine*. Milan: Feltrinelli, 1962.

This is an extreme externalist interpretation of the history of science. Important for our purposes are Rossi's discussions of "the idea of knowledge as construction," the image of God as clock-maker of "the book of nature," and the new interactions of science, technology, and religion. See, in particular, the chapters on the idea of scientific progress, the seventeenth century, Bacon, and the symbol of Prometheus.

Topics: *VI*, IV, VII, X Name: Bacon

1361 Rossi, Paolo, ed. *La Rivoluzione Scientifica da Copernico a Newton*. Turin: Loescher, 1976.

This is an anthology of sources, translated into Italian, with introductory essays by Rossi. In addition to the usual topics (revolutions in astronomy and physics), it also treats: the refutation of magic; scripture and geology; the infinity of God and of the world; and the plurality of worlds.

Topics: *VII*, X, XII Name: Newton

1362 Rossi, Sergio, ed. *Science and Imagination in XVIIIth-Century British Culture*. Milan: Edizioni Unicopli, 1987.

The essays by Brooke (0135), Giorello (0545), Hoskin (0733), and A. J. Smith (1459) are annotated separately in this bibliography.
Topic: *X*

1363 Rostvig, Maren-Sofie. *The Happy Man: Studies in the Metamorphoses of a Classical Ideal, 1600-1700*. Oslo: Akademisk, 1954.

Rostvig traces the classical motif of the "Happy Man" in seventeenth-century English poetry. Although not focusing directly on science and religion, the author notes various attitudes towards nature and spiritual ideals. The study is useful for studying the ancient-versus-moderns controversy—giving special attention to the "ancients" in the debate.

Topics: *X*, IV Names: Chudleigh, Cowley, Dryden, Finch, A., Habington, Marvel, Milton, Norris, Philips, Vaughan, H.

1364 Rousseau, George S. "Mysticism and Millenarianism: 'Immortal Dr. Cheyne'." In *Millenarianism and Messianism in English Literature and Thought, 1650-1800*, edited by Richard H. Popkin, 81-126. Leiden: Brill, 1988. See (1238). A slightly different version appears in *Hermeticism and the Renaissance*, edited by Ingrid Merkel and Allen G. Debus, 197-230. Washington, D.C.: Folger Shakespeare Library, 1988. London: Associated Universities Presses, 1988. See (1051).

This is a detailed intellectual biography of George Cheyne. It describes the changes in Cheyne's Newtonian, mystical, and millenarian ideas—and their relation to his activities as a physician. Rousseau is attempting to correct the views of those historians who have commented on only a portion of Cheyne's ideas and life. (He wishes to restore Cheyne to his early eighteenth-century stature as a fascinating and widely-known figure.)

Topics: *X*, VII, XI, XII Name: Cheyne

1365 Rousseau, George S. "Poiesis and Urania: the Relation of Poetry and Astronomy in the English Enlightenment." In *Science et Philosophie, XVIIe et XVIIIe Siecles*, 113-116. Paris: Albert Blanchard, 1971. See (1418).

In this brief talk, Rousseau warns against "the grave error of assuming that science *directly* affected literary artists." He uses

Pope's attendance at Whiston's coffee-house lectures (rather than reading Newton's works) as his example. This functionally is a brief abstract from *This Long Disease, My Life* (1143).
Topic: *X* Names: Pope, Whiston

1366 Rousseau, George S. "Science." In *The Eighteenth Century*, edited by Pat Rogers, 153-207. New York: Holmes and Meier, 1978. See (1346).
This survey (along with the whole book) is aimed at providing the context of English literature during the period. Rousseau argues that the impact of science was the most significant cultural development of the eighteenth century. He sees science, religion, and literature as all interrelated. About one-third of the essay is relevant to our period.
Topics: *X*, V, VII, XII Names: Locke, Newton, Pope, Swift

1367 Rousseau, George S. "'Wicked Whiston' and the Scriblerians: Another Ancients-Modern Controversy." In *Studies in Eighteenth Century Culture, Vol. 17*, edited by John W. Yolton and Leslie Ellen Brown, 17-44. East Lansing, Mich.: Colleagues Press, 1971.
Rousseau describes William Whiston's millenarian interpretation of comets as well as the satirical response to him and to his views put forward by the Scriblerians (especially Gay, Pope, and Swift). Whiston is presented as superstitious and fanatical—and as representing "the Ancients"; the Scriblerians are presented as "more discriminating, less superstitious"—and as representing "the Moderns."
Topic: *X* Names: Gay, Pope, Swift, Whiston

1368 Rousseau, George S., and Roy Porter, eds. *The Ferment of Knowledge: Studies in the Historiography of Eighteenth-Century Science*. Cambridge: Cambridge University Press, 1980.
The essays by Harre (0629), Schaffer (1404), and Shapin (1431) are annotated separately.
Topic: *I*

1369 Rudrum, Alan. "Thomas Vaughan's *Lumen de Lumine*: An Interpretation of Thalia." In *Literature and the Occult: Essays in*

Comparative Literature, edited by Luanne Frank, 234-243. Arlington, Texas: University of Texas at Arlington, 1977.

Rudrum proposes that Thalia represents nature transfigured, the goddess "Natura" Christianized, and nature as alchemists hoped to transform it. Vaughan hid this meaning in allegorical form in his writings just as he believed the spirit that infuses the world is hidden from the eyes of most people because of their spiritual incapacity.

Topic: *II*, X Name: Vaughan, T.

1370 Rudrum, Alan. *The Works of Thomas Vaughan*. Edited by Alan Rudrum. Oxford: Clarendon Press, 1984.

Vaughan (1621-1665) was a speculative alchemist, mystic, and sometime Anglican clergyman who is also known for his pamphlet war with Henry More. He wrote both under his own name and that of Eugenius Philalethes. Besides the 550 pages of Vaughan's writings, Rudrum provides: (1) a thirty-page biographical introduction; (2) a fifteen-page textual introduction; (3) a 148-page item-by-item commentary; and (4) a seven-page index to the scholarly apparatus. The whole effect is much like a modern (scholarly) biblical commentary.

Topics: *II*, X Names: More, Vaughan, H., Vaughan, T.

1371 Rudwick, Martin J. S. *The Meaning of Fossils: Episodes in the History of Paleontology*. New York: Science History, 1976. Reprint, Chicago: University of Chicago Press, 1985.

The "episodes" in the subtitle indicates that one of Rudwick's primary aims is to convey "a sense of period"—that is, a sense of the philosophical, theological, and social "coherence of the theories and activity of studying fossils." Thus, he tries to understand the theories, observations, and problems of men "in terms of their own view of the world" rather than that of the twentieth century. Chapter 2, dealing with the seventeenth century, thus involves biblical chronology and the story of the Flood. Rudwick argues that the Biblical elements played positive roles as well as posing problems for natural philosophy. For example, the theory of a diluvial origin of strata "focused attention on questions of stratification, and hence on the problems of reconstructing the history of the earth." This is an important, groundbreaking book.

Topics: *X*, VII Names: Burnet, T., Hooke, Lister, Ray, Woodward

1372 Rudwick, Martin J. S. "Senses of the Natural World and Senses of God: Another Look at the Historical Relation of Science and Religion." In *Sciences and Theology in the Twentieth Century*, edited by A. Peacocke, 241-261. Notre Dame, Ind.: University of Notre Dame Press, 1981.

Rudwick advocates enlarging the "strong program" in historical sociology of science—and treating science and religion on an equal basis. This means the elimination of any scientific triumphalism, taking seriously both input from "the externality of the natural world" and inputs from "externality characterized in theistic terms." Such a program would "do justice to the reality of personal purposes as well as collective social interests." While Rudwick's brief examples do not come from seventeenth-century England, his methodological suggestions are relevant and partially practiced by a contemporary generation of young historians.

Topics: *I*, IV

1373 Rukeyser, Muriel. *The Traces of Thomas Hariot*. New York: Random House, 1971.

This is a literary account of Harriot's life and times as interpreted through traces left to us in the twentieth century. Beginning from Ralegh's poetry, Rukeyser became fascinated with this "lost man who was great." Many direct quotes from primary documents, as well as from secondary sources, are interwoven into this poetic biography. Rukeyser's ability to bridge the humanities and science reflects the same ability in Harriot.

Topics: *X*, II, V, IX Names: Bacon, Donne, Harriot, Percy, Ralegh, Warner (1558)

1374 Rumsey, Peter Lockwood. *Acts of God and the People, 1620-1720*. Ann Arbor, Mich.: UMI Research Press, 1986.

Rumsey promises to bridge the gap between social and intellectual history; he does not succeed in building this bridge. However, the chapters on witchcraft, comets, and earthquakes do contain interesting observations. And his major theme—changes in the theological understanding of the relation of God's Providence to the natural world—is an important one.

Topics: *VII*, II, IV, V Names: Doolittle, Mather, C., Mather, I.

1375 Rupp, E. Gordon. *Religion in England, 1688-1791*. Oxford: Clarendon Press, 1986.

This is indirectly useful as a background survey. Rupp's historiographic approach could be described as "internalist history of Christianity" with major biographical and editorial propensities.
Topics: *V*, VII

1376 Russell, Colin A. *Cross-Currents: Interaction Between Science and Faith*. Grand Rapids, Mich.: W. B. Eerdmans, 1985.
This book is a survey for laypeople written from a Christian perspective. Russell avoids "conflict" approaches by taking for granted "the continuity between science and religion." (He uses a metaphor of two currents in a single river.) Our period is discussed in Chapters 4-6: "Converging Streams: Science and Biblical Ideology;" "Deepening Waters: The Scientific Revolution;" and "Harnessing the River: Making Use of Science." He essentially follows Hooykaas' lead in approach and in theses. Thus, science is seen as arising in the new climate resulting from the Reformation emphasis on biblical Christianity—and particularly as found in Calvinism. Russell's discussion of the uses of science includes questions concerning the purposes of natural theology and questions about nonconformists and technology. Against the Jacobs, he sees natural theologians as intending to defend biblical Christianity, not the social order. Yet, because of his "high view" of biblical Christianity, Russell also stresses critiques of natural theology.
Topics: *X*, III, IV, VI, VIII, XII Names: Bacon, Boyle, Newton, Ray

1377 Russell, Colin A., ed. *Science and Religious Belief: A Selection of Recent Historical Studies*. London: University of London Press, 1973.
This is an anthology of once-important articles and excerpts from books. The anthology was designed for a university course entitled "Science and Belief, from Copernicus to Darwin." The volume emphasizes seventeenth- and eighteenth-century England. The articles by A. R. Hall (0602), Kemsley (0873), Kubrin (0919), Merton (1056), and Sailor (1389) are annotated separately. The book also includes an excerpt from Burtt (0160).
Topic: *VII*

1378 Russell, G. A., ed. *The 'Arabick' Interest of the Natural Philosophers in Seventeenth-Century England*. Leiden: E. J. Brill, 1994.

This is an anthology of fifteen essays representing an interdisciplinary symposium organized in 1986. Very little in the book connects science and religion in any direct way. But the following indirect conclusions follow from various parts of the book: (1) the increase in attention paid to Arabic was caused by both religious and scientific/medical motivations; (2) the establishment of Arabic professorships helped lead to the institutional separation of science and religion (a structural secularization); (3) sacred chronology and alchemy were areas in which science, religion, and "Arabick" intersected. Russell's Introduction (1380) and essays by Newman (1110) and Russell (1379) are annotated separately in the present bibliography. Essays by Feingold, M. B. Hall, Holt, Mercier, Salmon, and Wakefield are not annotated here but do add to the cumulative effect in support of the above generalizations. For those interested in Islam and science in seventeenth-century England, this is one of only a few relevant secondary sources.

Topics: *X*, V, VII, XI Names: Greaves, Locke, Marsh, Pococke, Ussher

1379 Russell, G. A. "The Impact of the *Philosophus autodidactus*: Pococke, John Locke and the Society of Friends." In *The 'Arabick' Interest of the Natural Philosophers in Seventeenth-Century England*, edited by G. A. Russell, 224-265. Leiden: Brill, 1994. See (1378).

Russell argues that John Locke read and was influenced by a translation of Ibn Tufayl's *Hayy ibn Yaqzan*. This twelfth-century Arabic narrative describes how a child growing up alone on an island discovers (completely on his own) basic principles concerning natural science, God, and morality. The "Quaker Connection" (or the related activities of several Quakers) is part of Russell's proof and explanation of the larger thesis.

Topics: *VII*, IX, X Names: Locke, Pococke

1380 Russell, G. A. Introduction to *The 'Arabick' Interest of the Natural Philosophers in Seventeenth-Century England*, edited by G. A. Russell, 1-19. Leiden: Brill, 1994. See (1378).

Russell discusses in a general way the topic indicated by the title of the book and gives brief summaries of the essays included in the book. Science and religion are not connected in any direct way.

But Russell does note various parallels between scientific and religious factors in connection with the Arabic interest.
Topic: *X*

1381 Russell, Robert John, and others, eds. *Physics, Philosophy, and Theology: A Common Quest for Understanding*. Vatican City: Vatican Observatory, 1988.
The papers by Buckley (0150) and McMullin (1043) are annotated separately in the present bibliography.
Topic: *VII*

1382 Russo, Francois. "Catholicism, Protestantism, and the Development of Science in the Sixteenth and Seventeenth Centuries." Translated by D. Woodward. In *The Evolution of Science*, edited by Guy S. Metraux and Francois Crouzet, 291-320. New York: New American Library, 1963. See (1062).
This is a translation of (1384).
Topic: *III*

1383 Russo, Francois. "Les Etudes Newtoniennes d'Alexandre Koyre." *Arch. Philos.* 37(1974): 107-132.
This is a systematic analysis of Koyre's main work on Newton. Russo briefly treats the question of the relation of God and the world.
Topics: *XII*, I Name: Newton

1384 Russo, Francois. "Role Respectif du Catholicisme et du Protestantisme dans le Development des Sciences aus XVIe et XVIIe Siecles." *J. World Hist.* 3(1957): 854-880.
Russo aims to correct Hooykaas and Pelseneer by arguing that Catholicism was as receptive to science as Protestantism. He argues that both forms of Christianity were receptive but that science in the seventeenth century also shows an increasing independence from religion. The focus is on continental Catholics but the essay still is relevant to the debate regarding England.
Topic: *III*

1385 Ryan, Michael T. "The Diffusion of Science and the Conversion of the Gentiles in the Seventeenth Century." In *In the Presence of the Past*, edited by Richard T. Bienvenu and

Mordechai Feingold, 9-40. Dordrecht: Kluwer Academic, 1991. See (0094).

This is a wide-ranging and sophisticated survey of the relationship between natural philosophy and ideas about evangelism. While Ryan deals with the visions of a few Englishmen, continental Europeans predominate.

Topic: *IV*

1386 Ryle, Gilbert. "John Locke on the Human Understanding." In *John Locke: Tercentenary Addresses*, 15-28. London: Oxford University Press, 1933. Reprinted in *Locke and Berkeley: A Collection of Critical Essays*, edited by C. B. Martin and D. M. Armstrong, 14-39. Notre Dame, Ind.: University of Notre Dame Press, 1968.

This is an ahistorical appreciation of Locke and the *Essay*. Ryle recognizes the importance of both religion and natural philosophy for Locke. He asserts that Locke's greatness is that he distinguished between them: "he gave us . . . a theory of the sciences."

Topic: *VII* Name: Locke

1387 Sadler, Lynn Veach. "Relations Between Alchemy and Poetics in the Renaissance and Seventeenth Century, With Special Glances at Donne and Milton." *Ambix* 24(1977): 69-76.

Sadler argues that alchemy insinuated itself into poetic theory and technique. Alchemy and poetry, he finds, meet in their emphases on morality, medicine, secrecy, and religious fervor.

Topics: *X*, II, IV Names: Donne, Jonson, Milton

1388 Sailor, Danton B. "Cudworth and Descartes." *J. Hist. Ideas* 23(1962): 133-140.

Sailor discusses Cudworth's understanding of the theological implications of mechanism and his resulting opposition to Cartesianism.

Topic: *VII* Name: Cudworth

1389 Sailor, Danton B. "Moses and Atomism." *J. Hist. Ideas* 25(1964): 3-16. Reprinted in *Science and Religious Belief*, edited by Colin A. Russell, 5-19. London: University of London Press, 1973. See (1377).

Sailor here sets the tradition connecting Moses with atomism within the context of Renaissance humanism—but not very convincingly.

Topics: *VII*, II, IX

1390 Sailor, Danton B. "Newton's Debt to Cudworth." *J. Hist. Ideas* 49(1988): 511-518.

In this note, Sailor describes Newton's manuscript "out of Cudworth," now in the Clarke Library. In it, Newton excerpts portions of Cudworth's *True Intellectual System of the Universe*. Sailor notes the relation between these excerpts and Newton's other writings—especially as regards *prisca theologia*, ancient history, and atomism.

Topics: *XII*, II Names: Cudworth, Newton

1391 Sarasohn, Lisa T. "Motion and Morality: Pierre Gassendi, Thomas Hobbes and the Mechanical World-View." *J. Hist. Ideas* 46(1985): 363-379.

Sarasohn describes and compares the philosophies of Gassendi and Hobbes. In correlating a mechanistic physics with cosmology, theology, and human psychology, Gassendi argued for human free will while Hobbes was a materialist and determinist even in his view of human behavior.

Topic: *VII* Name: Hobbes

1392 Sarasohn, Lisa T. "A Science Turned Upside Down: Feminism and the Natural Philosophy of Margaret Cavendish." *Hunt. Lib. Q.* 47(1984): 289-307.

Sarasohn proposes that Cavendish fused an organic, living universe with an underlying (though ambivalent) feminism. She indicates that Cavendish's shift from atomism to an organic materialism might have been related to her Royalist political views. Sarasohn also discusses the following topics about Cavendish: the relations between her skepticism and her fideism; her views on the distant God and living feminine "nature"; her views on class and gender structures in the great chain of being. Above all, for Sarasohn, Cavendish may be seen as an isolated conservative person attempting to assault traditional authority.

Topic: *VII* Name: Cavendish

1393 Sarton, George. *A Guide to the History of Science*. Waltham, Mass.: Chronica Botanica, 1952.

The essay "Science and Tradition" that introduces this volume is a personal historiographical proclamation of the need for scientific specialists and technocrats to become aware of their history, of the need to reconcile science and the humanities, and of the need to consecrate a deeper interpretation of science to the "Good Life." The bibliography and lists of relevant organizations now are outdated.

Topic: *I*

1394 Saurat, Denis. *Milton et le materialisme Chretien en Angleterre*. Paris: Reider, 1928.

This is a rearranged version (minus the biographical sections) in French of *Milton, Man and Thinker*. See (1396).

Topic: *X* Name: Milton

1395 Saurat, Denis. *Milton, Man and Thinker*. New York: Dial Press, 1925. Revised ed., London: J. M. Dent and Sons, 1944.

This is a rather bombastic explication and appreciation of Milton's religious philosophy by a professor of French Literature. Milton is presented as a unitarian pantheist and a kabbalist (by one who seems to agree with these positions for the present day). Saurat asserts that Milton was interested in science but presents almost no scientific views of Milton. There is some explication of the metaphysical aspects of Milton's philosophy of nature. The book also includes a section on Robert Fludd and compares Milton to Fludd.

Topics: *X*, II, VII Names: Fludd, Milton

1396 Saurat, Denis. *La Pensee de Milton*. Paris: F. Alcan, 1920.

This is identical to the first half of *Milton, Man and Thinker*. See (1395).

Topic: *X* Name: Milton

1397 Saveson, J. E. "Descartes' Influence on John Smith, Cambridge Platonist." *J. Hist. Ideas* 20(1959): 258-263.

Saveson finds in Smith's sermons parallels to Descartes' philosophy. He notes especially: Smith's physiology of human and animal

sensation; and, in humans, the separation but interaction between body and soul.
Topic: *VII* Name: Smith

1398 Saveson, J. E. "Differing Reactions to Descartes Among the Cambridge Platonists." *J. Hist. Ideas* 21(1960): 560-567.
Saveson here points out theological and philosophical differences amongst the Cambridge Platonists (particularly over reactions to Descartes' cosmology).
Topic: *VII* Names: Cudworth, More, Smith

1399 Sawyer, Ronald C. "'Strangely Handled in All Her Lyms': Witchcraft and Healing in Jacobean England." *Journal of Social History* 22(1989): 461-485.
Sawyer uses the notebooks of the Rev. Richard Napier to examine the relation between illnesses and witchcraft accusations in a specific time and place. He finds that illnesses attributed to witchcraft were usually "psychological" and were those brought to Napier after "naturalistic explanations seemed unsuitable." That is, witchcraft accusations were part of the people's existential and social world and should be seen as part of their activist approach to seeking relief from the healer.
Topics: *XI*, II, IV Name: Napier

1400 Scarre, Geoffrey. "Tillotson and Hume on Miracles." *The Downside Review* 110(1992): 45-65.
Scarre analyzes Tillotson's empiricist attack on the Catholic doctrine of transubstantiation: since sense perception is the fundamental avenue to truth, transubstantiation is absurd. Likewise, the Catholic appeal to miracles to validate the Eucharist is contradictory because "the force of miracles presupposes the reliability of the senses." Scarre carries then this argument further—discussing Hume's and his own views.
Topic: *VII* Name: Tillotson

1401 Schaffer, Simon. "Comets & Idols: Newton's Cosmology and Political Theology." In *Action and Reaction*, edited by S. G. Brush and others, 206-231. Newark, Delaware: University of Delaware Press, 1993.

This is a study of two features of the hermeneutics of idolatry: (1) interpretations of Newton as an idol in the eighteenth century and its "fracturing" in the twentieth; and (2) Newton's own interpretation of idolatry, cometography, and their relationship. Schaffer argues that Newton created his own authoritative interpretations of comets between 1681 and 1684; and that he endowed them with divine significance as the principle transmitters of life and restorers of vitality. Likewise, during the same period, he developed his theory of the false hermeneutics of past corrupt political interests that resulted in errors both in natural philosophy and in scriptural interpretation.
Topics: *XII*, I, IV Name: Newton

1402 Schaffer, Simon. "Godly Men and Mechanical Philosophers: Souls and Spirits in Restoration Natural Philosophy." *Science in Context* 1(1987): 55-85.
Schaffer here shows how and why natural philosophers incorporated spirits and souls into their experimental labor and "extended these experimentally developed entities throughout the cosmos, both social and natural." Further, he argues that the "mechanization of the world picture" is not an adequate historiographical generalization. Nor is any intellectualist approach useful. Rather, a social and political analysis of "the ways of working in the community of natural philosophy" is necessary. That is, the laboratory—especially experiments in pneumatics—was the "safe space where the evidence for spirits could be displayed and managed." See (0072) for a response by Zev Bechler.
Topics: *IV*, I, VII, XII Names: Boyle, Hooke, Newton

1403 Schaffer, Simon. "Halley's Atheism and the End of the World." *Notes Rec. Royal Soc. London* 32(1977): 17-40.
Schaffer argues that Halley was theologically unorthodox but *not* an atheist. He was interested in the theological and biblical implications of natural philosophy but always with a strict scientific integrity. Manuscript sources, says Schaffer, show that Halley changed his position on the question of the end of the world. He did this not because of inconsistency or ecclesiastical pressure, however, but because of additional scientific evidence which he had discovered.
Topic: *IX* Name: Halley

1404 Schaffer, Simon. "Natural Philosophy." In *The Ferment of Knowledge*, edited by George S. Rousseau and Roy Porter, 55-91. Cambridge: Cambridge University Press, 1980. See (1368).

Schaffer has mainly to do with the post-1720 period and presents very little discussion of religion. But his analysis of three historiographic approaches to eighteenth-century natural philosophy raises important questions as to how the interplay between science and religion is to be understood. First, Schaffer discusses a dominant historiographic tradition which presents natural philosophy as a coherent, unified body of theory and practice—and which is limited to an essentially scientific discourse. This historiography assumes the unity of "Newtonian" tradition of matter theory. Schaffer also discusses revisionist historiography and critics of the dominant tradition who focus either: (1) on impersonal structures which politically (and religiously) produce and organize natural philosophy; or (2) on cosmology and the formulations of specific individuals—and which seem to derive from the disciplines of archaeology and anthropology. As examples, Schaffer uses a wide variety of historians and philosophers of science.

Topics: *I*, XII Name: Newton

1405 Schaffer, Simon. "Occultism and Reason." In *Philosophy, Its History and Historiography*, edited by A. J. Holland, 117-143. Dordrecht: D. Reidel, 1985.

This is a response to "Occultism and Philosophy in the Seventeenth Century," by MacDonald Ross (1353). Generally, Schaffer supports the views of Ross. In particular, he examines the role of spirits and souls in the theory and practice of natural philosophy, and the limits it put on the mechanical philosophy. Schaffer concludes with a discussion of the Leibniz-Clarke correspondence and the related debates over terms such as "miracles" and "occult."

Topics: *VII*, II Names: Boyle, Clarke, More

1406 Schaffer, Simon. "The Political Theology of Seventeenth-Century Natural Science." *Ideas and Production* 1(1983): 2-14.

In seeking to establish political authority and account for error and scientific heresy, Schaffer argues, the new philosophers drew upon their understandings of the politics of church history. After

showing their use of the image "priest of nature," Schaffer proposes that they developed two models for organizing the new science. In the "populist" (hermetic) model, nature could be encountered directly by individuals. The "technical" model, by contrast, placed a high value on technical skill and emphasized the origin of corruption and error by illegitimate political authorities. Newton is the prime example in this analysis. Finally, for Schaffer, "science is a politically constructed human practice which claims to be accessible to all in order to be accessible to few."
Topics: *IV*, V, XII Name: Newton

1407 Schaffer, Simon. "Wallification: Thomas Hobbes on School Divinity and Experimental Pneumatics." *Stud. Hist. Philos. Sci.* 19(1988): 275-298.
Schaffer reproduces a short manuscript of Hobbes and discusses the issues involved. Hobbes had begun a dispute with Ward and Wallis in the 1650s over both their natural philosophy and their apparent political use of priestcraft. The manuscript shows that Hobbes continued this controversy in the 1660s and attacked the experimental pneumatics of Boyle and Hooke—focusing on this "empty name condensation."
Topic: *VII* Name: Hobbes, Wallis

1408 Schankula, H. A. S. "Locke, Descartes, and the Science of Nature." *J. Hist. Ideas* 41(1980): 459-477. Reprinted in *Philosophy, Religion and Science in the Seventeenth and Eighteenth Centuries*, edited by John W. Yolton, 306-324. Rochester, N.Y.: University of Rochester Press, 1990. See (1720).
Against Aaron and Roth, Schankula argues that Locke was opposed to Descartes both in his epistemology and in his natural philosophy. Locke was an empiricist—not both an empiricist and rationalist. Schankula makes no explicit connections with regard to religion, theology, or God.
Topic: *VII* Name: Locke

1409 Scheurer, P. B., and G. Debrock, eds. *Newton's Scientific and Philosophical Legacy*. Dordrecht: Kluwer Academic, 1988.
The articles by Christianson (0208), Dobbs (0387), Feingold (0457), and Popkin (1242) are annotated separately.
Topic: *XII* Name: Newton

1410 Schiebinger, Londa. *The Mind Has No Sex? Women in the Origins of Modern Science*. Cambridge, Mass.: Harvard University Press, 1989.

This is a detailed analysis of the various levels and complications of the historical phenomenon of the exclusion of women and femininity from the sciences. Schiebinger focuses on England and Europe in the seventeenth and eighteenth centuries. She divides the problem into four constituent parts: institutional organizations; individual biographies of noblewomen and craftswomen; scientific definitions of female nature; and cultural meanings of gender. Issues relevant to this bibliography appear not only with respect to historiography generally but also in her treatments of: Margaret Cavendish; discussions of the relations between mind, body, and soul; and the Christian Neoplatonic basis for the iconography of the sciences and feminine muses.

Topics: *I*, IV, V, VI, VII, X Name: Cavendish

1411 Schilling, Bernard N., ed. *Essential Articles: For the Study of English Augustan Backgrounds*. Hamden, Conn.: Archon Books, 1961.

This anthology includes five relevant articles: Bond (0106), Howell (0741), and two by R. F. Jones (0836) and (0842). All are reprinted from the 1930s and 1940s and each of the four is annotated separately in this bibliography.

Topic: *X*

1412 Schirmer, W. F. "Das Problem des Religiosen Epos im 17. Jahrhundert in England." *Deutsche Vierteljahrschrift* 14(1936): 60-74.

Schirmer discusses religion and the new science as warring elements; he interprets *Paradise Lost* accordingly.

Topic: *X* Name: Milton

1413 Schmitt, Charles B. "*Prisca Theologia e Philosophia Perennis*: Due Temi del Rinascimento Italiana e la Loro Fortuna." In *Il Pensiero Italiano del Rinascimento e il Tempo Nostro*, 211-236. Florence: L. S. Olschki, 1970. Reprinted in *Studies in Renaissance Philosophy and Science*. London: Variorum Reprints, 1981.

This article is relevant to our field for the historical background which it provides. Schmitt describes sources possibly read by men in seventeenth-century England who believed in a *prisca theologia*. He traces the ideas from Clement of Alexandria through the Italian Renaissance to English natural philosophers.
Topics: *II*, XII Names: Cudworth, Newton

1414 Schmitt, Charles B. "Some Considerations on the Study of the History of Seventeenth-Century Science: Lessons from Helene Metzger." In *Etudes sur/Studies on Helene Metzger*, 23-33. Edited by Gideon Freudenthal. Leiden: Brill, 1990. See (0513). Originally this appeared in *Corpus: Revue de Philosophie* 8/9(1988).
Schmitt suggests that we follow Metzger's lead in understanding great seventeenth-century natural philosophers in their own context. This means immersing oneself in the period through wide reading in the primary literature and understanding scientific, magical, and religious terms as contemporaries did. Schmitt, however, thinks that Yates (1706-1709) and her followers present an unbalanced picture of sixteenth- and seventeenth-century scientists.
Topic: *I*

1415 Schofield, Robert E. *Mechanism and Materialism: British Natural Philosophy in an Age of Reason*. Princeton, N.J.: Princeton University Press, 1970.
Written by a historian of science, this study has a strongly scientific focus. It mostly deals with the post-1720 period but it includes pre-1720 sections on the Newtonian background and on the diffusion of Newtonianism. Doctrines of God and theology are acknowledged but treated as basically tangential to what really is important (hard science).
Topics: *VII*, XII Names: Clarke, Newton

1416 Schuler, Robert M. "Some Spiritual Alchemists of Seventeenth-Century England." *J. Hist. Ideas* 41(1980): 293-318.
This is a transcription and analysis of three alchemical manuscripts. The first is a document from an association of Royalist, Anglican alchemists, similar to Ashmole, with a meditative alchemy. The second is that of an orthodox Calvinist who linked his personal endeavor as an adept to his being one of the elect. The

third is that of John Everard—a radical who saw in material alchemy a symbol of the regeneration of the whole fallen world. Thus, alchemy is seen both to cross and to reflect a range of political and religious divisions.

Topic: *II* Names: Ashmole, Everard

1417 "Science and Religion." *Metascience* 1(1992): 31-46.

See (0959) by David C. Lindberg and others.

Topic: *I*

1418 *Science et Philosophie, XVIIe et XVIII3 Siecles*. XIIe Congress Internationale d'Histoire des Sciences. Actes. Tome III, B. Paris: Albert Blanchard, 1971.

Two brief essays, by Debus (0345) and Rousseau (1365), are annotated separately.

Topics: II, X

1419 Scupholme, A. C. "John Smith, a Cambridge Platonist." *Theology* 42(1941): 26-34.

This consists of brief biographical notes—and a summary of Smith's *Discourses*. Also see (1420).

Topic: *VII* Name: Smith

1420 Scupholme, A. C. "The Life and Opinions of Benjamin Whichcote." *Theology* 31(1935): 79-93.

Scupholme begins with a short biographical sketch of Whichcote. Then, in this and the following three articles, Scupholme discusses Cambridge Platonist views of reason, faith, nature and philosophy of nature, metaphysics, atheism, and immortality of the soul—but very little about the new science or Cambridge views of natural science itself. Scupholme addresses his articles to a general audience with a theological interest.

Topic: *VII* Name: Whichcote

1421 Scupholme, A. C. "The Light of Reason." *Theology* 38(1939): 259-268.

This is a continuation of Scupholme's summary of Culverwel's *Discourse of the Light of Nature* (1422). See also (1420).

Topic: *VII* Name: Culverwel

1422 Scupholme, A. C. "Nathaniel Culverwel—A Cambridge Platonist." *Theology* 38(1939): 196-206.

This consists of brief biographical notes—and a summary of Culverwel's *Discourse of the Light of Nature*. Also see (1420).

Topic: *VII* Name: Culverwel

1423 Sellin, Paul R. "The Last of the Renaissance Monsters: The *Poetical Institutions* of Gerardus Joannis Vossius and Some Observations on English Criticism." In *Anglo-Dutch Cross-currents in the Seventeenth and Eighteenth Centuries. Papers Read at a Clark Library Seminar, May 10, 1975*, by Paul R. Sellin and Stephen B. Baxter, 1-36. Los Angeles: William Andrews Clark Memorial Library, University of California, 1976.

This lecture sets Vossius' theory of poetics in the context of his synthesizing scheme of the distribution of the arts and sciences—and thereby in its relation to ethics, theology and the sciences. For Vossius, "poetry is a special kind of rhetoric that has few specific poetic principles of its own, but takes its essential form from its subject matter: the ethical *res* . . . and the empirical study of nature." Sellin also suggests relationships between Vossius and Bacon, Milton, and Dryden.

Topics: *X*, VII Names: Bacon, Dryden, Milton, Vossius

1424 *Seventeenth Century Studies Presented to Sir Herbert Grierson*. Oxford: Clarendon Press, 1938.

The essays by Bullough (0153), Garrod (0525), Metz (1063), and Willey (1683), are annotated separately.

Topic: *X*

1425 Shanahan, Timothy. "God and Nature in the Thought of Robert Boyle." *J. Hist. Phil.* 26(1988): 547-569. Reprinted in *Seventeenth-Century Natural Scientists*, edited by Vere Chappell, 123-145. New York: Garland, 1992. See (0205).

Shanahan argues against both deistic and occasionistic (read McGuire's) interpretations of Boyle's thought concerning the relation between God and nature. He argues for "a version of concurrentism expressed within the context of the mechanical philosophy." Shanahan places Boyle's thought in the intellectual context of discussions of these issues by medieval Christian and

Islamic theologians. Hence, he also is asserting that the intellectualist task is as important as the new social contextualist approaches to seventeenth-century natural philosophers.
Topic: *VII* Name: Boyle

1426 Shanahan, Timothy. "Theological Reasoning in Boyle's *Disquisition about Final Causes*." In *Robert Boyle Reconsidered*, edited by Michael Hunter, 177-192. Cambridge: Cambridge University Press, 1994. See (0759).
Shanahan considers Boyle's work on final causes from within "the adversarial context in which it was written" (that is, against both Epicureans and Cartesians). He discusses the structure of Boyle's argument, pointing out its complexity and sophistication. Boyle argued that careful empirical examination of natural phenomena does allow some final causes to be known. Caution is necessary, however, in deciding what phenomena may provide more probable conjectures. For example, arguments from the parts of animals are better than celestial mechanics. Overall, Boyle's purpose was apologetic: he believed that knowledge of final causes of nature is not necessary for a naturalist *qua* naturalist, but it is an incentive for acknowledging God by all rational human beings.
Topics: *VIII*, VII Name: Boyle

1427 Shapin, Steven. "The House of Experiment in Seventeenth-Century England." *Isis* 79(1988): 373-404.
This is a detailed study of the locations in which experimental science took place and of the role of these sites in the development of science. Shapin focuses on Boyle's and Hooke's rooms as well as those of the Royal Society. Boyle understood his private laboratory as a sacred space—and his activity therein is likened to the "solitude of the religious isolate." To legitimize an experiment, it had to be shown to reliable gentlemen who, through rites of passage, had entered a more public space.
Topic: *V* Names: Boyle, Hooke

1428 Shapin, Steven. "Licking Leibniz." *Hist. Sci.* 19(1981): 293-305.
This, simultaneously, is an essay review of and an historiographic dispute with A. R. Hall's *Philosophers at War* (0605). Shapin argues not only that the wider intellectual differences between

Leibniz and Newton are important but that the social and political differences between Leibnizians and Newtonians also should be explored.

Topics: *XII*, I, IV, VII Name: Newton

1429 Shapin, Steven. "Of Gods and Kings: Natural Philosophy and Politics in the Leibniz-Clarke Disputes." *Isis* 72(1981): 187-215.

Shapin first correlates the respective natural philosophies of Newton and Leibniz to different theological positions. He then correlates Newtonianism to various Whig politico-ideological views as represented by Clarke and other Anglican clerics. And he notes the use of Leibniz' philosophy by radical Whigs, freethinkers, and deists. Shapin says that he is not making any claims about the *intentions* of Newton and Leibniz. But, he says, there are these connections between political and moral order (on one side) and diverging notions of divine and natural order (on the other).

Topics: *IV*, V, VII, IX, XII Names: Bentley, Clarke, Newton, Toland

1430 Shapin, Steven. "Robert Boyle and Mathematics: Reality, Representation, and Experimental Practice." *Science in Context* 2(1988): 23-58.

Shapin argues that Boyle imposed strict limits on the use of mathematics in experimental practice for two basic reasons: "mathematical representations pointed to an improper ontology" and "mathematical means of persuasion were embedded" in an "immoral social order." The first reason was related to Boyle's nominalist theology. (Only particulars exist in the creation—and those exist by God's will.) The second reason was related to Boyle's concern to increase, not limit, the social accessibility of experimental knowledge.

Topics: *IV*, VII Name: Boyle

1431 Shapin, Steven. "Social Uses of Science." In *The Ferment of Knowledge*, edited by George S. Rousseau and Roy Porter, 93-139. Cambridge: Cambridge University Press, 1980. See (1368).

In this lengthy study of the historiography of the social "uses" of science, Shapin focuses on Newtonianism and matter theory as well as Enlightenment natural knowledge and eighteenth-century physiological thought. He criticizes traditional intellectualist historiography and favors revisionist contextualist writers. Shapin concludes that the (then) recent contextualist writers : (1) identify "an important role for social interests in scientific change or in sustaining scientific accounts;" (2) understand "use" to be "intimately associated with the production, judgment and institutionalization of science;" and (3) are "implicit anthropologists" in that they explain the "social work" which scientific theories do in legitimating or criticizing social and political arrangements. Among historians annotated in the present bibliography, Shapin criticizes the historiographic approaches of Gillespie, Koyre, Buchdahl, Guerlac, McGuire, and McMullin. He presents as most constructive the work of George Grinnell, J. R. Jacob and M. C. Jacob—as well as some of his own publications.

Topics: *I*, IV, V, XII Name: Newton

1432 Shapin, Steven. "Who Was Robert Hooke?" In *Robert Hooke: New Studies*, edited by Michael Hunter and Simon Schaffer, 253-285. Woodbridge, England: Boydell Press, 1989. See (0765).

Shapin discusses Hooke by drawing extensively from Hooke's diary. He shows that Hooke was dealt with by other members of the Royal Society "as a mechanic, as a tradesman, as a servant." According to Shapin, Hooke's religious life was minimal. Shapin contrasts Hooke with Boyle, the gentleman and Christian virtuoso, and indicates how the empirical science of the Royal Society related to each man's perceived code of conduct.

Topics: *V*, III, IV, VI Names: Boyle, Hooke

1433 Shapin, Steven, and Simon Schaffer. *Leviathan and the Air Pump: Hobbes, Boyle, and the Experimental Life*. Princeton: Princeton University Press, 1985.

This is a thorough examination of Boyle's experimental activity, particularly with the airpump, and Hobbes' objections to it. Writing historical sociology of science, Shapin and Schaffer show that a scientific method is "a means of regulating social interaction with the scientific community" and has wide political and social consequences. Part of the authors' description of Boyle's response to Hobbes includes the point that Boyle combined technical and theological arguments.

Topics: *IV*, VI, VII Names: Boyle, Hobbes

1434 Shapin, Steven, and Arnold Thackray. "Prosopography as a Research Tool in History of Science: The British Scientific Community, 1700-1900." *Hist. Sci.* 12(1974): 1-28.

Prosopography means the analysis of the common background characteristics of a group of persons by means of collective biographies. The intentions of the authors of this essay are not limited to the eighteenth and nineteenth centuries: they only use this period for their main examples. In doing the history of science in this way, they note that the following questions are important: Who are that "set of individuals who did original research into natural phenomena?" What are such persons called? How did they relate to their general culture? Shapin and Thackray also note that, in the seventeenth century, the "virtuosi" became "natural philosophers" (with both groups including clergymen). In the nineteenth century, they became "men of science."

Topics: *I*, V

1435 Shapiro, Barbara J. "Debate: Science, Politics and Religion." *Past and Present* 66(1975): 133-138.

Shapiro here argues against Mulligan's position—and especially against Mulligan's method presented in (1095). Shapiro reasserts the argument connecting the new science with Latitudinarianism.

Topic: *III*

1436 Shapiro, Barbara J. "Early Modern Intellectual Life: Humanism, Religion and Science in Seventeenth-Century England." *History of Science* 29(1991): 45-71.

This is both a systematic argument for a case and a proposal for a historiographic programme. Shapiro's case is that there were important parallels and connections between Erasmian humanism, latitudinarian religion, and the new science in early modern Europe (and especially in seventeenth-century England). Her programmatic proposal is that a more extensive examination of the English scientific revolution in the context of humanism should be adopted by intellectual historians generally. Shapiro does note many parallels as she makes a strong and interesting case for her views.

Topics: *VII*, I, III, X

1437 Shapiro, Barbara J. "History and Natural History in Sixteenth- and Seventeenth-Century England: An Essay on the Relationship between Humanism and Science." In *English Scientific Virtuosi in the 16th and 17th Centuries*, edited by Barbara Shapiro and Robert G. Frank, Jr., 3-55. Los Angeles: William Andrews Clark Memorial Library, University of California, 1979.

Shapiro argues that changing conceptions of evidence and proof brought historical thought into closer contact with the natural sciences. Among other things, these developments had effects upon late-seventeenth-century treatments of scriptural history and chronology. This essay has been superseded by Chapter 4 of Shapiro's book, *Probability and Certainty in Sixteenth-Century England* (1440).

Topics: *VII*, X

1438 Shapiro, Barbara J. *John Wilkins, 1614-1672: An Intellectual Biography*. Berkeley and Los Angeles: University of California Press, 1969.

Shapiro places Wilkins in the context of the philosophical and religious issues in seventeenth-century science. Some of her theses are rather questionable and her understanding of "intellectual" is rather narrow. But the book remains useful for studying Wilkens and for strictly "intellectual" connections between science and theology in our period.

Topics: *VII*, III, V Names: Sprat, Wilkins

1439 Shapiro, Barbara J. "Latitudinarianism and Science in Seven teenth-Century England." *Past and Present* 40(1968): 16-41. Reprinted in *The Intellectual Revolution of the Seventeenth Century*, edited by Charles Webster, 286-316. London: Routledge and Kegan Paul, 1974. See (1618).

Shapiro argues against the Puritanism thesis—and specifically against Merton and Stimson. She connects the rise of science to religious moderation or to a Latitudinarian religious position. Portions of this essay appear in Chapter 3 of Shapiro's 1983 book (1440).

Topic: *III* Name: Wilkins

1440 Shapiro, Barbara J. *Probability and Certainty in Seventeenth-Century England: A Study of the Relationship Between Natural*

Science, Religion, History, Law, and Literature. Princeton: Princeton University Press, 1983.

Shapiro's major thesis is that, among seventeenth-century English thinkers, assumptions and theories about the nature of truth and methods for attaining it were *changing*. The quest for certainty was given up (more or less simultaneously) by intellectuals in all of the fields indicated by the subtitle of the book. Concerning truth, human infallibility, matters of fact, evidence, sense observation, and credible testimony—Shapiro argues—criteria of high probability or "moral certainty" replaced criteria of demonstrable certainty as the test of what counts as knowledge. This book is not original, especially as regards religion and natural philosophy, but it provides a good starting point and overview for persons interested in the various topics found in it. There is a valuable and engaging chapter on witchcraft. Shapiro writes intellectual history in the narrow sense—with almost no reference to the social, political, or ecclesiastical contexts. Sometimes, her evidence seems highly selective and important religious groups are left out of her discussions. But the book is well-written and can be read by a general audience as well as by scholars.

Topic: *VII* Names: Bacon, Boyle, Glanvill, Hale, Locke, Sprat, Wilkins

1441 Shapiro, Barbara J. "Science, Politics, and Religion." *Past and Present* 66(1975): 133-138.

Shapiro here argues against Mulligan's statistical analysis (1095) and arguments concerning religion and the rise of science in seventeenth-century England. Within the context of a formal debate, Shapiro challenges both the propriety of Mulligan's method and the truth of her conclusions. Shapiro also defends her own view that there is a "linkage of latitudinarianism to science."

Topics: *III*, IV, V

1442 Shapiro, Barbara J. "The Universities and Science in Seventeenth Century England." *J. Brit. Stud.* 10(1971): 47-82.

Writing against Hill (0686, 0689, 0693) Shapiro argues that the universities showed a continuous interest in science throughout the century, that Puritan intervention did not significantly alter the pattern of scientific concerns, and that analysis of the Wadham group of the 1650s does little to lend support to the Puritanism thesis. After studying Oxford, Cambridge, and Harvard, she concludes: "The case for a causal connection between Puritanism and science is not supported by the data in the realm of education."

Topics: *V*, III

1443 Sharp, Linsay. "Walter Charleton's Early Life 1620-1659, and Relationship to Natural Philosophy in Mid-Seventeenth Century England." *Ann. Sci.* 30(1973): 311-340.

Sharp focuses on important details of Charleton's intellectual biography, especially in the 1650s. This includes the theological issues, as seen by Charleton, connected with Epicurean atomism.

Topic: *VII* Name: Charleton

1444 Shea, William R. "Introduction: Trends in the Interpretation of Seventeenth Century Science." In *Reason, Experiment, and Mysticism in the Scientific Revolution*, edited by M. L. Righini and William R. Shea, 1-17. New York: Science History, 1975. See (1329).

Shea introduces the essays in the book and raises the question "Are 'mystical' traditions and 'modern' elements in the Scientific Revolution antagonistic or, paradoxically, complementary extremes." Shea equates "mystical" with Hermetic and speaks only incidentally of God and religion in the usual sense. He assumes a retrospective historiography and tries to judge alchemy/hermeticism in terms of its positive contribution to scientific method.

Topics: *I*, II

1445 Shea, William R., ed. *Nature Mathematized: Historical and Philosophical Studies in Classical and Modern Natural Philosophy*. Dordrecht: D. Reidel, 1983.

The essays by Kirsanov (0880), McGuire (1030), and North (1143) are annotated separately in this bibliography.

Topics: *VII*, XII Name: Newton

1446 Sheppard, Harry J. "European Alchemy in the Context of a Universal Definition." In *Die Alchemie in der Europaischen Kultur und Wissenschaftsgeschichte*, edited by Christoph Meinel, 13-17. Wiesbaden, Germany: O. Harrassowitz, 1986.

Sheppard briefly proposes a single definition which covers three sometimes disputed aspects of alchemy. Namely, "Alchemy is the art of liberating parts of the cosmos from temporal existence and achieving perfection which, for metals is gold, and for man, longevity, then immortality and, finally, redemption. Material perfection was sought through the action of a preparation (Philosopher's Stone for metals; Elixir of Life for humans), while spiritual ennoblement resulted from some form of inner revelation or other enlightenment (Gnosis, for example, in Hellenistic and western practices)."
Topic: *II*

1447 Sherburn, George. "Pope and 'The Great Shew of Nature'." In *The Seventeenth Century*, edited by Richard F. Jones and others, 306-315. Stanford, Calif.: Stanford University Press, 1951. See (0845).
Sherburn says that nature is a great show, for Pope, resembling an opera. Pope viewed nature (and nature's God), however, not in small detail but as a universal whole. Pope intimated the millennium with his symbolic uses of astronomical images—wide visions in a spirit of cosmic reverence.
Topic: *X* Name: Pope

1448 Shirley, John W. "Sir Walter Ralegh and Thomas Harriot." In *Thomas Harriot: Renaissance Scientist*, 16-35. Edited by John W. Shirley. Oxford: Clarendon Press, 1974. See (1450).
This article focuses on the relationship between Ralegh and Harriot and contains a discussion of the charges of atheism directed against the "Northumberland circle."
Topics: *IX*, VII Names: Harriot, Ralegh

1449 Shirley, John W. *Thomas Harriot: A Biography*. Oxford: Clarendon Press, 1983.
This is a detailed and definitive biography of Harriot's life—if not of his thought. Shirley touches on Harriot's religious views and analyzes the accusations of atheism against Ralegh and Harriot. But he does little to place him among the natural philosophers of his time. Shirley basically sees Harriot as an independent scholar and, thus, undermines the image of a close "Northumberland Circle."

The first chapter usefully reviews the image of Harriot as known and interpreted since the early seventeenth century.

Topics: *IX*, V, VI Names: Harriot, Percy, Ralegh

1450 Shirley, John W., ed. *Thomas Harriot: Renaissance Scientist.* Oxford: Clarendon Press, 1974.

This is a collection of papers (originally presented in 1971) on various facets of Harriot's life and scientific activity . Two of them are relevant to our field and are annotated separately. See: Jacquot (0819) and Shirley (1448).

Topics: *VII*, IX Name: Harriot

1451 Shryock, Richard H. "Early American Immunology." In *Medicine in America*, by Richard H. Shryock, 252-258. Baltimore: Johns Hopkins Press, 1966.

This is a brief discussion of Mather's contributions to preventive medicine—contributions which were motivated by his concern for public welfare.

Topic: *XI* Name: Mather, C.

1452 Shryock, Richard H. *Medicine and Society in America: 1660-1860.* New York: New York University Press, 1960.

This is a survey of the beginnings of the medical profession, of medical thought and practice, and of understandings of health and disease. It is written prior to historiographical interest in Paracelsian medicine—but Shryock does note the role of well-read clergy. Using his retrospective criteria, Shryock can find no "creative" medical science in the seventeenth century.

Topic: *XI* Name: Mather, C.

1453 Shumaker, Wayne. "The Popularity of Renaissance Esotericism." *Bucknell Review* 20(Winter 1972): 45-52.

Shumaker proposes that (1) the belief that the oldest opinions are the most profound and (2) human irrationality both account for the popularity of the occult in the Renaissance. He also suggests that Protestantism restrained the popularity of esotericism in seventeenth-century England.

Topic: *II*

1454 Shumaker, Wayne. *Renaissance Curiosa.* (Medieval and Renaissance Studies, Vol. 8.) Binghamton, N.Y.: Center for Medieval and Early Renaissance Studies, 1982.

This work includes readable (even personal) summaries and explications of: John Dee's *Conversations with Angels* (published by Meric Casaubon in 1639); Girolamo Cardano's *Horoscope of Christ* (1555); Johannes Trithemius' *Stenographia* (written c. 1500 but published 1616); and George Dalgarno's *Ars Signorum* (a philosophical universal language published in 1661). Shumaker hopes that, by bringing attention to these "odd" and difficult texts, we may become aware of "axiomatic ideas wildly discontinuous" with our own.

Topics: *II*, X Names: Casaubon, M., Dalgarno

1455 Silverman, Kenneth. *The Life and Times of Cotton Mather.* New York: Harper and Row, 1984.

In this detailed and fair biography, Silverman concludes that Mather was the "first unmistakenly American figure in the nation's history." This generalization is exemplified in Mather's correspondence with the Royal Society: Mather emphasizes the American quality of his observations, reflecting both provincial discomfort and pride. Mather appears as a popularizer of the new science—which, however, remains a handmaiden to theology. While most of the book demonstrates Mather's mixed personality, the chapter on the inoculation controversy shows him at his best (wanting to do good).

Topics: *VII*, X, XI Names: Boylston, Mather, C.

1456 Skinner, Quentin. "Meaning and Understanding in the History of Ideas." *History and Theory* 8(1969): 3-53.

Skinner argues for the influence of social factors on intellectual history—"externalism" in the history of science.

Topic: *I*

1457 Skinner, Quentin. "Thomas Hobbes and the Nature of the Early Royal Society." *Hist. J.* 12(1969): 217-239.

Skinner here argues against several "misunderstandings" about the Royal Society. Two of these wrong views are that the Society was a "professional body" and that it stood strongly against "Hobbsian atheism."

Topics: *V*, IX Names: Boyle, Hobbes, Wallis

1458 Slack, Paul. *The Impact of the Plague in Tudor and Stuart England*. London: Routledge and K. Paul, 1985.

This is a study of the plague and of social responses to it. Slack deals with theological understandings of plague and changes in views about providence within the context of controversies about social and political control. It is a good discussion of the social control of religion and medicine.

Topics: *XI*, IV, V

1459 Smith, A. J. "Sacred Earth: The Advance of Science and the Scope of Imagination." In *Science and Imagination in XVIIIth-Century British Culture*, edited by Sergio Rossi, 359-368. Milan: Edizioni Unicopli, 1987. See (1362).

Smith contrasts the images of nature and of Eden in the writings of Henry Vaughan, Thomas Burnet, and Alexander Pope. He finds a "spiritual revolution"—from a sacred organism to a Mosaic scientific description to an aesthetic ideal. This progression illustrates a shift of thought and understanding which accompanied the advance of science and brought in the modern world.

Topic: *X* Names: Burnet, T., Pope, Vaughan, H.

1460 Smith, Constance I. "Richard Bentley and the Innate Idea of God: A Correction." *J. Hist. Ideas* 22(1961): 117-118.

Smith argues (against Yolton) that Bentley rejected the doctrine of innate ideas and that Bentley put forth a natural theology based on other grounds.

Topics: *VIII*, VII, XII Names: Bentley, Newton

1461 Snow, Adolph J. *Matter and Gravity in Newton's Physical Philosophy: a Study in the Natural Philosophy of Newton's Time*. London: Oxford University Press, 1926.

This is a once-important work which now is outdated.

Topics: *XII*, VII Name: Newton

1462 Sokolowski, R. "Idealization in Newton's Physics." In *Newton and the New Direction in Science*, edited by G. V. Coyne and others, 65-72. Vatican City: Specola Vaticana, 1988. See (0281).

This is a brief philosophical analysis of Newton's physics in terms of "the process of idealization" (which Sokolowski defines as

projecting concepts to their ideal limits). According to the argument, Newton's notion of God as an idealized mind is a logical correlative to his idealized physics.
Topics: *XII*, VII Name: Newton

1463 Solberg, Winton U. Introduction to *The Christian Philosopher,* by Cotton Mather, edited by Winton U. Solberg, xix-cxxxiv. Urbana, Ill.: University of Illinois Press, 1994.
In this extended introduction, Solberg thoroughly discusses the origins, sources, contents, reception, and significance of Mather's book. His primary thesis is that, though mainly derived from the works of others, it is the first American book of natural theology. For Solberg, Mather was "one of the most progressive thinkers in early America," a clergyman who enthusiastically adopted the new science. A useful "Biographical Registrar," a recapitulation of his sources, a biblical index, and a general index follow the text in this critical edition.
Topics: *VIII*, V, VII, X Names: Derham, Mather, C., Ray

1464 Solberg, Winton U. "Science and Religion in Early America: Cotton Mather's *Christian Philosopher.*" *Church History* 56(1987): 73-92.
This study of *The Christian Philosopher* emphasizes its place as the first statement in America of natural theology and of the design argument. This essay has been superseded by (1463).
Topics: *VIII*, VII Name: Mather, C.

1465 Solomon, Julie Robin. "From Species to Speculation: Naming the Animals with Calvin and Bacon." In *Women and Reason*, edited by Elizabeth D. Harvey and Kathleen Okruhlik, 77-113. Ann Arbor, Mich.: University of Michigan Press, 1992.
Solomon compares Calvin's exegesis of Genesis 2:18-22 with Bacon's interpretation. She finds Calvin's notion of human knowledge to be embodied; whereas Bacon distances "the domain of knowledge from that of desire." Bacon thus separates these elements in his interpretation of the passage. Solomon further relates each thinker's interpretation to his social, political, and sexual situation. For Bacon, she concludes, scientific activity (analogous to Adam's "naming") should be under rational control

and standardization; and that this reflects Bacon's view of kingship.
Topics: *X*, IV, VII Name: Bacon

1466 Solt, Leo F. "Anti-Intellectualism in the Puritan Revolution." *Church History* 24(1956): 306-316.
Solt shows that all "Puritans" did not have the same views respecting educational pursuits—including treatment of the sciences in the university curriculum. He focuses particularly on the anti-intellectualism of the army preachers during the revolution.
Topics: *IV*, III, V Name: Dell

1467 Solt, Leo F. "Puritanism, Capitalism, Democracy, and the New Science." *Am. Hist. Rev.* 63(1967): 18-29.
Solt here argues against the purported link between Puritanism and each of the other developments included in his title. He emphasizes the antilearning attitudes found in Puritanism.
Topics: *IV*, III

1468 Sommerville, C. John. "On the Distribution of Religious and Occult Literature in Seventeenth Century England." *Library* 29(1974): 221-225.
Sommerville argues that Keith Thomas (in 1541) exaggerates the sceptical and magical qualities of the English mentality in our period. He gives a detailed estimate of the numbers of Bibles and other religious books published and distributed in England intending to correct Thomas' use of such statistics.
Topics: *II*, X

1469 Sortais, Gaston. *La Philosophie Moderne depuis Bacon jusqu'a Leibniz*. 2 vols. Paris: P. Lethielleux, 1920-1922.
In Volume One, Sortais thoroughly examines Bacon's life and works. He accepts the thesis that Bacon's philosophy is a form of utilitarian empiricism motivated for the good of humankind. But he also indicates that all the separate lines of Bacon's philosophy converge on the "throne of truth," the first philosophy. Perhaps most useful for our purposes are Sortais' descriptions of the division of the philosophical sciences as discussed in *De Augmentis* and presented in *De Sapientia Veterum*. Bacon's division is constituted as: of God (or natural theology); of nature; and of man. In

Volume Two, Sortais discusses Gassendi and Hobbes, briefly touching on Gassendi's influence in England. After analyzing Hobbes' controversies with Descartes, Sortais presents Hobbes' ideas in the framework of the trilogy: matter, man, and citizen. A discussion of "humans as religious beings" appears as a part of the second section in the trilogy. Now seventy-five years old, Sortais' volumes are dated but they are still useful for their clear organization.

Topic: *VII* Names: Bacon, Cudworth, Hobbes

1470 Southgate, Beverley C. "'Cauterizing the Tumour of Pyrrhonism': Blackloism versus Skepticism." *J. Hist. Ideas* 53(1992): 631-645.

This is a study of the attempts by two English Catholic priests to defeat the "mitigated scepticism" which they believed was the prevailing orthodoxy among members of the Royal Society. Thomas White began a debate with Joseph Glanvill and then John Sergeant continued White's polemic. They claimed that their Catholic Aristotelian empiricism would reinstate both the certainty and the unity of scientific and religious truth.

Topic: *VII* Names: Glanvill, Sergeant, White

1471 Southgate, Beverley C. *"Covetous of Truth": The Life and Work of Thomas White, 1593-1676*. Dordrecht: Kluwer Academic, 1993.

Southgate here has written the definitive study of the Catholic controversialist and natural philosopher. In her opening chapters, she explores White's life, his reputation and his political controversies. She then turns to the intellectual context of his thought: scepticism and scholasticism. White's cosmology, physics and psychology are analyzed in the part appropriately entitled "Science Old and New." White was as enthusiastic about the new natural philosophy as about scholastic Catholicism; and Southgate finds him incorporating the new in the old. Southgate posits as her central theme that White argued for the conformity and complementarity of science and religion. Against several of his contemporaries, White refused to accept the limitations of scepticism, and saw science and religion as two mutually supportive routes to certainty. Southgate concludes that White hoped thereby to form a synthesis—a possible replacement for Thomism—the goal of which is absolute truth.

Topics: *VII*, IV Names: Boyle, Digby, Glanvill, Hobbes, Sargeant, White

1472 Southgate, Beverley C. "Excluding Sceptics: The Case of Thomas White, 1593-1676." In *The Sceptical Mode in Modern Philosophy: Essays in Honor of Richard H. Popkin*, edited by Richard A. Watson and James E. Force, 71-85. Dordrecht: M. Nijhoff, 1988.

Southgate pictures White as appearing to be "some latter-day Canute"—defying the inevitability of sceptical philosophies. He attempted to marry Catholic theology and a Copernican/ Aristotelian science to form an integrated antisceptical philosophy. In so doing, he engaged in controversies with Hobbes and Glanvill.

Topics: *VII*, IX Names: Glanvill, Hobbes, White

1473 Southgate, Beverley C. "'Forgotten and Lost': Some Reactions to Autonomous Science in the Seventeenth Century." *J. Hist. Ideas* 50(1989): 249-268.

Southgate revises the ancients versus moderns thesis by proposing a division between seventeenth-century advocates and opponents of autonomous science. After she outlines Bacon's proposals for "intellectual apartheid," she examines criticisms of autonomous science and the ideas of those who reasserted a synthesis of science, religion, and morality. She concludes with comments on the present relevance of these issues.

Topics: *VII*, IV, X Names: Bacon, Casaubon, M., Comenius, More

1474 Southgate, Beverley C. "'No Other Wisdom': Humanist Reactions to Science and Scientism in the Seventeenth Century." *The Seventeenth Century* 5(1990): 71-92.

Southgate here argues that the "Two Cultures" debate dates back to the seventeenth century. Not only was there a differentiation between science on the one hand and the arts and humanities on the other, but the arts and humanities (which includes religion) were devalued by arrogant advocates of scientism. This essay reads as a lawyer's brief—with a battery of seventeenth-century texts to support the case. But many of the quotes are presented out-of-context and the entire essay basically is a retrospective exercise.

The advocates of science are presented as the bad guys and the humanists (seeking cultural and religious synthesis) are presented as the good guys.

Topic: *X* Name: Comenius

1475 Spiller, Michael R. G. *'Concerning Natural Experimental Philosophie': Meric Casaubon and the Royal Society*. The Hague: M. Nijhoff, 1980.

Spiller proposes "to do justice" to Casaubon's *A Letter to Peter du Moulin* . . ., possibly the most intelligent seventeenth-century criticism of the new science. A facsimile of the letter and extracts from an unpublished manuscript, "On Learning," are included as appendices. The first chapter usefully describes Casaubon's life and works. The second summarizes the lines of attack on the Royal Society by the "conservative opposition." Spiller then analyzes each of Casaubon's issues. Casaubon associated the new philosophy with religious fanaticism, Epicurean atheism, Cartesianism, occult learning, and latitudinarianism. He saw humans as religious by nature, and he believed that the new philosophy did not take this sufficiently into account. "Usefulness" cannot "be confined to physical phenomena" according to Casaubon. Thus, science should not be cut off from religion and morality—and science should not be taken to be the paradigm for all kinds of understanding.

Topics: *V*, IV, VII Names: Casaubon, M., Stubbe, White

1476 Spinka, Matthew. *John Amos Comenius: That Incomparable Moravian*. Chicago: University of Chicago Press, 1943. Reissued with updated bibliography. Chicago: University of Chicago Press, 1955.

This is a solid biography presenting Comenius as a hero. The chapter on his experiences in England details the politics of the situation as well as the hopes and plans for a "pansophic" college.

Topics: *IV*, II Names: Comenius, Dury, Hartlib

1477 Sprunger, Keith L. *The Learned Doctor William Ames*. Urbana, Ill.: University of Illinois Press, 1972.

Sprunger combines a biographical study with an analysis of Ames' theology. He emphasizes the continental—especially the Dutch—features of Puritanism in England and New England. For our field, see especially Chapter 6, "Technometria: Prolegomena to Theology," which summarizes Ames' encyclopedic organization of all the "arts," discussing their nature, uses, and boundaries. (The six arts include Physics and Theology.) Sprunger notes the influence of Peter Ramus on Ames as well as Ames' influence on other Puritan thinkers and encyclopedists.

Topics: *VII*, III, X Name: Ames

1478 Spurr, John. "'Rational Religion' in Restoration England." *J. Hist. Ideas* 49(1988): 563-585.

Spurr describes the context and arguments of the theological controversies focusing on "rational religion" during the Restoration period. He notes the differing ways in which reason was understood and emphasizes the continuing significance of the traditional understanding of "right reason." He mentions but does not emphasize mathematical reasoning and natural religion based on scientific argument. The effect of Spurr's article is to imply that the new science did not have much impact on the debates.

Topics: *VII*, VIII, IX Names: Glanvill, Stillingfleet

1479 Spurr, John. *The Restoration Church of England, 1646-1689.* New Haven, Conn.: Yale University Press, 1991.

Spurr's survey includes no direct discussion of science. But the book is a good summary of the theological views of Anglican clergy during the period. Some of these clergy appear in the literature regarding science and religion—including Barrow, Patrick, Stillingfleet, Tillotson, Ward, and Wilkins. The effect of Spurr's book is to challenge indirectly some of the received historiography connecting science and "rationality," science and deism, and—most indirectly—science, religion, and socioeconomic ideology.

Topics: *V*, IV, IX, X Names: Barrow, Evelyn, Patrick, Stillingfleet, Tillotson, Ward, Wilkins

1480 Squadrito, Kathleen M. "Locke's View of Dominion." *Environmental Ethics* 1(1979): 255-262.

Squadrito argues that Locke's interpretation of Genesis in his educational writings mitigates the political interpretation of dominion over nature and the laboring classes in the *Second Treatise*. The one implies responsible stewardship, while the other leads to technological abuses.

Topics: *VI*, IV, X Name: Locke

1481 Stace, W. T. *Religion and the Modern Mind*. London: Macmillan, 1953.

In the tradition of Burtt (0160) and Randall (1285), Stace describes the rise of modern science and what he sees as its "negative" consequences for religion. He openly states that his emphasis is on the logical connections between ideas and not on their history. Unfortunately, a distortion of seventeenth-century science and religion results (especially with respect to Newton).

Topics: *VII*, XII Name: Newton

1482 Stahlman, William. "Astrology in Colonial America." *Wm. Mary Q.* 13(1956): 551-563.

Stahlman tests the general thesis that Colonial American intellectual endeavors reflected those of England by comparing numbers of astrological books printed and distributed from the 1640s through the eighteenth century. In America, almanacs were the main source of astrological data. Several contemporaries, including clergy, spoke about the wide influence of astrology—but Stahlman decides that his findings are inconclusive.

Topics: *II*, IV Name: Lilly

1483 Stangl, Walter. "Mutual Interaction: Newton's Science and Theology." *Perspectives on Science and Christian Faith* 43(1991): 82-91.

This is a rather superficial summary of ways in which Newton's science and theology interacted—and a call for inspiration today from Newton's example. Stangl argues that Newton's antitrinitarianism was not a positive heresy: it was merely part of the anti-metaphysical bias of his scientific mentality.

Topics: *XII*, VII, IX Name: Newton

1484 Starkman, Miriam K. *Swift's Satire on Learning in "A Tale of a Tub."* Princeton: Princeton University Press, 1950.

Starkman argues that Swift satirized the presumptions of utilitarianism and universalism proclaimed by the new science. She also suggests that Swift discredited Catholicism by relating it to science and nonconformity and by associating it with occultism. For Swift, these abuses in religion and scientific learning formed modernity and progress—his main targets. Yet Swift's satire, Starkman

concludes, represents a "prodigiously skillful espousal of a lost cause."
Topics: *X*, IV Name: Swift

1485 Staudenbaur, C. A. "Galileo, Ficino, and Henry More's *Psychathanasia*." *J. Hist. Ideas* 29(1968): 565-578.
Staudenbaur proposes that More wrote *Psychathanasia* with a copy of Ficino's *Theologia Platonica de Immortalitate Animorum* in one hand and a quill in the other. Moreover, he did so while reading Galileo's defense of the Copernican theory.
Topic: *VII* Name: More

1486 Staudenbaur, C. A. "Platonism, Theosophy, and Immaterialism: Recent Views of the Cambridge Platonists." *J. Hist. Ideas* 35(1974): 157-169.
This is an essay review of four books on the Cambridge Platonists. Staudenbaur gives: brief and positive notices of edited books by Cragg (0285) and Patrides; a medium-length, slightly negative review of Roberts' book on Whichcote (1333); and an extended, most negative review of Hutin's book on Henry More (0777). Science or natural philosophy is not discussed directly.
Topics: *VII*, II Names: More, Whichcote

1487 Steadman, John M. "Islamic Tradition and That Divelish Engin." *Hist. Ideas News* 4(1958): 39-41.
Steadman connects Milton to the interesting traditional idea that the Devil invented the cannon.
Topics: *X*, VI Name: Milton

1488 Stearns, Raymond Phineas. "Colonial Fellows of the Royal Society of London, 1661-1788." *Notes Rec. Royal Soc. London* 8(1951): 178-246. Earlier versions appeared in *Wm. Mary Q.* 3rd ser. 3(1946): 208-268 and *Osiris* 8(1948): 73-121.
Stearns gives brief sketches of the circumstances of the elections of the colonial fellows (including clergy) from 1661 to 1788. He includes a few English fellows associated with the colonies along with a discussion of statutes relating to admission.
Topic: *V* Names: Mather, C., Winthrop

1489 Stearns, Raymond Phineas. "The Relations Between Science and Society in the Seventeenth Century." In *The Restoration of the Stuarts, Blessing or Disaster?*, 67-75. Washington, D.C.: Folger Library, 1960.

This is a general lecture touching briefly on our topic. It has been superseded and is of little use.

Topic: *V*

1490 Stearns, Raymond Phineas. *Science in the British Colonies of America*. Chicago: University of Illinois Press, 1970.

This is a comprehensive survey; Stearns is not analytical, but he is thorough. The book contains studies of the relations between the Royal Society and the colonies, of Fellows who were clergy-scientists, and of colonial institutions. It covers all of the colonies, even the West Indies. Stearns is not particularly sympathetic to the theological dimension—but he does recognize that science had a prominent place in the theology of the period. Alchemy is not mentioned at all.

Topics: *VII*, V Names: Bannister, Byrd, Clayton, Josselyn, Mather, C., Mather, I., Petiver, Ray, Robie, Sloane, Winthrop

1491 Stebbins, Sara. *Maxima in Minima: Zum Empirie- und Autoritatsverstandnis in der physikotheologishen Literatur der Fruhaufklarung*. Frankfurt: P. D. Lang, 1980.

Stebbins mainly deals with physico-theology on the continent and after 1720; but she does include some discussion of William Derham, Nathaniel Grew, and John Ray. She sees physico-theology as a theologically oriented form of nature study which is based in a need to reconcile observations of natural phenomena with traditional religious beliefs. The result, according to Stebbins, was a secularization of religious belief—but also a sacralization of empirical knowledge.

Topics: *VIII*, VII Names: Derham, Grew, Ray

1492 Stein, Howard. "Newtonian Space-Time." *Texas Quarterly* 10(1967): 174-200. Reprinted in *The Annus Mirabilis of Sir Isaac Newton, 1666-1696*, edited by Robert Palter, 258-284. Cambridge, Mass.: MIT Press, 1970. See (1201).

Stein, a philosopher, argues that Newton's conception of absolute space was a logically necessary part of his mechanics. Newton's

theological and metaphysical views about absolute space were consistent with but in no way the basis of his mechanics.
Topics: *XII*, VII Name: Newton

1493 Stein, Howard. "On the Notion of Field in Newton, Maxwell, and Beyond." In *Historical and Philosophical Perspectives of Science*, edited by Roger H. Struewer, 264-287. Minneapolis: University of Minnesota Press, 1970. See (1514).
Stein investigates the "concepts and principles loosely associated with the notion of field" (while recognizing that "field" itself is a recent term). His case study is Newton on the gravitational field. For Stein, this "field" is describable in terms of Newton's doctrines of God and creation.
Topics: *XII*, VII Name: Newton

1494 Steneck, Nicholas H. "Greatrakes the Stroker: The Interpretations of Historians." *Isis* 73(1982): 161-177.
Steneck argues against James Jacob's thesis which uses the Greatrakes incident to illustrate the social tension between religious moderates and extremists. He says that "the contention that the advance of science is based on social foundations" is a "new orthodoxy." Against this, Steneck finds the variety of opinions regarding natural philosophy among the supporters of Greatrakes to be more important than conflicting ideologies. And, by examining in chronological detail the healings and responses of the people involved, he argues for the "complexity of human records."
Topics: *IV*, I, XI Names: Boyle, Greatrakes, More

1495 Stephen, Leslie. *History of English Thought in the Eighteenth Century*. 2 vols. London: Smith, 1876. 3rd ed., New York: G. Putnam's Sons; London: Smith, Elder, 1902.
This once classic survey is now outdated but some historians still refer to it. See the chapters in Volume 1 on deism and on the continuity between the Latitudinarians and deism.
Topics: *VII*, VIII, IX Names: Clarke, Collins, A., Locke, Toland, Whiston, Wollaston

1496 Stephens, James. *Francis Bacon and the Style of Science*. Chicago: University of Chicago Press, 1975.

Stephens approaches Bacon's writings through an examination of his rhetorical style. In Chapter 4, "Fable-making as a Strategy of Style," he investigates Bacon's use of ancient fables and parables. Using these, Bacon constructs a new myth of science and ensures that only the "sons of science" can appreciate the true meanings of the delightful images. Stephens concludes by describing the Christian imagery of *New Atlantis*. Through his own style, Bacon is seen as undivided—both poet and one of the fathers of modern science.
Topics: *X*, VII Name: Bacon

1497 Stephens, Michael D., and Gordon W. Roderick. "The First Phases of the Dissenting Academies Movement in England and Wales and the Teaching of Science." *Technikgeschichte* 43(1976): 206-212.
In this brief article, the authors argue that the dissenting academies set up by clergy ejected from the Church of England after 1662 had a strong emphasis on science, as well as on classics and theology. They also attracted Anglicans and supplied "much of the science which backed up" the Industrial Revolution.
Topics: *V*, III Name: Morton

1498 Stern, William T. "The Influence of Leyden on Botany in the Seventeenth and Eighteenth Centuries." *Brit. J. Hist. Sci.* 1(1962): 137-158.
According to Stern, the University of Leyden's policy of toleration allowed English and Arminian dissenters to have a university education and to spread the teachings of their professors.
Topic: *V*

1499 Stewart, Larry. "Samuel Clarke, Newtonianism, and the Factions of Post-Revolutionary England." *J. Hist. Ideas* 42(1981): 53-72. Reprinted in *Philosophy, Religion and Science in the Seventeenth and Eighteenth Centuries*, edited by John W. Yolton, 520-539. Rochester, N.Y.: University of Rochester Press, 1990. See (1720).
Stewart argues that the High Church anti-Newtonian critique focused on Clarke's theology of the Trinity, with its Newtonian metaphysical basis. Stewart analyzes both Clarke's view and Roger North's papers. For North, a nonjuror and Cartesian, Newton's

understanding of gravitational attraction and Newton's distinction between absolute and relative terms resulted in an occult philosophy, a heterodox theology, and social and political factions.
Topics: *IV*, VII, IX, XII Names: Clarke, Hickes, Newton, North

1500 Stewart, M. A. Introduction to *Selected Philosophical Papers of Robert Boyle*, edited by M. A. Stewart, xi-xxxi. Manchester, England: Manchester University Press, 1979.
Besides a brief biography, the introduction contains paragraph summaries of the papers included in this volume. Stewart's suggestion that Boyle's scientific and theological interests converged is demonstrated in the selections.
Topic: *VII* Name: Boyle

1501 Stimson, Dorothy. "Ballad of Gresham College." *Isis* 18(1932): 103-117.
Stimson here edits a version of a ballad from the 1660s. She argues that the ballad was an early work of Joseph Glanvill. And she presents it as early evidence concerning the Royal Society.
Topics: *V*, III Name: Glanvill

1502 Stimson, Dorothy. "Comenius and the Invisible College." *Isis* 23(1935): 373-388.
Stimson argues against the thesis that the group around Comenius (when he visited England in 1641-42) was influential in the formation of the "Invisible College." The only person common to both groups was Theodore Haak (and he was not a central figure in either). Stimson also sees differences in age and in scientific method between the groups. She notes a possible Puritan influence in the formation of the Invisible College.
Topics: *V*, III, IV Names: Comenius, Haak, Hartlib, Wilkins

1503 Stimson, Dorothy. "Dr. Wilkins and the Royal Society." *J. Mod. Hist*. 3(1931): 539-563.
Stimson's early article focuses on the activities of John Wilkins in the creation and ongoing work of the Royal Society. In particular, she focuses on his universal language project. She also outlines Wilkins' life and ecclesiastical positions.
Topics: *V*, III, X Name: Wilkins

1504 Stimson, Dorothy. "Puritanism and the New Philosophy in Seventeenth-Century England." *Bulletin of the Institute of the History of Medicine* 3(1935): 321-334. Excerpted in *Puritanism and the Rise of Modern Science: The Merton Thesis*, edited by Bernard I. Cohen, 151-158. New Brunswick, N.J.: Rutgers University Press, 1990. See (0246).

This is a work which appeared three years before Merton's famous monograph (1057) and helped start the controversy. Stimson emphasizes "moderate Puritanism" as a movement which helped further the new philosophy. She bases her evidence on a statistical analysis of the early Royal Society and other groups. She also includes New England in her argument.

Topics: *III*, V

1505 Stimson, Dorothy. *Scientists and Amateurs: A History of the Royal Society*. New York: H. Schuman, 1948.

The first six chapters cover the seventeenth century. Among other topics, Stimson touches on the Puritanism question, notes the bishops among the original members, and discusses Sprat's apologetic defense of the Society. The title of this book reflects Stimson's view that one can distinguish between true scientists and amateur virtuosi. This book (purposely) is written on a rather popular level.

Topics: *V*, III Names: Boyle, Evelyn, Oldenburg, Sprat, Wilkins

1506 Stock, R. D. *The Holy and the Daemonic from Sir Thomas Browne to William Blake*. Princeton, N.J.: Princeton University Press, 1982.

Against "whig" oriented historians, philosophers, and literary critics, Stock argues that religious experience in the seventeenth and eighteenth centuries should be seen as including a sense of mystery. The first half of the book deals with figures who lived in our time frame. Stock's revisionist positions include the following. Donne's famous remark about the new philosophy is a lament for humans continuing to drift into egoism rather than for the passing of the old cosmology. Dryden and Thomas Browne were not reactionary fideists—but poets who simultaneously promoted the numinous elements of religion and accepted human reason. Apologists for witchcraft were at least as open-minded and scientifically sophisticated as the sceptics. In general, Stock plays down any

immediate impact of the new science on the religious beliefs of the writers he studies. But, in the second half of the book, he describes such an impact occurring later in the eighteenth century.
Topics: *X*, I, II Names: Browne, T., Casaubon, M., Defoe, Donne, Dryden, Glanvill, More, Swift, Webster

1507 Stone, Lawrence. "The Educational Revolution in England, 1560-1640." *Past and Present* 28(1964): 41-80.
Stone connects the expansion of educational facilities to the rise of the new science. And he connects Puritanism and Baconianism as forces in the development of scientific courses at all levels of education.
Topics: *V*, III, IV

1508 Straker, Stephen M. "What is the History of Theories of Perception the History Of?" In *Religion, Science, and Worldview*, edited by Margaret J. Osler and Paul Lawrence Farber, 245-273. Cambridge: Cambridge University Press, 1985. See (1184).
As a general historiographical argument, Straker proposes that the history of theories of perception and cognition be far more central to the history of science than heretofore has been the case. Such an approach would help both historians and philosophers to deal adequately with Koyre's tragic vision of "the significance of the Newtonian synthesis." Koyre leaves the answer as a riddle: the solution to the riddle of the universe becomes the riddle of the modern mind. Straker suggests that perhaps the scientific revolution as a transformation of the comprehension of the "self" is the answer to this riddle. And, he suggests, in the seventeenth century a voluntarist God was necessary in maintaining coherence between the perceiving self and the world.
Topics: *I*, VII

1509 Strauss, Emil. *Sir William Petty: Portrait of a Genius*. London: Bodley Head, 1954. Glencoe, Ill.: Free Press, 1954.
This biography may be seen as a particular example of the thesis relating Puritanism to the rise of science. Emphasis is given to Petty's interest in anatomy and medicine; recognition also is given to his continuing use of quantitative analysis in social, economic, and political as well as in scientific areas. Strauss notes Petty's

Puritan principles, especially his valuing of productive labor. Petty's political and social ideas paralleled those of some radicals. But his charm and shrewdness allowed him to seek, unsuccessfully, to serve Restoration monarchs as well.

Topics: *III*, IV, XI Names: Aubrey, Charles II, Evelyn, Graunt, Hartlib, Hobbes, Petty, Southwell, Worsley

1510 Stromberg, Roland N. *Religious Liberalism in Eighteenth-Century England*. Oxford: Oxford University Press, 1954.

Although Stromberg places little emphasis on the new science, this book includes good discussions on the Arians and on deism in the period after 1690.

Topics: *IX*, VII Names: Chubb, Clarke, Collins, A., Locke, Toland, Whiston

1511 Strong, Edward W. "Newton and God." *J. Hist. Ideas* 13(1952): 147-167.

Strong acknowledges the importance of God for Newton; but he then compartmentalizes Newton's thought so that he can argue against the relevance of God for Newton's science. "God was irrelevant to works written by a scientist in fulfillment of a scientific purpose."

Topics: *XII*, VII, VIII Name: Newton

1512 Strong, Edward W. "Newton's Mathematical Way." *J. Hist. Ideas* 12(1951): 90-110. Reprinted in *Roots of Scientific Thought*, edited by Philip P. Wiener and Aaron Noland, 412-432. New York: Basic Books, 1958. See (1670).

Newton's philosophy of science is presented as both mathematical/rationalist and as experimental/empiricist. Strong also argues that God is a "third-level" hypothesis for Newton and therefore not part of his science. Finally, Strong attempts (rather unsuccessfully) to show that Newton viewed absolute space and time as mathematical terms (mere postulates) and not as physical concepts.

Topics: *XII*, VII Name: Newton

1513 Strong, Edward W. *Procedures and Metaphysics: A Study in the Philosophy of Mathematical-Physical Science in the Sixteenth and Seventeenth Centuries*. Berkeley: University of California Press, 1936.

This is a sustained argument in book form criticizing many of the theses of E. A Burtt (see 0160). It is almost entirely continental in subject matter but it still is relevant to English developments. Strong interprets early modern science in terms of his philosophy of "the mathematical way" (that is, his division between "the metaphysical use of mathematics and the procedure of mathematics proper").

Topics: *VII*, I

1514 Struewer, Roger H., ed. *Historical and Philosophical Perspectives of Science.* (Vol. 5 in Minnesota Studies in the Philosophy of Science.) Minneapolis: University of Minnesota Press, 1970.

This anthology includes several relevant articles discussing historiographic principles in the history of science. The papers by Buchdahl (0143), Hesse (0674), Laudan (0935) and (0936), McMullin (1041), Stein (1493), and Thackray (1539) are annotated separately.

Topic: *I*

1515 Such, Jan. "Newton's Fields of Study and Methodological Principia." In *Isaac Newton's Philosophiae Naturalis Principia Mathematica: 15-17 October 1987, Lublin, Poland*, edited by W. A. Kaminski, 113-125. Teaneck, N.J.: World Scientific, 1988.

Such constructs a taxonomy of Newton's fields of study consisting of: (1) the classical physical sciences; (2) Baconian Studies; and (3) Parascientific Studies. He analyzes these in an ahistorical way and concludes that Newton used scientific method in his scientific studies but used "parascientific (or pseudoscientific) studies in alchemy and theosophy." This essay is of little historical use.

Topics: *XII*, VII Name: Newton

1516 Sullivan, Robert E. *John Toland and the Deist Controversy.* Cambridge, Mass.: Harvard University Press, 1982.

This is a solid study of deism and Toland in relationship to Socinianism and extremely liberal Anglicanism. Sullivan may go too far in emphasizing the similarities between deism, Toland's views, and latitudinarianism. There is little emphasis on the role of science or natural philosophy.

Topics: *IX*, VII Names: Blount, Clarke, Collins, A., Harley, Herbert, E., Locke, Stillingfleet, Tillotson, Tindal, Toland

1517 Sutton, John. "Religion and the Failures of Determinism." In *The Uses of Antiquity*, edited by Stephen Gaukroger, 25-51. Dordrecht: Kluwer Academic, 1991. See (0530).

Sutton discusses Renaissance and seventeenth-century views of materialism, determinism, and free will. He notes the parallels between predestinarian theology and deterministic philosophy of nature. He emphasizes the inconsistencies and failures of every position. "To trace a path from Pico . . . to . . . Marston and Webster is to document not an inflation of hopes for dominion over the natural world, but rather a loss of confidence in the possibility of control over even human affairs."

Topics: *VII*, X Names: Cudworth, Hobbes, Webster

1518 Svendsen, Kester. "Cosmological Lore in Milton." *ELH* 9(1942): 198-223.

See following annotation (1519).

Topic: *X* Name: Milton

1519 Svendsen, Kester. *Milton and Science*. Cambridge, Mass.: Harvard University Press, 1956.

Svendsen sees the popular science of the Renaissance (and not the new science) as most important in Milton's works. He analyzes several popular encyclopedias of science from the Middle Ages and the Renaissance—and emphasizes their influence on Milton.

Topic: *X* Name: Milton

1520 Svendsen, Kester. "Milton and the Encyclopedias of Science." *Stud. Philol.* 39(1942): 303-327.

See the previous annotation (1519).

Topic: *X* Name: Milton

1521 Svendsen, Kester. "Satan and Science." *Bucknell Review* 9(1960): 130-142.

Svendsen here presents *Paradise Lost* as including the traditional idea that the advance of science and technology has had Satanic connections.

Topics: *X*, VI Name: Milton

1522 Syfret, R. H. "The Origins of the Royal Society." *Notes Rec. Royal Soc. London* 5(1948): 75-137.

Syfret argues that (among various likely sources which contributed to the founding of the Royal Society) it was Theodore Haak "who first suggested the meetings" in London in 1645. Syfret frames Haak's motives and larger activities within the pansophic schemes of Hartlib, Dury, and Comenius. And she indicates the Christian utopian dimension of these schemes.

Topics: *V*, III, IV Names: Boyle, Comenius, Dury, Haak, Hartlib

1523 Syfret, R. H. "Some Early Critics of the Royal Society." *Notes Rec. Royal Soc. London* 8(1950): 20-64.

Here Syfret summarizes selected works and passages from the late 1660s into the early eighteenth century. She argues that the satirists or "men of wit" were more successful in presenting the Royal Society in a negative light then were the serious critics such as Meric Casaubon and Henry Stubbe. In addition to the usual arguments regarding the religious consequences of the new philosophy, Syfret discusses the more subjective issue of blasphemy. The satirists, she writes, presented the virtuosi as lowering themselves in their study of insects and other trivia—and, by extension, as affronting the dignity of God.

Topics: *V*, IV, X Names: Addison, Casaubon, M., Crosse, Glanvill, Stubbe

1524 Syfret, R. H. "Some Early Reactions to the Royal Society." *Notes Rec. Royal Soc. London* 8(1950): 207-258.

Syfret here summarizes the arguments of early (1660s) critics of the Royal Society and Sprat's defense of it. She organizes the essay according to Sprat's classification of potential critics: the universities; the Church; the physicians; the tradesmen; the nobility and gentry. About twenty-five pages are given over to religious issues and to the religious dimension of university-related criticisms. Syfret criticizes previous historians who had interpreted the age in terms of the warfare metaphor—but then puts forward her own theory as to why the religious critics of the Royal Society "were justified in their fears."

Topics: *V*, IV, IX, X Names: South, Sprat

1525 Sykes, Norman. *From Sheldon to Secker: Aspects of English Church History, 1660-1768*. Cambridge: Cambridge University Press, 1958.

This is a series of lectures on Anglican history. In Chapter 5, "True Religion and Sound Learning," Sykes surveys the theological currents with particular attention to Latitudinarianism. He briefly touches on the challenge of the new science. The entire book is interesting for its perspective and for background.

Topics: *VII*, VIII, IX Names: Burnet, G., Wilkins

1526 Sypher, G. W. "Similarities between the Scientific and the Historical Revolutions at the End of the Renaissance." *J. Hist. Ideas* 26(1965): 353-368.

Sypher compares the methodologies of Francis Bacon and Lancelot du Voison de la Popeliniere, a contemporary French historian. They held, he argues, similar views on the relations between science (or history) and religion. Both "disengaged theology from historical and scientific theory."

Topics: *VII*, X Name: Bacon

1527 Tagart, Edward. *Locke's Writings and Philosophy Historically Considered and Vindicated from the Charge of Contributing to the Scepticism of Hume*. London: Longmans, Brown, Green, and Longmans, 1855. Reprinted, New York: Garland, 1984.

A metaphysician and admirer of Locke sets his interpretation within his own philosophical debates (of the later eighteenth and early nineteenth centuries). Tagart argues that Gassendi was "the true intellectual parent of Locke." Tagart's exegesis of Locke's religious ideas is solid—but kept completely separate from natural philosophic issues. This work is apologetic but interesting and well-written.

Topic: *VII* Name: Hobbes, Locke

1528 Talmor, Sascha. "Glanvill and Hume." *Durham University Journal* 72(1980): 183-193. A shorter version appears as "Glanvill and Locke," in *Glanvill*, by Sascha Talmor, 52-72. Oxford: Pergamon Press, 1981. See (1529).

This essay concerns Glanvill's (and Hume's) views on causality and the supernatural.

Topic: *VII* Name: Glanvill

1529 Talmor, Sascha. *Glanvill: The Uses and Abuses of Scepticism*. Oxford: Pergamon Press, 1981.

Talmor sees Glanvill as consistent throughout his career; moreover, his philosophical, scientific, and religious views were of a piece. She emphasizes that, for Glanvill, human knowledge is limited because of the Fall of Adam. Human knowledge of the soul and mind is especially obscure and mysterious. Thus, Glanvill's scepticism could be used by theologians as well as by "secular" thinkers. Talmor also argues that, in historical context, Glanvill's belief in witches and the supernatural makes sense. Finally, she discusses Glanvill as an example of the change in literary style that took place in seventeenth-century England under the influence of the new science.

Topics: *VII*, II, IX, X Names: Glanvill, Locke

1530 Talmor, Sascha. "Locke and Glanvill: A Comparison." *Locke Newsletter* 9(1978): 101-120. A similar version appears as "Glanvill and Locke," in *Glanvill*, by Sascha Talmor, 37-51. Oxford: Pergamon Press, 1981. See (1529).

Talmor argues that, because of his Platonism, Glanvill's differences from Locke are significant. This is especially to be noted with respect to Glanvill's discussion of the human soul.

Topic: *VII* Names: Glanvill, Locke

1531 Tarnas, Richard. *The Passion of the Western Mind: Understanding the Ideas That Have Shaped Our World View*. N.Y.: Harmony Books, 1991.

This is a well-written survey of "Western Civilization" from Socrates to the present based on a wide reading of the sources. Tarnas traces the development of successive "world views of the West's mainstream high culture, focusing on the crucial sphere of interaction between philosophy, religion, and science." Tarnas mainly writes with an extreme, retrospective approach based on his own evolutionary philosophy of history; but, as a minor supplement, he sometimes presents past writers' views in their own terms. He discusses the relationships of science and religion in terms of concord, compromise, and conflict—with an overall emphasis on "The Triumph of Secularism." About fifty pages are given over specifically to our period—but there are also sections on relevant background topics such as "The Christian World View"

and medieval, Renaissance, and Reformation factors. "Passion" for Tarnas seems to refer to some archetypal set of unconscious factors; this survey does not present intellectual history as passionate in the usual sense. Sometimes, in fact, the "Western mind" becomes so reified in the discussion (having a logic of its own) that it seems to lose all connection to actual human beings.
Topics: *VII*, III, X, XII Names: Bacon, Locke, Newton

1532 Taube, Mortimer. *Causation, Freedom and Determinism*. London: G. Allen and Unwin, 1936.
This is marginal but does include a discussion of Hobbes.
Topics: *VII*, IX Name: Hobbes

1533 Taylor, E. G. R. "The English Worldmakers of the Seventeenth Century and their Influence on the Earth Sciences." *Geographical Review* 38(1948): 104-112.
See the following annotation (1534).
Topics: *VII*, X Names: Burnet, T., Hooke, Keill, Ray, Whiston, Woodward

1534 Taylor, E. G. R. "The Origin of Continents and Oceans: A Seventeenth Century Controversy." *Geographical Journal* 116(1950): 193-198.
In his two articles, Taylor relates the birth of the geological sciences to theories controverted in late-seventeenth-century England. The first essay treats theories of the origin of the world and the second essay focuses on theories of the origin of the oceans, continents, and mountains. Taylor's approach is based on a retrospective historiography with "the new experimental philosophy" as the progressive element. But he does (dutifully) describe the importance of theological factors in his discussions.
Topics: *VII*, X Names: Burnet, T., Hooke, Keill, Ray, Whiston, Woodward

1535 Teich, Mikulas, and Robert Young, eds. *Changing Perspectives in the History of Science: Essays in Honour of Joseph Needham*. London: Heinemann, 1973.
This anthology includes several important articles on historiographic, alchemical and societal topics. See the entries under

Buchdahl (0141), Cohen (0240), Hesse (0675), Rattansi (1298), Webster (1627), and Young (1725).
Topics: *I*, II, IV

1536 Tellkamp, August. *Das Verhaltnis John Lockes zur Scholastik.* (Veroffentlichungen des Katholischen Instituts fur Philosophie Albertus-Magnus-Akademie zu Koln, Band 2, Heft 2.) Munster in Westfalen, Germany: Aschendorff 1927.
This is a short book of 120 pages which focuses on Locke, scholasticism, and the English universities in the seventeenth century. Tellkamp includes no direct discussion of natural philosophy, but he does focus on epistemology, metaphysics, and theology.
Topics: *VII*, V, X Name: Locke

1537 Thackray, Arnold. *Atoms and Powers: An Essay on Newtonian Matter-Theory and the Development of Chemistry.* Cambridge, Mass.: Harvard University Press, 1970.
Thackray describes the impact of Newtonianism on eighteenth-century chemistry with a special focus on matter theory. (And he argues that Newtonian writings had a deep and pervasive influence on eighteenth-century scientific thought.) He considers a variety of philosophical, experimental, polemical, nationalistic, and theological factors for each writer or group that he considers. For the period up to the 1720s, Thackray studies Newton, various Newtonians, and Robert Greene. (Over 100 pages cover the period into the 1720s.) Although he does not explicate the connection in much detail, Thackray repeatedly notes the importance of theological considerations for British natural philosophy in the eighteenth century.
Topics: *VII*, XII Names: Cheyne, Greene, Gregory, Hales, Harris, Keill, Newton, Pitcairne

1538 Thackray, Arnold. "'Matter in a Nut-Shell': Newton's *Opticks* and Eighteenth-Century Chemistry." *Ambix* 15(1968): 29-53.
This, in effect, is a summary of Thackray's book (1537) published two years later. "Matter in a Nut-Shell" refers to the doctrine that "all the solid matter in the solar system might be contained within a nut-shell."
Topics: *VII*, XII Names: Clarke, Newton

1539 Thackray, Arnold. "Science: Has Its Present Past a Future?" In *Historical and Philosophical Perspectives of Science*, edited by Roger H. Struewer, 112-133. Minneapolis: University of Minnesota Press, 1970. See (1514).

Thackray both summarizes past historiographical positions and proposes that there should be greater diversity in approaches to their topics by historians of science. (Although he does not deal directly with issues concerning possible relations between science and religion, Thackray's suggestions have been realized to some extent in subsequent studies involving those relations.) Also see Thackray's "Comment" on Mary Hesse's paper, pp. 160-162, in the same volume.

Topic: *I*

1540 Thiel, Udo. "Cudworth and Seventeenth-Century Theories of Consciousness." In *The Uses of Antiquity*, edited by Stephen Gaukroger, 79-99. Dordrecht: Kluwer Academic, 1991. See (0530).

In the course of making other points, Thiel discusses Cudworth's philosophy of nature and the possible relations between material and immaterial causes. Cudworth's understanding of "plastic nature" as both immaterial and unconscious constituted an interesting argument against Cartesian dualism. Cudworth's understanding of God and of the human soul are also mentioned.

Topic: *VII* Name: Cudworth

1541 Thomas, Keith. *Religion and the Decline of Magic: Studies in Popular Beliefs in Sixteenth and Seventeenth Century England*. London: Weidenfeld and Nicolson; New York: Scribner, 1971

The purpose of this massive volume is "to make sense of the systems of belief which were current in sixteenth- and seventeenth-century England, but which no longer enjoy much recognition today." Thus, Thomas includes chapters on magic (including healing and alchemy), astrology, prophecy, witchcraft, and other allied beliefs. Because of their close relationship to religion, he also discusses the impact of the reformation and of popular religion. He poses the speculation that Protestantism attempted to separate the magic found in medieval Catholicism from religion. But, as he shows, the practices persisted in other forms (e.g., "cunning men" and "wise women"). Thomas poses several explanations for the decline of magic, but he does not seem completely satisfied with any of them. For example, there is the empiricism of the new natural philosophy, and the increased control of the environment. But magic appears to have declined before technology substituted for it. He appears to place greatest emphasis on the "new faith in the potentialities of human initiative" embodied by the aspirations of the new philosophers. He is, however, more confident about the gulf that deepened "between the educated classes and the lower strata of the rural population." And he believes that the new natural philosophy certainly played a role in this sociological phenomenon.

Topics: *II*, III, IV, VI, XI Names: Ashmole, Aubrey, Bacon, Booker, Gadbury, Hopkins, Lilly, Perkins

1542 Thorndike, Lynn. "The Attitude of Francis Bacon and Descartes towards Magic and Occult Science." In *Science, Medicine and History*, edited by E. Ashworth Underwood, Vol. 1, 451-454. London: Oxford University Press, 1953. See (1580).

Thorndike argues with examples that Bacon retained many elements of the occult traditions.

Topic: *II* Name: Bacon

1543 Thorndike, Lynn. *History of Magic and Experimental Science*. Vols. 7-8. New York: Columbia University Press, 1958.

Generally, Thorndike assumes that there is no significant relationship between science and religion. He does see an "admixture" of scientific and nonscientific elements in seventeenth-century England. His volumes are of little use for historical interpretation; but they are still valuable for bibliographic references and quotations from primary texts.

Topics: *II*, I, VII

1544 Thorner, Isidor. "Ascetic Protestantism and the Development of Science and Technology." *Am. J. Sociol.* 58(1952): 25-33.

Thorner reinforces de Candolle and Merton by analyzing Protestant and Catholic countries.

Topics: *III*, VI

1545 Thrower, Norman J. W., ed. *Standing on the Shoulders of Giants: A Longer View of Newton and Halley. Essays Com-*

memorating the Tercentenary of Newton's "Principia" and the 1985-1986 Return of Comet Halley. Berkeley and Los Angeles: University of California Press, 1990.

See the essays by Dobbs (0385), Force (0493), Genuth (0534), and Kubrin (0921) annotated separately in this bibliography.

Topics: *VII*, XII Names: Halley, Newton

1546 Thulesius, Olav. *Nicholas Culpepper: English Physician and Astrologer*. New York: St. Martin's Press, 1992.

Thulesius sketches the life and works of this self-trained astrologer and physician. The chapters of his book deal with: Culpepper's life; his writings on astrology, midwifery, alchemy and herbs; and his later influence. Thulesius indicates that we know a lot about what Culpepper wrote but very little about him as a person. Though he recognizes Culpepper's importance as an influential figure among his contemporaries, Thulesius writes in a very present-minded way (looking for Culpepper's lasting achievements).

Topics: *XI*, II Name: Culpepper

1547 Tielsch, Elfriede W. "The Secret Influence of the Ancient Atomistic Ideas and the Reaction of the Modern Scientist under Ideological Pressure." *Hist. Euro. Ideas* 2(1981): 339-348.

Tielsch proposes a six-category typology for the history of science as regards scientists since the Renaissance era who were under pressure to conceal the true origins of their ideas. He uses ancient atomism as a test case.

Topics: *I*, IV, VII

1548 Todd, Richard. *The Opacity of Signs*. Columbia, Mo.: University of Missouri Press, 1986.

Todd's basic thesis is that George Herbert's interpretation of his own experience of relationship to God is analogous to the reader's act of interpretation of Herbert's verse. Within that framework, the chapter on "Providence" presents Herbert's reading of the book of nature. Todd discusses Herbert's views on how fallen humans must strive towards a prelapsarian state by deciphering creation and also on how humans are priests in and for creation.

Topic: *X* Name: Herbert, G.

1549 Tolles, Frederick B. *Meeting House and Counting House: The Quaker Merchants of Colonial Philadelphia*. Chapel Hill, N.C.: University of North Carolina Press, 1948.

In the chapter "Votaries of Science," Tolles proposes that the religious ethos of the Quakers (on both sides of the Atlantic) led to an interest in science on their part. This can be seen in their educational system, in the careers of Quaker doctors, and in the activities of James Logan. (A short portion of this chapter is reprinted in Chapter 4 of Tolles' *Quakers and the Atlantic Culture*, New York, 1960.)

Topics: *VII*, III, V, XI Names: Logan, Penn

1550 Torrance, Thomas F. "The Making of the 'Modern' Mind from Descartes and Newton to Kant." In *Transformation and Convergence in the Frame of Knowledge*, by Thomas F. Torrance, 1-61. Grand Rapids, Mich.: W. B. Eerdmans, 1984.

Torrance argues that the "modern" mind was shaped by two "fateful forces": (1) the transfer of intelligibility from an independent contingent reality to "the human pole of the knowing relationship"; and (2) the belief that humans can understand only what we make. Within this framework, Torrance focuses on Newton's understanding of the relation of the universe to God. This article serves as background to Torrance's program of exploring the interrelation of methods in theology and science. His style is difficult but his knowledge of the primary sources in both areas make his writings worthwhile.

Topics: *VII*, XII Name: Newton

1551 Toulmin, Stephen. *Cosmopolis: The Hidden Agenda of Modernity*. New York: Free Press, 1990.

This book combines a philosophical autobiography with both historical and historiographical approaches to the question of modernity. The first three chapters are relevant to our themes as they are focused to a large extent on the seventeenth century and its interpreters. In particular, Toulmin finds Donne grieving for cosmopolis (the playful coherence between the order of nature and that of society). And from his discussion of Newton comes an outline of the principal elements of modernity, including the creator God's relation to nature. While the creators of the modern world view hoped "to frame all their questions in terms that rendered them

independent of context," Toulmin hopes to "recontextualize" them. Thus he places the development of the vision of a clear and stable society and nature in the context of the political/social/theological turmoil of the seventeenth century.
Topics: *IV*, I, X, XII Names: Donne, Newton

1552 Toulmin, Stephen. "Criticism in the History of Science: Newton on Absolute Space, Time, and Motion." *Philos. Rev.* 68(1959): 1-29 and 203-227.
This case study from the extreme "internalist" phase of Toulmin's career is directed toward "a happier reading of Newton's famous scholium" on absolute space, time, and motion. His happier reading is that Newton did not believe in the reality of absolute space and that there were no significant connections between space and theological issues in Newton's view. This is a forced interpretation, as even the later Toulmin himself might recognize.
Topics: *XII*, VII Name: Newton

1553 Toulmin, Stephen. "From Form to Function: Philosophy and History of Science in the 1950's and Now." *Daedalus* 106(1977): 143-162.
This is an excellent review of change in his discipline from the 1950s to the mid-1970s. Toulmin elaborates on the change from internalist "rational reconstructions" of scientific achievements to interdisciplinary, contextual investigations of problems and issues. Although nothing is said regarding religion, one can see its relevance to Toulmin's argument in this essay (and one finds Toulmin including it in his later works).
Topic: *I*

1554 Toulmin, Stephen. "Nature and Nature's God." *J. Rel. Ethics* 13(1985): 37-52.
This is a response to James Gustafson's proposed ethics as a revival of stoic thought. In a section on historical background, Toulmin suggests that natural philosophers in the seventeenth century shared two assumptions: "the deanimation of nature" and the associated separation of God the Author of Creation from Nature, its object.
Topic: *VII*

1555 Toulmin, Stephen. "Seventeenth Century Science and the Arts." In *Seventeenth Century Science and the Arts*, edited by Hedley Howell Rhys, 3-28. Princeton, 1961. See (1328).

Toulmin discusses his general theory of how science could have affected culture and how culture could have affected science. He seems generally to accept Lovejoy's leading ideas or presuppositions—but argues against science reflecting the *zeitgeist* of the surrounding culture. He does not discuss religion directly.

Topics: *IV*, I

1556 Toulmin, Stephen, and June Goodfield. *The Discovery of Time*. New York: Harper and Row, 1965.

English natural philosophers are discussed in Chapter 4 of this survey of the development of modern historical inference. They propose the insight that conflicts were not between science and religion—but within science itself. But they retrospectively suggest that their retention of biblical chronology "blinded" seventeenth-century natural philosophers to the geological and cosmological evidence.

Topics: *X*, VII Names: Burnet, T., Hooke

1557 Trengrove, Leonard. "Newton's Theological Views." *Ann. Sci.* 22(1966): 277-294.

This article is concerned mainly with Newton's published biblical studies—especially his *Observations upon the Prophecies*.

Topics: *XII*, X Name: Newton

1558 Trevor-Roper, Hugh R. *Catholics, Anglicans and Puritans: Seventeenth-Century Essays*. Chicago: University of Chicago Press, 1988.

This collection includes two essays, one on James Ussher and one on Nicholas Hill, which are relevant to our field. See (1561) and (1562).

Topic: *VII*

1559 Trevor-Roper, Hugh R. *The Crisis of the Seventeenth Century: Religion, the Reformation and Social Change*. New York: Harper and Row, 1968.

This is the same book as (1564), *Religion, the Reformation, and Social Change*, 1967.

Topic: *IV*

1560 Trevor-Roper, Hugh R. "The European Witch-craze of the Sixteenth and Seventeenth Centuries." In *Religion, the Reformation and Social Change*, by Hugh R. Trevor-Roper, 90-192. London: Macmillan, 1967. See (1564).

This long essay deals mainly with the continental situation and with a large period of time. But the essay is relevant as background for seventeenth-century Britain. Trevor-Roper sees the witch-craze as part of a total ideological pattern reinforced by the post-reformation orthodoxy, the Counter-Reformation, and neo-Aristotelianism as well as by important social-psychological factors. Natural philosophy is part of this larger ideological pattern.

Topics: *II*, IV

1561 Trevor-Roper, Hugh R. "James Ussher, Archbishop of Armagh." In *Catholics, Anglicans, and Puritans*, by Hugh R. Trevor-Roper, 120-165. Chicago: University of Chicago Press, 1988. See (1558).

This is a summary of Ussher's life and thought. Trevor-Roper emphasizes his anti-Catholicism and his encyclopedic historical scholarship (both used in defense of the Elizabethan settlement of the Church of England.) Trevor-Roper also discusses Ussher's best-known project—his chronology of the ancient world beginning with the creation in 4004 B.C. (Although Trevor-Roper does not discuss science, Ussher's biblical chronology influenced most theories in seventeenth-century discussions of "Genesis and geology.")

Topics: *X*, VII Name: Ussher

1562 Trevor-Roper, Hugh R. "Nicholas Hill, the English Atomist." In *Catholics, Anglicans, and Puritans*, by Hugh R. Trevor-Roper, 1-39. Chicago: University of Chicago Press, 1988. See (1558).

Trevor-Roper reconstructs Hill's life and thought—and attempts to correct previous writers. Hill is a little-known figure who was a follower of Bruno and a convert to Roman Catholicism. His book, *Philosophia Epicurea* (1601), was influential later in the century.
Topic: *VII* Name: Hill

1563 Trevor-Roper, Hugh R. "The Paracelsian Movement." In *Renaissance Essays*, by Hugh R. Trevor-Roper, 149-199. Chicago: University of Chcago Press; London: M. Secker and Warburg, 1985. See (1566).

This essay mainly deals with the continental story. But pages 178-180 and 185-195 focus on seventeenth-century England. Trevor-Roper emphasizes the essential Neoplatonism of Paracelsianism and does a good job of describing it as a total ideology. The author argues that there was no necessary connection between Paracelsianism and Protestantism or Puritanism: but the way in which the Counter-Reformation developed in an anti-Platonic manner led the Paracelsian movement to become effectively Protestant; and the particular polarization of English politics and religion in the reign of Charles I led it to become effectively Puritan in England.
Topics: *XI*, II, IV, V Names: Comenius, Hartlib, Webster

1564 Trevor-Roper, Hugh R. *Religion, the Reformation and Social Change*. London: Macmillan, 1967.

This is an anthology of nine essays with various theses about religion, politics, and society in seventeenth-century Europe (including England and Scotland). Most have been published elsewhere; three are annotated separately in this bibliography. See: (1560), (1565), and (1568). This is the same book as (1559), *The Crisis of the Seventeenth Century*, 1968.
Topic: *IV*

1565 Trevor-Roper, Hugh R. "The Religious Origins of the Enlightenment." In *Religion, the Reformation and Social Change*, by

Hugh R. Trevor-Roper, 193-236. London: Macmillan, 1967. See (1564).

Trevor-Roper postulates a general thesis covering all of Europe over a 300-year period. He sees the orthodox Reformation parties and the Counter-Reformation as all inimical to the development of the Enlightenment. (And he sees the new philosophy or science as part of the Enlightenment.) For him, Arminianism, Socinianism, heresy, and the liberal-laicized traditions (of Erasmus, Bacon, Grotius and others) provide the true religious roots of the Enlightenment.

Topics: *VII*, III, IX Name: Bacon

1566 Trevor-Roper, Hugh R. *Renaissance Essays*. Chicago: University of Chicago Press; London: M. Secker and Warburg, 1985.

This anthology includes two essays, "The Paracelsian Movement" and "Robert Burton and *The Anatomy of Melancholy*," which are not published elsewhere and which are annotated separately in this bibliography. See (1563) and (1567).

Topic: *XI*

1567 Trevor-Roper, Hugh R. "Robert Burton and *The Anatomy of Melancholy*." In *Renaissance Essays*, by Hugh R. Trevor-Roper, 239-274. Chicago: University of Chicago Press; London: M. Secker and Warburg, 1985. See (1566).

This is a rambling summary and commentary on Burton and his most famous work. Trevor-Roper presents Burton as, among other things, an amateur physician and a rather indifferent Anglican clergyman. Several pages focus on Burton's views on the medical and super-natural causes of melancholy and the related cures.

Topics: *X*, XI Name: Burton

1568 Trevor-Roper, Hugh R. "Three Foreigners: The Philosophers of the Puritan Revolution." In *Religion, the Reformation and Social Change*, by Hugh R. Trevor-Roper, 237-293. London: Macmillan, 1967. See (1564). An earlier version appeared in *Encounter*, 14(Feb. 1961): 3-20.

In this extended article, Trevor-Roper argues that "the real philosophers, the only philosophers, of the English Revolution" were Hartlib, Dury, and Comenius. Their version of Baconianism correlated with the country party's interests in political and religious decentralization and laitization. They thus provided the center of the intellectual world that surrounded Cromwell. The group these men gathered later became members of the Royal Society. But Trevor-Roper does not see the three foreigners as scientists; and he argues that there was only an indirect and gradual effect of the Pansophical dream on the institutionalization of science.

Topics: *IV*, V Names: Bacon, Comenius, Dury, Hartlib

1569 Trinterud, L. J. "A.D. 1689: The End of the Clerical World." In *Theology in Sixteenth- and Seventeenth-Century England*, by L. J. Trinterud and W. S. Hudson, 25-51. Los Angeles: William Andrews Clark Memorial Library, University of California, 1971.

Trinterud suggests that the Anglican clerics who turned to natural philosophy and natural theology did so "to find a way out of England's problems" (i.e., to stabilize society). He sets this thesis within an overarching examination of how the five largest religious groups dealt with issues arising from religious pluralism.

Topics: *IV*, VII, VIII

1570 Trompf, Garry W. "On Newtonian History." In *The Uses of Antiquity*, edited by Stephen Gaukroger, 213-249. Dordrecht: Kluwer Academic, 1991. See (0530).

This is an excellent discussion of Newton's views on chronology and ancient history. Trompf unites not only Newton's biblical, chronological, and theological work—but connects all of these with Newton's *prisca* and natural philosophical outlook. A key factor here, besides Newton's understanding of God, is the role of Noah in his theologico-historical scheme. Trompf builds upon Manuel and goes beyond both Manuel (0984) and Westfall (1640). He has mastered both the published and manuscript sources: he analyzes them in Newton's own terms (and without any false surprise at Newton's methods or results). He concludes that Newton's "was no secular mind" and "his work bespeaks as much, if not more, of traditionalisms of the past than of liberalisms of the future."

Topics: *XII*, X Name: Newton

1571 Turnbull, George H. "George Stirk, Philosopher by Fire (1628?-1665)." *Pub. Colonial Soc. Mass. Trans.* 38(1947-1951): 219-251.

This paper is a biographical study of Stirk (or Starkey) which emphasizes his alchemical and medical experiments. Turnbull also argues that the anonymous alchemist Eirenaeus Philalethes should be identified as Stirk. In a useful appendix, Turnbull lists works definitely written by Stirk and works probably written by Stirk.

Topics: *II*, XI Names: Hartlib, Starkey

1572 Turnbull, George H. *Hartlib, Dury, and Comenius: Gleanings from Hartlib's Papers*. Liverpool: University of Liverpool Press, 1947.

Turnbull reconstructs portions of the lives of Hartlib, Dury, and Comenius from the (then) newly found papers of Hartlib. He revises his own previously published views (1574). This work contains long quotes and appendices of documents by or about each man named in the title.

Topic: *IV* Names: Comenius, Dury, Hartlib

1573 Turnbull, George H. *Samuel Hartlib, with Special Regard to His Relations with J. A. Comenius*. London: Oxford University Press, 1919.

This work is superseded by Turnbull's later work (see 1572).

Topic: *IV* Names: Comenius, Hartlib

1574 Turnbull, George H. "Samuel Hartlib's Influence on the Early History of the Royal Society." *Notes Rec. Royal Soc. London* 10(1953): 101-130.

This article includes much biographical detail regarding Hartlib's acquaintances and correspondence. Turnbull argues against R. H. Syfret as regards the influence of Comenius, Haak, and Hartlib in the formation of the Royal Society. He downplays Hartlib's scientific significance *because* of Hartlib's religious and utopian motives. But Turnbull's spadework on Hartlib has been extremely useful to later scholars who have emphasized the connections between natural philosophy, religion, and utopian reform.

Topics: *V*, IV Names: Comenius, Haak, Hartlib

1575 Tuveson, Ernest Lee. *Millennium and Utopia: A Study in the Background of the Idea of Progress*. Berkeley and Los Angeles: University of California Press, 1949.

Tuveson sees the idea of progress as intertwined with both science and religion. About one-third of the book deals with seventeenth-century England. This book is especially good on the ideas of Providence, the millennium, and their relation to his topic in our period. It also is relevant for the ancients versus moderns controversy.

Topics: *IV*, VII, X Names: Boyle, Burnet, T., Mede

1576 Tuveson, Ernest. "Space, Deity and the 'Natural Sublime'." *Mod. Lang. Q.* 12(1951): 20-38.

Tuveson traces the transformation whereby infinite space and "bigness" in nature became symbols of deity and, thereby, became "sublime." According to the author, More's writings about infinite space and Locke's argument for God's existence formed the intellectual basis. Burnet and (especially) Addison fitted the new sensibility and theory together into a new aesthetic.

Topics: *X*, VII Names: Addison, Burnet, T., Locke, More

1577 Tuveson, Ernest Lee. "Swift and the World-Makers." *J. Hist. Ideas* 11(1950): 54-74.

Tuveson proposes that, beginning in the 1690s, one of the central issues in the relation of science to religion was the debate over the writings of the "World-makers" and their "scientific" interpretations of the events in Genesis. At issue, finally, was the sense of progress and the testing of scripture by reason—a new "world spirit" that Swift regarded as degenerate.

Topic: *X* Names: Burnet, T., Swift, Temple, W., Whiston, Wood ward

1578 Tyacke, Nicholas. "Science and Religion at Oxford before the Civil War." In *Puritans and Revolutionaries*, edited by D. Pennington and K. Thomas, 73-93. Oxford: Clarendon Press, 1978. See (1219).

Tyacke presents evidence that science at Oxford did not decline or stagnate during the Laud's chancellorship of the late 1620s and the

1630s; and he argues that *therefore* no positive correlation exists between Puritanism and the rise of science.
Topics: *V*, III Names: Bainbridge, Briggs, Greaves, Laud

1579 Ullmann-Margalit, Edna. *The Scientific Enterprise*. The Bar-Hillel Colloquium: Studies in the History, Philosophy, and Sociology of Science, Vol. 4. Dordrecht: Kluwer Academic, 1992.
See the annotations for essays by Dobbs (0382), Popkin (1240), Vickers (1587), and Westfall (1637).
Topics: *VII*, II, XII Name: Newton

1580 Underwood, E. Ashworth, ed. *Science, Medicine and History*. 2 vols. London: Oxford University Press, 1953.
The essays by Miller (1070) and Thorndike (1542) are annotated separately.
Topic: *XI*

1581 Underwood, T. L. "Quakers and the Royal Society of London in the Seventeenth Century." *Notes Rec. Royal Soc. London* 31(1976): 133-150.
Underwood sets himself the goal of testing "Puritanism-and-the-rise-of-science" hypotheses by focusing on Quakers. He concludes with mixed results—claiming to have found evidence both to support and to undermine a connection between Quakerism and the rise of science. Underwood argues that only two Fellows of the Royal Society had any significant connection to the Society of Friends.
Topics: *III*, V Names: Conway, Finch, J., Haistwell, Penn

1582 Van de Wetering, Maxine. "Moralizing in Puritan Natural Science: Mysteriousness in Earthquake Sermons." *J. Hist. Ideas* 3(1982): 417-438.
Van de Wetering traces the providentialist interpretation of earthquakes in the sermons of American Puritans in the seventeenth century. She argues that the Puritan tendency to moralize rationally about strange natural events and to dwell on secondary causes precluded "any true sense of the mysterious." Moralizing, not science, "forced the diminution of the mystery of the universe."
Topics: *VII*, X Names: Doolittle, Mather, I.

1583 Van Leeuwen, Henry G. *The Problem of Certainty in English Thought: 1630-1690*. The Hague: Martinus Nijhoff, 1963. 2nd. ed., The Hague: Martinus Nijhoff 1970.

Van Leeuwen argues that the development of the theory of knowledge characteristic of seventeenth-century English science arose from theological controversies. He reviews Catholic and Protestant polemicists who previously had used an epistemological scepticism which could not be philosophically answered. He sees the empirical philosophy as a commonsense constructive scepticism which issued from the destructive scepticism of the religious controversies. What for religious philosophers was a practical means for obtaining the most probable knowledge which the evidence supports became, for the natural philosophers, their empirical method. One weakness of Van Leeuwen's approach is that he often treats seventeenth-century English writers as if they were pragmatists 250 years ahead of their time.

Topics: *VII*, XII Names: Bacon, Boyle, Chillingworth, Glanvill, Locke, Newton, Tillotson, Wilkins

1584 Van Pelt, Robert Jan. "The Utopian Exit of the Hermetic Temple; or, A Curious Transition in the Tradition of the Cosmic Sanctuary." In *Hermeticism and the Renaissance*, edited by Ingrid Merkel and Allen G. Debus, 400-423. Washington, D.C.: Folger Shakespeare Library, 1988. London: Associated Universities Presses, 1988. See (1051).

This is an interesting exploration of the symbolisms of the Hermetic-Christian temple and of utopia. In the seventeenth century, Van Pelt proposes, the architectural images of the temple at Jerusalem show the development of the symbol of an exit from the Holy of Holies by which the saved were to return to earth to form a new utopian society.

Topics: *II*, IV, VI, X Names: Bacon, Herbert, G.

1585 Vassilieff, A. "H. More, Newton, et Berkeley." In *Atti del V Congresso Internazionale di Filosofia*, International Congress of Philosophy, 1924., 1045-1049. Naples: Perrella, 1924.

This is a brief note on the spiritual character of absolute space, its historical background and its critics.

Topics: *XII*, VII Names: More, Newton

1586 Vickers, Brian. "Analogy Versus Identity: The Rejection of Occult Symbolism, 1580-1680." In *Occult and Scientific Mentalities in the Renaissance*, edited by Brian Vickers, 95-163. Cambridge: Cambridge University Press, 1984. See (1591).

Vickers contends, at great length, "that the occult and scientific traditions can be differentiated . . . in terms of goals, methods, and assumptions." The key distinction is found in the use of language: for the scientific tradition, there is a clear distinction between words and things; for the occult tradition, there is no such distinction. Vickers examines theories of language (e.g., Adamic language) as well as how natural philosophers used language. Though Vickers claims initially to put persons along a spectrum, his practice seems to make the categories clear-cut and rigid.

Topics: *X*, II Names: Bacon, Fludd, Hobbes

1587 Vickers, Brian. "Critical Reactions to the Occult Sciences During the Renaissance." In *The Scientific Enterprise*, edited by Edna Ullman-Margalit, 43-92. Dordrecht: Kluwer Academic, 1993. See (1579).

Against Yates (1709) and others, Vickers argues that astrology, as an example of an occult science, was the subject of an articulate critique throughout the Renaissance. He concludes with a brief discussion of a few figures from the mid-to-late-seventeenth century. His view here, as in other of his writings, is that "occult sciences" were rhetorical devices—not sciences.

Topics: *II*, X Names: More, Ray

1588 Vickers, Brian, ed. *Essential Articles for the Study of Francis Bacon*. Hamden, Conn.: Archon Books, 1968.

This is an anthology of important reprinted articles, with an introduction by Vickers (1589). Articles found in the present bibliography include: Bullough (0153); Harrison (0636); Hesse (0673); R. F. Jones (0839); Kocher (0891); Lemmi (0942); Prior (1267); and Whitaker (1658).

Topics: *VII*, IV, X Name: Bacon

1589 Vickers, Brian. Introduction to *Essential Articles for the Study of Francis Bacon*, edited by Brian Vickers, vii-xxiii. Hamden, Conn.: Archon Books, 1968. See (1588).

This is a survey of Bacon scholarship with a focus on the essays included in the anthology—and a description of what yet needs to be done (in 1968). Vickers emphasizes the continuity of approach, for Bacon, across the variety of disciplines; and he notes "the ever-present religious basis" in all of Bacon's work.
Topics: *VII*, I, X Name: Bacon

1590 Vickers, Brian. Introduction to *Occult and Scientific Mentalities in the Renaissance*, edited by Brian Vickers, 1-55. Cambridge: Cambridge University Press, 1984. See (1591).
In his lengthy introduction, Vickers presents historiographic background as well as summarizing and commenting on the essays in the volume. He suggests that further studies in Renaissance occult mentalities should make use of anthropological research.
Topics: *I*, II

1591 Vickers, Brian, ed. *Occult and Scientific Mentalities in the Renaissance*. Cambridge: Cambridge University Press, 1984.
Annotated separately in our bibliography are: an essay (1586) as well as the Introduction (1590) by Vickers; and essays by Clark (0219), Mulligan (1097), Rees (1320), Westfall (1642), and Westman (1654).
Topic: *II*

1592 Wade, Gladys I. "Thomas Traherne as 'Divine Philosopher'." *Hibbert J.* 32(1934): 400-408.
This is a brief appreciation of Traherne and of his platonic philosophy. Wade compares him to several Cambridge Platonists and notes (in passing) his philosophy of nature.
Topic: *X* Names: Cudworth, Traherne

1593 Wagner, Fritz. "Church History and Secular History as Reflected by Newton and His Time." *History and Theory* 8(1969): 97-111.
Wagner foresees the separate disciplines of church history and secular history becoming more interdependent because both have a transcendent reference point: metahistorical (i.e., hermeneutical) and metaphysical assumptions. The author seems to be guided by continental more than by Anglo-American discussions. He proposes 1700 as an approximate time of when the disciplines began to be

separated. He uses Newton and especially Desaguliers as examples or symptoms of the division. But, he also argues, the whole of Newton does not quite fit the division (given his theological and alchemical manuscripts). Indeed, the two types of history could share in the kind of theological studies which Newton undertook.
Topics: *I*, XII Names: Desaguliers, Newton

1594 Walker, Daniel P. *The Ancient Theology: Studies in Christian Platonism from the Fifteenth to the Eighteenth Century*. Ithaca, N.Y.: Cornell University Press, 1972.

Important for our period and topic areas are the Introduction and Chapters 5 and 7. These are excellent studies on Edward, Lord Herbert of Cherbury, and on the Chevalier Ramsay. Combining biography and analysis of specific writings, Walker shows how each interwove natural philosophy (ancient and contemporary) with religion: star symbolism and astronomy in Herbert's case and a Newtonian interpretation of God's *sensorium* in Ramsay's. Theologically, Herbert's "common notions" and Ramsay's emphasis on the Chinese as a source for the ancient true theology indicate their willingness to transcend the boundaries of Christendom. This is a valuable, early contribution to the study of the *prisca* tradition.
Topics: *II*, VIII, IX, X XII Names: Herbert, E., Newton, Ramsay

1595 Walker, Daniel P. *The Decline of Hell: Seventeenth-Century Discussions of Eternal Torment*. London: Routledge and Kegan Paul, 1964.

Generally, this work deals with purely theological discussions and theories of hell among selected seventeenth-century writers. If anything, it may show how little impact the new science could have on some theological discussions. However, Walker does take up some writers who were integrating their theological and natural philosophic concepts—including Cudworth, More, Glanvill, Burnet, and Whiston. This book also has relevant material concerning platonism and *prisca* traditions.
Topics: *VII*, II, X Names: Burnet, T., Conway, Cudworth, Glanvill, More, Whiston

1596 Walker, Daniel P. "The Elusive Rosicrucians." *Hist. Sci.* 11(1973): 306-310.

This is an essay review of Yates' *The Rosicrucian Enlightenment* (1709). Walker reviews the work positively and adds some of his own judgments concerning the need for further research. He emphasizes the magical, alchemical, and religiously heterodox factors in the Rosicrucian movement.

Topics: *II*, IX

1597 Walker, Daniel P. "Francis Bacon and *Spiritus*." In *Science, Medicine and Society in the Renaissance*, edited by Allen G. Debus, Vol. 2, 121-130. New York: Science History, 1972. See (0357).

Walker examines the complexities of Bacon's distinction between the "rational soul" and Spiritus. The former is directly infused into humans by God while the latter makes up the irrational soul which is derived from the elements and is shared by other animals. This brief article asks the right questions but has been superseded by the work of Graham Rees.

Topics: *VII*, XI Name: Bacon

1598 Walker, Daniel P. "Medical Spirits and God and the Soul." In *Spiritus*, edited by M. Fattori and M. Bianchi, 223-244. Rome: Edizioni dell' Ateneo, 1984. See (0446).

Walker proposes that confusions between medical spirits and the Christian meanings of spirit resulted in "audacious conceptions" of the soul and unorthodox theologies of God. Henry More and Ralph Cudworth are his seventeenth-century English examples—although he also raises some questions about Newton. In particular, More is castigated by Walker for his tendency to pile up "coextensive entities all doing the same thing at the same time" (e.g., God as infinitely extended, the incorporeal Spirit of Nature, and the aethereal but corporeal spirits).

Topics: *XI*, VII, XII Names: Cudworth, More, Newton

1599 Walker, Daniel P. *Spiritual and Demonic Magic: from Ficino to Campanella*. London: Warburg Institute, University of London, 1958. Reprinted, Notre Dame, Ind.: University of Notre Dame Press, 1975.

Primarily, this is a study of fifteenth- and sixteenth-century continental theories of magic (broadly interpreted). Towards the end, there is an excellent four-page note on Bacon. Walker includes Bacon among the followers of Telesio in his conception

of the human spirit. And he suggests that Bacon dislikes magic because it is too easy.
Topics: *II*, XI Name: Bacon

1600 Wallace, Dewey D. "Socinianism, Justification by Faith, and the Sources of John Locke's *The Reasonableness of Christianity*." *J. Hist. Ideas* 45(1984): 49-66. Reprinted in *Philosophy, Religion and Science in the Seventeenth and Eighteenth Centuries*, edited by John W. Yolton, 152-169. Rochester, N.Y.: University of Rochester Press, 1990. See (1720).
There is no direct connection to science here: but, indirectly, the article serves to undercut theses which connect the new science either to Locke's *The Reasonableness of Christianity* or to deism. Wallace argues: (1) that "justification" was the principal subject of Locke's treatise; (2) that Locke was a sincere Anglican of the Latitudinarian type; and (3) that Locke was neither a Socinian nor a Deist. This is a good article for the religious background to our field.
Topics: *IX*, X Name: Locke

1601 Wallace, Karl R. *Francis Bacon on the Nature of Man*. Urbana, Ill.: University of Illinois Press, 1967.
This is an exploration of Bacon's understanding of the six faculties of the whole person: understanding; reason; memory; imagination; will; and appetite. Wallace finds Bacon to have invented a "psychology of discovery"—and not a new kind of induction. This "psychology" (though Wallace prefers to use seventeenth-century terms) has implications both for theology and for science. For example, the imagination is the faculty through which God has communicated directly with humans. And, Wallace says of Bacon, the spirit is the instrument of the rational soul and the faculties are modes of spirit activity.
Topic: *VII* Name: Bacon

1602 Wallace, William A. "Newton's Early Writings: Beginning of a New Direction." In *Newton and the New Direction in Science*, edited by G. V. Coyne and others, 65-72. Vatican City: Specola Vaticana, 1988. See (0281).
The author describes and summarizes those parts of Newton's "Trinity notebook" which reflect his reading of Aristotelian textbooks by Ioannes Magirus and Daniel Stahl. Wallace argues that

these notes from his early years show a certain Aristotelianism on Newton's part which should be noted (along with the new direction in science associated with his name). This Aristotelian factor in Newton's thought includes several theological elements—including the notion that nature is God's instrument.

Topics: *XII*, VII, VIII Name: Newton

1603 Walton, Michael T. "Boyle and Newton on the Transmutation of Water and Air from the Root of Helmont's Tree." *Ambix* 27(1980): 11-18.

In this brief note, Walton suggests that Boyle drew ideas about the transmutation of water from van Helmont's experiments, and that Newton was influenced by Boyle in his discussion of the tails of comets. Finally, for Boyle and Newton, a similar understanding of divine creation played an important role in their ideas about transmutation.

Topics: *VII*, XII Names: Boyle, Newton

1604 Warch, Richard. *School of the Prophets: Yale College, 1701-1740*. New Haven, Conn.: Yale University Press, 1973.

This book provides a good discussion of the early years of Yale—including theological disputes. Chapters on the curriculum detail the subjects covered and note lectures on various schools of natural philosophy. In this regard, Warch sees a slow shift from Aristotelianism, to Cartesianism, to Newtonian physics. He suggests that a tension between natural philosophy and revealed theology, though "latent," produced a shift in mood toward an emphasis on "the great Establisher."

Topics: *V*, VII, XII Names: Ames, Cutler, Mather, C., Pierson

1605 Ward, Robert. "What Forced by Fire: Concerning Some Influences of Chemical Thought and Practice Upon English Poetry." *Ambix* 23(1976): 80-95.

Ward notes instances of the use of chemical and alchemical imagery in English poetry with religious and scientific themes from the sixteenth through the twentieth centuries.

Topic: *X* Names: Donne, Herbert, G.

1606 Warhaft, Sidney. "The Providential Order in Bacon's New Philosophy." *Studies in Literary Imagination* 4(1971): 49-64.

Warhaft argues that Bacon thoroughly believed that God providentially created and ordered nature and would continue to do so to the end of time. This leads to a discussion of Bacon's view of the role of humans in searching after the knowledge of nature combined with charity. For Bacon, Warhaft indicates, nature was created for human benefit but not for human exploitation.

Topics: *VII*, IV, VI Name: Bacon

1607 Watson, Patricia A. *The Angelical Conjunction: The Preacher-Physicians of Colonial New England*. Knoxville, Tenn.: University of Tennessee Press, 1991.

This is a study of the large group of Congregational ministers in Colonial New England who were trained in and practiced medicine. Watson describes their theological framework for understanding and diagnosing illness. He finds that they combined Galenic and Paracelsian (including alchemical) medical practices. He also sociologically analyzes their dual profession—particularly their finances and job security.

Topics: *XI*, II, V Names: Bulkeley, Culpepper, Lee, Mather,C., Palmer, Wigglesworth

1608 Webb, C. J. J. *Studies in the History of Natural Theology*. Oxford: Clarendon Press, 1915.

Basically, this has to do with ancient and medieval philosophers—but it includes a short chapter on Herbert of Cherbury. Webb, a philosopher, approaches the material in a rather ahistorical way. He shows a tangential interest in natural philosophy. The long "Introduction" includes an interesting distinction between natural theology and revealed theology.

Topics: *VIII*, VII Name: Herbert, E.

1609 Webb, Suzanne S. "Raleigh, Hariot, and Atheism in Elizabethan and Early Stuart England." *Albion* 1(1969): 10-18.

Webb finds little real evidence for Raleigh's reputed atheism. She finds his association with Harriot to be the primary source "of the calumny which accrues to Raleigh." She is arguing against the thesis that there was a tightly organized "coterie of maverick intellectuals, generally referred to as the School of the Night." And she sees Raleigh both as a Christian and as similar to Bacon in his approach to nature.

Topics: *IX*, VII Names: Harriot, Ralegh

1610 Weber, Max. *The Protestant Ethic and the Spirit of Capitalism*. Translated by Talcott Parsons. New York: Scribner, 1930. Translated from *Gesammelte Aufsatze zur Religionssoziologie*. Vol. 1. Tubingen: J. C. B. Mohr, 1920.

In a long footnote (no. 145 on p. 249) Weber suggested a link between Protestantism and science—even specifying, among other forms, seventeenth-century Puritan Christianity. Weber's work as a whole, though outside our topical bounds, was a springboard for the work of Robert K. Merton (1057).

Topic: *III*

1611 Webster, Charles. "The Authorship and Significance of *Macaria*." *Past and Present* 56(1972): 34-48. Reprinted in *The Intellectual Revolution of the Seventeenth Century*, edited by Charles Webster, 369-385. London: Routledge and Kegan Paul, 1974. See (1618).

Webster argues that the Utopian tract *Macaria* was written by Gabriel Plattes within a Puritan context; and he shows the importance of studying the English scientific movement in its larger social and intellectual context. He argues against the restrictive historiography of A. R. Hall. This essay later was incorporated into (1616).

Topics: *IV*, I, III Names: Hartlib, Plattes

1612 Webster, Charles. "Benjamin Worsley: Engineering for Universal Reform from the Invisible College to the Navigation Act." In *Samuel Hartlib and Universal Reformation*, edited by Mark Greengrass, 213-235. Cambridge: Cambridge University Press, 1994. See (0573).

This is a brief biographical study. Webster shows how, in Worsley's checkered political career, engineering was an instrument of reform. Worsley thereby exemplifies the interplay between spiritual motivations and practical endeavors in the universal religious vision of the Hartlib circle. This pattern of vision, motivations, and endeavors resulted in both the failures and the successes of the members of the circle.

Topics: *VI*, III, IV Names: Hartlib, Worsley

1613 Webster, Charles. "The College of Physicians: 'Solomon's House' in Commonwealth England." *Bull. Hist. Med.* 41 (1967): 393-412.

Webster takes his thesis from Walter Charleton's view that there was a close parallel between the College of Physicians and Bacon's Solomon's House. In a portion of *The Immortality of the Human Soul*, Charleton describes the activities of the College—which is quoted in full in this article. Webster then goes on to propose that the College of Physicians and the Royal Society had complementary scientific outlooks and personnel.

Topics: *V*, IV, XI Names: Bacon, Charleton

1614 Webster, Charles. "English Medical Reformers of the Puritan Revolution: A Background to the 'Society of Chymical Physitians'." *Ambix* 14(1967): 16-41.

Webster shows that the Puritan Revolution allowed medical institutions which claimed a monopoly—the College of Physicians in particular—to come under suspicion. Utopian authors proposed, among other reforms, that priests be allowed to practice medicine. Moreover, the Revolution provided the climate for new chemical, Paracelsian, and hermetic therapies. The "Society of Chymical Physitians" was a brief attempt after the Restoration to continue that tradition as an alternative to the authority of the College of Physicians.

Topics: *XI*, II, III, IV, V Names: Hartlib, Rand, Starkey

1615 Webster, Charles. *From Paracelsus to Newton: Magic and the Making of Modern Science*. The Eddington Memorial Lectures, 1980. Cambridge: Cambridge University Press, 1982.

Webster demonstrates "an important degree of contiguity between the worldviews of the early sixteenth and late seventeenth centuries." Against the then-received historiographical tradition that there is a perfect correlation between the rise of science and the decline of magic, Webster was one of the historians who posed a less radical epistemological shift and greater diversity in the worldview of the Scientific Revolution. His method is to explore three "test cases": prophecy, spiritual magic and demonic magic. Though large portions of these chapters cover the ideas of continental Europeans (especially Paracelsus), there are insights into developments in England which have since been elaborated upon by other historians. "Throughout the Scientific Revolution," he writes, "Christian eschatology provided an undiminished incentive for science, if not a primary motivating factor." Likewise, the Baconian reform of natural magic, not its elimination, was the "guiding principle" for organized scientific activity. And the new natural philosophers (especially Newton) saw themselves as magus figures. Finally, in the last section, Webster finds the new natural philosophers "dragged along with the tide" rather than leaders in the decline of witchcraft.

Topics: *I*, II, XII Names: Hartlib, Newton, Whiston

1616 Webster, Charles. *The Great Instauration: Science, Medicine and Reform, 1626-1660*. London: Duckworth, 1975.

This is a massive book (630 pages) which thoroughly examines the social and intellectual context of science, technology, medicine, education, Puritan religion, and reform during the middle decades of the seventeenth century. Webster sees a complicated but close relationship between Puritan/Protestant views of a this-worldly millennialism and the rise of the new science. The book can be dry and hard to read: a possible approach is to read the last chapter ("Conclusions") for the main theses and orientation—and then read other sections selectively. It includes good discussions of Bacon, Harvey, Comenius, Hartlib, Dury, Wilkins, Boyle, and many other individuals. Webster includes coverage of the various natural philosophic traditions but is especially thorough as regards Baconianism and "the new science." He emphasizes the context of science rather than the content of science. There is extensive discussion of all the major institutions relevant to science, medicine, education, and social reform. He also gives significant attention to technology and agriculture. Basically, Webster agrees with Merton and Hill in support of the connection between Puritanism and the rise of science. This is a valuable book even for those who do not agree with every thesis that Webster puts forward.

Topics: *IV*, III, V, VI, X, XI Names: Bacon, Beale, Biggs, Boate, A., Boate, G., Boyle, Briggs, Chamberlen, Charleton, Child, Comenius, Dell, Dury, Hartlib, Harvey, Hubner, Plattes, Webster, Wilkins, Winstanley, Wren

1617 Webster, Charles. "Henry Power's Experimental Philosophy." *Ambix* 14(1967): 150-178.

This is a biographical study of Power's life, philosophy, and work. Webster shows that Power's experimental philosophy was eclectic and that his activities were carried on independently of the Royal Society. The motivation for his activities (e.g., his microscopy) was "to pay homage to the creator."

Topics: *VII*, XI Names: Boyle, Browne, T., Power

1618 Webster, Charles, ed. *The Intellectual Revolution of the Seventeenth Century*. London: Routledge and Kegan Paul, 1974.

This anthology is a collection of 28 articles from *Past and Present* (originally published from 1953 to 1973). Half of the articles are relevant to this bibliography. The collection focuses: (1) on the Protestantism or Puritanism-and-the-rise-of-science debate; and (2) on science, society, and social ideals. Annotated separately are a good introduction by Webster (1619) and articles by: 'Espinasse (0436); Hill (0689), (0693), (0697), and (0698); Kearney (0862) and (0863); Mason (0996); Mulligan (1095); Rabb (1279) and (1280); Shapiro (1439); Webster (1611); and Whitteridge (1666).

Topics: *IV*, III

1619 Webster, Charles. Introduction to *The Intellectual Revolution of the Seventeenth Century*, edited by Charles Webster, 1-22. London: Routledge and Kegan Paul, 1974. See (1618).

This amounts to a review of selected scholarly literature concerning the revolutionary aspects of seventeenth-century England. These include political theory, radical political movements, millenarianism, Puritanism, and the new science. As he reviews key issues related to these topics, Webster defends contextual (not "anachronistic") historiography in general—and externalist (not "restrictive") historiography of science in particular.

Topics: *IV*, I, III

1620 Webster, Charles. Introduction to *Samuel Hartlib and the Advancement of Learning*, by Charles Webster, 1-72. Cambridge: Cambridge University Press, 1970.

The lengthy introduction to a collection of documents by Hartlib essentially is an intellectual biography of Hartlib and his circle.

Webster focuses on proposals for educational reform as the basis for social, religious, and scientific advancement. Although several of the proposals had the backing of some parliamentarians during the Interregnum, Hartlib's vision remained unfulfilled.

Topics: *IV*, V Names: Bacon, Comenius, Dury, Hartlib, Milton, Pell

1621 Webster, Charles. "New Light on the Invisible College: The Social Relations of English Science in the Mid-17th Century." *Trans. Royal Hist. Soc.* 24(1974): 19-42.

Webster uses Boyle's early biography to reconstitute the membership of the "Invisible College." He argues that it referred neither to the Hartlib group nor to the Oxford predecessors of the Royal Society. Rather, it centered in Lady Ranelagh's home and Benjamin Worsley's laboratory; and it consisted primarily of "Anglo-Irish intellectuals associated with the Boyle family." As such, it focused on experimental problems that had immediate economic application (e.g., husbandry). Thus, in the conditions of 1646, it "was one expression of the debate on religious, political and social affairs within the puritan movement." This work is only summarily incorporated into Webster's book (1616).

Topics: *V*, III, IV, VI Names: Boyle, Ranelagh, Worsley

1622 Webster, Charles. "The Origins of the Royal Society: Review Essay." *Hist. Sci.* 6(1967): 106-128.

This is a wide-ranging and sustained criticism of Margery Purver's book (1269).

Topic: *V*

1623 Webster, Charles. "Puritanism, Separatism, and Science." In *God and Nature: Historical Essays on the Encounter between Christianity and Science*, edited by David C. Lindberg and Ronald Numbers, 192-217. Berkeley and Los Angeles: University of California Press, 1986. See (0954).

Webster summarizes the arguments for a connection between English science and Puritanism in the revolutionary period. He notes especially the relevance of Puritan separatism while he usefully points out the diversity of Puritan science. In response to the critics of the thesis, Webster here carefully chooses his terms and phraseology.

Topic: *III*

1624 Webster, Charles. "Richard Towneley (1629-1707), The Towneley Group, and Seventeenth-Century Science." *Transactions of the Historic Society of Lancashire and Cheshire* 118(1966): 51-76. Excerpt reprinted in *Puritanism and the Rise of Modern Science*, edited by I. Bernard Cohen, 51-76. New Brunswick, N.J.: Rutgers University Press, 1990. See (0246).

This is a study of a lesser-known Catholic natural philosopher and his associates in Lancashire. Webster notes that case-by-case studies such as this provide exceptions to broad correlations between science and religion in this period. This case shows both that not all natural philosophers interested in the new science were Puritans and that there was not a single "Catholic" approach to science.

Topics: *IV*, III Name: Towneley

1625 Webster, Charles. *Utopian Planning and the Puritan Revolution: Gabriel Plattes, Samuel Hartlib, and the "Macaria."* Oxford: Wellcome Unit for the History of Medicine, 1979.

This is a facsimile of the utopian tract *Macaria* (printed in 1641) with introduction and notes. Repeating his usual themes—and amounting to an appendix to *The Great Instauration*—Webster connects Puritanism, Baconianism, and the Hartlib circles to science, technology, and social reform. He argues that intellectuals and craftsmen together sought to bring about the Kingdom of God. He sees Plattes as the author of the anonymous tract.

Topics: *IV*, III, VI Names: Hartlib, Plattes

1626 Webster, Charles. "Water as the Ultimate Principle in Nature: The Background to Boyle's Skeptical Chymist." *Ambix* 13(1966): 96-107.

Webster here examines the Willow Tree experiment and the Water Culture experiment as they influenced Boyle. He touches on Izaak Walton's biblical basis for claiming the supremacy of water among the elements.

Topics: *VII*, X Names: Boyle, Browne, T., Walton

1627 Webster, Charles. "William Dell and the Idea of University." In *Changing Perspectives in the History of Science*, edited by

Nicholas Teich and Robert Young, 110-126. London: Heinemann, 1973. See (1535).

William Dell was a radical Puritan and Master of Caius College, Cambridge, from 1649 to 1660. Webster's study shows a division within Puritanism regarding education. Dell wished to transform education and society in order that people receive the Holy Spirit. Though not an active experimenter himself, Dell was an advocate for the sciences as useful to "the new millennial social order which had no place for the ordained ministry or scholastic education."

Topics: *IV*, III, V   Name: Dell

1628 West, J. F. *The Great Intellectual Revolution*. London: J. Murray, 1965.

West briefly surveys the intellectual revolution of the seventeenth century—leading to a present worldview characterized by specialized studies. He emphasizes the separation of metaphor and truth.

Topics: *X*, VII, XII   Names: Bacon, Hobbes, Locke, Newton

1629 West, Muriel. "Notes on the Importance of Alchemy to Modern Science in the Writings of Francis Bacon and Robert Boyle." *Ambix* 9(1961): 102-114.

For West, alchemy is a "physical science" and does not deal with metaphysics. Its basic subject matter must be understood as chemistry. She argues that this also is the view of Bacon and Boyle.

Topic: *II*   Names: Bacon, Boyle

1630 Westfall, Richard S. "The Changing World of the Newtonian Industry." *J. Hist. Ideas* 37(1976): 175-184.

This is an essay review (with humorous asides) of selected works by Whiteside, Cohen, McGuire, Rattansi, Dobbs, and Manuel. Westfall's theme is that Newtonian studies since 1960 have been moving away from Newton the ideal modern scientist and ideal citizen to Newton the seventeenth-century man. The latter approach includes Neoplatonism, alchemy, and religion as significant factors. Westfall emphasizes the importance of previously unpublished manuscripts as a key to the revised picture of Newton.

Topics: *XII*, I, II   Name: Newton

1631 Westfall, Richard S. "Charting the Scientific Community." In *Trends in the Historiography of Science*, edited by Kostas

Gavrogin and others, 1-14. Dordrecht: Kluwer Academic, 1993.

Westfall summarizes his present research project of exploring the "parameters of the social existence of those engaged in the study of nature during the sixteenth and seventeenth centuries." He has analyzed 630 "scientists" taken from the *Dictionary of Scientific Biography*—using a number of categories, including religion and profession.

Topics: *I*, V

1632 Westfall, Richard S. *Force in Newton's Physics: The Science of Dynamics in the Seventeenth Century*. London: McDonald; New York: American Elsevier, 1971.

Westfall mainly focuses on the basic problems of seventeenth-century mechanics without connection to religion or theology. But he does set the development of seventeenth-century dynamics within the context of competing philosophies of nature and the respective metaphysical considerations. And Chapter 7, "Newton and the Concept of Force," includes an important discussion of God in Newton's natural philosophy. Westfall sees Newton's concept of God as relevant to his theory of matter, to his theory of absolute space, to his theories concerning an aether, and to his postulation of universal gravitation. Newton's idea of God was foundational to his entire conception of nature and it freed him to develop a quantitative approach to the regularities of nature (including the concept of force). Westfall also argues that his idea of God allowed Newton to reconcile the mechanical tradition and the Pythagorean tradition without being thwarted by a demand for causal explanations in terms of impact mechanics. Newton's quantitative dynamics—including forces of attraction and repulsion—was possible because of his belief in a Divine Medium which is not mechanical. This part of the book (both the main text and the extensive chapter endnotes) is important because it is not reproduced in Westfall's biographies of Newton.

Topics: *XII*, VII Name: Newton

1633 Westfall, Richard S. "The Influence of Alchemy on Newton." In *Science, Pseudo-Science, and Society*, edited by Marsha P. Hanen and others, 145-169. Waterloo, Ont.: Wilfrid Laurier University Press, 1980. See (0623).

This is a chronological study of Newton's interest in alchemy. Westfall argues, hesitantly, that alchemy stimulated Newton's mind to consider attractions as physical forces.
Topics: *XII*, II, VII Name: Newton

1634 Westfall, Richard S. "Isaac Newton: A Sober, Thinking Lad." *Review* (Indiana Univ.) 18(1975): 17-33.
This is an earlier draft of Chapter 2 of *Never at Rest* (1640).
Topic: *XII* Name: Newton

1635 Westfall, Richard S. "Isaac Newton in Cambridge: The Restoration University and Scientific Creativity." In *Culture and Politics From Puritanism to the Enlightenment*, edited by Perez Zagorin, 135-164. Berkeley and Los Angeles: University of California Press, 1980.
Westfall here describes the state of corruption of Cambridge University in the late seventeenth century. Westfall relates this state of affairs to the patronage system in general—and to the patronage system of the Church of England in particular. He argues that the university gave no direct support to science or any scientific community during this period. (This material is *not* reproduced in *Never at Rest*.)
Topics: *XII*, V Name: Newton

1636 Westfall, Richard S. "Isaac Newton, Religious Rationalist or Mystic?" *Review of Religion* 22(1958): 155-170.
Westfall argues effectively against categorizing Newton as a mystic—especially of a Boehmenistic sort.
Topics: *XII*, II Name: Newton

1637 Westfall, Richard S. "Isaac Newton: Theologian." In *The Scientific Enterprise*, edited by Edna Ullman-Margalit, 223-239. Dordrecht: Kluwer Academic, 1993. See (1579).
This article is an excellent summary of Westfall's understanding of Newton's theology. He distinguishes between Newton's "traditional" published position on natural theology and his unorthodox Arian doctrines found in manuscript form. Westfall also relates Newton's study of the prophecies to his unorthodox, secretive positions and sets his secretiveness in its institutional context. Westfall concludes that Newton's thought was one chapter in the

great, possibly tragic shift from a Christian civilization to a scientific one.
Topics: *XII*, VII, VIII, X, IX Name: Newton

1638 Westfall, Richard S. "Isaac Newton's 'Theologiae Gentilis Origines Philosophicae'." In *The Secular Mind: Transformations of Faith in Modern Europe*, edited by W. Warren Wager, 15-34. New York: Holmes and Meier, 1982.
Westfall calls attention to a theological manuscript of Newton's ("The Philosophical Origins of Gentile Theology") which never has been published but which sheds much light on Newton's heterodox theological views. Newton's extremely "low" Christology, his view of the history of religion as the continuous corruption of true religion, and his view of revelation all are described—but in a way that is debatable.
Topics: *XII*, VII Name: Newton

1639 Westfall, Richard S. *The Life of Isaac Newton*. Cambridge: Cambridge University Press, 1993.
This is a reduced version of *Never at Rest* (1640). Aimed at a general audience, it has less technical material and no footnotes. It is 340 pages long instead of 900. The main theses and historiography remain the same, of course, and the effective date of publication remains 1980 (except for a few bibliographic references).
Topic: *XII* Names: Barrow, Bentley, Conduitt, Fatio, Locke, Newton, Whiston

1640 Westfall, Richard S. *Never at Rest: A Biography of Isaac Newton*. Cambridge: Cambridge University Press, 1980.
This is certainly the definitive biography of Newton and probably will remain so for many years to come. The book runs to over 900 pages and is well-done in its basic historiography, in its organization, and in its detail. Westfall also is a good storyteller and the book makes for interesting reading. It meets the highest standards of scholarship in terms of Westfall's thorough mastery of published and unpublished primary sources and in terms of his mastery of the secondary literature. Westfall is an outstanding historian of science—and the book's strength lies especially in this dimension. But over one hundred pages of the book are given over to Newton's religion; and about thirty pages relate Newton's science and

religion to each other. Westfall basically organizes the biography in chronological order and emphasizes issues of development and stability in Newton's thought. He places Newton nicely within his historical context. He gives some psychological interpretations but generally disagrees with Manuel's neo-Freudian explanations. Westfall provides a full account of Newton's Arianism and other heretical opinions. He presents Newton as extremely pious (albeit in an heretical way) whose religious views and passion significantly influenced his natural philosophic views. Some of these influences have to do with Newton's views on gravity, absolute space, atomism, active principles in the material universe, the mechanical philosophy in general and, of course, the creation of the universe. Likewise, Westfall sees Newton's science as influencing his theological views. Newton's biblical studies were influenced by his empirical and mathematical orientations; and his natural theology was influenced by his natural philosophy. For Westfall, Newton's understanding of God's dominion and his views of ancient philosophers seem to have been mutual to his natural philosophy and to his theology. Westfall does not include his earlier extreme and retrospective interpretations concerning Newton's relation to deism, atheism, and Enlightenment religious rationalism. He also deals extensively with Newton's alchemy and with Newton's mixed reactions to Neoplatonism. This book is required reading for all students of Newton and for most students of science and religion in our period.

Topics: *XII*, II, IV, VII, IX, X Names: Barrow, Bentley, Clarke, Conduitt, Fatio, Locke, More, Newton, Whiston

1641 Westfall, Richard S. "Newton and Absolute Space." *Arch. Int. d'Hist. Sci.* 17(1964): 121-132.

Westfall here discusses Newton's reasons for postulating absolute space and (among other things) the relationship between absolute space and God in Newton's philosophy of nature. He also argues that Newton's view of God liberated him from an exclusive concern with the construction of mechanical models of causal explanation. This essay largely is incorporated in (and superseded by) the same author's *Force in Newton's Physics* (1632).

Topics: *XII*, VII Name: Newton

1642 Westfall, Richard S. "Newton and Alchemy." In *Occult and Scientific Mentalities in the Renaissance*, edited by Brian Vickers, 315-335. Cambridge: Cambridge University Press, 1984. See (1591).

Regarding the relation between Newton's alchemy and his physics, Westfall seeks to defend a position between the extremes taken by I. B. Cohen and A. R. Hall on the one hand and David Castillejo on the other. He concludes that Newton's theory of "force embodies the enduring influence of alchemy" upon his scientific thought.

Topics: *XII*, II Name: Newton

1643 Westfall, Richard S. "Newton and Christianity." In *Religion, Science, and Public Policy*, by Frank T. Birtel, 79-95. New York: Crossroad, 1987.

This essay includes: (1) a good summary of Newton's natural theology and of its place in his natural philosophy; (2) a good description of Newton's Christology based on the unpublished manuscripts; and (3) a defense of Westfall's previously published interpretations of Newton's relationship to later developments in European civilization.

Topics: *XII*, VIII, X Name: Newton

1644 Westfall, Richard S. "Newton and the Hermetic Tradition." In *Science, Medicine and Society in the Renaissance*, edited by Allen G. Debus, Vol. 2, 183-198. New York: Science History, 1972. See (0357). Reprinted in Vol. 11 of *Articles on Witchcraft, Magic and Demonology*, edited by Brian P. Levack, 217-232. New York: Garland, 1992. See (0945).

This is a proposal about the ways in which Newton synthesized Hermetic and mechanical modes of thought. (Westfall later modified his position on this matter.)

Topics: *XII*, II, VII Name: Newton

1645 Westfall, Richard S. "Newton's Scientific Personality." *J. Hist. Ideas* 48(1987): 551-570.

This article includes several pages on Newton as theologian and argues that "this endeavor was a central aspect of his personality." It basically recapitulates what Westfall has written elsewhere.

Topics: *XII*, X Name: Newton

1646 Westfall, Richard S. "Newton's Theological Manuscripts." In *Contemporary Newtonian Research*, edited by Zev Bechler, 129-143. Dordrecht: D. Reidel, 1982. See (0071).

This is a summary of Newton's theological writings with a helpful (66-item) "Checklist of Newton's Theological Manuscripts." Westfall also summarizes his own well-known interpretation of the relation of religion and theology to Newton's science. His theses include: (a) there is no theological influence on Newton's science; (b) but there is a "religious" influence (which he leaves unexplained); (c) there definitely is a scientific influence both on Newton's theology and with respect to European civilization generally; (d) this influence has to do with the displacement of traditional biblical revelation by natural religion.

Topics: *XII*, VII, VIII, X Name: Newton

1647 Westfall, Richard S. "The Rise of Science and the Decline of Orthodox Christianity: A Study of Kepler, Descartes, and Newton." In *God and Nature: Historical Essays on the Encounter between Christianity and Science*, edited by David C. Lindberg and Ronald Numbers, 218-237. Berkeley and Los Angeles: University of California Press, 1986. See (0954).

With respect to religion in Western civilization, what was the ultimate effect of the scientific revolution? Westfall argues that, while it did not logically entail the decline of Christianity and religion, the scientific revolution inevitably did raise challenges to central aspects of received Christianity. Newton, explicitly and implicitly for Westfall, is a central figure in this development.

Topics: *VII*, X, XII Name: Newton

1648 Westfall, Richard S. "Robert Hooke, Mechanical Technology, and Scientific Investigation." In *The Uses of Sciences in the Age of Newton*, edited by John G. Burke, 85-110. Berkeley and Los Angeles: University of California Press, 1983. See (0157).

Westfall here uses Hooke's technological inventions as a test case for internalist versus contextualist history of science. Westfall specifically questions whether the Baconian religio-utilitarian goals regarding applied science actually bore major consequences.

Topics: *I*, IV, VI Name: Hooke

1649 Westfall, Richard S. "The Role of Alchemy in Newton's Career." In *Reason, Experiment, and Mysticism in the Scientific Revolution*, edited by M. L. Righini Bonelli and William R. Shea, 189-232 and 305-316. New York: Science History, 1975. See (1329).

At a time when the appropriate place of alchemy in the history of science was hotly debated, Westfall attempted to pin down the exact place of alchemy in Newton's natural philosophy. Religion is only indirectly connected to natural philosophy in this discussion. Generally, this essay has been superseded by Westfall's biography of Newton and by the writings of Dobbs.

Topics: *XII*, II, VII Name: Newton

1650 Westfall, Richard S. *Science and Religion in Seventeenth-Century England*. New Haven, Conn.: Yale University Press, 1958. Second ed. with new preface, Ann Arbor, Mich.: University of Michigan Press, Ann Arbor Paperbacks, 1973.

Based on research done in 1950-1952, this was an important contribution at the time when it was first published. The young Westfall writes loosely within the tradition of W. E. H. Lecky, Andrew White, and Franklin Baumer. In the thought of the seventeenth-century English virtuosi, Westfall emphasizes what he sees as the ultimate conflict between science and religion and what he sees as the superficial harmony attributed to them. In his running commentary on the seventeenth-century writers, he treats the proposed reconciliations between science and religion not only as superficial, but as artificial, arbitrary, weak, inconsistent and failed. He sees no influence of religion on the rise of modern science and postulates a *negative* influence of science on Christianity. Philosophical and theological topics which he discusses include: God as creator; providence, miracles, and natural law; reason and faith/revelation; natural theology; the mechanical philosophy; materialism; atomism; deism; and atheism. Even in his early work, Westfall includes an impressively broad selection of seventeenth-century natural philosophers and uses primary sources extensively. Westfall's early book has been superseded by the works of Force, Klaaren, and Popkin—and, to a certain extent, by the later Westfall. The second edition is unchanged from the first, except for a new preface in which Westfall defends a few of his most questionable theses from the first edition.

Topics: *VII*, V, VIII, IX, X, XII Names: Barrow, Boyle, Browne, T., Charleton, Digby, Glanvill, Grew, Hooke, Locke, Mapletoft, Newton, Ray, Ross, Sydenham, Wilkins

1651 Westfall, Richard S. "The Scientific Revolution of the Seventeenth Century: the Construction of a New World View." In *The Concept of Nature*, edited by John Torrance, 63-93. Oxford: Clarendon Press, 1992.

Westfall proposes that the "scientific revolution" is a true and useful term for describing the new view of physical reality that was developed in the seventeenth century. From his vast reading of the texts of the period, Westfall skillfully describes the revolutionary changes in the conception of nature under four headings: "nature was quantified; it was mechanized; it was perceived to be other; it was secularized." It is the last of these historiographical generalizations that is most relevant to this bibliography. By secularization, Westfall means that "the cord that had bound the study of nature in Europe to Christianity" had been severed. Secularization is most clearly seen in the transformation of the locus of authority for truth. And natural theology is seen as a manifestation of this secularization.

Topics: *I*, VII, VIII

1652 Westfall, Richard S. "Short-writing and the State of Newton's Conscience, 1662." *Notes Rec. Royal Soc. London* 18(1963): 10-16.

This article is important for those interested in the emotional side of Newton's religion. It has been incorporated directly into Westfall's biography of Newton and indirectly into those of other biographers.

Topic: *XII* Name: Newton

1653 Westman, Robert S. "Magical Reform and Astronomical Reform: The Yates Thesis Reconsidered." In *Hermeticism and the Scientific Revolution*, edited by Robert S. Westman and J. E. McGuire, 5-91. Los Angeles: William Andrews Clark Memorial Library, University of California, 1977. See (1655).

Westman critically examines a portion of Yates thesis that the Hermetic tradition contributed to the justification of the Copernican hypothesis as a physical theory by interpreting it as a "magical symbol." While he mainly discusses Bruno, Westman also points out that Fludd (who also is one of Yates' examples) rejected the Copernican theory. Westman concludes that Hermetic thinkers had a diversity of responses to Copernicanism, that most derived their positions from other astronomers of their generation, and that Bruno was a unique thinker.

Topic: *II* Name: Fludd

1654 Westman, Robert S. "Nature, Art, and Psyche: Jung, Pauli, and the Kepler-Fludd Polemic." *Scientific Mentalities in the Renaissance*, edited by Brian Vickers, 177-229. Cambridge: Cambridge University Press, 1984. See (1591).

Westman both reconstructs the Kepler-Fludd debate and discusses Pauli's relation to Jungian psychology. Regarding the former, Fludd's rhetoric and epistemology differ from that of Kepler. Yet the central concern for both "lay in the problem of establishing the connections between pictures, words, and things." For Fludd, visual images retrieve the original creation in Genesis and thereby unite the soul with God. For Kepler, mathematical "symbolization is an activity by which the soul matches intelligibles coeternal with God to sensibles." In connection with this essay, see also the essay by Pauli (1212).

Topics: *II*, VII, X Name: Fludd

1655 Westman, Robert S., and J. E. McGuire. *Hermeticism and the Scientific Revolution*. Los Angeles: William Andrews Clark Memorial Library, University of California, 1977.

See the separate annotations for the essays by Westman (1653) and McGuire (1025).

Topics: *II*, I

1656 Whewell, William. *History of the Inductive Sciences*. 3 vols. London: J. W. Parker, 1837. *History of the Inductive Sciences from the Earliest to the Present Time*. 3rd ed. 3 vols. London: J. W. Parker, 1847.

This is a pioneering work in the history of science. It is outdated, of course, but it is still of value for those who are interested in the historiography of our field.

Topics: *VII*, I

1657 Whinney, Margaret. *Wren*. London: Thames and Hudson, 1971.
This is a very readable book by an acknowledged expert on Wren. Whinney provides descriptions and illustrations of small churches in London, of St. Paul's, and of other buildings (such as Greenwich Hospital). She notes the theological basis of Wren's emphases on pulpit and altar. She is opinionated but appreciative—especially of Wren's ability to negotiate in the midst of changing and sometimes difficult circumstances.
Topics: *VI*, IV Name: Wren

1658 Whitaker, Virgil K. *Francis Bacon's Intellectual Milieu*. Los Angeles: William Andrews Clark Memorial Library, 1962. Reprinted in *Essential Articles for the Study of Francis Bacon*, edited by Brian Vickers, 28-49. Hamden, Conn.: Archon Books, 1968. See (1588).
Whitaker argues that Bacon, in his natural philosophy, was much more a man of his times than has often been supposed. Whitaker supports this view by discussing: Bacon as Renaissance encyclopedist; Bacon's conventional ideas; and Bacon's distrust of human intelligence. The author suggests that "the real key to Bacon's thought was his religion" and that "the effects of Calvinist views may reasonably be found at the center of Bacon's thought."
Topics: *VII*, III, X Name: Bacon

1659 White, Andrew D. *A History of the Warfare of Science with Theology in Christendom*. New York: D. Appleton, 1896. Reprinted. New York: Dover, 1960.
As the classic statement of the "warfare thesis," White's volumes themselves are under increasing scrutiny. Now seen as a product of a time of disciplinary independence and professionalization, White's work covers a wide variety of fields—from biology and astronomy to anthropology and history, and from medicine and psychology to political science and biblical higher criticism. Discussions of seventeenth-century English material may be found particularly in the chapters on geology, meteors, chemistry and physics, and medicine. White hoped both to show the positive establishment of independent disciplines and to clear away what he considered to be irrational theology so that "religion pure and undefiled" could be "a blessing to humanity."
Topics: *VII*, I, X, XI

1660 White, Lynn. "The Context of Science." In *Machina Ex Deo: Essays in the Dynamism of Western Culture*, by Lynn White, 95-105. Cambridge, Mass.: MIT Press, 1968.

White proposes that any concentration of intellectual and social energies in science depends on the social and emotional context in which the science is done. In particular, "natural theology uses the motivational basis of late medieval and early modern science."

Topics: *I*, IV, VIII

1661 White, R. J. *Dr. Bentley: A Study in Academic Scarlet*. London: Eyre and Spottiswoode, 1965.

White's flowery biographical study focuses on Bentley at Cambridge University as Regis Professor of Divinity, as Master of Trinity College, and as involved in college politics. But two chapters are devoted to Bentley's Boyle Lectures of 1692. White is as interested in recovering Bentley's critical classical studies as his theology or his popularization of Newtonian natural philosophy.

Topics: *V*, IV, VIII, X, XII Names: Bentley, Newton

1662 Whitebrook, J. C. "Dr. John Stoughton the Elder." *Transactions of the Congregational Historical Society* 6(1913-1915): 89-107 and 177-184.

This is an old biographical essay on an English minister with Puritan leanings. He was a father-figure to Ralph Cudworth. Stoughton sometimes is used as an example of a Puritan with both millenarian and Baconian ideas.

Topics: *IV*, III, VII, X Names: Cudworth, Stoughton

1663 Whitehead, Alfred North. *Science and the Modern World: Lowell Lectures, 1925*. New York: Macmillan, 1925.

This work includes Whitehead's generalizations concerning science and metaphysics in the seventeenth century. Without giving evidence or argumentation, Whitehead was one of the first to note the connection between a voluntaristic doctrine of God and early modern natural philosophy.

Topics: *VII*, I

1664 Whitla, William. *Sir Isaac Newton's Daniel and the Apocalypse with an Introductory Study of the Nature and the Cause of*

Unbelief, of Miracles and Prophecy. London: J. Murray, 1922.

This introductory study exposits some of Newton's biblical views from the vantage point of a modern-day millenarian and biblical literalist.

Topics: *XII*, X Name: Newton

1665 Whittaker, E. Jean. *Thomas Lawson, 1630-1691: North Country Botanist, Quaker and Schoolmaster*. York, England: Sessions, 1986.

This is a biography of a Quaker botanist, schoolmaster, and polemicist. Converted by George Fox, he became a follower and correspondent of John Ray.

Topic: *VII* Names: Fell, Fox, Lawson, Ray

1666 Whitteridge, Gweneth. "William Harvey: A Royalist and No Parliamentarian." *Past and Present* 30(1964): 104-109. Reprinted in *The Intellectual Revolution of the Seventeenth Century*, edited by Charles Webster, 182-188. London: Routledge and Kegan Paul, 1974. See (1618).

Whitteridge argues against Hill (0697) regarding Harvey's political views and the ideological implications of Harvey's science. She also brings up the topic of the human soul and relevant biblical texts.

Topics: *IV*, III Name: Harvey

1667 Whyte, Alexander. "Appreciation and Introduction." In *Sir Thomas Browne: An Appreciation with Some of the Best Passages of the Physician's Writings Selected and Arranged by Alexander Whyte*, by Alexander Whyte, 11-45. London: O. Anderson and Ferrier, 1898. Reprinted, Port Washington, N.Y.: Kennikat Press, 1971.

This is a rhetorical appreciation of Browne with special emphasis on the literary excellence of his writings. It is superficial by our later standards. But it is interesting in that this essay from the 1890's stresses the harmony (rather than the warfare) between science, medicine, and religion in Browne's person and views.

Topics: *X*, XI Name: Browne, T.

1668 Wiener, Philip P. "The Experimental Philosophy of Robert Boyle." *Philos. Rev.* 41(1932): 594-609.

Wiener explicates and analyzes Boyle's methodology and philosophy of nature. He argues that Boyle misunderstood Aristotelianism and that there are metaphysical lacunae in Boyle's thinking. He treats Boyle's theological ideas—his "pious discourses on things above reason"—as separate from (and possibly inconsistent with) Boyle's experimental philosophy. This early essay has been superseded by later analyses of Boyle's thought.

Topic: *VII* Name: Boyle

1669 Wiener, Philip P. "Some Problems and Methods in the History of Ideas." *J. Hist. Ideas* 22(1961): 531-548.

Wiener provides a useful and insightful analysis of the problems and approaches to intellectual history in the early 1960s. Both religious and scientific ideas are used as examples. He concludes with a still relevant series of historiographical questions.

Topic: *I*

1670 Wiener, Philip P., and Aaron Noland, eds. *Roots of Scientific Thought: A Cultural Perspective.* New York: Basic Books, 1957.

This collection of articles is reprinted from the *Journal of the History of Ideas*. The articles by Johnson (0832) and Strong (1512) are listed separately in the present bibliography.

Topic: *VII*

1671 Wilbur, Earl Morse. *A History of Unitarianism in Transylvania, England, and America*. Cambridge, Mass.: Harvard University Press, 1952.

This is a survey history which is helpful for understanding religious rationalism in our period. Wilbur makes clear: (1) that deism and atheism should be strictly differentiated from various Christian heresies; and (2) that science had no stereotypical relationship with the various forms of unitarianism.

Topics: *IX*, VII Names: Biddle, Clarke

1672 Wildiers, N. Max. *The Theologian and His Universe: Theology and Cosmology from the Middle Ages to the Present*. New York: Seabury Press, 1982.

Wildiers uses secondary sources to discuss a few seventeenth-century Englishmen in his survey of the impact on theology of the scientific study of the universe.

Topics: *VII*, XII Name: Newton

1673 Wiley, Margaret L. *The Subtle Knot: Creative Scepticism in Seventeenth-Century England*. London: Allen and Unwin, 1952.

Wiley describes skepticism mainly in religion and literature with some discussion of the significance of the new science. She has full chapters on Thomas Browne, Glanvill, and others.

Topics: *X*, VII Names: Browne, T., Glanvill, Norris, Whichcote

1674 Wilkinson, Ronald Sterne. "The Alchemical Library of John Winthrop, Jr." *Ambix* 13(1966): 139-186.

This is the second part of (1675).

Topic: *II* Name: Winthrop

1675 Wilkinson, Ronald Sterne. "The Alchemical Library of John Winthrop, Jr. (1606-1676) and His Descendants in Colonial America." *Ambix* 11(1963): 33-51 and *Ambix* 13(1966): 139-186.

In the first installment of this essay, Wilkinson describes Winthrop's varied interests and his attempts to recreate the associations with other alchemists which led to the acquisition of the library. That library is catalogued in the second installment of the essay. The library exhibits Winthrop's wide-ranging reading of the various kinds of alchemical writings.

Topic: *II* Name: Winthrop

1676 Wilkinson, Ronald Sterne. "George Starkey, Physician and Alchemist." *Ambix* 11(1963): 121-152.

Although little is said about religion, this is a good introductory study of Starkey's life and writings—and of the controversies which they initiated.

Topic: *II* Names: Hartlib, Starkey, Thomson

1677 Wilkinson, Ronald Sterne. "The Hartlib Papers and Seventeenth-Century Chemistry: Part I." *Ambix* 15(1968): 54-69.

See (1678).

Topics: *II*, V Names: Boyle, Hartlib

1678 Wilkinson, Ronald Sterne. "The Hartlib Papers and Seventeenth-Century Chemistry: Part II, George Starkey." *Ambix* 17(19-70): 85-110.

In these two articles, Wilkinson discusses Hartlib, Boyle, the Invisible College, and the related ideas with respect to manuscripts which had recently been made available.

Topics: *II*, V Names: Hartlib, Starkey

1679 Wilkinson, Ronald Sterne. "'Hermes Christianus:' John Winthrop, Jr. and Chemical Medicine in Seventeenth-Century New England." In *Science, Medicine and Society in the Renaissance*, edited by Allen G. Debus, Vol. 1, 221-241. New York: Science History, 1972. See (0357).

Wilkinson presents Winthrop as "that most charitable Christian" who was not only a governor of Connecticut and an alchemist but also "New England's foremost physician." He details Winthrop's chemical medicine and indicates his "paracelsian compromise." The title is drawn from Cotton Mather's name for Winthrop and indicates the blending of traditions in Winthrop's life.

Topics: *XI*, II Name: Winthrop

1680 Willey, Basil. *The Eighteenth Century Background: Studies on the Idea of Nature in the Thought of the Period*. London: Chatto and Windus, 1940.

This is a well-written book concentrating on changing conceptions of "Nature" in eighteenth-century England. Willey basically is interested in the effects of science on religion with examples such as the flood, the Fall of Adam, and miracles. This book is still good for beginning students of literature.

Topics: *X*, VII Names: Burnet, T., Derham, Ray, Swift

1681 Willey, Basil. *The Seventeenth Century Background: Studies in the Thought of the Age in Relation to Poetry and Religion*. London: Chatto and Windus, 1934.

Willey discusses how the new conceptions of the real, the true, and causality undermined vital religion and "told against poetry." Although his disparaging of seventeenth-century philosophy tends to distort the historical picture, Willey's book remains a well-written work with valuable insights.

Topics: *X*, VII Names: Bacon, Browne, T., Cudworth, Glanvill, Herbert, E., Hobbes, Locke, Milton, More, Smith, Sprat, Whichcote

1682 Willey, Basil. "The Touch of Cold Philosophy." In *The Seventeenth Century*, edited by Richard F. Jones and others, 369-376. Stanford, Calif.: Stanford University Press, 1951. See (0845).

Here are broad generalizations about the pervasive effect of the Scientific Revolution and the New Philosophy on seventeenth-century English culture—with special attention to poetry and religion.

Topics: *X*, IV, VII

1683 Willey, Basil. "The Turn of the Century." In *Seventeenth Century Studies Presented to Sir Herbert Grierson*, 372-393. Oxford: Clarendon Press, 1938. See (1424). Reprinted in *The Eighteenth Century Background*, by Basil Willey, 1-26. London: Chatto and Windus, 1940. See (1681).

Willey here discusses the changing conceptions of "Nature" around the year 1700.

Topic: *X*

1684 Williams, Arnold. *The Common Expositor: An Account of the Commentaries on Genesis 1527-1633*. Chapel Hill, N.C.: University of North Carolina Press, 1948.

Williams analyzes Genesis commentaries and related works by subjects. The chapters most relevant to our field are those on "The Corruption of the World and the Flood" and "Science and Pseudo-Science." While much of the material is drawn from sixteenth-century continental Europe, some of the material has to do with interpretations by seventeenth-century Englishmen.

Topic: *X* Names: Browne, T., Milton

1685 Williamson, George. "Milton and the Mortalist Heresy." *Studies in Philology* 32(1935): 553-579. Reprinted in *Seventeenth Century Contexts*, by George Williamson, 148-177. London: Faber and Faber, 1960. See (1689).

Williamson argues that Milton was influenced by Epicurus and Hobbes in his view that the soul dies with the body. For Milton,

both the body and the soul are later resurrected.
Topics: *X*, IX Names: Hobbes, Milton

1686 Williamson, George. "Mutability, Decay, and Seventeenth-Century Melancholy." *ELH* 2(1935): 121-151. Reprinted in *Seventeenth Century Contexts*, by George Williamson, 9-41. London: Faber and Faber, 1960. See (1689).
Williamson here connects the new astronomy of the late sixteenth and early seventeenth centuries to poetic views of the decay of the world. A resulting melancholy and various religious responses are also discussed.
Topics: *X*, IV Names: Bacon, Browne,T., Burton, Donne, Goodwin, Hakewill

1687 Williamson, George. "The Restoration Revolt against Enthusiasm." *Stud. Philol.* 30(1933): 571-603. Reprinted as in *Seventeenth Century Contexts*, by George Williamson, 202-239. London: Faber and Faber, 1960. See (1689).
Williamson writes that seventeenth-century views concerning rhetoric and literary imagination were intertwined with views on religion, science, and "social consequences." The plainer style of the new science was favored over the rhetoric and sermonic enthusiasm of the Interregnum. Dryden represents the movement towards a new style.
Topics: *X*, III, IV Names: Bacon, Dryden, Hobbes, Sprat

1688 Williamson, George. "Richard Whitlock, Learning's Apologist." In *Seventeenth Century Contexts*, by George Williamson, 178-201. London: Faber and Faber, 1960. See (1689).
This is a rambling essay about Whitlock, an M.D. and Anglican priest, who wrote *Zootomia* (a volume of essays published in 1654). Williamson presents Whitlock as a Baconian and as a skeptical rationalist. Williamson compares Whitlock to Thomas Browne and to Joseph Glanvill.
Topics: *X*, IV, XI Names: Browne, T., Glanvill, Whitlock

1689 Williamson, George. *Seventeenth-Century Contexts*. London: Faber and Faber, 1960. Revised ed., Chicago: University of Chicago, 1969.

This volume includes four relevant articles by Williamson annotated in the previous four entries (1685-1688) of this bibliography. The articles may be found in either edition with the same pagination.
Topic: *X*

1690 Wilson, Catherine. "Visual Surface and Visual Symbol: The Microscope and the Occult in Early Modern Science." *J. Hist. Ideas* 49(1988): 85-108. Reprinted in *Philosophy, Religion and Science in the Seventeenth and Eighteenth Centuries*, edited by John W. Yolton, 85-108. Rochester, N.Y.: University of Rochester Press, 1990. See (1720).

Wilson relates various seventeenth-century attitudes towards the microscope to the various schools of natural philosophy. She correlates a pro-microscope position to the mathematico-corspuscularian philosophy and an anti-microscope position both to Aristotelian and to Paracelsian schools of thought. Wilson treats theology as a related factor within the larger complex of issues which were addressed in the controversies over the microscope.
Topics: *VII*, II Names: Bacon, Hooke

1691 Wilson, Frank Percy. *Seventeenth Century Prose: Five Lectures*. Berkeley and Los Angeles: University of California Press, 1960.

This short book is without specific theses concerning science and religion. But it touches on various relevant topics (including the idea of progress) and individuals (including Thomas Browne). The style is anecdotal and descriptive rather than argumentative. Wilson suggests that changes in prose style during the seventeenth century (including changes in the style of sermons) were part of a complex process. The triumph of the plain style was a revolution which should not "be attributed merely, or even chiefly, to the new rationalism and the successes of experimental science."
Topics: *X*, IV Names: Barrow, Browne, T., Donne, Dryden, Ross

1692 Winnett, Arthur Robert. *Peter Browne: Provost, Bishop, Metaphysician*. London: S.P.C.K., 1974.

This is a readable biography of the Irish/Anglican bishop who was a well-known philosopher in his own day. Browne was a strong

critic of Socinianism and of deism—especially that of Toland. Natural philosophy is of indirect relevance in his case.
Topics: *VII*, IX Name: Browne, P.

1693 Winnett, Arthur Robert. "Were the Deists 'Deists'?" *Church Q. Rev.* 161(1960): 70-77.
Winnett makes a distinction between what he calls "historical Deism" and what he calls "philosophical Deism." It is quite unwarranted, he shows, to ascribe to the Deists in general (historical deists) the views commonly called Deism (philosophical deism of the absentee God).
Topics: *VII*, IX

1694 Winship, Michael P. "Prodigies, Puritanism, and the Perils of Natural Philosophy: The Example of Cotton Mather." *William and Mary Quarterly*, 51(1994): 92-105
Winship argues that Mather refrained from discussing prodigies after 1690 but not because of a change in his own natural philosophy. Rather, he became aware of limitations on legitimate speech by "the natural philosophers whose good opinion he craved." He could be "both a good Newtonian and a Puritan." Yet, as a learned Puritan, he found that he had to leave prodigies and other divine wonders to the "vulgar."
Topic: *VII*, X Name: Mather, C.

1695 Winslow, Ola E. *A Destroying Angel: The Conquest of Smallpox in Colonial Boston*. Boston: Houghton-Mifflin, 1974.
This is a good popular history. It begins with chapters on the unprofessional background of medicine and on the dual roles of pastor-physicians. The central chapters contain an interesting narrative of the controversy over inoculation. Winslow assumes that this "conquest" by scientific medicine is an example of the reshaping of American culture from a new center: from "God's hands" to a "new power in man's hands." She also finds in the combined efforts of Cotton Mather and Zabdiel Boylston "a rare human partnership."
Topics: *XI*, X Names: Boylston, Douglas, Mather, C., Montagu

1696 Wojcik, Jan W. "The Theological Context of Boyle's *Things above Reason*." In *Robert Boyle Reconsidered*, edited by

Michael Hunter, 139-155. Cambridge: Cambridge University Press, 1994. See (0759).

Wojcik sets *Things above Reason* in the context of controversies over the limits or competency of reason and the relation between predestination and free will among Anglicans and nonconformists. He notes, however, that Boyle's purpose is irenical. Because reason is limited, humans cannot understand God's revelation completely. Hence, Boyle hoped to preclude dogmatic pronouncements about things beyond reason. Wojcik concludes that theological voluntarism provided the unity of Boyles' thought. God willed limits on human knowledge—both of theological truth and in natural philosophy. Boyle thus was not a predecessor of deism because God and God's will is the starting point for all of his thought.

Topics: *VII*, X Names: Baxter, Boyle, Ferguson, Glanvill, Howe, Owen

1697 Wolf, Edwin. *The Library of James Logan of Philadelphia, 1674-1751*. Philadelphia: Library Company of Philadelphia, 1974.

This work includes Logan's notes on Newton's *Chronology of Ancient Kingdoms Amended*.

Topics: *X*, XII Name: Logan

1698 Woodfield, Richard. "Hobbes on the Laws of Nature and the Atheist." *Renaiss. Mod. Stud.* 15(1971): 34-43.

Woodfield discusses the place of God in the philosophy of Hobbes.

Topics: *IX*, VII Name: Hobbes

1699 Wooton, David. "New Histories of Atheism." In *Atheism from the Reformation to the Enlightenment*, edited by Michael Hunter and David Wooton, 13-53. Oxford: Clarendon Press, 1982.

This is a methodological survey of present discussions of atheism (and deism) in early modern Europe. In the "new histories," science is seen as playing only a small and indirect role in relationship to atheism. (By encouraging certain theological arguments and through the development of probability theory, Wooton says, the Scientific Revolution had such an indirect role.)

Topics: *IX*, I, VII

1700 Wormhoudt, Arthur. "Newton's Natural Philosophy in the Behmenistic Works of William Law." *J. Hist. Ideas* 10(1949): 411-429.

Wormhoudt agrees with Hobhouse that Newton was not influenced by the theosophy of Jacob Boehme.

Topics: *XII*, II Name: Newton

1701 Wright, John P. "Locke, Willis, and the Seventeenth-Century Epicurean Soul." In *Atoms, Pneuma, and Tranquility*, Margaret J. Osler, 239-258. Cambridge: Cambridge University Press, 1991. See (1179).

Wright argues that Locke extended the Epicurean conception of the material soul as the principle of life. He proposes that Locke was influenced by Thomas Willis' medical lectures on the two souls: the rational soul ("a particle of divine breath") and the lower soul (which is common to humans and animals). Locke, however, appears to have gone farther in suggesting a unified human being on the basis that God superadded thought to matter.

Topics: *VII*, XI Names: Locke, Willis

1702 Wright, Peter W. G. "On the Boundaries of Science in Seventeenth-Century England." In *Sciences and Cultures*, edited by Everett Mendelsohn and Yehuda Elkana, 77-100. Dordrecht: D. Reidel, 1981.

Wright proposes a social-constructivist perspective on seventeenth-century natural philosophy. He uses the Webster-Ward debate and Sprat's *History of the Royal Society* as examples. The image of "science," he argues, was formed by political, religious, and ideological struggles.

Topics: *IV*, I, V Names: Sprat, Ward, Webster

1703 Wybrow, Cameron, ed. *Creation, Nature, and Political Order in the Philosophy of Michael Foster (1903-1959): The Classical Mind Articles and Others, with Modern Critical Essays*. Lewiston, N.Y.: E. Mellen Press, 1992.

This is a very useful collection of Foster's writings and responses by others to his work. It includes a biographical introduction by the editor and a bibliography of Foster's published writings. Foster's relevant articles (0501) and (0502) are annotated separately

in the present bibliography, as are essays by Gruner (0579), Oakley (1157), and Patrick (1209).
Topic: *VII*

1704 Yan, Kangnian. "On Isaac Newton's Ideas of Gravitation and God." *Hist. Scientiarium* (34)1988: 43-56.
The thesis here is that Newton "transformed from a deist into a rather thoroughgoing mechanical materialist." Yan concludes this after outlining what he understands to be Newton's positions on gravitation, action at a distance, and God's relation to or identification with nature.
Topics: *XII*, VII Name: Newton

1705 Yates, Frances A. *The Art of Memory*. Chicago: University of Chicago Press, 1966.
This is a history of the art of memory from the ancient Greeks to Bruno and his followers. The last three chapters of the book are relevant to this bibliography. In them, Yates suggests that Fludd's theatre memory system took specific form in the Globe theatre and she hints at possible transformations of the art by philosophers in the seventeenth century.
Topics: *II*, VI, X Name: Fludd

1706 Yates, Frances A. *Giordano Bruno and the Hermetic Tradition*. London: Routlege and Kegan Paul; University of Chicago Press, 1964.
This book basically deals with topics outside our place and time; but it is very influential and has been a rather controversial work even for historians of seventeenth-century English thought. In her concluding chapter on the Robert Fludd controversies, Yates proposes that the Hermetic tradition provided "a new direction of the will towards the world" that lay behind the emergence of modern science.
Topics: *II*, VII Names: Casaubon, I., Cudworth, Fludd, More

1707 Yates, Frances A. "The Hermetic Tradition in Renaissance Science." *Art, Science, and History in the Renaissance*, edited by Charles S. Singer, 255-274. Baltimore, Md.: Johns Hopkins Press, 1967. Reprinted in Vol. 11 of *Articles on*

Witchcraft, Magic, and Demonology, edited by Brian P. Levack, 233-252. New York: Garland, 1992. See (0945).

This essay provides a good brief description of the Hermetic tradition and its significance for interpretation in the history of science.

Topics: *II*, I

1708 Yates, Frances A. *The Occult Philosophy in the Elizabethan Age*. London: Routlege and Kegan Paul, 1979.

Yates deals almost totally with occult traditions in the fifteenth and sixteenth centuries. But she concludes by reaffirming her belief that Bacon's community in *New Atlantis* was a Christian cabalistic community and that Bacon's movement for the advancement of scientific learning was continuous with the Rosicrucian movement.

Topics: *II*, IV Name: Bacon

1709 Yates, Frances A. *The Rosicrucian Enlightenment*. London: Routlege and Kegan Paul, 1972.

Yates focuses on the idea of Rosicrucianism on the continent and in England from the late sixteenth century through the seventeenth century. In this tradition, she says, two symbols persisted: (1) the image of God as the Great Architect; and (2) the expectation of a coming Enlightenment. She concludes that the Hermetic tradition did not lose its force but remained "in the background of the minds" of the new natural philosophers. Yates' controversial theses appear to be good for the earlier seventeenth century. Her suggestions for the later part of the century have been criticized for being weakly supported by the evidence.

Topics: *II*, IV, VI Names: Ashmole, Bacon, Comenius, Fludd, Hartlib, James I

1710 Yates, Frances A. *Shakespeare's Last Plays: A New Approach*. London: Routlege and Kegan Paul, 1975.

In the chapter on *The Tempest*, Yates argues that the play should be interpreted from the perspective of the Hermetic tradition. In other words, attention should be paid to its assumed "philosophy of nature with religious and reforming undercurrents" and to the related practitioners.

Topics: *X*, II Name: Shakespeare

1711 Yates, Frances A. *Theatre of the World*. London: Routlege and Kegan Paul; Chicago: University of Chicago Press, 1969.

In this third volume of a series beginning with *Giordano Bruno and the Hermetic Tradition*, Yates expands on her suggestions made in *The Art of Memory*. She argues that London theatres of the late-sixteenth and early-seventeenth centuries were adaptations of the image of the ancient theatre made under the influence of the Hermetic tradition. And she closely analyzes the architectural ideas and relations of Robert Fludd and Inigo Jones. Thus, she concludes, the Globe Theatre was a "cosmic theatre, a religious theatre"—a microcosm within which players "enacted the drama of the life of man."

Topics: *VI*, II, X Names: Fludd, Jones

1712 Yolton, Jean S., ed. *A Locke Miscellany: Locke Biography and Criticism for All*. Bristol, England: Thoemmes, 1990.

Here are odds and ends mainly addressed to the general reader. The collection includes three relevant reprints by Ashcraft (0035), Broad (0126), and Gregory (0576).

Topic: *VII* Name: Locke

1713 Yolton, John W. Introduction to *The Locke Reader: Selections from the Works of John Locke*, edited by John W. Yolton, 1-9. Cambridge: Cambridge University Press, 1977.

Yolton gives introductory remarks stressing the wide range of Locke's interests and his tendency to support and express new or radical ideas in all areas of thought. Locke's views of religion and of the new science are briefly discussed along these lines.

Topic: *VII* Name: Locke

1714 Yolton, John W. *John Locke and the Way of Ideas*. London: Oxford University Press, 1956.

Here Yolton traces the developments and summarizes the contemporary responses to Locke's *Essay concerning Human Understanding*. He deals with natural philosophy implicitly and the various theological positions explicitly as he discusses the debates over innate ideas, skepticism, substance, and thinking matter. Included are excellent summaries and a review of a wide range of primary sources. The interpretive and judgmental theses are not so strong;

Yolton confuses the various unorthodox theological positions (deistic, socinian, unitarian, etc.). But this book remains useful.

Topics: *VII*, IX Names: Browne, P., Collins, A., Gastrell, Locke, Stillingfleet, Toland

1715 Yolton, John W. *John Locke: Problems and Perspectives: A Collection of New Essays*. London: Cambridge University Press, 1969.

This is a collection of thirteen essays, six of which are relevant to our topic and are listed elsewhere in this bibliography. See Aarsleff (0002-0003), Ashcraft (0034), Axtell (0049), Leyden (0949), and Yolton (1721).

Topic: *VII* Name: Locke

1716 Yolton, John W. *Locke: An Introduction*. Oxford: B. Blackwell, 1985.

This could be entitled "An Introduction to Locke's Philosophy." There is some but not much attention to the historical context of Locke's thinking. There is no direct discussion of the relation between science and theology for Locke; but there is discussion about God, theology, and natural science done separately.

Topic: *VII* Name: Locke

1717 Yolton, John W. *Locke and the Compass of Human Understanding: A Selective Commentary on the 'Essay'*. Cambridge: Cambridge University Press, 1970.

This is a philosophical analysis of and commentary on *An Essay concerning Human Understanding*. There is rather extensive discussion of the relation of the new science to Locke's epistemology. Generally, there are only tangential references to religion and theology (though there is a brief analysis of Locke's views of the resurrection body). The commentary basically is aimed at graduate students and it is set within the context of Yolton's continuing debates with other Locke scholars.

Topic: *VII* Names: Boyle, Locke, Stillingfleet

1718 Yolton, John W. "Locke and the Seventeenth-Century Logic of Ideas." *J. Hist. Ideas* 16(1955): 431-452.

Yolton relates epistemological problems of the time to religion and to the new philosophy. He presents Locke within the seventeenth-century movement that attempted to reconcile the corpuscular theory with traditional theology.

Topic: *VII* Names: Locke, More

1719 Yolton, John W. "Locke's Unpublished Marginal Replies to John Sergeant." *J. Hist. Ideas* 12(1951): 528-559.

Yolton interweaves Locke's unpublished notes with a detailed analysis of Sergeant's published objections to Locke's *Essay*. The issues are categorized under the headings of method, perception, and substance. The basis of Sergeant's objections lies in his realist epistemology and Aristotelian theological metaphysics. Thus, Locke responds to Sergeant's concern that Locke's skeptical positions threaten both science and theology.

Topic: *VII* Names: Locke, Sergeant

1720 Yolton, John W., ed. *Philosophy, Religion and Science in the Seventeenth and Eighteenth Centuries*. Rochester, N.Y.: University of Rochester Press, 1990.

This is a collection of articles reprinted from the *Journal of the History of Ideas*—nineteen of which are listed separately in this bibliography. The "Introduction" is brief and unoriginal. See Barnow (0061), Biddle (0093), Farr (0443), Force (0488) and (0489), Guerlac (0582), M. C. Jacob (0815), Jolley (0833), Laudan (0937), Osler (1183), Perl (1220), Rogers (1340), (1342), and (1343), Schankula (1408), Stewart (1499), Wallace (1600), Wilson (1690), and Yost (1723).

Topics: *VII*, XII Names: Boyle, Locke, More, Newton

1721 Yolton, John W. "The Science of Nature." In *John Locke: Problems and Perspectives*, edited by John W. Yolton, 183-193. London: Cambridge University Press, 1969. See (1715). An expanded version appears as Chapter 2 in *Locke and the Compass of Human Understanding*, by John W. Yolton, 44-75. Cambridge: Cambridge University Press, 1970. See (1717).

Yolton here relates Locke's theory of knowledge to the new science and argues that, indeed, Locke "gave a philosophical foundation for the new science." Tangential references to God, angels, and spirit show how such were included in Locke's overall theory of knowledge. (Yolton also tries in this article to clarify and

support his position in an ongoing debate with Mandelbaum and Laudan.)
Topic: *VII* Name: Locke

1722 Yolton, John W. *Thinking Matter: Materialism in Eighteenth-Century Britain*. Minneapolis: University of Minnesota Press, 1983.
Yolton postulates a "thinking matter controversy" and approaches it as "the story of reactions to Locke's suggestion." Various forms of materialism and immaterialism are described as well as the implications concerning: man as automaton; matter as inert or active; space and extension; and human physiology. About one-half of the book (about 100 pages) is relevant to the pre-1720 period. Yolton states repeatedly that religion and theology are important for understanding the historical context; and he says that the implications of these philosophical (and scientific) issues were significant for theology. But he does not specify that context or those implications in any detail or depth. As is typical of Yolton, he summarizes an impressive array of primary sources.
Topics: *VII*, IX, XII Names: Clarke, Collins, A., Cudworth, Ditton, Layton, Locke, Newton

1723 Yost, R. M., Jr. "Locke's Rejection of Hypotheses about Submicroscopic Events." *J. Hist. Ideas* 12(1951): 111-130. Reprinted in *Philosophy, Religion and Science in the Seventeenth and Eighteenth Centuries*, edited by John W. Yolton, 251-270. Rochester, N.Y.: University of Rochester Press, 1990. See (1720).
Yost argues that Locke opposed the use of hypotheses concerning submicroscopic events (including atomism or corpuscularianism). He provides no direct discussion of theology or religion: but Locke's methodological position regarding natural philosophy is related implicitly to his theological views and to questions about the separation of science and theology.
Topic: *VII* Name: Locke

1724 Young, Davis A. *Christianity and the Age of the Earth*. Grand Rapids, Mich.: Zondervan, 1982.
Chapter 2, "Geological Investigations to 1750," briefly summarizes seventeenth-century theories about fossils and the deluge.
Topics: *X*, VII Names: Burnet, T., Hooke, Whiston, Woodward

1725 Young, Robert F. "The Historiographic and Ideological Contexts of the Nineteenth Century Debate on Man's Place in Nature." In *Changing Perspectives in the History of Science*, edited by Nicholas Teich and Robert Young, 344-438. London: Heinemann, 1973. See (1535).

This is an articulate critique of many positions in the historiography of science by a Marxist making a trumpet-call for a new "radical historiography." Sections III and VI of Young's long essay are relevant to seventeenth- and eighteenth-century studies.

Topics: *I*, IV

1726 Young, Robert Fitzgibbon. *Comenius in England*. Oxford: Oxford University Press, 1932.

This is a collection of primary documents related to Comenius' visit to England in 1641-42. The introduction sketches the interest in forming scientific societies and colleges in the early seventeenth century—and the influence of Comenius' encyclopedic vision on these movements. Young also notes the education of Native Americans in the British colonies.

Topics: *IV*, V Names: Comenius, Winthrop

1727 Youngren, William H. "Generality, Science and Poetic Language in the Restoration." *ELH* 35(1968): 158-187.

Youngren questions the thesis of R. F. Jones and others that in our period mathematical symbolism determined how words should function and made a universe inimical to poetry. He finds continuity in discussions about poetry throughout the seventeenth and into the eighteenth century. Critics distinguished between the knowledge imparted by scientific prose and the knowledge imparted by poetry (and, indirectly, religion). As a moral teacher, the "lively examples" of poetry made it a clearer and more general moral teacher.

Topic: *X* Names: Dennis, Dryden, Hobbes, Sprat

1728 Zafiropulo, Jean, and Catherine Monod. *Sensorium Dei, dans l'Hermetisme et la Science*. (Collection d'Etudes Anciennes.) Paris: Belles Lettres, 1976.

In this 370-page survey from ancient Greece to Einstein and de Broglie, over 200 pages are devoted to Newton. The authors emphasize the importance of alchemy, esoteric tradition, and theology in Newton's natural philosophy. They relate the "Sensorium Dei"

not only to his metaphysics but also to universal gravitation. The title and theses are intriguing but the scholarship is secondhand. The book is based on selected (and not the complete) secondary literature. There are almost no references to Leibniz or Clarke! This work has been superseded by the writings of McGuire, Rattansi, Dobbs, and other works in English.
Topics: *VII*, II, XII Name: Newton

1729 Zeitz, Lisa M. "Natural Theology, Rhetoric, and Revolution: John Ray's Wisdom of God, 1691-1704." *Eighteenth Century Life* 18(Feb. 1994): 120-133.
Zeitz argues that the rhetoric of Ray's work "provided a cultural model of cooperation and consensus in post-Revolutionary England." She examines both the design argument and the expansions of the text (from 249 pages in 1691 to 464 pages in 1704). Both editions reflected a broad appeal and an inclusive sense of community and shared effort.
Topics: *VIII*, IV, X Name: Ray

1730 Zetterberg, J. Peter. "Echoes of Nature in Solomon's House." *J. Hist. Ideas* 43(1982): 179-193.
Zetterberg disputes Rossi's assertion that Bacon rejected the doctrine that art imitates nature. He shows that Bacon believed that human art and industry should be an activity which is patterned after God's creation. But, he says, Bacon did reject speculative metaphysical claims about the relation of nature to art.
Topics: *VII*, VI Name: Bacon

1731 Zilsel, Edgar. "The Genesis of the Concept of Physical Law." *Philos. Rev.* 51(1942): 245-279.
Zilsel discusses various possibilities for explaining the genesis of physical law—from biblical and classical authors to the end of the seventeenth century. Developments in Christian theology, the "quantitative rules of the early capitalistic artisans," and changes in the structure of the political state are seen by Zilsel as factors in the development of the concept of physical law.
Topics: *IV*, III, VII, X

1732 Zilsel, Edgar. "The Sociological Roots of Science." *Am. J. Sociol.* 47(1941-1942): 544-562.

Zilsel traces the rise of science to the breakdown of medieval class barriers. This essay basically is outside the scope of seventeenth-century England, but it has influenced some writers in our field.
Topic: *III*

1733 Zimmerman, Robert. "Henry More und die verte Dimension des Raumes." *Sitzungsberichte der Philosophisch-Historischen Classe der Kaiserlichen Akademie der Wissenschaften* 98 (1881): 403-448.
This includes a rather full account of More's natural philosophic views. But the same material can be found in later English works (such as J. T. Baker, Burtt, Koyre, and Snow).
Topic: *VII* Name: More

1734 Zolla, Elemire. *Le Meraviglie della Natura: Introduzione all'Alchimia*. Milan: Bompiani, 1975.
Although she treats alchemists in our period only tangentially, we are including this work because Zolla treats alchemical symbolism so symbolically and so completely. In her massive volume, she comparatively relates alchemy to the Hindu, Christian, and Islamic traditions.
Topics: *II*, X

1735 Zycinski, J. "The Rise and Fall of Methodological Positivism in Newton's *Principia*." In *Newton and the New Direction in Science*, edited by G. V. Coyne and others, 73-83. Vatican City: Specola Vaticana, 1988. See (0281).
Zycinski is a philosopher. He argues that, in the *Principia*, Newton separated physics from metaphysics and theology in a methodological way but not ontologically or psychologically. Another way to put this is that metaphysics and theology were important to Newton's context of scientific discovery but were irrelevant in his context of scientific justification. Zycinski pushes his view so far as radically to separate alchemy from Newton's "methodological positivism" and to say that Napoleon's famous question to Laplace would have been formulated a century earlier if Newton's contemporaries had read the *Principia* carefully. Zycinski also argues that eighteenth- and nineteenth-century commentators misunderstood *both* Newton's positivism and his somewhat Platonic theology. Zycinski makes some good points despite his overstatements.
Topics: *XII*, VIII Name: Newton

A Bibliographic List of Bibliographies

See also the following annotated works which include significant bibliographies: Brooke (0132); Brooks (0137); I. B. Cohen (0234); Debus (0359); *Dictionary of National Biography* (0374); *Dictionary of the History of Ideas* (0375); M. Hunter (0752), (0759), and (0763); M. Hunter and Schaffer (0765); A. Jacob (0791); Kargon (0857); Laudan (0938); N. H. Nelson (1108); Sarton (1393); Westfall (1640) and (1650).

1736 *American Studies: An Annotated Bibliography*. Edited by Jack Salzman on behalf of the American Studies Association. Cambridge: Cambridge University Press, 1986.
Topic: *X*

1737 *Annual Bibliography of British and Irish History*. (Royal Historical Society in association with the Institute of Historical Research. Writings on British History.) Brighton, England: Harvester Press, 1976-1987. Tokyo: Harvester Wheatsheaf, 1988-1989. Oxford: Oxford University Press, 1990 and following. Before 1976, this title was *Annual Bibliography of British History*. London: Institute for Historical Research, University of London, 1934-1975.)
Topic: *X*

1738 *Annual Bibliography of English Language and Literature*. (Modern Humanities Research Association.) London: W. S. Maney and Sons, 1920 and following.
Topic: *X*

1739 Armitage, Christopher M. *Sir Walter Ralegh: An Annotated Bibliography*. Chapel Hill. University of North Carolina Press, 1987.
Topics: *X*, IX Name: Ralegh

1740 Attig, John C. *The Works of John Locke: A Comprehensive Bibliography from the Seventeenth Century to the Present*. Westport, Conn.: Greenwood Press, 1985.
Topic: *VII* Name: Locke

1741 Bateson, F. W., ed. *The Cambridge Bibliography of English Literature: Vol. 2, 1660-1800*. Cambridge: Cambridge University Press, 1940.
Topic: *X*

1742 Bechtle, Thomas C., and Mary F. Riley. *Dissertations in Philosophy Accepted at American Universities, 1861-1975*. New York: Garland, 1978.
Topic: *VII*

1743 Brewster, John W., and Joseph A. McLeod. *Index to Book Reviews in Historical Periodicals*. Metuchen, N.J.: Scarecrow Press, 1972-1979.

1744 Brush, Stephen G., ed. *Resources for the History of Physics*. Hanover, N.H.: University Press of New England, 1972.
Topic: *VII*

1745 Bush, Douglas, ed. *English Literature of the Earlier Seventeenth Century, 1600-1660. Vol. 5 of The Oxford History of English Literature*. 2nd. ed. Oxford: Clarendon Press, 1962.
Topic: *X*

1746 Carabelli, Giancarlo. *Tolandiana: Materiali Bibliografici per lo Studio dell'Opera e della Fortuna di John Toland (1670-1722)*. 2 volumes. Florence: Nuova Italia, Italy, 1975-1978.
Topics: *VII*, IX Name: Toland

1747 Christophersen, H. O. *A Bibliographical Introduction to the Study of John Locke*. Oslo, Norway: Jacob Dybwad, 1930. Reprint-

ed, New York: B. Franklin, 1968
Topics: *VII*, X Names: Locke, Stillingfleet, Toland

1748 Crane, Ronald S., ed. *English Literature, 1660-1800: A Bibliography of Modern Studies*. 6 vols. Princeton: Princeton University Press, 1950 and following.
Topic: *X*

1749 "Critical Bibliography of the History of Science and Its Cultural Influences." *Isis*. (Volumes 4-79) 1913-1988.

1750 Crocker, Robert. "A Bibliography of Henry More." In *Henry More (1614-1687) Tercentenary Studies*, edited by Sarah Hutton, 219-248. Dordrecht: Kluwer, 1990. See (0782).
Topics: *VII*, II

1751 Davies, Godfrey, ed. *Bibliography of British History: Stuart Period, 1603-1714*. Oxford: Clarendon, 1928.

1752 DeGeorge, Richard T. *A Guide to Philosophical Bibliography and Research*. New York: Appleton-Century-Crofts, 1971.
Topic: *VII*

1753 Doumato, Lamia. *Sir Christopher Wren and St. Paul's Cathedral*. Monticello, Ill.: Vance Bibliographies, 1979.
Topic: *VI* Name: Wren

1754 Duarte, Francisco Jose. *Bibliografia: Euclides, Arquimedes, Newton*. Caracas: N.p., 1967.
Topic: *XII* Name: Newton

1755 "English Literature, 1660-1800: A Current Bibliography." *Philol. Q*. Edited by Ronald S. Crane and others. Appeared annually from 1926-1975.
Topic: *X*

1756 Ferrer Benimeli, Jose. *Bibliografia de la Masoneria: Introduccion Historico-Critica*. 2nd ed. Madrid: Fundacion Universitaria Espanola, 1978.
Topic: *IX*

1757 Fulton, John F. *A Bibliography of the Honourable Robert Boyle, F.R.S.* 2nd ed. Oxford: Clarendon Press, 1961.
Topic: *VII*, X Name: Boyle

1758 Garcia, Alfred. *Thomas Hobbes: Bibliographie Internationale de 1620 a 1986*. Caen: Centre de Philosophie Politique et Juridique, Universite de Caen, 1986.
Topic: *VII* Name: Hobbes

1759 Gillett, Charles R. *Catalogue of the McAlpin Collection of British History and Theology*. 5 vols. New York: Union Theological Seminary, 1927-1930.
Topic: *X*

1760 Grose, Clyde Leclare. *A Select Bibliography of British History, 1660-1760*. Chicago: University of Chicago Press, 1957. Reprinted, New York: Octagon Books, 1967.

1761 Guerry, Herbert, ed. *A Bibliography of Philosophical Bibliographies*. Westport, Conn.: Greenwood Press, 1977.
Topic: *VII*

1762 Guffey, George Robert. *Traherne and the Seventeenth-Century Platonists: 1900-1966*. (Elizabethan Bibliographies Supplements, XI.) London: Nether Press, 1969.
Topics: *X*, VII Names: Cudworth, More, Norris, Smith, Traherne, Whichcote

1763 Hall, Roland, and Roger Woolhouse. *Eighty Years of Locke Scholarship: A Bibliographical Guide*. Edinburgh: Edinburgh University Press, 1983.
Topic: *VII* Name: Locke

1764 Hall, Roland, and Roger Woolhouse. "Forty Years of Work on John Locke (1929-1969)." *Philos. Q.* 20(1970): 258-268 and 394-396.
Topic: *VII* Name: Locke

1765 Henrey, Blanche. *British Botanical and Horticultural Literature Before 1800: Comparing a History and Bibliography of*

Botanical and Horticultural Books Printed in England, Scotland, and Ireland from the Earliest Times Until 1800. London: Oxford University Press, 1975.

Topic: *X*

1766 Herzenberg, Caroline L. *Women Scientists from Antiquity to the Present: An Index.* West Carroll, Conn.: Locust Hill Press, 1986.

1767 Hubner, Jurgen. *Der Dialog zwischen Theologie und Naturwissenschaft: Ein Bibliographischer Bericht.* (*Forschungen und Berichte der Evangelischen Studiengemeinschaft*, 41.) Munich: Kaiser, 1987.

Topic: *VII*

1768 Huckabay, Calvin. *John Milton: An Annotated Bibliography, 1929-1968.* Rev. ed. Pittsburgh, Pa.: Duquesne University Press, 1969.

Topic: *X* Name: Milton

1769 *Isis Cumulative Bibliography: A Bibliography of the History of Science Formed from Isis Critical Bibliographies 1-90, 1913-1965.* Edited by Magda Whitrow. 6 vols. London: Mansell, 1971-1984.

Topics: *VII*, I, II, IV, V, VI, XI, XII

1770 *Isis Cumulative Bibliography, 1966-1975: A Bibliography of the History of Science Formed from Isis Critical Bibliographies 91-100 Indexing Literature Published from 1965 through 1974.* 2 vols. Edited by John Neu. London: Mansell, 1980-1985.

Topics: *VII*, I, II, IV V, VI, XI, XII

1771 *Isis Cumulative Bibliography, 1976-1985: A Bibliography of the History of Science Formed from Isis Critical Bibliographies 101-110 Indexing Literature Published from 1975 through 1984.* 2 vols. Edited by John Neu. Boston: G. K. Hall, 1989.

Topics: *VII*, I, II, IV, V, VI, XI, XII

1772 *Isis Current Bibliography of the History of Science and Its Cultural Influences. Isis.* (Volumes 80 and following.) 1989 and fol-

lowing.
Topics: *VII*, I, II, IV, V, VI, XI, XII

1773 Jacobs, P. M. *History Theses, 1901-70: Historical Research for Higher Degrees in the Universities of the United Kingdom.* London: University of London, Institute of Historical Research, 1976.

1774 Keeler, Mary Frear, ed. *Bibliography of British History: Stuart Period, 1603-1714*. 2nd ed. Oxford: Clarendon Press, 1970.

1775 Keynes, Geoffrey. *A Bibliography of Sir Thomas Browne*. Oxford: Clarendon Press, 1968.
Topics: *X*, XI Name: Browne, T.

1776 Keynes, Geoffrey. *John Ray, 1627-1705: A Bibliography, 1660-1970*. Amsterdam: G. Th. van Heusden, 1976.
Topics: *VII*, VIII Name: Ray

1777 Kies, Cosette N. *The Occult in the Western World: An Annotated Bibliography*. Hamden, Conn.: Library Professional, 1986.
Topic: *II*

1778 Knight, David M. *Natural Science Books in English, 1600-1900*. London: Batsford, 1972.
Topic: *VII*

1779 Knight, David M. *Sources for the History of Science, 1660-1914*. Ithaca, N.Y.: Cornell University Press, 1975.
Topics: *VII*, I

1780 Kren, Claudia. *Alchemy in Europe: A Guide to Research*. New York: Garland, 1990.
Topic: *II*

1781 Macey, Samuel L. *Time: A Bibliographical Guide*. New York: Garland, 1991.
Topic: *VII*

1782 Meynell, G. G. *A Bibliography of Dr. Thomas Sydenham (1624-1689)*. Folkestone, England: Winterdown Books, 1990.
Topics: *VII*, XI Name: Sydenham

1783 Mitcham, Carl, and Jim Grote. "Select Bibliography of Theology and Technology." Pp. 325-502 in *Theology and Technology: Essays in Christian Analysis and Exegesis*. Edited by Carl Mitcham and Jim Grote. Lanham, Md.: University Press of America, 1984.
Topic: *VI*

1784 Morrill, J. S. *Seventeenth-Century Britain, 1603-1714*. (Critical Bibliographies in Modern History.) Folkestone, England: Dawson, 1980.

1785 Neu, John, ed. *Isis Cumulative Bibliography, 1966-1975*, and *Isis Cumulative Bibliography, 1976-1985*. See (1770) and (1770).

1786 Pighetti, Clelia. "Cinquant' anni di Studi Newtoniani (1908-1959)." *Rivista Critica di Storia della Filosofia* (1960): 181-203 and 295-318.
Topics: *XII*, VII, X Name: Newton

1787 Porter, Roy. *The Earth Sciences: An Annotated Bibliography*. New York: Garland, 1983.
Topic: *VII*

1788 Porter, Roy, and Kate Poulton. "Geology in Britain, 1660-1800: A Selective Biographical Bibliography." *Journal of the Society for the Bibliography of Natural History* 9(1978): 74-84
Topics: *VII*, X

1789 Porter, Roy, and Kate Poulton. "Research in British Geology, 1660-1800: A Survey and Thematic Bibliography." *Annals of Science* 34(1977): 33-42.
Topic: *VII*

1790 "Principales Publications D'Alexandre Koyre." In *Melanges Alexandre Koyre*, Vol 1, xiii-xvii, and (repeated) Vol. 2, xv-

xix. Paris: Hermann 1964. See (1046).
Topics: *VII*, XII

1791 Pritchard, Alan. *Alchemy: A Bibliography of English-Language Writings*. London: Routledge and Kegan Paul, 1980.
Topic: *II*

1792 Rider, K. J. *History of Science and Technology: A Select Bibliography for Students*. London: Library Association, 1970.
Topics: *VII*, VI

1793 Rubin, Davida. *Sir Kenelm Digby, F.R.S., 1603-1665: A Bibliography Based on the Collection of K. Garth Huston, Sr.* San Francisco: J. Norman, 1991.
Topics: *VII*, II, XI Name: Digby

1794 Sacksteder, William. *Hobbes Studies (1879-1979): A Bibliography*. Bowling Green, Ohio: Philosophical Documentation Center, Bowling Green State University, 1982.
Topics: *VII*, IX Name: Hobbes

1795 Sarjeant, William A. S. *Geologists and the History of Geology: An International Bibliography from the Origins to 1978*. 5 vols. New York: Arno Press, 1980.
Topic: *VII*

1796 Schatzberg, Walter. "Relations of Literature and Science: A Bibliography of Scholarship." *Clio* 4(1974): 73-93.
Topic: *X*

1797 Schatzberg, Walter. "Relations of Literature and Science: A Bibliography of Scholarship, 1975-1976." *Clio* 7(1977): 135-155.
Topic: *X*

1798 Schatzberg, Walter. "Relations of Literature and Science: A Bibliography of Scholarship, 1976-1977." *Clio* 8(1978): 135-155.
Topic: *X*

1799 Schatzberg, Walter. "Relations of Literature and Science: A Bibliography of Scholarship, 1977-1978." *Clio* 9(1979): 111-132.
Topic: *X*

1800 Schatzberg, Walter. "Relations of Literature and Science: A Bibliography of Scholarship, 1978-1979." *Clio* 10(1980): 57-84.
Topic: *X*

1801 Schuler, Robert M. *English Magical and Scientific Poems to 1700: An Annotated Bibliography*. New York: Garland, 1979.
Topics: *X*, II

1802 Schuler, Robert M. "English Scientific Poetry, 1500-1700: Prolegomena and Preliminary Check List." *The Papers of the Bibliographic Society of America* 69(1975): 482-502.
Topics: *X*, II, VIII

1803 Stathis, James J. *A Bibliography of Swift Studies, 1945-1965*. Nashville: Vanderbilt University Press, 1967.
Topic: *X* Name: Swift

1804 Sutherland, James Runcieman. *English Literature of the Late Seventeenth Century. Vol. 6 of The Oxford History of English Literature*. Oxford: Clarendon Press, 1962.
Topic: *X*

1805 Thorton, John L., and R. I. J. Tully. *Scientific Books, Libraries and Collectors: A Study of Bibliography and the Book Trade in Relation to Science*. 3rd revised ed. London: Library Association, 1971.

1806 Tobin, James E. *Eighteenth Century English Literature and Its Cultural Background: A Bibliography*. New York: Fordham University Press, 1939.
Topic: *X*

1807 Wallis, Peter, and Ruth Wallis. *Newton and Newtoniana: 1672-1975: A Bibliography*. Folkestone, England: Dawson, 1975.
Topics: *XII*, II, VII, X Name: Newton

1808 Watson, George, ed. *The Cambridge Bibliography of English Literature: Vol. 2, 1660 1800. Supplement*. Cambridge, 1957. 2nd Edition. Cambridge: Cambridge University Press, 1965.
Topic: *X*

1809 Watson, George, ed. *The New Cambridge Bibliography of English Literature, Vol. 2, 1660-1800*. Cambridge: Cambridge University Press, 1971.
Topic: *X*

1810 Whitrow, Magda, ed. *Isis Cumulative Bibliography: A Bibliography of the History of Science Formed from Isis Critical Bibliographies 1-90, 1913-1965*. See (1769).

1811 Wikelund, Philip. "Restoration Literature: An Annotated Bibliography." *Folio* 19(1954): 135-155.
Topic: *X*

1812 Winder, Marianne. "A Bibliography of the Writings of Walter Pagel." In *Science, Medicine and Society in the Renaissance*, edited by Allen G. Debus, Vol. 2, 289-326. New York: Science History, 1972. See (0357).
Topics: *XI*, II

1813 Wing, Donald G. *Short Title Catalogue of Books Printed in England, Scotland, Ireland, Wales, and British America, and of English Books Printed in Other Countries, 1641-1700*. 2nd ed., revised and enlarged. 3 vols. New York: Index Committee of the Modern Language Association of America, 1972-1988.

1814 Witschi-Bernz, Astrid. *Bibliography of Works in the Philosophy of History, 1500-1800*. (Beiheft 12 of *History and Theory: Studies in the Philosophy of History*.) Middletown, Conn.: Wesleyan University Press, 1972.
Topic: *I*

1815 Yolton, Jean S., and John W. Yolton. *John Locke: A Reference Guide*. Boston: G. K. Hall, 1985.
Topics: *VII*, X Name: Locke

A Bibliographic List of Doctoral Dissertations

1816 Abromitis, Lois I. "William Gilbert as Scientist: The Portrait of a Renaissance Amateur." Brown University, 1977.
Topic: *XI* Name: Gilbert

1817 Acworth, R. "La Philosophie de John Norris, 1657-1712." University of Paris, 1977.
Topic: *VII* Name: Norris

1818 Adams, Charles V. "An Introduction to the *Divine Dialogues* of Henry More." University of Cincinnati, 1934.
Topics: *VII*, X Name: More

1819 Adamson, Ian R. "The Foundation and Early History of Gresham College London, 1596-1704." Cambridge University, 1975.
Topic: *V*

1820 Anderson, Fulton H. "The Influence of Contemporary Science on Locke's Method and Results." University of Toronto, 1920.
Topic: *VII* Name: Locke

1821 Anderson, Paul Russell. "Science in Defense of Liberal Religion." Columbia University, 1934.
Topic: *VII* Names: Cudworth, More, Smith, Whichcote

1822 Ashworth, William B., Jr. "The Sense of the Past in English Scientific Thought of the Early Seventeenth Century: The Impact of the Historical Revolution." University of Wisconsin-Madison, 1975.
Topics: *X*, VII

1823 Axtell, James L. "The Educational Writings of John Locke: A Critical Edition." Cambridge University, 1966.
Topic: *VII* Name: Locke

1824 Baker, John Tull. "A Historical and Critical Examination of English Space and Time Theories from H. More to Bishop Berkeley." Columbia University, 1932.
Topics: *VII*, XII Names: Barrow, Clarke, Locke, More, Newton, Smith

1825 Barnett, Francis J. "'The Exantlation of Truth': Ways of Knowing in the Works of Sir Thomas Browne." University of Southwestern Louisiana, 1980.
Topics: *X*, VII, XI Name: Browne, T.

1826 Bates, Donald G. "Thomas Sydenham: The Development of his Thought, 1666-1676." Johns Hopkins University, 1975.
Topic: *XI* Name: Sydenham

1827 Bazeley, Deborah Taylor. "An Early Challenge to the Precepts and Practice of Modern Science: The Fusion of Fact, Fiction, and Feminism in the Works of Margaret Cavendish, Duchess of Newcastle (1623-1673)." University of California, San Diego, 1990.
Topics: *IV*, X Name: Cavendish

1828 Beck, Daniel A. "Miracle and the Mechanical Philosophy: The Theology of Robert Boyle in Its Historical Context." Notre Dame University, 1986.
Topic: *VII* Name: Boyle

1829 Beier, Lucinda McCray. "Sufferers and Healers: Health Choices in Seventeenth Century England." Lancaster University, 1984.
Topic: *XI*

1830 Bennett, J. A. "Studies in the Life and Work of Sir Christopher Wren." Cambridge University, 1973.
Topic: *VI* Name: Wren

1831 Biddle, John C. "John Locke on Christianity: His Context and His Text." Stanford University, 1972.
Topics: *VII*, IX, X Name: Locke

1832 Birken, William J. "The Fellows of the Royal College of Physicians of London, 1603-1643: A Social Study." University of North Carolina, Chapel Hill, 1977.
Topics: *V*, XI

1833 Boas (Hall), Marie. "Robert Boyle and the Corpuscular Philosophy: A Study of Theories of Matter in the Seventeenth Century." Cornell University, 1949.
Topic: *VII* Name: Boyle

1834 Bowen, Mary E. C. "'This Great Automaton, the World': The Mechanical Philosophy of Robert Boyle, F.R.S." Columbia University, 1976.
Topic: *VII* Name: Boyle

1835 Bowles, Geoffrey. "The Place of Newtonian Explanation in English Popular Thought, 1687-1727." Oxford University, 1977.
Topics: *XII*, VII, X Name: Newton

1836 Bradford, Edward B., Jr. "Creation, Contingency, and Early Modern Science: The Impact of Voluntaristic Theology on Seventeenth Century Natural Philosophy." Indiana University, 1984.
Topic: *VII*

1837 Bradish, Norman Conyers. "John Sergeant, a Seventeenth Century Critic of Locke." Northwestern University, 1932.
Topic: *VII* Names: Locke, Sargeant

1838 Brooks, Richard S. "The Relationships between Natural Philosophy, Natural Theology and Revealed Religion in the Thought

of Newton and Their Historiographic Relevance." Northwestern University, 1976.
Topics: *XII*, I, VII, VIII, X Name: Newton

1839 Brown, Cedric C. "The Early Works of Henry More." University of Reading, 1968.
Topics: *VII*, X Name: More

1840 Bruneteau, C. "John Arbuthnot (1667-1735) et les Idees au debut au XVIIIe Siecle." University of Paris, 1973.
Topic: *XI* Name: Arbuthnot

1841 Budick, Sanford. "Dryden's *Religio Laici*: A Study in Context and Meaning." Yale University, 1967.
Topic: *X* Name: Dryden

1842 Burns, Norman Thomas. "The Tradition of Christian Mortalism in England: 1530-1660." University of Michigan, 1967.
Topics: *VII*, X Names: Browne, T., Hobbes, Milton

1843 Burns, William E., Jr. "An Age of Wonders: Prodigies, Providence, and Politics in England, 1580-1727." University of California at Davis, 1994.
Topics: *IV,* VII, XII Name: Newton

1844 Burtt, Edwin Arthur. "The Metaphysics of Sir Isaac Newton: An Essay on the Metaphysical Foundations of Modern Science." Columbia University, 1925.
Topics: *XII*, VII, VIII Name: Newton

1845 Caldwell, Wayne T. "'Affliction Then Is Ours': George Herbert's *The Temple* as an Anatomy of Religious Melancholy." Duke University, 1973.
Topic: *X* Name: Herbert, G.

1846 Canavan, Thomas L. "Madness and Enthusiasm in Burton's *Anatomy of Melancholy* and Swift's *Tale of a Tub*." Columbia University, 1973.
Topics: *X*, XI Names: Burton, Swift

1847 Carrithers, David W. "Joseph Glanvill and Pyrrhonic Scepticism: A Study in the Revival of the Doctrines of Sextus Empiricus in Sixteenth-Century and Seventeenth-Century Europe." New York University, 1972.
Topic: *VII* Name: Glanvill

1848 Carroll, J. T. "Dryden and the Great Chain of Being." National University of Ireland, 1961.
Topic: *X* Name: Dryden

1849 Caudill, R. L. "Some Literary Evidence of the Development of English Virtuoso Interests in the Seventeenth Century, with Particular Reference to the Literature of Travel." Oxford University, 1975.
Topic: *X*

1850 Cobb, Joann P. "Jonathan Swift and Epistemology: A Study of Swift's Satire on Ways of Knowing." St. Louis University, 1975.
Topics: *X*, VII Name: Swift

1851 Cook, Harold J. "The Regulation of Medical Practice in London Under the Stuarts, 1607-1704." University of Michigan, 1981.
Topics: *XI*, V

1852 Coudert, Allison.
See Gottesman, Allison Coudert.

1853 Craig, George A. "Umbra Dei: Henry More and the Seventeenth-Century Struggle for Plainness." Harvard University, 1947.
Topic: *X* Name: More

1854 Crocker, Robert. "An Intellectual Biography of Henry More (1614-1687)." Oxford University, 1986.
Topics: *VII, II* Name: More

1855 Crouch, Laura E. "The Scientist in English Literature: Domingo Gonsales (1638) to Victor Frankenstein (1817)." University of

Oklahoma, 1975.
Topic: *X*

1856 Crowley, M. E. "The Notion of Nature in the Corpuscular Philosophy of Robert Boyle." Marquette University, 1970.
Topic: *VII* Name: Boyle

1857 Dahrendorf, Walter. "Lockes Kontroverse mit Stillingfleet und ihre Bedeutung fur seine Stellung zur Anglikanischen Kirche." University of Hamburg, 1932.
Topic: *VII* Names: Locke, Stillingfleet

1858 Davis, Edward B., Jr. "Creation, Contingency and Early Modern Science: the Impact of Voluntaristic Theology on Seventeenth-Century Natural Philosophy." Indiana University, 1984.
Topic: *VII* Name: Boyle

1859 Deason, Gary B. "The Philosophy of a Lord Chancellor: Religion, Science, and Social Stability in the Work of Francis Bacon." Princeton Theological Seminary, 1977.
Topics: *VII*, IV Name: Bacon

1860 Dick, Steven J. "Plurality of Worlds and Natural Philosophy: An Historical Study of the Origins of Belief in Other Worlds and Extra Terrestial Life." Indiana University, 1977.
Topics: *X*, VII

1861 Dobbs, Betty Jo Teeter. "The Foundation of Newton's Alchemy, or 'The Hunting of the Greene Lyon'." University of North Carolina, 1973.
Topics: *XII*, II, VII Name: Newton

1862 Drumin, William A. "The Corpuscular Philosophy of Robert Boyle: Its Establishment and Verification." Columbia University, 1973.
Topic: *VII* Name: Boyle

1863 Dyche, E. I. "The Life and Works, and Philosophical Relations, of John (Janus Junius) Toland, 1670-1722." University of Southern California, 1944.
Topics: *VII*, IX Name: Toland

1864 Eamon, William C. "Books of Secrets and the Empirical Foundations of English Natural Philosophy, 1550-1650." University of Kansas, 1977.
Topics: *VII*, II

1865 Elmer, Peter. "Medicine, Medical Reform and the Puritan Revolution." University of Swansea, 1980.
Topics: *XI*, II, III Name: Hart

1866 Emerson, Roger L. "English Deism 1670-1755: An Enlightenment Challenge to Orthodoxy." Brandeis University, 1962.
Topics: *IX*, IV, VII Names: Blount, Collins, Tindal, Toland

1867 Estes, Leland L. "The Role of Medicine and Medical Theories in the Rise and Fall of the Witch Hunts in England." University of Chicago, 1985.
Topics: *XI*, II, IV

1868 Evans, Robert Rees. "John Toland's Pantheism: A Revolutionary Ideology and Enlightenment Philosophy." Brandeis University, 1965.
Topics: *IX*, IV, VII Name: Toland

1869 Farrell, Sister Maureen. "The Life and Works of William Whiston." University of Manchester, 1973.
Topics: *IX*, VII, XII Name: Whiston

1870 Feingold, Mordechai. "Science, Universities, and Society in England, 1580-1640." Oxford Uniersity, 1981.
Topics: *V*, IV

1871 Figala, Karin. "Die 'Kompositionshierarchie' der Materie Newtons quantitative Theorie und Interpretation der qualitativen Alchemie." Technische Universitat, Munich, 1977.
Topics: *XII*, II Name: Newton

1872 Fisher, Mitchell S. "Robert Boyle, Devout Naturalist: A Study in Science and Religion in the Seventeenth Century." Columbia University, 1946.
Topics: *VII*, VIII Name: Boyle

1873 Fishman, Joel H. "Edward Stillingfleet, Bishop of Worcester (1635-1699): Anglican Bishop and Controversialist." University of Wisconsin-Madison, 1977.
Topics: *VII*, VIII Name: Stillingfleet

1874 Force, James E. "Whiston Controversies: The Development of 'Newtonianism' in the Thought of William Whiston." Washington University, 1977.
Topics: *IX*, VII, X, XII Names: Newton, Whiston

1875 Galbraith, Kenneth J. "Henry More's *Divine Dialogues*: A Critical Analysis." University of North Carolina, 1969.
Topics: *VII*, X Name: More

1876 Gascoigne, John. "'The Holy Alliance': The Rise and Diffusion of Newtonian Natural Philosophy and Latitudinarian Theology within Cambridge from the Restoration to the Accession of George II." Cambridge University, 1980.
Topics: *XII*, IV, V, VIII Name: Newton

1877 Gentilcore, Roxanne M. "The Classical Tradition and American Attitudes towards Nature in the Seventeenth and Eighteenth Centuries." Boston University, 1991.
Topics: *X*, VII

1878 Genuth, Sara Schechner. "From Monstrous Signs to Natural Causes: The Assimilation of Comet Lore into Natural Philosophy." Harvard University, 1988.
Topics: *VII*, X

1879 Gottesman, Allison Coudert. "Francis Mercury van Helmont: His Life and Thought." London University, 1972.
Topic: *II* Name: Helmont

1880 Gouk, Penelope M. "Music in the Natural Philosophy of the Early Royal Society." London University, 1982.
Topics: *VII*, XII Names: Hooke, Newton

1881 Griffith, Richard R. "Science and Pseudo-Science in the Imagery of John Dryden." Ohio State University, 1957.
Topic: *X* Name: Dryden

1882 Haber, Francis C. "Revolution in the Concept of Historical Time: A Study in the Relationship between Biblical Chronology and the Rise of Modern Science." Johns Hopkins University, 1957.
Topics: *X*, VII

1883 Habicht, Hartwig. "Joseph Glanvill, ein spekulatuver Denker im England des XVII. Jahrhunderts: Eine Studie uber des fruhwissenschaftliche Weltbild." University of Zurich, 1936.
Topics: *VII*, V Names: Cudworth, Glanvill, More

1884 Hamlin, Howard Phillips, Jr. "A Critical Evaluation of John Locke's Philosophy of Religion." University of Georgia, 1972.
Topics: *VII*, VIII Name: Locke

1885 Hammil, Carrie E. "The Celestial Journey and the Harmony of the Spheres in English Literature, 1300-1700." Texas Christian University, 1972.
Topic: *X*

1886 Haring, Lee. "Henry More's *Psychoanthanasia* and *Democritus Platonissans*: A Critical Edition." Columbia University, 1961.
Topics: *VII*, II Name: More

1887 Harrison, Charles T. "The Ancient Atomists and English Humanism in the Seventeenth Century." Harvard University, 1932.
Topics: *VII*, X Names: Bacon, Boyle, Hobbes, More

1888 Hefelbower, Samuel G. "The Relation of John Locke to English Deism." Harvard University, 1912.
Topics: *IX*, VII Name: Locke

1889 Hellegers, Desiree E. M. "The Politics of Redemption: Science, Conscience and Poetry from John Donne and Francis Bacon to Anne Finch." University of Washington, 1993.
Topics: *X*, IV Names: Bacon, Conway, Donne

1890 Henry, John C. "Matter in Motion: The Problem of Activity in Seventeenth-Century English Matter Theory." Open Univer-

sity, 1983.
Topics: *VII*, XII Name: Newton

1891 Hofstadter, Albert A. "Locke and Scepticism." Columbia University, 1935.
Topics: *VII*, IX Name: Locke

1892 Huebsch, Daniel A. "Ralph Cudworth: ein Englischer Religionsphilosoph des Siebenzehnten Jahrhunderts." University of Jena, 1904.
Topics: *VII*, VIII Name: Cudworth

1893 Huffman, William H. "Robert Fludd: The End of an Era." University of Missouri-Columbia, 1977.
Topics: *VII*, II Name: Fludd

1894 Hunt, Deray Louis. "Some of the Relations Between Descartes and Locke." University of Toronto, 1940.
Topic: *VII* Name: Locke

1895 Hunter, Michael. "The Place of John Aubrey in Intellectual History." Oxford University, 1974.
Topic: *X*

1896 Hurlbutt, Robert Harris, III. "Science and Theology in Eighteenth-Century England." University of California-Berkeley, 1953.
Topics: *VII*, VIII, IX, XII Name: Newton

1897 Hvolbek, Russell H. "Seventeenth-Century Dialogues: Jacob Boehme and the New Sciences." University of Chicago, 1984.
Topics: *II*, VII

1898 Iliffe, R. C. "'The Idols of the Temple': Isaac Newton and the Private Life of Anti-Idolatry." Cambridge University, 1989.
Topics: *XII*, X Name: Newton

1899 Innes, David C. "Francis Bacon, Christianity, and the Hope of Modern Science." Boston College, 1992.
Topics: *VII*, X Name: Bacon

1900 Jacob, Margaret C. "The Church and the Boyle Lectures: the Social Context of the Newtonian Natural Philosphy." Cornell University, 1969.
Topics: *IV*, V, VII, VIII, IX, XII Names: Bentley, Newton, Toland

1901 Jones, Rex F. "Genealogy of a Classic: *The English Physitian* of Nicholas Culpepper." University of California at San Francisco, 1984.
Topics: *XI*, X Name: Culpepper

1902 Kaplan, Barbara B. "The Medical Writings of Robert Boyle: Medical Philosophy in Mid-Seventeenth Century England." University of Maryland, 1979.
Topics: *XI*, VII Name: Boyle

1903 Kargon, Robert Hugh. "Science and Atomism in England: From Hariot to Newton." Cornell University, 1964.
Topics: *VII*, V, VIII, IX, XII Names: Boyle, Cavendish, Charleton, Hariot, Hobbes, Newton, Percy

1904 Karlsen, Carol F. "The Devil in the Shape of a Woman: The Witch in Seventeenth Century New England." Yale University, 1980.
Topics: *II*, X

1905 Kelly, John T. "Practical Astronomy During the Seventeenth Century: A Study of Almanac-Makers in America and England." Harvard University, 1977.
Topics: *VI*, VII

1906 Kemerling, Garth Leroy. "John Locke and Mind/Body Dualism." University of Iowa, 1974.
Topic: *VII* Name: Locke

1907 King, Bruce. "Dryden's Treatment of Ideas and Themes in His Dramatic Works, with Some Reference to the Intellectual Movement of His Time." Leeds University, 1959.
Topic: *X* Name: Dryden

1908 Kite, Jon B. "A Study of the Works and Reputation of John Aubrey, with Emphasis on his *Brief Lives*." University of California-Santa Barbara, 1977.
Topic: *X* Name: Aubrey

1909 Klawitter, George A. "The Poetry of Henry More." University of Chicago, 1981.
Topics: *X*, VII Name: More

1910 Koppel, Richard M. "English Satire on Science, 1660-1750." University of Rochester, 1978.
Topic: *X*

1911 Krakowski, Edouard. "Les Sources Medievales de la Philosophie de Locke." University of Paris, 1915.
Topic: *VII* Name: Locke

1912 Kronemeyer, Ronald J. "Matter and Meaning: Dualism in the Thought of Robert Boyle, Isaac Newton, and John Ray." Kent State University, 1978.
Topics: *VII*, XII Names: Boyle, Newton, Ray

1913 Kubrin, David. "Providence and the Mechanical Philosophy: The Creation and Dissolution of the World in Newtonian Thought. A Study of the Relations of Science and Religion in Seventeenth-Century England." Cornell University, 1968.
Topics: *VII*, VIII, X, XII Names: Burnet, T., Keill, Newton, Whiston

1914 Lai, Tyrone Tai Lun. "Infinitesimals and the Infinite Universe: A Study of the Relation between Newton's Science and His Metaphysics." University of California at San Diego, 1972.
Topics: *XII*, VII Name: Newton

1915 Langton, Larry B. "Milton, J. A. Comenius, and Hermetic Natural Philosophy." University of Wisconsin, Madison, 1977.
Topics: *II*, VII, X Names: Comenius, Milton

1916 Loney, Roy P. "Faith, Reason, and Natural Philosophy in Sir Thomas Browne's *Urn Burial* and *The Garden of Cyrus*." University of Colorado, 1974.
Topics: *X*, VII Name: Browne, T.

1917 Lowery, W. R. "John Milton, Henry More, and Ralph Cudworth: A Study in Patterns of Thought." Northwestern University, 1970.
Topics: *X*, VII Names: Cudworth, Milton, More

1918 Luecke, Richard Henry. "God and Contingency in the Philosophies of Locke, Clarke, and Leibniz." University of Chicago, 1955.
Topics: *VII*, VIII, XII Names: Clarke, Locke

1919 MacDonald, Michael. "Madness and Healing in Seventeenth-Century England." Stanford University, 1979.
Topic: *XI* Name: Napier

1920 Mackinnon, Flora I. "The Philosophical Writings of Henry More." Toronto University, 1924.
Topic: *VII* Name: More

1921 Martz, William J. "Dryden's Religious Thought: A Study of *The Hind and the Panther* and Its Background." Yale University, 1957.
Topic: *X* Name: Dryden

1922 Massa, Daniel P. "Giordano Bruno and Sixteenth and Seventeenth Century English Writers with Particular Reference to the Works of Henry More." University of Edinburgh, 1975.
Topics: *VII*, X Name: More

1923 Mattern, Ruth Marie. "Locke on the Essence and Powers of the Soul." Princeton University, 1975.
Topic: *VII* Name: Locke

1924 McCarthy, Paul J. "A Doctor's Language of Devotion: The Occult Sciences in the Works of Thomas Browne." Michigan State University, 1974.
Topics: *II*, X, XI Name: Browne, T.

1925 McLachlan, Patricia. "Scientific Professionals in the Seventeenth Century." Yale University, 1968.
Topics: *X*, V

1926 Merton, Robert K. "Sociological Aspects of Scientific Development in Seventeenth-Century England." Harvard University, 1935.
Topics: *IV*, III

1927 Meyer, Marilyn. "Ralph Cudworth's Philosophical System." Columbia University, 1952.
Topics: *VII*, VIII Name: Cudworth

1928 Michalec, Gerald J. "Metaphysics and Realist Theories of Space in the Seventeenth Century." University of Pittsburgh, 1977.
Topic: *VII*

1929 Miles, Rogers. "The Clerical Virtuosi of the Royal Society: 1663-1687." Princeton University, 1987.
Topics: *V*, VII, VIII, X

1930 Milton, John. "The Influence of the Nominalist Movement on the Thought of Bacon, Boyle, and Locke." London University, 1982.
Topic: *VII* Names: Bacon, Boyle, Locke

1931 Mintz, Samuel I. "The Hunting of Leviathan: Seventeenth-Century Reactions to the Materialism and Moral Philosophy of Thomas Hobbes." Columbia University, 1958.
Topics: *VII*, V, IX Name: Hobbes

1932 Moore, John Thomas. "Locke's Concept of Faith." University of Kansas, 1971.
Topics: *VII*, X Name: Locke

1933 Morawetz, Bruno. "The Epistemology of John Norris." University of Toronto, 1964.
Topics: *VII*, X Name: Norris

1934 Nicolson, Marjorie Hope. "The Life and Works of Henry More: A Study in Cambridge Platonism." Columbia University, 1920.
Topics: *VII*, X Name: More

1935 Ornstein, Martha. "The Role of Scientific Societies in the Seventeenth Century." Columbia University, 1913.
Topic: *V*

1936 Osler, Margaret J. "John Locke and Some Philosophical Problems in the Science of Newton and Boyle." Indiana University, 1968.
Topics: *VII*, XII Names: Boyle, Locke, Newton

1937 Oster, Malcolm. "Nature, Ethics, and Divinity: The Early Thought of Robert Boyle." Oxford University, 1990.
Topics: *VII*, X Name: Boyle

1938 Overman, Ronald J. "Theories of Gravity in the Seventeenth Century." Indiana University, 1974.
Topic: *VII*

1939 Page, Leroy E. "The Rise of the Diluvial Theory in British Geological Thought." University of Oklahoma, 1963.
Topics: *VII*, X

1940 Pegg, Barry M. "Optimistic and Pessimistic Attitudes to Generation and Corruption in Selected Literature and Scientific Texts, 1590-1660." University of Wisconsin-Madison, 1977.
Topics: *X*, XI

1941 Penrose, Stephen B. L. "The Reputation and Influence of Francis Bacon in the Seventeenth Century." Columbia University, 1934.
Topics: *VII*, X Name: Bacon

1942 Perez-Ramos, A. "Francis Bacon's Idea of Science and the Maker's Knowledge Tradition." Cambridge University, 1982.
Topics: *VII*, X Name: Bacon

1943 Perkins, James C. "Some Aspects of the Religious Thought of John Locke." Duke University, 1956.
Topic: *VII* Name: Locke

1944 Porter, Roy S. "The Making of the Sciences of Geology in Britain, 1660-1815." Cambridge University, 1974.
Topics: *VII*, X

1945 Pratt, Minnie B. "Sir Thomas Browne and the Hermetic Maze: The Structure of the *Religio Medici*." University of North Carolina, Chapel Hill, 1979.
Topics: *XI*, II, X Name: Browne, T.

1946 Prior, Moody. "Joseph Glanvill and the New Science." University of Chicago, 1930.
Topics: *VII*, X Name: Glanvill

1947 Reimann, Hugo. "Henry Mores Bedeutung fur die Gegenwart." University of Basel, 1941.
Topics: *VII*, X Name: More

1948 Robins, Harry F. "The Cosmology of Paradise Lost: A Reconsideration." Indiana University, 1951.
Topics: *X*, VII Name: Milton

1949 Roche, J. J. "Thomas Harriot's Astronomy." Oxford University, 1977.
Topic: *VII* Name: Harriot

1950 Rogers, G. A. J. "John Locke and the Scientific Revolution: A Study of the *Essay Concerning Human Understanding* in Relation to Seventeenth Century Science." University of Keele, 1972.
Topic: *VII* Name: Locke

1951 Romagosa, Sister Edward O. "A Compendium of the Opinions of John Dryden." Tulane University, 1959.
Topic: *X* Name: Dryden

1952 Ruffner, J. A. "The Background and Early Development of Newton's Theory of Comets." Indiana University, 1966.
Topics: *XII*, VII Name: Newton

1953 Sabre, Susan L. "The Separation of the Divine from the Natural: A Perspective on the Thought of Francis Bacon, 1584-1609." University of North Carolina at Chapel Hill, 1977.
Topics: *VII*, X Name: Bacon

1954 Sailor, Danton B. "Ralph Cudworth: Forlorn Hope of Humanism in the Seventeenth Century." University of Illinois, 1956.
Topics: *VII*, X Name: Cudworth

1955 Schaffer, Simon. "Newtonian Cosmology and the Steady State." Cambridge University, 1980.
Topics: *XII*, IV, VII Name: Newton

1956 Schiebinger, Londa L. "Women and the Origins of Modern Science." Harvard University, 1984.
Topics: *IV*, V

1957 Schnorrenberg, John M. "Anglican Architecture, 1558-1662: Its Theological Implications and Its Relation to the Continental Background." Princeton University, 1964.
Topic: *VI*

1958 Schuchard, Marsha K. Manatt. "Freemasonry, Secret Societies, and the Continuity of the Occult Traditions in English Literature." University of Texas, Austin, 1975.
Topics: *X*, II, IV

1959 Shapiro, Barbara. "John Wilkins, 1614-1672." Harvard University, 1966.
Topics: *VII*, III, V Names: Sprat, Wilkins

1960 Sharp, L. G. "Sir William Petty and Some Aspects of Seventeenth-Century Natural Philosophy." Oxford University, 1977.
Topics: *VII*, IV Name: Petty

1961 Shaw, Ann Laura. "Hooker, Tillotson, Locke, Toland, and Tindal: Questions of Authority in Assent." University of Michigan, 1975.
Topics: *VII*, IX, X Names: Locke, Tillotson, Tindal, Toland

1962 Shay, Cari Lee Gabiou. "The Transmutation of Alchemy into Science and Political Thought." University of Oregon, 1974.
Topics: *II*, IV, VII, XII Names: Locke, Newton

1963 Sherrer, G. B. "Francis Mercury van Helmont: A Study of his Personality and Influence." Cleveland University, 1937.
Topics: *II*, VII Name: Helmont

1964 Shute, Michael N. "Earthquakes and Early American Imagination: Decline and Renewal in Eighteenth-Century Puritan Culture." University of California, Berkeley, 1977.
Topics: *VII*, X

1965 Singer, Thomas C. "Sir Thomas Browne and 'The hieroglyphical schools of the Egyptians': A Study of the Renaissance Search for the Natural Language of the World." Columbia, University, 1985.
Topics: *X*, II, XI Name: Browne, T.

1966 Smith, Donna S. "Tudor and Stuart Midwifery." University of Kentucky, 1980.
Topics: *XI*, II

1967 Smyth, Marina B. "Understanding the Universe in Seventeenth-Century Ireland." University of Notre Dame, 1984.
Topics: *VII*, X

1968 Snow, Adolph J. "Matter and Gravity in Newton's Physical Philosophy: A Study in the Natural Philosophy of Newton's Time." Columbia University, 1926.
Topics: *XII*, VII Name: Newton

1969 Sonnichsen, Charles L. "The Life and Works of Thomas Sprat." Harvard University, 1931.
Topics: *IV*, V, X Name: Sprat

1970 Southgate, Beverly C. "The Life and Work of Thomas White (1593-1676)." London University, 1979.
Topics: *VII*, IV Name: White

1971 Staudenbaur, Craig Anthony. "The Metaphysical Thought of Henry More: Its Sources and Development." Johns Hopkins University, 1961.
Topic: *VII* Name: More

1972 Stewart, Larry. "Whigs and Heretics: Science, Religion, and Politics in the Age of Newton." University of Toronto, 1978.
Topics: *IV*, VII, IX, XII Name: Clarke, Hickes, Newton, North

1973 Sullivan, Robert E. "John Toland and the Deist Controversy: A Study in Adaptation." Harvard University, 1977.
Topics: *IX*, VII Name: Toland

1974 Svendsen, Kester. "Milton's Use of Natural Science, with Special Reference to Certain Encyclopedias of Science in English." University of North Carolina, 1940.
Topic: *X* Name: Milton

1975 Taffee, James G. "Milton, the Boyles, and Their Circle." Indiana University, 1960.
Topics: *VII*, X Names: Boyle, Milton

1976 Tamny, Martin. "The Early Epistemological Thought of Isaac Newton." City University of New York, 1976.
Topics: *XII*, VII Name: Newton

1977 Tang, Michael. "The Intellectual Context of John Wilkins' *Essay Towards a Real Character and a Philosophical Language*." University of Wisconsin-Madison, 1975.
Topics: *VII*, X Name: Wilkins

1978 Teague, B. C. "The Origins of Robert Boyle's Philosophy." Cambridge University, 1971.
Topic: *VII* Name: Boyle

1979 Thackray, Arnold W. "The Newtonian Tradition and Eighteenth-Century Chemistry." Cambridge University, 1966.
Topics: *VII*, XII Name: Newton

1980 Tihinen, Paul E. "The Transition in the Treatment of the Body-Soul Relationship: A Study of Juan Huarte, Robert Burton, and Rene Descartes." Miami University, 1978.
Topics: *VII*, X Name: Burton

1981 Traister, Barbara H. "Heavenly Necromancy: The Figure of the Magician in Tudor and Stuart Drama." Yale University, 1973.
Topics: *X*, II Name: Shakespeare

1982 Turner, C. E. A. "The Puritan Contribution to Scientific Education in the Seventeenth Century in England." London University, 1952.
Topics: *X*, III

1983 Tyler, Glenn E. "The Influence of Calvinism on the Development of Early Modern Science in England and America." University of Minnesota, 1951.
Topics: *III*, VII

1984 Van Leeuwen, Hendrik Gerrit. "The Problem of Certainty in English Thought from Chillingworth to Locke." University of Iowa, 1961.
Topic: *VII* Names: Chillingworth, Locke

1985 Wagenblass, John H. "Lucretius and the Epicurean Tradition in English Poetry." Harvard University, 1946.
Topic: *X* Names: Blackmore, Pope

1986 Wagner, Joseph B. "Samuel Butler's Satire of the Hermetic Philosophers." Kent State University, 1974.
Topics: *X*, II Name: Butler

1987 Waller, M. J. "Joseph Glanvill and the Seventeenth Century Reaction against Enthusiasm." University of St. Andrews, 1968.
Topics: *VII*, IV Name: Glanvill

1988 Wegman, Nora J. "Argument and Satire: the Christian Response to Deism, 1670-1760." Northwestern University, 1967.
Topics: *VII*, IX, X Names: Bentley, Clarke, Norris, Swift, Toland

1989 Weir, C. "Francis Mercury van Helmont: His Life and His Position in the Intellectual History of the Seventeenth Century." Harvard University, 1941.
Topics: *II,* V, VII Name: Helmont

1990 Westfall, Richard S. "Christianity and the Virtuosi." Yale University, 1955.
Topics: *VII*, VIII, IX, XII Names: Boyle, Newton

1991 Westfall, T. M. "Sir Thomas Browne's Revisions of the *Pseudodoxia Epidemica*: A Study in the Development of his Mind." Princeton University, 1938.
Topics: *XI*, VII, X Name: Browne, T.

1992 Whitman, Julie. "Cotton Mather and Jonathan Edwards: Philosophy, Science, and Puritan Theology." Indiana University, 1993.
Topics: *VII*, X Name: Mather, C.

1993 Whitney, Charles C. "Bacon's Modernity: From Literature to Science." City University of New York, 1977.
Topics: *X*, VII Name: Bacon

1994 Willard, Thomas S. "The Life and Work of Thomas Vaughan." University of Toronto, 1978.
Topics: *X*, II Name: Vaughan, T.

1995 Wojcik, Jan W. "Robert Boyle and the Limits of Reason: A Study in the Science and Religion in Seventeenth-Century England." University of Kentucky, 1992.
Topics: *VII*, VIII Name: Boyle

1996 Wright, D. G. "The Theology of John Locke." Edinburgh University, 1938.
Topics: *VII*, VIII Name: Locke

1997 Wright, Peter W. G. "Astrology in Mid-Seventeenth-Century England: A Sociological Analysis." University of London, 1983.
Topics: *IV*, II

1998 Yolton, John W. "John Locke and the Way of Ideas: An Examination and Evaluation of the Epistemological Doctrines of John Locke's *Essay Concerning Human Understanding*." Oxford University, 1952.

Topic: *VII* Name: Locke

1999 Yost, Robert M., Jr. "Locke's Theory of the External World." Harvard University, 1948.

Topic: *VII* Name: Locke

2000 Zetterberg, J. Peter. "'Mathematical Magic' in England: 1550-1650." University of Wisconsin-Madison, 1977.

Topics: *VII*, II

Topical Indexes

For subject descriptions of each of these categories, please see the "Descriptions of Topical Categories," pp. xvii-xxix.

Topic I: Historiography

Topic II: The Magical, Alchemical, and *Prisca* Traditions

pard (1446); Sherrer (1963); Shumaker (1453-1454); Singer (1965); D. Smith (1966), Sommerville (1468); Spinka (1476); Stahlman (1482); Staudenbaur (1486); Stock (1506); Talmor (1529); Teich and Young (1535); Thomas (1541); Thorndike (1542-1543); Thulesius (1546); Traister (1981); Trevor-Roper (1560, 1563); Turnbull (1571); Ullmann-Margalit (1579); Van Pelt (1584); Vickers (1586-1587, 1590-1591); J. Wagner (1986); Walker (1594-1596, 1599); Wallis and Wallis (1807); P. Watson (1607); Webster (1614-1615); Weir (1989); M. West (1629); R. S. Westfall (1630, 1633, 1636, 1640, 1642, 1644, 1649); Westman (1653-1654); Westman and McGuire (1655); Wilkinson (1674-1679); Willard (1994); C. Wilson (1690); Winder (1812) Wormhoudt (1700); P. Wright (1997); Yates (1705-1711); Zafiropulo and Zolla (1734); Zetterberg (2000).

Topic III: Protestantism and the Rise of Modern Science

Abraham (0004); Agassi (0008); P. Allen (0020); Ashworth (0036); Aspelin (0039); Bainton (0053); Baron (0062); Becker (0075); Ben-David (0078-0079); M. Berman (0086); Brauer (0114); J. H. Brooke (0132); R. Brooks (0137); Burnham (0158); J. W. Carroll (0191); G. Clark (0214); Jon Clark and others (0216); Clucas (0227); H. F. Cohen (0230); I. B. Cohen (0234, 0242-0244, 0246); Cole (0249); Conant (0257); Cook (0258); Cope (0264); Cunningham (0305); Deason (0332-0333); Debus (0356); de Candolle (0363); Dillenberger (0377); Duffin and Strickland (0404); Eisenstein (0418-0419); Elmer (0423, 1865); 'Espinasse (0436); Feingold (0454, 0458); Feuer (0465); Fisch (0475); Fleming (0481); H. Foster (0500); M. Foster (0501-0502); Frank (0505); Funkenstein (0518); C. George (0535); George and George (0536); Gillespie (0544); Goldman (0552); Greaves (0564-0566); Gruner (0579); A. R. Hall (0596, 0600, 0602); B. Hall (0611); Haller (0620); Hans (0625); Hessen (0676); Heyd (0679); Hill (0683, 0686, 0688-0691, 0693-0694, 0697-0698); Hooykaas (0715, 0721-0723); Hornberger (0728); Houghton (0739-0740); M. Hunter (0756, 0760, 0763); K. Hutchison (0775); J. R. Jacob (0799-0800, 0804); Jacob and Jacob (0805; 0807-0808); M. C. Jacob (0812); Jeske (0827); Johnson (0832); R. F. Jones (0835, 0838, 0840-0841, 0844); Kaiser (0855); Kearney (0862-0864, 0866); A. Keller (0867); Kemsley (0873); M. D. King (0879); Klaaren (0885); Knappen (0887); Kocher (0892); Kuhn (0922); Lash (0931); Lilley (0951); Manuel (0987, 0989); Mason

Topic IV: Christianity, Social Ideals, Ideology, and Science

Topic V: Social Institutions, Science, and Christianity

Topic VI: Religion, Technology, Architecture, and the Environment

Topic VII: Theology, Philosophy, and Science

Topic VIII: Natural Theology and Natural Philosophy

Topic IX: Heretical Christianity, Deism, and Atheism

Topic X: Science, the Bible, and Literature

Topic XI: Religion and Medicine

Topic XII: Newtonian Studies

Index of Persons from the Historical Period

The following items are included in each record: the name as used in our indexing; the person's full name; the dates of birth and death or when the person flourished; a nutshell identification in terms of science, religion, and fame; and relevant authors and item numbers from our bibliography. F.R.S. is an abbreviation for "Fellow of the Royal Society of London" and D.P.S. is an abbreviation for "Fellow of the Dublin Philosophical Society."

There are some works which significantly treat so many seventeenth- and eighteenth-century men and women that we have not attempted to index them. These include: *Dictionary of National Biography* (0374), M. Hunter (0760), the bibliographies in general, and especially the *Isis* bibliographies (1769-1772). In this same class is *The Dictionary of Scientific Biography*, edited by Charles Coulston Gillespie, 18 volumes, which will appear in the next volume of the present bibliography. Please consult these works in addition to the following index.

Abercromby: David Abercromby (died 1701-1702?).
Scottish physician; Protestant apologist and former Jesuit; philosopher
E. Davis (0324).

Addison: Joseph Addison (1672-1719).
Essayist; journalist; political official.
M. Bush (0166); 'Espinasse (0436); Fleischmann (0480); R. Harris (0630); Hollander (0703); Hoyles (0744); Kocher (0892); Milic (1069); Nicolson (1125, 1129, 1131, 1137); Syfret (1523); Tuveson (1576).

Allen. Thomas Allen (1542-1632).
Mathematician; astrologer.
Feingold (0456).

Ames: William Ames (1576-1633).
Puritan minister and theologian.
P. Miller (1072-1073); Morgan (1081); Morison (1083); Sprunger (1477); Warch (1604).

Andrewes: Lancelot Andrewes (1555-1626).
Anglican bishop; scholar; theologian; preacher.
H. Davies (0322); Mitchell (1077).

Arbuthnot: John Arbuthnot (1667-1735).
Scottish physician who settled in London; essayist; F.R.S.
Bruneteau (1840); Nicolson (1141).

Ardeme: James Ardeme (1636-1691).
Anglican cleric; F.R.S.
Miles (1067).

Ashe: St. George Ashe (1658?-1718).
Anglican bishop; natural philosopher; D.P.S.; F.R.S.
Hoppen (0725).

Ashmole: Elias Ashmole (1617-1692).
Alchemist; astrologer; Royalist public official; antiquarian; F.R.S.
Coudert (0274); P. Curry (0306); Debus (0334, 0343); Hartley 0642; M. Hunter (0753, 0755); Josten (0849, 0851); McCaffery (1009); Read (1308); Schuler (1416); Thomas (1541); Yates (1709).

Aubrey: John Aubrey (1626-1697).
Antiquarian; topographer; F.R.S.
O. L. Dick (0372); 'Espinasse (0437); M. Hunter (0755, 1895); Kite (1908); Powell (1256); Strauss (1509); Thomas (1541).

Austen: Ralph Austen (died 1676).
Planter of fruit trees; writer.
Otten (1189).

(1584); Vickers (1586, 1588-1589); Walker (1597, 1599); K. R. Wallace (1601); Warhaft (1606); Webster (1613, 1616, 1620); J. F. West (1628); M. West 1629; Whitaker (1658); Whitney (1993); Willey (1681); Williamson (1686-1687); C. Wilson (1690); Yates (1708-1709); Zetterberg (1730).

Backhouse: William Backhouse (1593-1662).
Alchemist; friend of Ashmole.
Josten (0851).

Bainbridge: John Bainbridge (1582-1643).
Astronomer; physician; Oxford professor.
Feingold (0456); Tyacke (1578).

Bannister: John Bannister (1650-1692).
Naturalist; Oxford Fellow in England; Anglican cleric in Virginia.
Ewan and Ewan (0440); Stearns (1490).

Barlow: Thomas Barlow (1607-1691).
Anglican bishop; Calvinist; friend of Boyle.
M. Hunter (0749).

Barrow: Isaac Barrow (1630-1677).
Anglican cleric; theologian; Royalist; Cambridge professor; mathematician; F.R.S.
J. T. Baker (0055, 0057, 1824); Burtt (0160); Capek (0180); Christianson (0207); Dobbs (0380); Feingold (0452, 0454); Gascoigne (0527); D. Greene (0569); A. R. Hall (0599); M. C. Jacob (0817); McAdoo (1008); McColley (1011); Miles (1067); Mitchell (1077); Osmond (1185); Reedy (1314-1315); Rivers (1330); Spurr (1479); R. S. Westfall (1639-1640, 1650); F. P. Wilson (1691).

Baxter: Richard Baxter (1615-1691).
Puritan (later dissenting) minister; theologian.
Cope (0266); H. Davies (0322); J. Henry (0660); R. K. Merton (1057); Mitchell (1077); Nuttall (1152); Rattansi (1297); Wojcik (1696).

Beale: John Beale (1608-1683).
Anglican cleric; follower of Comenius; friend of Boyle; F.R.S.
J. R. Jacob (0798, 0800, 0804); Miles (1067); Webster (1616).

Bentley: Richard Bentley (1662-1742).
Anglican cleric; theologian; Vice-Chancellor of Cambridge University; Newtonian; Boyle Lecturer; classical scholar; F.R.S.
Bentley (0160); I. B. Cohen (0233); Dahm (0311); Engdahl (0434); Fleischmann (0480); Garrod (0525); Gascoigne (0526); Guerlac (0583); J. C. Henry (0661, 0666); Horne (0731); Horstmann 0732; Hoskin (0733, 0737); Hutton (0786); M. C. Jacob (0809, 0817, 1900); Kerzberg (0874); Koyre (0903, 0907, 0909); McAdoo (1008); Metzger (1064); P. Miller (1071); More (1079); O'Higgins (1165); Olson (1171); Redwood (1313); Ronan (1348); Shapin (1429); C. I. Smith (1460); R. S. Westfall (1639-1640); Wegman (1988); R. White (1661).

Berkeley: George Berkeley (1685-1753).
Not included in this volume; it is planned that the early part of his career will be included in the next volume.

Biddle: John Biddle (1615-1662).
Socinian writer; schoolmaster.
Wilbur (1671).

Bidloo: Goverd Bidloo (1649-1713).
Dutch Mennonite medical practitioner; professor; administrator.
Cook (0259).

Biggs: Noah Biggs (fl. 1650s).
Puritan iatrochemist; medical reformer.
Debus (0353); Webster (1616).

Blackall: Offspring Blackall (1654-1716).
Anglican bishop; theologian; Boyle Lecturer.
Redwood (1313).

Blackmore: Richard Blackmore (c.1650-1729?).
Physician; philosophical poet.
Cook (0259); Wagenblass (1985).

Blith: Walter Blith (fl. 1640s).
Agriculturalist; member of Hartlib circle.
Haber (0592); Otten (1189).

Blount: Charles Blount (1654-1693).
Deist; political philosopher.
D. C. Allen (0016); Colie (0253); Emerson (0425, 1866); Empson (0430); Harth (0638); J. R. Jacob (0798); Orr (1178); Popkin (1227, 1243); Redwood (1312-1313); Reventlow (1326); Sullivan (1516).

Boate, A.: Arnold Boate (1606-1653).
Natural historian; physician.
Webster (1616).

Boate, G.: Gerard Boate (1604-1650).
Natural historian; physician.
Webster (1616).

Booker: John Booker (1602-1667).
Astrologer.
Capp (0182); Thomas (1541).

Boyle: Robert Boyle (1627-1691).
Natural philosopher; chemist; Anglican lay theologian; F.R.S.
Aarsleff (0002; Agassi (0008-0009); D. C. Allen (0017); F. Anderson (0023); F. Baumer (0068); Beck (1828); M. Boas (1833); Bodemer (0104); Bowen (1834); Bredvold (0116); J. H. Brooke (0132); R. M. Burns (0159); Burtt (0160); Carre (0188); Centore (0203); Clericuzio (0222); I. B. Cohen (0231); Colie (0252); Cope (0264, 0266); Coudert (0274, 0277); Crowley (1856); Cunningham (0305); Dahm (0311); E. Davis (0324, 0327, 1858); Dear (0330); Deason (0332); Debus (0338, 0340, 0342, 0362); Dijksterhuis (0376); Dillenberger (0377); Dobbs (0380, 0383); Drumin (1862); Duffy (0405); Easlea (0414); Emerton (0428); 'Espinasse (0436-0437); Fisch (0474-0476); M. S. Fisher (0478, 1872); Fulton (1758); Gibson (0538); R. Greene (0570); Guibbory (0585); A. R. Hall (0599, 0606); M. B. Hall (0612-0614); T. Hall (0619); C. Harrison (0633, 1887); Harth (0638); Hartley (0642); Harwood (0644); Heimann (0654); J. Henry

Brattle, W.: William Brattle (1662-1717).
New England minister; younger brother of Thomas Brattle.
Corrigan (0272).

Brigden: Zechariah Brigden (1639-1662).
New England minister; almanac compiler.
Morison (1084).

Briggs: Henry Briggs (1561-1631).
Mathematician; Gresham College professor; physician; Anglican lay theologian.
Feingold (0456); Hill (0686); Johnson (0832); Kearney (0863); Tyacke (1578); Webster (1616).

Brightman: Thomas Brightman (1562-1607).
Biblical scholar; nonseparating Puritan.
Hill (0696).

Broghill: Roger Boyle, Baron Broghill (1621-1679).
Brother of Robert Boyle; statesman; soldier; writer.
J. R. Jacob (0796).

Browne, E.: Edward Browne (1642-1708).
Physician; President of the College of Physicians; Son of T. Browne; F.R.S.
Finch (0471-0472).

Browne, P.: Peter Browne (1665-1735).
Philosopher; Irish Anglican bishop.
D. Berman (0084); K. Craven (0293); Winnett (1692); John Yolton (1714).

Browne, T.: Sir Thomas Browne (1605-1682).
Physician; writer; Anglican lay theologian.
D. C. Allen (0017); Ardolino (0029); Barnett (1825); Blau (0097); Bodemer (0104); Bredvold (0116); N. T. Burns (1842); Chalmers (0204); Clucas (0226); Cragg (0283); W. Dunn (0409); Finch (0471-0472); Fisch (0475-0476); E. George (0537); V. Harris (0631); C. Harrison (0632); Heninger (0658); Hollander (0703); A. Howell (0742); Huntley 0769-0770; G. Keynes (1775); Leroy

(0944); Loney (1916); McAdoo (1008); McCarthy (1924); E. S. Merton (1052); Nathanson (1104); Pratt (1945); Raven (1300); Singer (1965); Stock (1506); Webster (1617, 1626); R. S. Westfall (1650); T. M. Westfall (1991); Whyte (1667); Wiley (1673); Willey (1681); Williams (1684); Williamson (1686, 1688); F. P. Wilson (1691).

Buckmaster: Thomas Buckmaster (1532-1599).
Anglican cleric; physician; astrologer.
Capp (0182).

Bulkeley: Gersham Bulkeley (1636-1713).
New England Congregational minister; physician.
Watson (1607).

Burnet, G.: Gilbert Burnet (1643-1715).
Anglican bishop; latitudinarian; F.R.S.
S. C. Carpenter (0186); Griffin (0577); M. Hunter (0748-0749); Lasky (0932); Miles (1067); Mitchell (1077); Sykes (1525).

Burnet, T.: Thomas Burnet (1635?-1715).
Anglican cleric; theologian; natural historian.
D. C. Allen (0017); J. H. Brooke (0133); Collier(0254); Gordon Davies (0319); Dobbs (0383); Engdahl (0434); Farrell (0444); Force (0497); Froom (0514); Gascoigne (0528); Glacken (0548); Gould (0560); Haber (0591); Heimann (0654); M. Hunter (0763); M. C. Jacob 0817; M. C. Jacob and Lockwood (0818); Kubrin (1913); Macklem (0975); Mandelbrote (0980); Nicolson (1129); Ogden (1163); Porter (1251, 1253); Redwood (1313); Roger (1337-1338); P. Rossi (1355, 1359); Rudwick (1371); A. J. Smith (1459); Taylor (1533-1534); Toulmin and Goodfield (1556); Tuveson (1575-1577); Walker (1595); Willey (1680); D. Young (1724).

Burton: Robert Burton (1577-1640).
Anglican cleric; author of the "Anatomy of Melancholy."
Canavan (1846); Harth (0640); Miner (1075); Nicolson (1137); Tihinen (1980); Trevor-Roper (1567); Williamson (1686).

Butler: Samuel Butler (1612-1680).
Poet; satirist.
Miner (1075).

Byrd: William Byrd II (1674-1744).
Virginia planter; official; writer; F.R.S.
Ewan and Ewan (0440); Stearns (1490).

Camden: William Camden (1551-1623).
Antiquarian; historian.
Feingold (0456).

Carleton: George Carleton (1559-1628).
Anglican bishop; Calvinist; critic of astrology.
D. C. Allen (0018).

Caroline: Caroline of Anspach (1683-1737).
Princess of Wales/Queen of England; friend of Leibniz; friend of Clarke and Newton.
Aiton (0012); Alexander (0015).

Casaubon, I.: Isaac Casaubon (1559-1614).
Classical scholar; theologian; prebend of Canterbury, 1610-1614.
Yates (1706).

Casaubon, M.: Meric Casaubon (1599-1671).
Classical scholar; Anglican cleric; son of Isaac Casaubon.
Heyd (0680); M. Hunter (0763); Notestein (1147); Shumaker (1454); Southgate (1473); Spiller (1475); Stock (1506); Syfret (1523).

Cavendish: Margaret, Lady Cavendish, Duchess of Newcastle (1623-1673). Natural philosopher; friend and patron of philosophers; writer; Royalist.
Atherton (0041); Bazeley (1827); Blaydes (0099); Ferguson (0462); D. Grant (0561); Kargon (0857, 1903); Nicolson (1137); Perry (1221); Phillips (1223); Sarasohn (1392); Schiebinger (1410).

Chamber: John Chamber (1546-1604).
Anglican cleric; Greek scholar; critic of astrology.
D. C. Allen (0018).

Chamberlen: Peter Chamberlen or Chamberlain (1601-1683).
Physician; known for his changing political and religious views.
Hill (0687); Webster (1616).

Charles I: Charles I (1600-1649).
King of England, 1625-1649.
H. Davies (0322); Gough (0556); Hill (0686); Lasky (0932); Petersson (1222).

Charles II: Charles II (1630-1685).
King of England, 1660-1685; F.R.S.
Hartley (0642); Mendelsohn (1047); Petersson (1222); Strauss (1509).

Charleton: Walter Charleton (1619-1707).
Physician; natural philosopher; F.R.S.
Bodemer (0104); Debus (0338); Gelbart (0532); Guerlac (0580); Harth (0638); J. Henry (0665); K. Hutchinson (0776); Kargon (0857), (0859, 1903); Mayo (1000); McGuire (1023); McGuire and Tamny (1035); Mendelsohn (1047); Mulligan (1097); Oakley (1155); Olson (1170); Osler (1181); Pacchi (1190); Rattansi (1294); Rodney (1334); Sharp (1443); Webster (1613, 1616); R. S. Westfall (1650).

Chauncy: Charles Chauncy (1592-1672).
Puritan minister; President of Harvard.
Morison (1083).

Cherbury: Lord Herbert of Cherbury.
See Herbert, E.

Cheyne: George Cheyne (1671-1743).
Physician; Mathematician; Newtonian; mystic.
Bowles (0109); Coulter (0279); Fleischmann (0480); Guerrini 0584; McGuire (1020); Metzger (1064); Rousseau (1364); Thackray (1537).

(1499, 1972); Stromberg (1510); Sullivan (1516); Thackray (1538); Wegman (1988); R. S. Westfall (1640); Wilbur (1671); John Yolton (1722).

Clayton: John Clayton (1657-1725).
Anglican cleric in Virginia; naturalist; physician; F.R.S.
Berkeley and Berkeley (0081); Ewan and Ewan (0440); Stearns (1490).

Coleman: Benjamin Coleman (1673-1747).
New England minister; Harvard fellow.
Corrigan (0272).

Coles: William Coles (1626-1662).
Herbalist; naturalist.
Otten (1189).

Coley: Henry Coley (1633-1707).
Mathematician; astrologer.
Capp (0182).

Colliber: Samuel Colliber (fl. 1718-1737).
English philosophical theologian.
J. T. Baker (0058).

Collins, A.: Anthony Collins (1676-1729).
Deist philosopher; reputed atheist.
Attfield (0044); D. Berman (0082-0083, 0085); Colie (0253); Cranston (0289); Emerson (1866); *Encyclopedia of Unbelief* (0433); Fabrio (0441); J. Ferguson (0461); Force (0488); Hefelbower (0651); Horstmann (0732); M. C. Jacob (0817); Motzo Dentice de Accadia (1090); O'Higgins (1165-1166); Reventlow (1326); Rodney (1335); Stephen (1495); Stromberg (1510); Sullivan (1516); John Yolton (1714, 1722).

Collins, J.: John Collins (1625-1683).
Mathematician; F.R.S.
Osmond (1185).

Comenius: Johannes Amos Comenius (1592-1671).
Bohemian educational and utopian reformer; Moravian minister; resident in England, 1641-1642.
Batten (0064); Bernal (0087); Capkova (0181); Duncan (0407); Greaves (0565); Hill (0686); Langton (1915); Lasky (0932); Manuel (0987); Phillips (1223); Rattansi (1295); Southgate (1473-1474); Spinka (1476); Stimson (1502); Syfret (1522); Trevor-Roper (1563, 1568); Turnbull (1572-1574; Webster (1616, 1620); Yates (1709); R. Fitz. Young (1726).

Compton: Henry Compton (1632-1713).
Anglican Bishop.
Lang (0928).

Conduitt: John Conduitt (1688-1737).
Newton biographer; Newton's nephew by marriage; Master of the Mint.
R. S. Westfall (1639-1640).

Conway: Anne, Viscountess Conway (1631-1679).
Philosopher; friend and correspondent of various philosophers; Quaker.
S. Brown (0139); Coudert (0275, 0278); Duffy (0405); Gabbey (0519); A. R. Hall (0599); Hellegers (1889); Hirst (0701); Hutin (0777); Hutton (0783); Levine (0946); Loptson (0963); Merchant (1048-1049); Nicolson (1119); Nuttall (1153); Popkin (1247); (1249); Underwood (1581); Walker (1595).

Coste: Pierre Coste (1668-1747).
Huguenot; translator of works of Locke and Newton.
Axtell (0047).

Cotes: Roger Cotes (1682-1716).
Mathematician; astronomer; Newtonian.
A. R. Hall (0603); Koyre (0897, 0900, 0909); More (1079); R. S. Westfall (1640).

Cotta: John Cotta (1575?-1650).
Physician; author.
S. Clark (0219).

Cotton: John Cotton (1584-1652).
Dean of Emmanuel College, Cambridge; New England Puritan minister.
Hornberger (0728); P. Miller (1072-1073).

Cowley: Abraham Cowley (1618-1667).
Poet; playwright; satirist; Royalist government official; philosopher; physician; Fellow of and apologist for the Royal Society.
Eurich (0438); Hinman (0700); Hollander (0703); Kassler (0860); Rostvig (1363).

Coxe: Thomas Coxe (1615-1685).
Physician; F.R.S.
Cunningham (0305).

Cromwell: Oliver Cromwell (1599-1658).
Lord Protector (1653-1658).
Cope (0267); Lasky (0932); Nuttall (1152); Petersson (1222).

Crosse: Robert Crosse (1605-1683).
Anglican cleric; Puritan; critic of Joseph Glanvill and the Royal Society.
Syfret (1523).

Cudworth: Ralph Cudworth (1617-1688).
Philosopher; Cambridge Platonist; Anglican clergyman; F.R.S.
Aaron (0001); Albee (0013); P. R. Anderson (0025, 1821); Armstrong (0032); Aspelin (0040); Atherton (0041); Bercovitch (0080); Bodemer (0104); Carlini (0184); Carre (0188-0189); Cassirer (0201); Colie (0251); Cragg (0284-0285); de Pauley (0368); Fisch (0476); Fleischmann (0480); Gabbey (0520); Glacken (0548); Gregory (0575); Guerlac (0580); Guffey (1762); Gysi (0589-0590); Habicht (0593, 1883); A. R. Hall (0599); Harth (0640); J. Henry (0663); Hoopes (0714); Huebsch (1892); W. Hunter (0768); Hutchison (0775); A. Jacob (0792); M. C. Jacob (0817); Korshin (0894); Lamprecht (0926); Leyden (0948); Lowery (1917); McAdoo (1008); McGuire (1025, 1034); Merchant (1048); Meyer (1927); Mijuskovic (1066); Mintz (1076); Muirhead (1092); Nicolson (1118); Olson (1170); Popkin (1227, 1243); Quinn (1272); Raven (1302); Redwood (1313); Rivers

(1330); Rodney (1334, 1336); Sailer (1388, 1390, 1954); Saveson (1398); Schmitt (1413); Sortals (1469); Sutton (1517); Thiel (1540); Wade (1592); Walker (1595, 1598); Whitebrook (1662); Willey (1681); Yates (1706); John Yolton (1722).

Culpepper: Nicholas Culpepper (1616-1654).
Puritan reformer in England; Congregational minister in New England; physician; astrologer.
Capp (0182); Hill (0687); Nagy (1101); Poynter (1258); Thulesius (1546); Watson (1607).

Culverwel: Nathanael Culverwel (1618?-1651).
Philosopher; Cambridge Platonist; Anglican clergyman.
de Pauley (0368); Guffey (1762); Harth (0638); Scupholme (1421-1422).

Cutler: Timothy Cutler (1684-1765).
Rector of Yale; convert to Anglicanism.
Warch (1604).

Dalgarno: George Dalgarno (1626?-1687).
Scottish educator.
Shumaker (1454).

Defoe: Daniel Defoe (1660-1731).
Essayist; journalist; novelist; government agent.
R. Harris (0630); Stock (1506).

Dell: William Dell (1606?-1664).
Radical university reformer; Chaplain in the New Model Army.
Greaves (0565); Hill (0690); Solt (1466); Webster (1616, 1627).

Dennis: John Dennis (1657-1734).
Literary critic; playwright.
Youngren (1727).

Derham: William Derham (1657-1735).
Theologian; Newtonian; Boyle Lecturer; F.R.S.
Atkinson (0042); Attfield (0043); Coleman (0250); Collier (0254); Dahm (0311); Engdahl (0434); Genuth (0533); Glacken (0548);

Heyd (0679); M. C. Jacob (0817); Murdin (1100); Raven (1301); Solberg (1463); Stebbins (1491); Willey (1680).

Desaguliers: J. T. Desaguliers (1683-1744).
Newtonian natural philosopher.
Hans (0625); M. C. Jacob (0812); Quinn (1272); Wagner (1593).

Dickinson: Edmund Dickinson (1624-1707).
Physician; alchemist.
Collier (0254).

Digby: Kenelm Digby (1603-1655).
Natural Philosopher; alchemist; Catholic; F.R.S.
A. Baumer (0065); Bodemer (0104); Dobbs (0380, 0393-0395); Drake (0400); Feingold (0456); Finch (0472); C. Harrison (0632); Hartley (0642); Holmyard (0706); Huntley (0769); L. King (0878; Leroy (0944); Nicolson (1119); Petersson 1222; Rubin (1793); Southgate (1471); R. S. Westfall (1650).

Ditton: Humphrey Ditton (1675-1715).
Mathematician; dissenting minister; theologian.
John Yolton (1722).

Donne: John Donne (1573-1631).
Poet; dean of St. Paul's; raised a Roman Catholic.
D. C. Allen (0017); Bethel (0090); Bredvold (0117); Bush (0165); Clucas (0226); Coffin (0228); Cunnar (0304); Drake (0400); Guibbory (0585); Guthke (0588); Haydn (0646); Hayes (0647); Hellegers (1889); Hirst (0701); Hollander (0703); Kocher (0892); Macklem (0975); Mazzeo (1001, 1003, 1007); Miner (1075); Mitchell (1077); Nicolson (1117, 1124, 1130, 1137); Otten (1189); Pagel (1191); Patrides (1210); M. Ramsay (1282-1283); P. Rossi (1359); Rukeyser (1373); Sadler (1387); Stock (1506); Toulmin (1551); Ward (1605); Williamson (1686); F. P. Wilson (1691).

Doolittle: Thomas Doolittle (1632-1707).
New England minister.
Rumsey (1373); Van de Wetering (1582).

Douglas: William Douglas (1691?-1752).
New England physician; writer.
Winslow (1695).

Dryden: John Dryden (1631-1700).
Poet; dramatist; satirist; convert to Roman Catholicism; F.R.S.
Armistead (0030); Bredvold (0115-0116); Brett (0120); Budick (0151, 1841); J. T. Carroll (1848); Cope (0267); Empson (0430-0431); 'Espinasse (0436); Fleischmann (0480); Griffith (1881); Guibbory (0585); Harth (0638-0639); Hollander (0703); B. King (1907); Korshin (0894); Lasky (0932); Martz (1921); Miner (1074-1075); Romagosa (1951); Rostvig (1363); Sellin (1423); Stock (1506); Williamson (1687); F. P. Wilson (1691); Youngren (1727).

Dunster: Henry Dunster (1609-1658/59).
President of Harvard; Puritan minister.
Hornberger (0730); Morison (1083, 1085).

Duport: James Duport (1606-1679).
Master of Magdalene College, Cambridge; Professor of Greek; Royalist.
Costello (0273); Feingold (0454).

Dury: John Dury (1596-1680).
Puritan intellectual; utopian reformer; member of the Hartlib circle.
Batten (0064); Clucas (0225; Hill (0686); J. R. Jacob (0804); Lasky (0932); Popkin (1249); Spinka (1476); Syfret (1522); Trevor-Roper (1568); Turnbull (1572); Webster (1616, 1620).

Evelyn: John Evelyn (1620-1706).
Diarist; writer on architecture, gardening, and numismatics; son-in-law of Sir Thomas Browne; F.R.S.
Cope (0264); H. A. L. Fisher (0477); Fulton (0515); Glacken (0548); Hartley (0642); Houghton (0739-0740); M. Hunter (0756, 0763); M. C. Jacob (0810, 0815, 0817); Katz (0861); Mayo (1000); Ollard (1169); Otten (1189); Spurr (1479); Stimson (1505); Strauss (1509).

Everard: John Everard (1575-1650?).
Alchemist; mystic; member of the Family of Love.
Schuler (1416).

Falkland: Lucius Cary, Lord Falkland (1610?-1643).
Royalist; patron of the "Falkland Circle" at Great Tew.
Hayward (0648).

Fatio: Nicolas Fatio de Duillier (1664-1753).
Swiss mathematician; Newtonian; Protestant enthusiast; F.R.S.
Christianson (0207); Dobbs (0383); Heyd (0679); M. C. Jacob (0816); R. S. Westfall (1639-1640).

Fell: Margaret Fell (1614-1702).
Quaker activist.
Whittaker (1665).

Felton: Edmond Felton (fl. 1640's).
Inventor; Puritan.
Raylor (1307).

Ferguson: Robert Ferguson (died 1714).
Jacobite activist; religious controversialist.
Wojcik (1696).

Finch, A.: Anne Finch, Countess of Winchelsea (1666-1720).
Poet.
Rostvig (1363); Hellegers (1889).

Finch, J.: John Finch (1626-1682).
Friend of Henry More; brother of Lady Anne Conway; M.D.; F.R.S.
Nicolson (1119); T. L. Underwood (1581).

Firmin: Giles Firmin, Jr. (1615-1697).
Physician; Puritan minister.
Marks and Beatty (0992).

Flamsteed: John Flamsteed (1646-1719).
Anglican cleric; natural philosopher; astronomer; F.R.S.

Forbes (0484); Hoppen (0725); M. Hunter (0761); Miles (1067); More (1079); Murdin (1100); Ronan (1348).

Fludd: Robert Fludd (1574-1637).
Physician; natural philosopher; Rosicrucian.
Cafiero (0174); Collier (0254); Copenhaver (0269); J. B. Craven (0292); Debus (0335-0338, 0340-0342, 0344, 0346-0348, 0350-0352, 0354-0355; 0361); Eamon (0412); Emerton (0429); Favre (0450); French (0509); Godwin (0550); Heninger (0658); Hirst (0701); Huffman (0745-0746, 1893); Hutin (0778); Josten (0850); Pacchi (1190); Pagel (1191, 1193); Pauli (1212); Peuckert (1218); Saurat (1395); Vickers (1586); Westman (1653-1654); Yates (1705-1706, 1709, 1711).

Foley: Samuel Foley (1655-1695).
Anglican bishop; D.P.S.
Hoppen (0725).

Folkes: Martin Folkes (1690-1754).
President of the Royal Society; antiquary.
Force (0495).

Foster, J.: John Foster (1648-1681).
New England printer; astronomer.
M. B. Hall (0618).

Foster, W.: William Foster (died 1643).
Anglican cleric.
J. B. Craven (0292); Huffman (0746).

Fox: George Fox (1624-1691).
Founder of the Society of Friends.
Clarke (0220); H. Davies (0322); Hill (0687); Nuttall (1152-1153); Raistrick (1281); Whittaker (1665).

Gadbury: John Gadbury (1627-1704).
Physician; astrologer.
Capp (0182); P. Curry (0306); Thomas (1541).

Gale: Thomas Gale (1635?-1702).
Anglican cleric; F.R.S.
Miles (1067).

Gannse: Joachim Gannse (fl. 1600).
Jewish mining technologist; possible model for Bacon.
Feuer (0464).

Garth: Sir Samuel Garth (1661-1719).
Physician; member of the Kit-Cat Club; minor poet.
Fleischmann (0480).

Gastrell: Francis Gastrell (1662-1725).
Anglican bishop; theologian; Boyle lecturer.
Redwood (1313); John Yolton (1714).

Gauden: John Gauden (1605-1662).
Anglican bishop.
R. F. Jones (0840).

Gay: John Gay (1685-1732).
Poet; Playwright; Essayist.
Nicolson (1141); Rousseau (1367).

Gellibrand: Henry Gellibrand (1597-1636).
Natural philosopher; mathematician; Puritan.
Feingold (0456); F. Johnson (0832); Kearney (0863); Pumfrey (1269).

Gerard: John Gerard (1545-1612).
Herbalist.
Otten (1189).

Gilbert: William Gilbert (1540-1603).
Natural philosopher; physician.
Abromitis (1816); Debus (0347); Feingold (0456); Hooykaas (0720).

Glanvill: Joseph Glanvill (1636-1680)
Anglican cleric; natural philosopher; Fellow and publicist of the Royal Society.

Armistead (0030); Aspelin (0039); Bredvold (0116); R. M. Burns (0159); Carrithers (1847); S. Clark (0219); Cope (0263, 0266); Coudert (0277); Edelin (0416); Eurich (0438); D. Greene (0569); Griffin (0577); Habicht (0593); A. R. Hall (0599); Habicht (1883); Harth (0640); W. Howell (0743); M. Hunter (0762-0763); J. R. Jacob (0798); Jobe (0828); R. F. Jones (0835-0836, 0838, 0842); E. F. Keller (0871); L. King (0878); Lamprecht (0926); Laudan (0934); Lecky (0939); McAdoo (1008); Miles (1067); Mintz (1076); Mitchell (1077); Notestein (1147); Olson (1172); Ormsby-Lennon (1176); Popkin (1230, 1233, 1236-1237); Prior (1268, 1946); Shapiro (1440); Southgate (1470-1472); Spurr (1478); Stimson (1501); Stock (1506); Syfret (1523); Talmor (1528-1530); Van Leeuwen (1583); Walker (1595); Waller (1987); R. S. Westfall (1650); Wiley (1673); Willey (1681); Williamson (1688); Wojcik (1696).

Glisson: Francis Glisson (1597-1677).
Physician; Cambridge professor; anatomist; F.R.S.
Clericuzio (0222); Frank (0505); French (0509); J. Henry (0665).

Goad: John Goad (1616-1689).
Roman Catholic; astrologer; schoolmaster.
P. Curry (0306).

Godwin: Francis Godwin (1562-1633).
Anglican bishop; anonymous author of *A Man in the Moone*.
Guthke (0588); Knight (0889); Nicolson (1121, 1137-1138).

Goodman: Godfrey Goodman (1583-1656).
Anglican bishop; argued for the "decay" of nature.
V. Harris (0631); R. F. Jones (0835); Mazzeo (1005).

Graham: George Graham (1673-1751).
Clock and instrument maker; Quaker-turned-Anglican; F.R.S.
Raistrick (1281).

Graunt: John Graunt (1620-1674).
Statistician; F.R.S.
Strauss (1509).

Greatrakes: Valentine Greatrakes (1629-1683).
Irish healer; Puritan soldier and official.
Duffy (0405); J. R. Jacob (0798, 0804); Kaplan (0856); Nicolson (1119); Steneck (1494).

Greaves: John Greaves (1602-1652).
Astronomer; Egyptologist.
G. A. Russell (1378); Tyacke (1578).

Greene: Robert Greene (1678?-1730).
Natural philosopher; Anglican theologian; Fellow of Clare College, Cambridge.
Heimann and McGuire (0655); Thackray (1537).

Greenwood: Isaac Greenwood (1702-1745).
Harvard professor; natural philosopher; mathematician.
Hornberger (0730).

Gregory: David Gregory (1661-1708).
Scottish mathematician; astronomer; Newtonian natural philosopher.
Casini (0196); Thackray (1537).

Greville: Fulke Greville, First Baron Brooke (1554-1628).
Politician; writer; patron of Bacon; layman with Puritan tendencies.
Buckley (0146); Bullough (0153).

Grew: Nehemiah Grew (1641-1712).
Physician; natural philosopher; botanist; F.R.S.
Attfield (0043); Bolam (0105); Collier (0254); Stebbins (1491); R. S. Westfall (1650).

Groenevelt: Joannes Groenevelt (1640-1710?).
Dutch immigrant to England; physician; surgeon; Calvinist.
Cook (0260-0261).

Haak: Theodore Haak (1605-1690).
German-born Calvinist; scholar/translator; lived in England from 1625; Anglican deacon; member of the Hartlib circle; F.R.S.
Kemsley (0873); Stimson (1502); Syfret (1522); Turnbull (1574).

Habington: William Habington (1605-1654).
Roman Catholic poet.
Rostvig (1363).

Haistwell: Edward Haistwell (c.1658-1709).
Quaker; F.R.S.
Underwood (1581).

Hakewill: George Hakewill (1578-1649).
Anglican cleric; writer; opponent of the decay-of-the-world theory.
Baron (0062); Glacken (0548); V. Harris (0631); Hill (0686); R. F. Jones (0835); Mazzeo (1005); Williamson (1686).

Hale: Sir Matthew Hale (1609-1676).
Lord Chief Justice.
D. C. Allen (0017); Attfield (0043, 0045); J. Henry (0665); Shapiro (1440).

Hales: Stephen Hales (1677-1761).
Anglican minister; Newtonian natural philosopher.
Thackray (1537).

Hall (b.1575): John Hall (1575-1635).
Physician; Shakespeare's son-in-law.
Nagy (1101).

Hall (b.1627): John Hall (1627-1656).
Puritan educational writer and reformer; poet; member of Hartlib circle.
Freeman (0508); Greaves (0565).

Halley: Edmund Halley (1656-1742).
Newtonian; Natural philosopher/astronomer; Clerk of the Royal Society; F.R.S.
Albury (0014); Dobbs (0383); Genuth (0534); Hoskin (0738); Kubrin (0921); Murdin (1100); Ronan (1348); Schaffer (1403); Thrower (1545).

Harley: Robert Harley, Earl of Oxford (1661-1724).
Politician; book collector.
Sullivan (1516).

Harrington: James Harrington (1611-1677).
Political philosopher.
K. Craven (0293).

Harriot: Thomas Harriot or Hariot (1560-1621).
Natural philosopher; mathematician.
Feingold (0456); Feuer (0464); J. Henry (0668); M. Hunter (0757); Jacquot (0819-0820); Kargon (0857-0858, 1903); North (1144); Roche (1949); Rukeyser (1373); Shirley (1448-1450); Webb (1609).

Harris: John Harris (1667?-1719).
Lexicographer; Newtonian natural philosopher; Boyle Lecturer; F.R.S.
Bowles (0108); Hutton (0786); M. C. Jacob (0817); Thackray (1537).

Hart: James Hart (died 1633).
Puritan physician.
Elmer (0423, 1865).

Hartlib: Samuel Hartlib (c.1600-1662).
Social and educational reformer; writer.
Batten (0064); Hartlib (0181); Clucas (0225, 0227); J. Davis (0328); Dobbs (0380); Eamon (0411); Eurich (0438); Greaves (0565); Greengrass (0573); Haber (0592); A. R. Hall (0606); M. B. Hall (0613); Hill (0686, 0696); Houghton (0740); J. R. Jacob (0804); R. F. Jones (0835); Kittredge (0883); Lasky (0932); Maddison (0976); Manuel (0987); R. K. Merton (1057); Newman (1111, 1114); O'Brien (1158); Parker (1205); Pumfrey (1269); Raylor (1307); Spinka (1476); Stimson (1502); Strauss (1509); Syfret (1522); Trevor-Roper (1563, 1568); Turnbull (1571-1574); Webster (1611-1612, 1614-1616, 1620, 1625); Wilkinson (1676-1678); Yates (1709).

Harvey: William Harvey (1578-1657).
Physician; anatomist; physician to James I and Charles I.
A. Baumer (0065); Bodemer (0104); I. B. Cohen (0240); Debus (0347, 0355); Feingold (0456); French (0509); T. Hall (0619); Hill (0697-0698); Huffman (0746); W. Hunter (0768); Kassler

(0860); Pagel (1191 1192, 1196-1200); Webster (1616); Whitteridge (1666).

Hawksmoor: Nicholas Hawksmoor (1661-1736).
Architect.
Downes (0398).

Haydocke: Richard Haydocke (c.1578-c.1642).
Physician; translator; engraver; Puritan.
Holtgen (0707).

Helmont: Francois Mercury van Helmont (1614?-1699).
Flemish physician; philosopher; student of Kabbala; son of Jean Baptiste van Helmont; moved to England in 1670; Anne Conway's physician.
S. Brown (0139); Clericuzio (0222); Coudert (0275-0276, 0278); Debus (0335, 0338); Gottesman (1879); A. R. Hall (0599); Hutin (0777, 0779); Loptson (0963); Merchant (1048-1050); Nicolson (1119, 1133); Nuttall (1153); Sherrer (1963); Weir (1989).

Herbert, E.: Edward, Lord Herbert of Cherbury (1583-1648).
Writer; poet; philosopher; Anglican proto-deist.
Bedford (0076); Crous (0303); *Encyclopedia of Unbelief* (0433); Harth (0638); Hutcheson (0772); Motzo Dentice de Accadia (1090); Orr (1178); Popkin (1233, 1243); Preus (1260); Reventlow (1326); M. Rossi (1354); Sullivan (1516); Walker (1594); Webb (1608); Willey (1681).

Herbert, G.: George Herbert (1593-1633).
Anglican cleric; poet; brother of Edward, Lord Herbert of Cherbury.
Caldwell (1845); Otten (1189); Todd (1548); Van Pelt (1584); Ward (1605).

Heydon: Sir Christopher Heydon (died 1623).
Astrologer; member of Parliament.
D. C. Allen (0018); Capp (0182).

Hickes: George Hickes (1642-1715).
Nonjuring cleric; scholar.
L. Stewart (1499, 1972).

Highmore: Nathaniel Highmore (1613-1685).
Physician; Oxford medical writer.
A. Baumer (0065); Bodemer (0104).

Hill: Nicholas Hill (c.1570-c.1620).
Roman Catholic recusant; natural philosopher.
Jacquot (0819); McColley (1014); Trevor-Roper (1562).

Hobbes: Thomas Hobbes (1588-1679).
Philosopher.
H. Baker (0054); Barnouw (0061); D. Berman (0085); Brandt (0112); Bredvold (0115-0116); Brett (0120); K. C. Brown (0138); N. T. Burns (1842); Burtt (0160); Carre (0188-0189); Colie (0251); Cope (0266); Crous (0303); *Encyclopedia of Unbelief* (0433); Feingold (0453); Finocchiaro (0473); Fisch (0475); Fleischmann (0480); Flew (0482); Freudenthal (0511-0512); Funkenstein (0518); Garcia (1758); Gibson (0538); Glover (0549); Hacking (0594); A. R. Hall (0599); C. Harrison (0632-0633, 1887); Hayward (0648); Hazard (0649); N. Henry (0669); Hill (0686); Hinman (0700); Hoopes (0714); Horstmann (0732); M. Hunter (0755, 0763); A. Jacob (0791); J. R. Jacob (0798); M. C. Jacob (0812, 0817); Kargon (0857, 1903); Kassler (0860); Kearney (0865); Krook (0918); Laird (0925); Lange (0929); Lasky (0932); Lasswitz (0933); Loptson (0963); Mazzeo (1005); McGuire and Tamny (1035); Merchant (1048); Mintz (1076, 1931); Motzo Dentice de Accadia (1090); Mulligan (1097); Nicolson (1126); Oakeshott (1154); Olson (1172); Paul (1211); Petersson (1222); Popkin (1233-1235); Redwood (1313); Reventlow (1326); G. A. J. Rogers (1341); M. Rossi (1354); P. Rossi (1358); Sacksteder (1794); Sarasohn (1391); Schaffer (1407); Shapin and Schaffer (1433); Skinner (1457); Sortais (1469); Southgate (1471-1472); Strauss (1509); Sutton (1517); Tagart (1527); Taube (1532); Vickers (1586); J. F. West (1628); Willey (1681); Williamson (1685, 1687); Woodfield (1698); Youngren (1727).

Hobbs: William Hobbs (fl. 1700-1710).
Folk natural philosopher.
Porter (1252, 1255).

Hoby: Lady Margaret Hoby (1571-1633).
Diarist; lay medical practitioner.
Nagy (1101).

Holdsworth: Richard Holdsworth (1590-1649).
Master of Emmanuel College; theologian.
Costello (0273).

Hooke: Robert Hooke (1635-1703).
Natural philosopher; F.R.S.
Beier (0077, 1829); Birkett and Oldroyd (0095); Centore (0203); Gordon Davies (0319-0321); E. Davis (0327); 'Espinasse (0437); H. A. L. Fisher (0477); Gouk (1880); Haber (0591); A. R. Hall (0599); M. B. Hall (0613); T. Hall (0619); Hartley (0642); Harwood (0643); J. Henry (0667); Hoppen (0725); M. Hunter (0751-0755); M. Hunter and Schaffer (0765); Kearney (0865); L. King (0878); Kubrin (0921); Maddison (0976); Murdin (1100); Oldroyd (1167); Porter (1253); Ronan (1348); Rudwick (1371); Schaffer (1402); Shapin (1427-1432); Taylor (1533-1534); Toulmin and Goodfield (1556); R. S. Westfall (1648-1650); C. Wilson (1690); D. Young (1724).

Hopkins: Matthew Hopkins (fl. 1640s).
Witch finder.
Thomas (1541).

Horrox: Jeremiah Horrox (1619-1641).
Astronomer; Anglican cleric.
Orchard (1175).

Houghton: John Houghton (1645-1705).
London merchant; Tory pamphleteer; F.R.S.
J. R. Jacob (0801).

How: William How (1620-1656).
Naturalist.
Raven (1300).

Howe: John Howe (1630-1705).
Anglican cleric; writer.
Wojcik (1696).

Howell: James Howell (1594?-1666).
Royalist official; essayist.
Lasky (0932).

Hubner: Joachim Hubner (fl. 1640s).
Protestant refugee in England; member of the Hartlib circle.
Capkova (0181).

Hutchinson: John Hutchinson (1674-1737).
Anglican theologian; opponent of Newtonianism.
Cantor (0179).

Hutton: John Hutton (died 1712).
Physician; F.R.S.
Cook (0259).

Hyrne: Henry Hyrne (fl. 1670s).
Schoolmaster; Latin scholar; author of a treatise on the tides; correspondent of More.
Gabbey (0522).

James I: James I of England or James VI of Scotland (1566-1625).
King of Scotland (1567-1625); King of England (1603-1625); poet and writer.
Hill (0686); Huffman (0746); Notestein (1147); Rattansi (1289); Yates (1709).

Johnson: Thomas Johnson (1604-1644).
Apothecary; naturalist.
Raven (1300).

Jolly: Thomas Jolly or Jollie (1629-1703).
Dissenting minister.
Harley (0627).

Jones: Inigo Jones (1573-1652).
Architect.
Lang (0928); Yates (1711).

Jonson: Ben Jonson (1572-1637).
Playwright.
Johanssen (0830); Sadler (1387).

Josselin, R.: Ralph Josselin (1616-1683).
Anglican cleric.
Beier (0077, 1829); Nagy (1101).

Josselyn, J.: John Josselyn (1608-1675).
Puritan traveler and writer of natural history; English friend of John Winthrop and John Cotton.
Beier (0077, 1829); Stearns (1490).

Keill: John Keill (1671-1721).
Mathematician; Newtonian natural philosopher.
Force (0497); M. Hunter (0763); Kubrin (1913); Taylor (1533-1534); Thackray (1537).

Keogh: John Keogh (1650?-1725).
Anglican cleric; mathematician.
Hoppen (0725).

King: William King (1650-1729).
Irish/Anglican bishop; philosopher; D.P.S.
J. T. Baker (0058); D. Berman (0084); Garrod (0525); Hoppen (0725); Horne (0731).

Laney: Benjamin Laney (1591-1675).
Anglican bishop; F.R.S.
Miles (1067).

Laud: William Laud (1573-1645).
Archbishop of Canterbury; leading opponent of Calvinism.
H. Davies (0322); Tyacke (1578).

Lawson: Thomas Lawson (1630-1691).
Quaker naturalist.
Raistrick (1281); Whittaker (1665).

Layton: Henry Layton (1632-1705).
Writer of anonymous theological tracts.
John Yolton (1722).

Lee: Samuel Lee (1625-1691).
New England Congregational minister; antiquarian.
Hornberger (0729); Watson (1607).

LeFevre: Nicaise LeFevre (died 1669).
French neoplatonic alchemist; resident in England in the 1660s.
Debus (0354); Mendelsohn (1047).

Leverett: John Leverett (1662-1724).
President of Harvard; F.R.S.
Fiering (0467).

Lilly: William Lilly (1602-1681).
Astrologer; almanac writer.
Briggs (0125); Capp (0182); P. Curry (0306); Josten (0849); McCaffery (1009);N. Nelson (1108); D. Parker (1204); Stahlman (1482); Thomas (1541).

Lister: Martin Lister (1638?-1712).
Physician; natural philosopher; F.R.S.
Raven (1301); Rudwick (1371).

Lloyd, D.: David Lloyd (1635-1692).
Anglican cleric; biographer.
Duffy (0405); Kaplan (0856).

Lloyd, W.: William Lloyd (1627-1717).
Anglican bishop; Latitudinarian.
M. C. Jacob (0810).

Locke: John Locke (1632-1704).
Philosopher; lay theologian; physician; F.R.S.
Aaron (0001); Aarsleff (0002-0003); Acton (0005); Aiton (0012); F. Anderson (0023, 1820); Armstrong (0032); Ashcraft (0034-0035); Aspelin (0037-0039); Attfield (0045); Attig (1740); Axtell

Yolton (1713-1722, 1998); Yolton and Yolton (1815); Yost (1723, 1999).

Lockyer: Lionel Lockyer (died 1672).
Medical practitioner.
Newman (1111).

Logan: James Logan (1674-1751).
Pennsylvania merchant and government official; Quaker; natural philosopher; mathematician.
Raistrick (1281); Tolles (1549); Wolf (1697).

Lydiat: Thomas Lydiat (1572-1646).
Anglican cleric; astronomer.
Feingold (0456).

Mace: Thomas Mace (1619?-1709?).
Anglican cleric; music theorist.
Hollander (0703).

Maclaurin: Colin Maclaurin (1698-1746).
Scottish mathematician; Newtonian.
Odom (1162).

Maier: Michael Maier (1568-1622).
German physician; alchemist; mystic; Rosicrucian.
J. B. Craven (0291); Favre (0450); Huffman (0746).

Makin: Bathsua Makin (fl. 1612-1672).
School teacher; educational reformer; sister-in-law of John Pell.
Phillips (1223).

Mandeville: Bernard de Mandeville (1670-1733).
Dutch physician; satirist.
Davie (0315).

Mapletoft: John Mapletoft (1631-1721).
Physician; natural philosopher; Anglican cleric; F.R.S.
R. S. Westfall (1650).

Marsh: Narcissus Marsh (1638-1713).
Anglican bishop, mathematician; Lord Justice of Ireland; associated with the founding of the Royal Dublin Society.
K. Craven (0293); Hoppen (0725); G. A. Russell (1378).

Marvell: Andrew Marvell (1621-1678).
Poet; political writer; politician.
Rostvig (1363).

Mary II: Mary II (1662-1694).
Queen of England, Scotland, and Ireland, 1689-1694.
Cook (0259).

Masham: Lady Abigail Masham (died 1734).
Courtier; philosopher.
Atherton (0041).

Mather, C.: Cotton Mather (1663-1728).
New England Congregational minister; natural philosopher.
Beall (0069; Beall and Shryock (0070); Breitwieser (0118); Genuth (0533); Hornberger (0727-0729); Jeske (0827); G. Jones (0834); Kittredge (0882-0884); Marks and Beatty (0992); Middlekanff (1065); P. Miller (1073); Morison (1083-1085); Rumsey (1374); Shryock (1451-1452); Silverman (1455); Solberg (1463-1464); Stearns (1488, 1490); Warch (1604); Watson (1607); Winship (1694); Winslow (1695).

Mather, I.: Increase Mather (1639-1723).
New England Puritan minister and theologian; President of Harvard.
M. B. Hall (0617-0618); Middlekanff (1065); P. Miller (1073); Morison (1083-1085); Oakley (1155); Rumsey (1374); Stearns (1490); Van de Wetering (1582).

Mayow: John Mayow (1640-1679).
Physician; medical philosopher; physiologist; F.R.S.
T. Hall (0619); J. Henry (0665).

Mede: Joseph Mede (1586-1638).
Biblical scholar; interpreter of the prophecies of the Apocalypse; student of astrology.

Feingold (0456); Hill (0696); Hutton (0784); Iliffe (0788); Kochavi (0890); Popkin (1247, 1249); Tuveson (1575).

Melton: John Melton (died 1640).
Politician; writer; critic of astrology.
D. C. Allen (0018).

Merret: Christopher Merret or Merrett (1614-1695).
Physician; naturalist.
Raven (1300).

Middleton: Thomas Middleton (1570?-1627).
Playwright.
Johanssen (0830).

Milton: John Milton (1608-1674).
Poet; essayist; Puritan; lay theologian.
Acton (0005); D. C. Allen (0017); H. Baker (0054); Batten (0064); Brett (0120); Burns (1842); Bush (0165); K. Craven (0293); W. C. Curry (0308-0309); Drake (0400); Duncan (0407); Fisch (0475); Gilbert (0540-0541); Giorello (0545); Guibbory (0585); Guthke (0588); C. Harrison (0632); N. Henry (0669); Hill (0686); Hirst (0701); Hoopes (0714); Huckaby (1768); W. Hunter (0766-0767); Korshin (0894); Langton (1915); Lasky (0932); Lovejoy (0964-0966); Lowery (1917); McColley (1010, 1013, 1016); McLachlan (1038); Nicolson (1121, 1126-1128, 1135); Orchard (1174-1175); Otten (1189); Patrides (1210); Quinn (1272); Robins (1948); Rostvig (1363); Sadler (1387); Saurat (1394-1396); Schirmer (1412); Sellin (1423); Steadman (1487); Svendsen (1518-1521, 1974); Taffee (1975); Webster (1620); Willey (1681); Williams (1684); Williamson (1685).

Molyneux, S.: Samuel Molyneux (1689-1728).
Naturalist; son of William Molyneux; D.P.S.
Hoppen (0725).

Molyneux, T.: Thomas Molyneux (1661-1733).
D.P.S.; F.R.S.
Hoppen (0725).

Molyneux, W.: William Molyneux (1656-1698).
Natural philosopher; D.P.S. (founder); F.R.S.
Cranston (0289); Hoppen (0725).

Montagu: Lady Mary Wortley Montagu (1689-1762).
Poet; letter writer; friend of Pope and Swift; introduced into England inoculation for smallpox.
Winslow (1695).

Moore: John Moore (1646-1714).
Anglican bishop; Latitudinarian.
M. C. Jacob (0817).

Moray: Robert Moray (1608-1673).
Scottish lord; F.R.S.
M. Hunter (0751).

More: Henry More (1614-1687).
Cambridge Platonist; Anglican cleric; F.R.S.
Aaron (0001); C. Adams (1818); D. C. Allen (0016); Almond (0021); P. R. Anderson (0025, 1821); Armistead (0030); J. T. Baker (0055-0057, 1824); Biarnais (0091); Bodemer (0104); Boylan (0110); C. C. Brown (1839); S. Brown (0139); Budick (0151); Bullough (0154); Burnham (0158); Burtt (0160); Bush (0164); Carlini (0184); Carre (0188); Cassirer (0201); Clucas (0226); L. Cohen (0248); Colie (0251); Cope (0266); Coudert (0275, 0277-0278); Cragg (0284-0285); G. Craig (1853); K. Craven (0293); Cristofolini (0294); Crocker (0295-0296, 1854); de Pauley (0368); Dobbs (0380, 0383); Duffy (0405); Edelin (0416); Fabro (0441); Feilchenfeld (0451); Fiering (0467); Fisch (0475); Fleischmann (0480); Force (0496); Froom (0514); Funkenstein (0516, 0518); Gabbey (0519-0523); Galbraith (1875); Gascoigne (0527); E. George (0537); Gibson (0538); R. Greene (0570); Guffey (1762); Guinsburg (0586); Guthke (0588); Habicht (0593, 1883); A. R. Hall (0598-0599); M. B. Hall (0612); Haring (1886); C. Harrison (0632, 1887); Harth (0640); J. Henry (0660-0663); Heyd (0680); Hirst (0701); Hoopes (0714); Hoyles (0744); W. Hunter (0768); Hurlbutt (0771); Hutin (0777, 0779-0780); Hutton (0781-0786); Iliffe (0788); A. Jacob (0790-0792); J. R. Jacob (0798); M. C. Jacob (0817); Jammer (0823-0824);

Kaplan (0856); E. F. Keller (0871); Klawitter (1909); Kochavi (0890); Korshin (0894); Koyre (0903, 0909); Koyre and Cohen (0913); Lamprecht (0926); Levine (0946); Lichtenstein (0950); Loptson (0963); Lovejoy (0964); Lowery (1917); Mackinnon (0974, 1920); Marks (0991); Massa (1922); McAddo (1008); McColley (1011, 1016); McGiffert (1018); McGuire (1023-1025); McGuire and Rattansi (1034); McGuire and Tamny (1035); Mijuskovic (1066); P. Miller (1073); Mintz (1076); Mitchell (1077); Mulligan (1097); Nicolson (1118-1119, 1121-1122, 1127, 1129, 1135, 1934); Notestein (1147); Pacchi (1190); Popkin (1247, 1249); Power (1257); Prior (1268); Quinn (1272); Rattansi (1298); Raven (1302); Redgrove (1309); Reimann (1324, 1947); G. A. J. Rogers (1339, 1342); Rudrum (1370); Saveson (1398); Schaffer (1405); Soutgate (1473); Staudenbaur (1485-1486, 1971); Steneck (1494); Stock (1506); Tuveson (1576); Vassilieff (1585); Vickers (1587); Walker (1595, 1598); Westfall (1640); Willey (1681); Yates (1706); John Yolton (1718, 1720); Zimmerman (1733).

Morley: George Morley (1597-1684).
Anglican bishop; member of the Falkland Circle; F.R.S.
Hayward (0648).

Morton: Charles Morton (1627-1698).
In England, Puritan minister and schoolmaster; in New England, Puritan minister and Harvard professor; natural philosopher.
P. Miller (1073); Morison (1083); I. Parker (1205); Stephens and Roderick (1497).

Mullin: Allen Mullin or Mulines (died 1690).
F.R.S.
Hoppen (0725).

Napier: Richard Napier or Napper (1559-1634).
Anglican cleric; physician; astrologer.
MacDonald (1919); Sawyer (1399).

Nedham: Marchamont Nedham (1620-1678).
Parliamentarian pamphleteer; critic of the College of Physicians; iatrochemical physician; journalist.
L. King (0878); Lasky (0932).

Newcastle: First Duchess of Newcastle.
See Cavendish.

Newton: Sir Isaac Newton (1642-1727).
Natural philosopher; mathematician; alchemist; President of the Royal Society; (privately Arian) lay theologian.
See the listings in Topical Category XII—almost all of which include Newton in the listing of names.

Norris: John Norris (1657-1711).
Anglican cleric; philosopher; the last Cambridge Platonist; poet.
Aaron (0001); Acworth (1817); Fairchild (0442); Guffey (1762); Hoyles (0744); Mijuskovic (1066); Morawetz (1933); Rostvig (1363); Wegman (1988); Wiley (1673).

North: Roger North (1653-1734).
Lawyer; writer.
L. Stewart (1499, 1972).

Nye: Stephen Nye (1648-1719).
Socinian theologian.
Reedy (1317).

Oldenburg: Henry Oldenburg (1615-1677).
Natural philosopher; Secretary of the Royal Society.
Easlea (0413); 'Espinasse (0437); Hartley (0642); M. Hunter (0752, 0756, 0758); J. R. Jacob (0804); Kaplan (0856); Maddison (0976).

Oughtred: William Oughtred (1575-1660).
Anglican cleric; mathematician.
Feingold (0456).

Overton: Richard Overton (fl. 1642-1663).
Pamphleteer; leader of the Leveller movement.
Hill (0699); Mendelsohn (1047); Mosse (1088).

Owen: John Owen (1616-1683).
Puritan theologian; Vice-Chancellor of Oxford; chaplain to Cromwell.
Nuttall (1152).

Palmer: Thomas Palmer (1666-1743).
New England Congregational minister; physician.
Watson (1607).

Parish: Mary Parish (1630-1703).
Medium; counselor; alchemist.
J. K. Clark (0215).

Parker: Samuel Parker (1640-1688).
Natural philosopher.
Harth (0640); R. F. Jones (0836); Ormsby-Lennon (1176); G. A. J. Rogers (1342).

Parkinson: John Parkinson (1567-1650).
Naturalist; apothecary; writer of garden books.
Otten (1189); Raven (1300).

Partridge: John Partridge (1644-1715).
Cobbler; astrologer and almanac writer; the object of a satirical joke by Swift.
Capp (0182); P. Curry (0306); McCaffery (1009).

Patrick: Simon Patrick (1625-1707).
Anglican bishop; Latitudinarian.
Gabbey (0520); Griffin (0577); McAdoo (1008); Spurr (1479).

Payne: John Payne (1596-1651).
Anglican cleric; natural philosopher; friend of bishops and scholars.
Feingold (0453).

Peacham: Henry Peacham (1576?-1643?).
Author; painter; engraver; mathematician.
Houghton (0739).

Pearson: John Pearson (1613-1686).
Anglican bishop; F.R.S.
Miles (1067).

Pell: John Pell (1611-1685).
Mathematician; Anglican/Puritan cleric; member of Hartlib circle; F.R.S.
Miles (1067); Webster (1620).

Pemberton: Ebenezer Pemberton (1671-1717).
New England Puritan minister.
Corrigan (0272).

Penn: William Penn (1644-1718).
Founder and Governor of Pennsylvania; Quaker; F.R.S.
Tolles (1549); Underwood (1581).

Pepys: Samuel Pepys (1633-1703).
Public official; diarist; F.R.S.
Beier (0077, 1829); Fulton (0515); Nicolson (1132); Ollard (1169).

Percy: Henry Percy, Earl of Northumberland (1564-1632).
Natural Philosopher; called the "Wizard Earl" because of his scientific experiments.
Kargon (0857-0858, 1903); Rukeyser (1373); Shirley (1449).

Perkins: William Perkins (1558-1602).
Puritan minister and theologian.
P. Miller (1073); Morgan (1081); Thomas (1541).

Petiver: James Petiver (1663-1718).
Naturalist; entomologist; apothecary; F.R.S.
Stearns (1490).

Pett: Sir Peter Pett (1630-1699).
Friend of Boyle; lawyer and Irish office-holder; F.R.S.
J. R. Jacob (0801-0802, 0804).

Petty: William Petty (1623-1687).
Physician; public official; D.P.S.; F.R.S.
Eurich (0438); H. A. L. Fisher (0477); J. Henry (0665); Hoppen (0725); Houghton (0740); R. F. Jones (0835); R. K. Merton (1057); Sharp (1960); Strauss (1509).

Philips: Mrs. Katherine Philips (1631-1664).
Poet; used the pseudonym "Orinda."
Rostvig (1363).

Pierson: Abraham Pierson (1645-1707).
Rector of Yale.
Warch (1604).

Pitcairne: Archibald Pitcairne (1652-1713).
Scottish physician; Newtonian natural philosopher.
Emerson (0427); Guerrini (0584); L. King (0877); Thackray (1537).

Plattes: Gabriel Plattes (1590s-1643?).
Experimental philosopher; utopian reformer and tract writer; member of the Hartlib circle.
J. C. Davis (0328); Haber (0592); Webster (1611, 1616, 1625).

Plot: Robert Plot (1640-1696).
Writer of natural history.
M. Hunter (0755); Porter (1253).

Pococke: Edward Pococke (1604-1691).
Oriental scholar.
G. A. Russell (1378-1379).

Pope: Alexander Pope (1688-1744).
Poet; writer.
Clifford (0223-0224); Davie (0315); Fairchild (0442); Fleischmann (0480); R. W. Harris (0630); Humphreys (0747); R. F. Jones (0837); W. P. Jones (0846); Korshin (0894); Lovejoy (0964); Macklem (0975); Mazzeo (1001); Nicolson (1125, 1129, 1131); Nicolson and Rousseau (1141); Rousseau (1365-1367); Sherburn (1447); A. J. Smith (1459); Wagenblass (1985).

Pound: James Pound (1669-1724).
Chaplain to the East India Company; astronomer.
Murdin (1100).

Ranelagh: Katherine, Lady Ranelagh (died 1691).
Sister of Robert Boyle; patron of natural philosophers.
M. B. Hall (0613); J. R. Jacob (0796, 0804); Maddison (0976); Pilkington (1224); Webster (1621).

Raphson: Joseph Raphson (died 1715 or 1716).
Mathematician; natural theologian; F.R.S.
Koyre (0903).

Ray: John Ray (1627-1705).
Anglican cleric; naturalist; natural philosopher; F.R.S.
Attfield (0043, 0045); Collier (0254); Gordon Davies (0319); Ewan and Ewan (0440); Glacken (0548); M. Hunter (0755, 0763); Hutton (0786); Jenkins (0826); Keynes (1775); Kronemeyer (1912); McAdoo (1008); R. K. Merton (1057); Miles (1067); Ogden (1163); Olson (1170); Pedersen (1214); Porter (1253); Raven (1300-1303); Rudwick (1371); C. Russell (1376); Solberg (1463); Stearns (1490); Stebbins (1491); Taylor (1533-1534); Vickers (1587); R. S. Westfall (1650); Whittaker (1665); Willey (1680); Zeitz (1729).

Read: Alexander Read (also Reid; Rhead) (1586?-1614).
Surgeon; medical writer; anatomical lecturer.
French (0509).

Richardson: Alexander Richardson (1565-1621).
Cambridge scholar; popularizer of Ramus.
P. Miller (1073); Newman (1111).

Robie: Thomas Robie (1689-1729).
New England astronomer; almanac publisher.
Kilgour (0876); Stearns (1490).

Robinson: Tancred Robinson (1657?-1748).
Physician; naturalist; F.R.S.
Raven (1301).

Roe: Sir Thomas Roe (1581-1644).
Member of Parliament; economic planner; supporter of Hartlib.
Batten (0064).

Ross: Alexander Ross (1591-1654?).
Anglican cleric; Aristotelian controversialist.
Finch (0472); K. Hutchinson (0774); McColley (1013, 1015); R. S. Westfall (1650); Wilson (1691).

Salusbury: Thomas Salusbury (died 1666).
English translator and biographer of Galileo.
Drake (0400).

Saunders: Richard Saunders (1613-1675).
Physician; astrologer.
Capp (0182).

Sergeant: John Sergeant (1622-1707).
Philosopher; Roman Catholic cleric; critic of Locke.
Bradish (0111, 1837); Cooney (0262); Southgate (1470); John Yolton (1719).

Shakespeare: William Shakespeare (1564-1616).
Playwright; poet.
Haydn (0646); Traister (1981); Yates (1710).

Sheldon: Gilbert Sheldon (1598-1677).
Archbishop of Canterbury, 1663-1677; F.R.S.
Hayward (0648); Miles (1067).

Sherley: Thomas Sherley (1638-1678).
Physician to Charles II.
Debus (0362).

Sherman: John Sherman (died 1671).
Anglican cleric; historian of Jesus College, Cambridge.
Levine (0946).

Sibbald: Sir Robert Sibbald (1641-1722).
Scottish virtuoso; physician.
Emerson (0427).

Sloane: Sir Hans Sloane (1660-1753).
President of the Royal Society.
Raven (1301); Stearns (1490).

Smith: John Smith (1618-1652).
Anglican cleric; Cambridge Platonist.
P. R. Anderson (0025, 1821); J. T. Baker (0057, 1824); Cragg (0284-0285); de Pauley (0368); Guffey (1762); Mijuskovic (1066); Saveson (1397-1398); Scupholme (1419); Willey (1681).

Smyth: Edward Smyth (1665-1720).
Anglican bishop; D.P.S.; F.R.S.
Hoppen (0725).

South: Robert South (1634-1716).
Anglican cleric; Public Orator of the University of Oxford; critic of the Royal Society.
Syfret (1524).

Southwell: Robert Southwell (1635-1702).
Statesman and diplomat; President of Royal Society.
Strauss (1509).

Sprat: Thomas Sprat (1635-1713).
Anglican bishop; Fellow of and publicist for the Royal Society.
Agassi (0008); Bredvold (0115); Cope (0265, 0268); Dear (0330); Harwood (0643); M. Hunter (0751-0752, 0756, 0763); J. R. Jacob (0798, 0800, 0802-0804); M. C. Jacob (0817); Kemsley (0873); R. K. Merton (1057); Miles (1067); Mitchell (1077); Ormsby-Lennon (1176); Purver (1270); Shapiro (1438, 1440, 1959); Sonnichsen (1969); Stimson (1505); Syfret (1524); Willey (1681); Williamson (1687); Wright (1702); Youngren (1727).

Sprigg: William Sprigg (fl. 1652-1695).
Puritan; Fellow of Lincoln College, Oxford.
Greaves (0565).

Stanley: Thomas Stanley (1625-1678).
Poet; writer; F.R.S.
C. Harrison (0632); Heninger (0658); Kroll (0916).

Starkey: George Starkey or George Stirk (1628?-1665).
Alchemist; used the pseudonym Eirenaeus Philalethes; physician.

Hutin (0778); Mendelsohn (1047); Newman (1111-1114); Turnbull (1571); Webster (1614); Wilkinson (1676, 1678).

Stillingfleet: Edward Stillingfleet (1635-1699).
Anglican bishop; Latitudinarian; opponent of Locke.
Carlini (0184); R. T. Carroll (0192); Christophersen (1747); Collier (0254); Dahrendorf (1857); de Pauley (0368); Fishman (1873); Griffin (0577); Harth (0640); M. Hunter (0749); Hutton (0781, 0786); Leyden (0949); McAdoo (1008); Pacchi (1190); Popkin (1226); Reedy (1314-1315, 1317); Reventlow (1326); Rivers (1330); Spurr (1478)-1479; Sullivan (1516); Yolton (1714, 1717).

Stirk: George Stirk
See Starkey.

Stoughton: John Stoughton the Elder (died 1639).
Puritan minister; Baconian; member of the Hartlib circle.
Whitebrook (1662).

Stubbe: Henry Stubbe (1632-1676).
Social and political polemicist; opponent of the Royal Society; physician.
Cope (0263, 0267); Duffy (0405); M. Hunter (0756, 0763); J. R. Jacob (0793-0794, 0797-0798, 0804); Jacob and Jacob (0805); M. C. Jacob (0817); R. F. Jones (0835); Kaplan (0856); L. King (0878); Spiller (1475); Syfret (1523).

Stukeley: William Stukeley (1687-1765).
Antiquary; Anglican cleric; collector of Newtonian memoranda; F.R.S.
Hoskins (0738); Manuel (0984).

Swift: Jonathan Swift (1667-1745).
Irish writer; Anglican cleric; ecclesiastical and political pamphleteer.
Canavan (1846); Clifford (0223); Cobb (1850); Colie (0252); K. Craven (0293); Davie (0315); Fleischmann (0480); Force (0497); R. W. Harris (0630); Harth (0640); Hazard (0649); Hoppen (0725); R. F. Jones (0837); Landa (0927); Nicolson (1125,

1137); Nicolson and Mohler (1139-1140); Nicolson & Rousseau (1141); Quintana (1274); Reventlow (1326); Rousseau (1366-1367); Starkman (1484); Stathis (1803); Stock (1506); Tuveson (1577); Wegman (1988); Willey (1680).

Sydenham: Thomas Sydenham (1624-1689).
Physician; Puritan.
F. Anderson (0023); Aspelin (0038); Bates (1826); Coulter (0279); Cunningham (0305); L. King (0877-0878); Meynell (1782); Romanell (1347); R. S. Westfall (1650).

Symcotts: John Symcotts (1592-1662).
Physician.
Nagy (1101).

Tanner: John Tanner (1636-1715).
Physician; astrologer.
Capp (0182).

Taylor, J.: Jeremy Taylor (1613-1667).
Anglican bishop.
Hoopes (0714); Mitchell (1077); Mulligan (1097).

Taylor, Z.: Zachary Taylor (1653-1705).
Anglican cleric; pamphleteer; author of "The Surey Imposter."
Harley (0627).

Teackle: Thomas Teackle (died 1696 or 1697).
Anglican cleric in Virginia; collector of occult library.
Butler (0169).

Temple: Sir William Temple (1628-1699).
Diplomat; essayist—including "Upon Ancient and Modern Learning."
K. Craven (0293); Tuveson (1577).

Tenison: Thomas Tenison (1636-1715).
Archbishop of Canterbury; Boyle Lectures Trustee; F.R.S.
E. Carpenter (0185).

Thomson: George Thomson (fl. 1645-1679).
Physician; Royalist; Helmontian pamphleteer.
Clericuzio (0222).

Tillotson: John Tillotson (1630-1694).
Archbishop of Canterbury; latitudinarian theologian; F.R.S.
Bethell (0090); Brauer (0114); S. C. Carpenter (0186); Curtis (0310); Emerson (0425); Fiering (0467); D. Greene (0569); Griffin (0577); Harth (0638, 0640); Horstmann (0732); M. C. Jacob (0817); R. F. Jones (0842); Lang (0928); Locke (0962); May (0999); McAdoo (1008); McGiffert (1018); Mitchell (1077); O'Higgins (1165); Osmond (1185); Popkin (1244); Reedy (1314-1316); Reventlow (1326); Rivers (1330); Scarre (1400); Shaw (1961); Spurr (1479); Sullivan (1516); Van Leeuwen (1583).

Tindal: Matthew Tindal (1657-1733).
Deist philosopher.
Colie (0253); Emerson (1866); *Encyclopedia of Unbelief* (0433); Force (0485); Hefelbower (0651); M. C. Jacob (0817); Redwood (1313); Reventlow (1326); Shaw (1961); Sullivan (1516).

Toland: John Toland (1670-1722).
Deist; pantheist philosopher; political philosopher.
D. Berman (0084); Biddle (0093); R. M. Burns (0159); Carabelli (1746); Casini (0197); Christophersen (1747); Colie (0253); K. Craven (0293); Daniel (0313); *Encyclopedia of Unbelief* (0433); Dyche (1863); Emerson (1866); Evans (0439, 1868); Fabro (0441); Hazard (0649); Hefelbower (0651); Heinemann (0656-0657); Horstmann (0732); Hurlbutt (0771); J. R. Jacob (0798); M. C. Jacob (0814, 0817, 1900); Lange (0929); McGiffert (1018); Metzger (1064); Motzo Dentice de Accadia (1090); Nicholl (1116); Nourrisson (1148); Ogonowski (1164); O'Higgins (1165); Orr (1178); Popkin (1226); Redwood (1313); Reedy (1315, 1317); Reventlow (1326); Shapin (1429); Shaw (1961); Stephen (1495); Stromberg (1510); Sullivan (1516, 1973); Wegman (1988); J. W. Yolton (1714).

Tompion: Thomas Tompion (1638-1713).
Clock and instrument maker; Anglican patron of Quakers.
Raistrick (1281).

Topsell: Edward Topsell (died 1638).
Anglican cleric; naturalist.
Raven (1300).

Torporley: Nathaniel Torporley (1564-1632).
Mathematician.
Jacquot (0820).

Towneley: Richard Towneley (1629-1707).
Natural philosopher; Catholic layman.
Webster (1624).

Traherne: Thomas Traherne (1637-1674).
Poet; Anglican cleric; Anglican mystic.
Clucas (0226); Guffey (1762); C. Marks (0991); Otten (1189); Wade (1592).

Twisse: William Twisse (1578?-1648).
Anglican cleric; Puritan biblical scholar.
Popkin (1249).

Twysden: John Twysden or Twisden (1607-1688).
Physician; defender of the College of Physicians against Nedham.
L. King (0878).

Tymme: Thomas Tymme (died 1620).
Anglican cleric; Paracelsian apologist.
Johnson (0831); Debus (0341, 0352).

Ussher: James Ussher (1581-1656).
Irish Anglican archbishop; biblical scholar.
Barr (0063); Feingold (0456); J. R. Jacob (0804); Katz (0861); Reese and others (1322); G. A. Russell (1378); Trevor-Roper (1561).

Van Dyck: Sir Anthony Van Dyck (1599-1641).
Flemish painter; court painter to Charles I; alchemist.
Petersson (1222).

Vaughan, H.: Henry Vaughan (1622-1695).
Poet; twin brother of Thomas Vaughan.
Bethell (0090); Hill (0685); Hollander (0703); Holmes (0704); F. E. Hutchinson (0773); Llasera (0961); Nuttall (1153); Otten (1189); Rostvig (1363); Rudrum (1370); A. J. Smith (1459).

Vaughan, T.: Thomas Vaughan (1622-1666).
Alchemist; twin brother of Henry Vaughan.
Bullough (0154); Burnham (0158); F. E. Hutchinson (0773); E. F. Keller (0871); Korshin (0894); Guinsburg (0586); Hamilton (0622); Mendelsohn (1047); Mulligan (1097); Newman (1110-1111, 1115); Ormsby-Lennon (1176); Rudrum (1369-1370); Willard (1994).

Villiers: George Villiers, Duke of Buckingham (1592-1628).
Courtier and politician; friend and patron of Kenelm Digby.
Petersson (1222).

Vossius: Isaac Vossius (1618-1689).
Dutch polymath who lived in England, 1670-1689; Anglican cleric; son of G. J. Vossius; F.R.S.
Katz (0861); Popkin (1227); Sellin (1423).

Wadsworth: Benjamin Wadsworth (1669-1737).
New England Puritan minister; President of Harvard.
Corrigan (0272).

Wallis: John Wallis (1616-1703).
Anglican cleric; mathematician; natural philosopher; F.R.S.
Agassi (0008); Hartley (0642); J. R. Jacob (0798, 0804); Johnson (0832); Miles (1067); Murdin (1100); Oldroyd (1167); Schaffer (1407); Skinner (1457).

Walton: Izaak Walton (1593-1683).
Writer.
Webster (1626).

Walwyn: William Walwyn (fl. 1600-1649).
Leveller spokesman; physician; medical philosopher.
Hill (0687).

Ward: Seth Ward (1617-1689).
Astronomer; Anglican bishop; F.R.S.
Debus (0336, 0338, 0356); Greaves (0565); Miles (1067); Rattansi (1294); Rees (1319); Spurr (1479); P. Wright (1702).

Warner (1570): Walter Warner (1570-1642 or 1643).
Natural philosopher.
Henry (0665).

Warner (1558): William Warner (1558?-1609).
Poet; translator; lawyer.
Rukeyser (1373).

Warren: Erasmus Warren (died 1718).
Anglican cleric; critic of Thomas Burnet.
Collier (0254); Jacquot (0819); Ogden (1163).

Waterhouse: Edward Waterhouse (1619-1670).
Anglican cleric; F.R.S.
R. F. Jones (0840).

Watts: Isaac Watts (1674-1748).
Dissenting minister; theologian; hymn writer.
Hoyles (0744).

Webbe: Thomas Webbe (fl. 1640).
Mathematician; Navy official; Puritan; Ranter.
Mosse (1088).

Webster: John Webster (1610-1682).
Puritan minister; chaplain and surgeon in the New Model Army; pamphleteer; radical critic of the traditional universities.
Coudert (0277); Debus (0336, 0338, 0345, 0350, 0356); Greaves (0565); Hill (0687, 0699); Jobe (0828); E. F. Keller (0871); Notestein (1147); Ormsby-Lennon (1176); Rattansi (1294); Rees (1319); Stock (1506); Sutton (1517); Trevor-Roper (1563); Webster (1616); P. Wright (1702).

Wells: John Wells (died 1635).
Mathematician; Navy official; Puritan.
Johnson (0832).

Wharton (1617): Sir George Wharton (1617-1681).
Physician; anatomist; mathematician; astrologer; Royalist.
Capp (0182); Josten (0849).

Wharton (1653): Goodwin Wharton (1653-1704).
Government official; inventor; alchemist.
J. K. Clark (0215).

Whichcote: Benjamin Whichcote (1609-1683).
Cambridge Platonist; Vice-Chancellor of Cambridge.
P. R. Anderson (1821); Bullough (0154); Cassirer (0201); Cragg (0284-0285); de Pauley (0368); E. George (0537); R. Greene (0571-0572); Guffey (1762); Hoopes (0714); Levine (0946); Rivers (1330); Roberts (1333); Scupholme (1420); Staudenbaur (1486); Wiley (1673); Willey (1681).

Whiston: William Whiston (1667-1752).
Mathematician; theologian; Newtonian; Boyle Lecturer.
Briggs (0124); J. H. Brooke (0131); Carpenter (0185); Chistianson (0208); Collier (0254); Gordon Davies (0319); Duffy (0406); Farrell (0444, 1869); Ferguson (0459); Force (0485, 0487-0488, 0490-0493, 0495-0497, 0499, 1874); Froom (0514); Genuth (0533); Glacken (0548); Gould (0533); Hurlbutt (0771); M. C. Jacob (0817); Korshin (0894); Kubrin (1913); Manuel (0984); Metzger (1064); Nicolson (1141); O'Higgins (1165); Popkin (1244); Porter (1253); Redwood (1313); Roger (1337); Rousseau (1365, 1367); Stephen (1495); Stromberg (1510); Taylor (1533-1534); Tuveson (1577); Walker (1595); Webster (1615); R. S. Westfall (1639-1640); D. Young (1724).

White: Thomas White (1593-1676).
Roman Catholic priest; natural philosopher; opponent of Royal Society.
Cope (0266); Petersson (1222); Southgate (1470-1472, 1970); Spiller (1475).

Whitlock: Richard Whitlock (born 1616).
Anglican cleric; physician.
Williamson (1688).

Wigglesworth: Michael Wigglesworth (1631-1705).
New England Puritan minister; physician; poet.
Morison (1083); Newman (1111); P. Watson (1607).

Wilkins: John Wilkins (1614-1672).
Anglican bishop; natural philosopher; F.R.S.
Borges (0107); J. H. Brooke (0132); Burns (0159); Clauss (0221); Cope (0264); Deason (0331); S. Dick (0373); Drake (0400); Ferreira (0463); H. A. L. Fisher (0477); Gascoigne (0527); Greaves (0565); R. Greene (0572); Griffin (0577); Guthke (0588); Hacking (0594); C. Harrison (0632); Hartley (0642); Heyd (0679); Hooykaas (0720-0721); Hornberger (0729); M. Hunter (0752, 0756, 0763); J. R. Jacob (0804); Jacob and Jacob (0807); M. C. Jacob (0817); R. F. Jones (0835, 0842-0843); Knight (0888-0889); Locke (0962); Mason (0996); McColley (1012-1013, 1015); R. K. Merton (1057); Miles (1067); Mitchell (1077); Nicolson (1121, 1137-1138); Ormsby-Lennon (1176); Rattansi (1294); Rivers (1330); Jack Rogers and McKin (1345); P. Rossi (1355, 1359); Shapiro (1438-1440, 1959); Spurr (1479); Stimson (1503, 1505); Sykes (1525); Tang (1977); Van Leeuwen (1583); Webster (1616); R. S. Westfall (1650).

Willard: Samuel Willard (1639 or 1640-1707).
New England Puritan minister; Harvard professor.
P. Miller (1073).

William III: William III (1650-1702).
King of England, Scotland, and Ireland, 1689-1702.
Cook (0259).

Willis: Thomas Willis (1621-1675).
Physician; medical philosopher; F.R.S.
T. Brown (0140); Clericuzion (0222); Coulter (0279); T. Hall (0619); J. R. Jacob (0798); L. King (0877); Murdin (1100); J. Wright (1701).

Willughby: Francis Willughby (1635-1672).
Naturalist; F.R.S.
Raven (1301).

Wing: Vincent Wing (1619 1668).
Welsh astronomer; mathematician; astronomical writer.
Capp (0182); Fleming (0481).

Winstanley: Gerrard Winstanley (c.1609-after 1660).
Cloth merchant; radical Digger critic of the establishment.
Easlea (0414); Eurich (0438); Greaves (0565); Hill (0690, 0699); Lasky (0932); Mulder (1093); Mulligan (1097); Rosen (1350).

Winthrop: John Winthrop (1606-1676).
Puritan Governor of Connecticut; alchemist; F.R.S.
Black (0096); Kittredge (0883); G. Miller (1070); Morison (1085); Newman (1111); Stearns (1488, 1490); Webster (1616); Wilkinson (1674-1675, 1679); R. Fitz. Young (1726).

Wise: Thomas Wise (fl. 1700-1710).
Anglican cleric; Follower of Cudworth.
Redwood (1313).

Wollaston: William Wollaston (c. 1660-1724).
Natural theologian; reputed deist.
Burns (0159); Stephen (1495).

Wood: Anthony Wood (1632-1695).
Diarist; antiquary.
Hargreaves-Mawdsley (0626); M. Hunter (0755).

Woodward: John Woodward (1665-1728).
Physician; natural philosopher; geologist; F.R.S.
Collier (0254); Gordon Davies (0319); Farrell (0444); Gascoigne (0528); Glacken (0548); Porter (1253); Redwood (1313); Rudwick (1371); Taylor (1533-1534); D. Young (1724).

Woolston: Thomas Woolston (1670-1733).
Anglican cleric; Cambridge Fellow; radical theologian; reputed deist.
Burns (0159).

Worsley: Benjamin Worsley (1618?-1677).
Physician; merchant; Secretary and Treasurer of the Council for Trade and Plantations; member of the Hartlib circle.
Newman (1111); Strauss (1509); Webster (1621).

Index of Authors and Editors